Materials Engineering

Bonding, Structure, and Structure–Property Relationships

Designed for both one- and two-semester courses, this textbook provides a succinct and easy-to-read introduction to crystal structures and structure-property relations. By linking together the fundamentals of bond strength and the arrangement of atoms in space with the mechanical, optical, magnetic, and electrical properties that they control, students will gain an intuitive understanding of how different materials are suited to particular applications. The systematics of crystal structures are described for both organic and inorganic materials, with coverage including small molecular crystals, polymers, metals, ceramics, and semiconductors. Hundreds of figures and practice problems help students gain an advanced, 3D understanding of how structure governs behavior, and a wealth of examples throughout show how the underlying theory is translated into practical devices. With solutions, video lectures, and PowerPoints available online for instructors, this is an excellent resource for graduates and senior undergraduates studying materials science and engineering.

Susan Trolier-Mckinstry is the Steward S. Flaschen Professor of Ceramic Science and Engineering, Professor of Electrical Engineering, and Director of the Nanofabrication Facility at Pennsylvania State University. She was also the 2017 President of the Materials Research Society (MRS), and is a fellow of the IEEE, MRS, and the American Ceramic Society.

The late **Robert E. Newnham** was a professor in the Department of Materials Science and Engineering at Pennsylvania State University and a member of the National Academy of Engineering. He was the recipient of numerous awards, including the John Jeppson Medal from the American Ceramic Society and the Turnbull Lecturer Award from the Materials Research Society.

Materials Engineering

Bonding, Structure, and Structure–Property Relationships

SUSAN TROLIER-McKINSTRY
Pennsylvania State University

ROBERT E. NEWNHAM
Pennsylvania State University

CAMBRIDGE
UNIVERSITY PRESS

Shaftesbury Road, Cambridge CB2 8EA, United Kingdom

One Liberty Plaza, 20th Floor, New York, NY 10006, USA

477 Williamstown Road, Port Melbourne, VIC 3207, Australia

314–321, 3rd Floor, Plot 3, Splendor Forum, Jasola District Centre, New Delhi – 110025, India

103 Penang Road, #05–06/07, Visioncrest Commercial, Singapore 238467

Cambridge University Press is part of Cambridge University Press & Assessment,
a department of the University of Cambridge.

We share the University's mission to contribute to society through the pursuit of
education, learning and research at the highest international levels of excellence.

www.cambridge.org
Information on this title: www.cambridge.org/9781107103788

DOI: 10.1017/9781316217818

First published 2018

A catalogue record for this publication is available from the British Library

Library of Congress Cataloging-in-Publication data
Names: Trolier-McKinstry, Susan, 1965– author. | Newnham, Robert E. (Robert Everest), 1929–2009 author.
Title: Materials engineering : bonding, structure, and structure–property relationships / Susan
    Trolier-McKinstry, Pennsylvania State University, Robert Newnham, Pennsylvania State University.
Description: Cambridge, United Kingdom ; New York, NY, USA : Cambridge University Press, [2017] |
    Includes bibliographical references and index.
Identifiers: LCCN 2017024291 | ISBN 9781107103788 (hardback : alk. paper)
Subjects: LCSH: Materials.
Classification: LCC TA403 .T746 2017 | DDC 620.1/1–dc23 LC record available at
    https://lccn.loc.gov/2017024291

ISBN   978-1-107-10378-8   Hardback

Additional resources available at www.cambridge.org/trolier-mckinstry

This book is dedicated to the memory of Professor Robert E. Newnham, an extraordinary scientist, engineer, and educator.

For my family. Thank you for your patience and your love.

Susan Trolier-McKinstry

Contents

Foreword

Use of this Book for Different Classes

The intended audience is primarily undergraduate or beginning graduate students in materials science and engineering. That said, the book is written in such a way that it could be used as a text for a single class in materials to be taught to engineering students in other disciplines (civil, electrical, and mechanical, in particular).

Professor Newnham was a great believer in looking at structure models of crystals, dreaming he was an ångström (Å) high – walking around inside the crystals, hopping across grain boundaries and domain walls, pushing on the atomic bonds. Both he and I have thought about crystals, ceramics, metals, semiconductors, polymers, and their properties for many years. The ability to visualize crystal structures and the connectivity of strong and weak bonds in materials provides tremendous insight. Hence, this text is liberally illustrated with figures of crystal structures, all of which were drawn using CrystalMaker™. Numerous good drawing programs exist; students are strongly urged to explore both three dimensional hand models as well as computer packages so that they can develop an intuitive understanding of symmetry.

The modus operandi of materials research and development programs is to seek fundamental understanding while working toward an engineering goal and remaining alert for new applications. One of the best ways of introducing this line of thinking is to develop an understanding of the structure–property relationships.

I use this book to teach both undergraduate and graduate classes in Crystal Chemistry at Penn State.

The undergraduate course is a junior level class, intended to provide many of the underpinnings of the field to the student. The perspective begins from the structure and bonding of the solid, and uses this to provide students with an intuitive understanding of how materials are designed or chosen for particular applications.

Prerequisite knowledge includes:

- some chemistry (typically two semesters of College Chemistry)
- vector math
- some physics (particularly a class in waves, so that the ideas of phonons and photons are not new).

Course Objectives

(a) To identify important raw materials and minerals as well as their names and chemical formulas.

(b) To describe the crystal structure of important materials and to be able to build their atomic models.

(c)   To learn the systematics of crystal and glass chemistry.

(d)   To understand how physical and chemical properties are related to crystal structure and microstructure.

(e)   To appreciate the engineering significance of these ideas and how they relate to industrial products: past, present, and future.

Course Outcomes

(a)   Students should be able to write and balance chemical formulae for commercially important raw and engineered materials.

(b)   Students should be able to build important crystal structures and understand the impact of bond length, coordination, and symmetry on the resultant physical properties.

(c)   Given an initial chemistry, students should be able to apply Pauling's rules to determine anion and cation coordinations, and should be able to make intelligent suppositions about the resulting crystal structure. Similarly, on the basis of Zachariasen's rules, students should be able to assess the likelihood of easy glass formation in a particular materials system.

(d)   Students should understand the rules governing the stability of crystal structures as a function of temperature, pressure, and composition changes.

(e)   Students should understand the basic mechanisms controlling a wide variety of physical properties, and should be able to correlate this information with crystal structures to predict materials properties.

(f)   Students should begin to understand how materials are chosen and designed for particular engineering applications.

The best way of visualizing the atomic structures of materials is to build models. It is easy to see why graphite is a good lubricant, how slip occurs in metals, and why barium titanate develops a spontaneous polarization. Table F.1 lists the models built in the introductory course. Thirty hours of model building are accompanied by information discussing atomic coordination, chemical bonding, crystallographic symmetry, and structure–property relations. Raw material specimens and commercial products are circulated amongst the students to emphasize the usefulness of the materials being modeled. Of course, the structures selected can be modified to best suit particular disciplines: in classes taught to civil engineers, it is recommended that the cement structures alite, belite, and tobermorite be substituted for the last lab listed.

The first four laboratory sessions are devoted to basic structures illustrating the five principal types of chemical bonding. After that point, more sophisticated model building is pursued. It is important to move beyond the elementary crystal structures of rocksalt, diamond, and the metal structures. In building the models, we use the Orbit and Minit Molecular Building Systems™ utilizing flexible plastic straws and connectors, available from Cochranes of Oxford, Ltd. It is critical to utilize a building system that allows coordinations from 1 to 12.

Discussions of anisotropy develop naturally from crystal models. A simple example comes from the structures of graphite and boron nitride. Thermal conductivity coefficients are much higher parallel to the layers than in the perpendicular direction.

Table F.1 Crystal structures assembled in introductory crystal chemistry

Ionic solids	Halite and fluorite
Covalent crystals	Diamond, zincblende, and wurtzite
Metals	Copper (FCC), iron (BCC), and magnesium (HCP)
Molecular solids	Ice, pentacene, and coupling agents
Polymers	Polyethylene, polypropylene, polystyrene, and silicones
Structures with anisotropy	Rutile, graphite, and hexagonal BN
Octahedral coordinations	Brucite, gibbsite, and corundum
Classification of silicate structures	Isolated silicates, ring structures, single and double chains, zeolites
Layer silicates	Kaolinite, serpentine, talc, micas
Silica phases and stuffed derivatives	Cristobalite, tridymite, and their stuffed derivatives
Raw materials	Feldspars, beryl, and cordierite
Optical and electronic ceramics	Calcite and perovskite
Magnetic structures and defects	Spinels, β-alumina, edge and screw dislocations
Pb^{2+} and B^{3+} coordinations, and non-oxide ceramics	PbO, borax, and silicon nitride

Table F.2 Selected empirical rules of thumb of importance in materials science and engineering

- Goldschmidt's rules for atomic coordination changes at high temperature and pressure.
- Hume-Rothery and Brewer predictions for metallic structures and solid solutions.
- Line and Matthiessen rules for electrical conductivity of alloys.
- Mooser–Pearson rules for semiconductors.
- Pauling's rules for ionic coordination and crystal structures.
- Shewmon and Tamman rules for diffusion coefficients and annealing temperatures.
- Trouton formula for latent heat of evaporation and boiling point.
- Weidenmann–Franz law relating thermal and electrical conductivity of metals.
- Zachariasen's rules for oxide glass formation and structure.

This structure–property relationship can be further amplified with discussions of the correlations between thermal conductivity and chemical bonding, and with bond length.

"Rules of thumb" help put students in touch with reality, since most involve properties of practical importance. They provide the physical and chemical intuition required for back of the envelope calculations, and for rapid recall at technical meetings. Table F.2 lists several of these rules. Common sense ideas like these also facilitate discussion of more theoretical concepts, and provide a link to the advances in materials modeling.

For the graduate course, the outline is as follows.

Definition of Crystal Chemistry
Elements of Crystallography
    Symmetry
    Space Groups and Point Groups
    Coordination

Bonding
    Covalent Bonding
    Ionic Bonding
    Metallic Bonding
    Hydrogen Bonding
    Van der Waals Bonding
    Ionic and Atomic Radii
    Crystal Field Theory
    Molecular Orbital Theory
    Band Theory
Defects
Structure Prediction
    Pauling's Rules
    Bond Valence Sums
    $8 - N$ Rule
    Metal Structure Prediction
    Structure Field Maps
    Pressure–Temperature Variations
Important Structure Types
    Ionically Bonded Materials
    Metallically Bonded Materials
    Covalently Bonded Materials
    Small Molecule Crystals
    Polymers
    Glasses
Structure–Property Relations
    Neumann's Law
    Thermal Properties
    Electrical Conductivity
    Dielectric Properties
    Optical Properties
    Magnetism
    Mechanical Properties

This book will not describe two of the other pillars of materials science and engineering: processing and its link to microstructure, and materials characterization.

Acknowledgements

Portions of this book were adapted from previous works by Professor Robert E. Newnham. We gratefully acknowledge Elsevier for permission to adapt material from R. E. Newnham, "Phase diagrams and crystal chemistry," in *Phase Diagrams: Materials Science and Technology, Volume V*, Academic Press, NewYork (1978). We are similarly grateful to Springer-Verlag for permission to reuse and adapt material from R. E. Newnham, *Structure–Property Relations*, Springer-Verlag (1975).

Professor Newnham knew the name and chemical formula of all the minerals, and usually knew the crystal structure. He combined this knowledge with a breadth of understanding of the fundamental mechanisms responsible for various properties. As a result, he was often able to predict the likely material response from the chemical formula, and use this to solve engineering problems.

Bob received numerous prizes and honors over the course of his career, including the Franklin Medal for Electrical Engineering for development of the composite transducers now ubiquitous in medical ultrasound. One of the stories that he told over the years about the genesis of the idea was that he recognized that SbSI, a one-dimensional ferroelectric, had some of the key properties: strong bonding along the axis of the SbSI chains that produced a strong piezoelectric response in that direction; weak bonding between the chains that degraded the coupling to motion perpendicular to the chains. Then, it was simply a matter of artificially creating the desired connectivity in a material that is less electrically leaky. At this point, there are thousands to millions of people whose lives have been saved by this inspired application of structure–property relations.

It was Bob's hope as well as mine, that this book would inspire more generations of scientists to learn about structures and structure–property relations.

There are a great many people to thank for helping make this book a reality.

A small army of proofreaders has helped improve the text over the years – from students in MATSE 400 or MATSE 512 at Penn State, to Igor Levin, Cihangir Duran, Barry Scheetz, Della Roy, Paolo Colombo, Allison Beese, Michael Hickner, Robert Hickey, Wanlin Zhu, Thomas N. Jackson, Jon-Paul Maria, Jon Ihlfeld, Kim Trolier, and Michael Trolier. Thank you, thank you, thank you! Remaining errors are my responsibility.

This manuscript is strongly dependent on figures. The crystal structures were rendered in Crystal Maker™. Several people helped with figures along the way, including Ryan Haislmaier, Aileen McKinstry, Nathan McKinstry, and Herb McKinstry.

Finally, my family has been both enormously patient and supportive of the time it has taken to finish this. You all are my heroes.

Acknowledgements from the Family of Professor Robert Newnham

On behalf of Professor Robert Newnham's family, we would like to thank his friend and valued colleague, Professor Susan Trolier-McKinstry, for her work in carrying on our father's legacy. At the time of his death in 2009, Dad and Susan had written only about a quarter of this book, so it is very generous of her to share full co-author credit with my father on this final version, which is largely her work.

Susan's dedication does an excellent job in describing our father's scientific legacy. But there was much more to him than his work. First, he was a very kind and devoted family man, and is warmly remembered by his children, Randall and Rosemary, their spouses, Janet and Patrick, his grandchildren, Johnathan Robert, Henry Everest, and Eleanor Patricia, and a large extended family. He and our mother Patricia treated Dad's colleagues as family, too, including them in hikes, picnics, and family holiday celebrations. They enjoyed visiting his academic "children" and "grandchildren" all over the world. Second, he always approached his work – and his life – with a sense of imagination and fun, which we hope readers of this book will be able to share. For example, he opened the Preface of his first book with these words, to explain how he developed his love of structures as a child:

As a boy I loved to build model airplanes, not the snap-together plastic models of today, but the old-fashioned Spads and Sopwith Camels made of balsa wood and tissue paper. I dreamed of Eddie Rickenbacker and dogfights with the Red Baron as I sat there sniffing airplane glue. Mother thought I would never grow up to make an honest living, and mothers are never wrong. Thirty years later I sit in a research laboratory surrounded by crystal models and dream of what it would be like to be one Angstrom tall, to rearrange atoms with pick and shovel, and make funny things happen inside.[1]

Reading these words again, we can picture our father laughing as he wrote them, before turning to work on another of the carefully constructed crystal models which littered his office. We hope this book will inspire you to cultivate some of the same joy and wonder in the natural world that our father felt since he was a little boy.

Randall and Rosemary Newnham, July 2016

---

[1] Robert E. Newnham, *Structure–Property Relations*. Berlin: Springer-Verlag (1975).

1 Introduction to Bonding, Structure, and Structure–Property Relations

The goal of this book is to explore how a knowledge of bonding, crystal structure, and structure–property relations can be used in materials science and engineering to design and utilize materials whose properties match those required in a given application. This field is sometimes referred to as *crystal chemistry*. Crystals are groups of atoms that repeat periodically in space. A knowledge of how the atoms are arranged in space, and the strength of the bonds holding them together, in turn, provides direct insight into properties as diverse as refractive index, cleavage, thermal conductivity, ionic conductivity, and many others.

For many problems of interest in materials engineering, it is helpful to ask the following questions:

- What are the atoms involved and how are they arranged?
- How does this arrangement lead to certain mechanisms of electronic or atomic motion?
- How do these mechanisms give rise to the observed properties?
- How are these properties represented in equation form?
- How does symmetry modify the coefficients appearing in these equations?
- What controls the magnitudes of the coefficients? Are there useful trends and "Rules of Thumb"?
- Based on these structure–property relationships, when are exceptionally outstanding properties likely to develop?
- What are the important applications? How can engineers use these ideas?

Relationships between crystal structures and physical properties are described in this book, emphasizing the application of structure–property relations to engineering problems. Faced with the task of finding new materials with useful properties, the reader of this book should be able to use atomic/ionic radii, crystal fields, the arrangement of bonds, understanding of the bond type, and symmetry arguments as criteria for the materials selection process.

This field synthesizes a large amount of information, so that trends can be identified. It provides guidelines on:

- which materials might have interesting and useful properties,
- structure predictions as a function of temperature, pressure, and composition,
- crystalline solubility limits, and
- the nature and kinetics of solid state reactions.

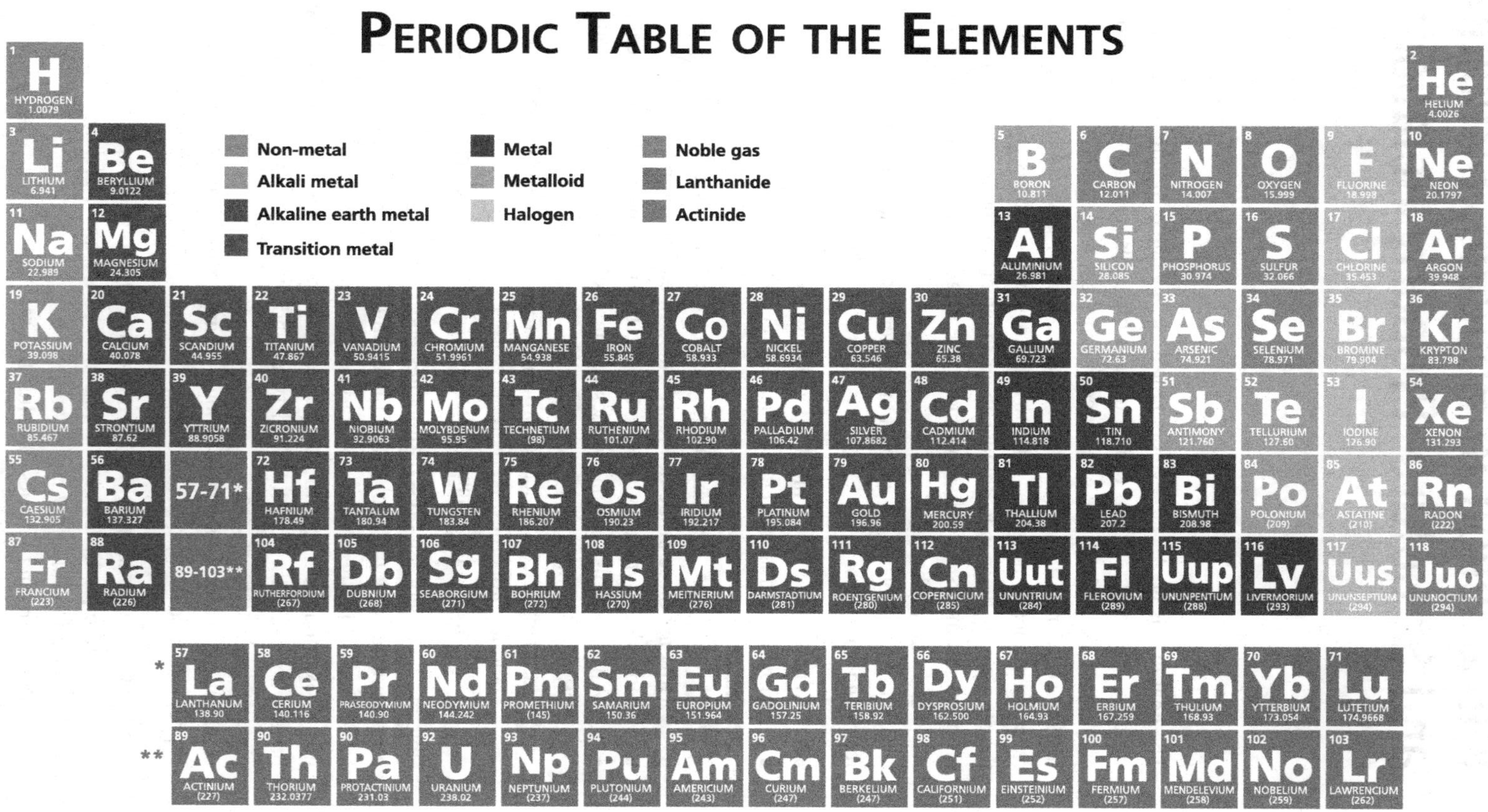
PERIODIC TABLE OF THE ELEMENTS
Non-metal
Alkali metal
Alkaline earth metal
Transition metal
Metal
Metalloid
Halogen
Noble gas
Lanthanide
Actinide
1 H HYDROGEN 1.0079
2 He HELIUM 4.0026
3 Li LITHIUM 6.941
4 Be BERYLLIUM 9.0122
5 B BORON 10.811
6 C CARBON 12.011
7 N NITROGEN 14.007
8 O OXYGEN 15.999
9 F FLUORINE 18.998
10 Ne NEON 20.1797
11 Na SODIUM 22.989
12 Mg MAGNESIUM 24.305
13 Al ALUMINIUM 26.981
14 Si SILICON 28.085
15 P PHOSPHORUS 30.974
16 S SULFUR 32.066
17 Cl CHLORINE 35.453
18 Ar ARGON 39.948
19 K POTASSIUM 39.098
20 Ca CALCIUM 40.078
21 Sc SCANDIUM 44.955
22 Ti TITANIUM 47.867
23 V VANADIUM 50.9415
24 Cr CHROMIUM 51.9961
25 Mn MANGANESE 54.938
26 Fe IRON 55.845
27 Co COBALT 58.933
28 Ni NICKEL 58.6934
29 Cu COPPER 63.546
30 Zn ZINC 65.38
31 Ga GALLIUM 69.723
32 Ge GERMANIUM 72.63
33 As ARSENIC 74.921
34 Se SELENIUM 78.971
35 Br BROMINE 79.904
36 Kr KRYPTON 83.798
37 Rb RUBIDIUM 85.467
38 Sr STRONTIUM 87.62
39 Y YTTRIUM 88.9058
40 Zr ZICRONIUM 91.224
41 Nb NIOBIUM 92.9063
42 Mo MOLYBDENUM 95.95
43 Tc TECHNETIUM (98)
44 Ru RUTHENIUM 101.07
45 Rh RHODIUM 102.90
46 Pd PALLADIUM 106.42
47 Ag SILVER 107.8682
48 Cd CADMIUM 112.414
49 In INDIUM 114.818
50 Sn TIN 118.710
51 Sb ANTIMONY 121.760
52 Te TELLURIUM 127.60
53 I IODINE 126.90
54 Xe XENON 131.293
55 Cs CAESIUM 132.905
56 Ba BARIUM 137.327
57-71*
72 Hf HAFNIUM 178.49
73 Ta TANTALUM 180.94
74 W TUNGSTEN 183.84
75 Re RHENIUM 186.207
76 Os OSMIUM 190.23
77 Ir IRIDIUM 192.217
78 Pt PLATINUM 195.084
79 Au GOLD 196.96
80 Hg MERCURY 200.59
81 Tl THALLIUM 204.38
82 Pb LEAD 207.2
83 Bi BISMUTH 208.98
84 Po POLONIUM (209)
85 At ASTATINE (210)
86 Rn RADON (222)
87 Fr FRANCIUM (223)
88 Ra RADIUM (226)
89-103**
104 Rf RUTHERFORDIUM (267)
105 Db DUBNIUM (268)
106 Sg SEABORGIUM (271)
107 Bh BOHRIUM (272)
108 Hs HASSIUM (270)
109 Mt MEITNERIUM (276)
110 Ds DARMSTADTIUM (281)
111 Rg ROENTGENIUM (280)
112 Cn COPERNICIUM (285)
113 Uut UNUNTRIUM (284)
114 Fl FLEROVIUM (289)
115 Uup UNUNPENTIUM (288)
116 Lv LIVERMORIUM (293)
117 Uus UNUNSEPTIUM (294)
118 Uuo UNUNOCTIUM (294)
*
57 La LANTHANUM 138.90
58 Ce CERIUM 140.116
59 Pr PRASEODYMIUM 140.90
60 Nd NEODYMIUM 144.242
61 Pm PROMETHIUM (145)
62 Sm SAMARIUM 150.36
63 Eu EUROPIUM 151.964
64 Gd GADOLINIUM 157.25
65 Tb TERIBIUM 158.92
66 Dy DYSPROSIUM 162.500
67 Ho HOLMIUM 164.93
68 Er ERBIUM 167.259
69 Tm THULIUM 168.93
70 Yb YTTERBIUM 173.054
71 Lu LUTETIUM 174.9668
**
89 Ac ACTINIUM (227)
90 Th THORIUM 232.0377
90 Pa PROTACTINIUM 231.03
92 U URANIUM 238.02
93 Np NEPTUNIUM (237)
94 Pu PLUTONIUM (244)
95 Am AMERICIUM (243)
96 Cm CURIUM (247)
97 Bk BERKELIUM (247)
98 Cf CALIFORNIUM (251)
99 Es EINSTEINIUM (252)
100 Fm FERMIUM (257)
101 Md MENDELEVIUM (258)
102 No NOBELIUM (259)
103 Lr LAWRENCIUM (262)

Fig. 1.1 Periodic table of the elements.

Learning crystal chemistry and crystal physics is an excellent way of developing an understanding of the structure–property relations underpinning materials science and engineering. While *crystal physics* determines which property coefficients are zero, based on symmetry arguments, *crystal chemistry* provides guidance on which property coefficients should be large or small.

Both of these fields, of course, require a strong understanding of crystal structures. Three-dimensional models and computer visualization software are both very helpful tools in understanding these. This book is extensively illustrated to facilitate an understanding of the three-dimensional nature of structures; the interested reader is strongly encouraged to seek out modeling kits, or one of the many programs that allow crystal structures to be plotted and rotated into various orientations.

The other part of crystal chemistry is *chemistry*. This textbook presumes a knowledge of the periodic table. There is a reasonably good chance that any material you make or utilize over the course of your career will be composed of elements on the periodic table. Therefore, it behooves any student in this field to have a firm grasp of the periodic table, including a working knowledge of the abbreviations for various elements, the location of different elements on the table, their electron configurations, and their common oxidation states. The periodic table itself is illustrated in Fig. 1.1 located on page 2.

2 Raw Materials

Common engineering materials are made from common minerals, which come mostly from common rocks. This is a comforting thought for materials scientists, for we will never run out of them.

2.1 Abundance of Elements

The continental crust is about 30 km thick, and is made up chiefly of silicates. Estimated abundances of the major elements in the continental crust are listed in Table 2.1. Seven of the eight common elements are cations. Oxygen is the *only* common element that is an anion. Hence, most minerals (and most ceramics) are oxides.

2.2 Common Minerals

The most abundant of the thousands of different minerals found in the Earth's crust are listed in Table 2.2. The Earth consists of three main parts: (1) a central iron–nickel *core*, (2) an intermediate oxide *mantle* composed of dense iron and magnesium silicates, and (3) a thin outer *crust* enriched in aluminosilicates, alkali elements, and calcium. The minerals used in making ceramics and metals come from this outer crust.

As expected, the list is dominated by silicates. In nearly all silicates, silicon is bonded to four oxygens, with a Si–O bond length of ~1.6 Å. As discussed later, silicate structures are often classified according to the way in which the SiO_4^{4-} tetrahedra link together. Among the common minerals, the feldspars and quartz are network silicates in which the tetrahedra link together via the oxygens on the corners to form a three-dimensional network. Micas and clays are layer silicates with strong bonding in two dimensions, and pyroxenes and amphiboles are chain silicates with one-dimensional linkages. The silicate tetrahedra in olivine are discrete; that is, they bond to other ions in the structure, rather than forming Si–O–Si linkages. In much the same way, the CO_3^{2-} groups in carbonate minerals are distinct. These structural differences have a strong influence on the chemical and physical properties of minerals.

Table 2.1 Major chemical elements in the Earth's crust and their abundance

Element	Weight%	Atomic%	Atomic #	Atomic wt. (amu)	Valence
Oxygen (O)	46.6	62.6	8	16.00	−2
Silicon (Si)	27.7	21.2	14	28.09	+4
Aluminum (Al)	8.1	6.5	13	26.98	+3
Iron (Fe)	5.0	1.9	26	55.85	+2, +3
Calcium (Ca)	3.6	1.9	20	40.08	+2
Sodium (Na)	2.8	2.6	11	22.99	+1
Potassium (K)	2.6	1.4	19	39.10	+1
Magnesium (Mg)	2.1	1.8	12	24.31	+2

Table 2.2 Mineral abundance in the continental crust

Mineral		Abundance
Feldspars		58%
Orthoclase	$KAlSi_3O_8$	
Albite	$NaAlSi_3O_8$	
Anorthite	$CaAl_2Si_2O_8$	
Pyroxenes and amphiboles		13%
Diopside	$CaMgSi_2O_6$	
Enstatite	$MgSiO_3$	
Tremolite	$Ca_2Mg_4Si_8O_{22}(OH)_2$	
Quartz	SiO_2	11%
Micas, clays		10%
Muscovite	$KAl_2(AlSi_3O_{10})(OH)_2$	
Kaolinite	$Al_2Si_2O_5(OH)_4$	
Carbonates		3%
Calcite	$CaCO_3$	
Dolomite	$CaMg(CO_3)_2$	
Olivines		3%
Forsterite	Mg_2SiO_4	
Other Minerals		2%

2.3 Feldspars

The crystal structure of sodium feldspar (albite = $NaAlSi_3O_8$) is illustrated in Fig. 2.1. As pointed out in Table 2.2, feldspars are by far the most abundant mineral family in the Earth's crust. Their chemical compositions are given by $(K_{1-x-y}Na_xCa_y)(Al_{1+y}Si_{3-y})O_8$, where x and y range between 0 and 1. The end members are the minerals orthoclase ($KAlSi_3O_8$), albite ($NaAlSi_3O_8$), and anorthite ($CaAl_2Si_2O_8$). Orthoclase also exists in slightly different high and low temperature forms known as sanidine and microcline.

The feldspar crystal structure consists of AlO_4^{5-} and SiO_4^{4-} tetrahedra linked together to form a three-dimensional framework, as shown in Fig. 2.1. Charge balance is maintained through the presence of alkali or alkaline earth cations in cavities in the tetrahedral framework.

(a)

(c)

(b)

(d)

Crystal System: Monoclinic
Lattice parameters
$a = 8.56$Å,
$b = 13.03$Å, $\beta = 116°$
$c = 7.18$Å,
Space Group: C2/m

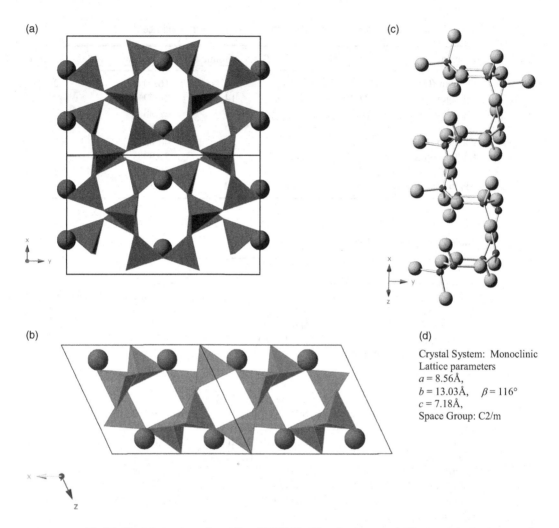

Fig. 2.1 Crystal structure of sanidine, $KAlSi_3O_8$. The aluminum and silicon tetrahedra (or atoms) are shown in a dark color, oxygen in large light spheres, and K as large dark spheres. (a) (001) projection, (b) (010) projection, (c) a portion of the aluminosilicate framework of the feldspars. Larger Na, Ca, or K ions fit into cavities within the framework. Si and Al atoms are bonded to four oxygens and each oxygen is bonded to two Si or one Si and one Al. Sanidine is the high temperature form of $KAlSi_3O_8$ in which aluminum and silicon are disordered on the tetrahedral sites, with an average bond length of 1.64 Å. Potassium is coordinated to nine oxygens at distances ranging from 2.70 to 3.13 Å. Sanidine is the prototype structure for the feldspars, an important family of minerals constituting almost 2/3 of the Earth's outer crust. The structures of most other feldspars are of lower symmetry than sanidine. The distortions can be traced to Al–Si ordering or to partial collapse around the large cation, or both.

The physical properties of the feldspars are readily understood in terms of the crystal structure. High hardness (6 on Mohs' scale) and high melting point (>800 °C) are related to the strong Al–O and Si–O bonds, which are cross-linked in all directions. Bonding is the strongest in the direction of the zig-zag chains shown in Fig. 2.1. Mechanical cleavage takes place on planes of lower bond density parallel to the chains.

Feldspars are generally pale in color because iron, the most common coloring agent in minerals, does not fit into the structure well.

2.4   Common Elements and Their Oxides

When discussing ceramics, materials scientists and engineers use both common names and mineral names.

Silicon dioxide (silica) occurs in nature as the mineral quartz (SiO_2). Second only to oxygen in abundance, silicon occurs in a very large number of minerals. Silicon single crystals form the basis of the vast majority of integrated circuits. Amorphous SiO_2 films are essential in that application for their excellent insulating properties and high dielectric breakdown strength. Silica glass (SiO_2), silicon nitride (Si_3N_4), and silicon carbide (SiC) are important industrial products. Optical fibers based on SiO_2 are at the heart of the high-speed telecommunications industry.

Aluminum is the third most common element in the Earth's crust. The principal ore is bauxite, an impure mixture of diaspore (AlOOH) and gibbsite ($Al(OH)_3$). Aluminum oxide (alumina) occurs in nature as the mineral corundum (α-Al_2O_3). Ruby and sapphire are gem varieties of corundum, where the colors are introduced by transition metal impurities. Alumina ceramics are noted for their strength, hardness, and high melting point. China clay (kaolinite = $Al_2Si_2O_5(OH)_4$) is used in making whitewares. Kyanite (Al_2SiO_5) is an aluminosilicate used in refractories. As a metal, aluminum is widely used in packaging, and in applications where its light weight is an asset, as in aeronautical engineering.

Iron is the heaviest of the common elements. Steel (FeC_x) is a strong intermetallic alloy hardened by interstitial carbon atoms. Hematite (Fe_2O_3) and magnetite (Fe_3O_4) are ore minerals used for steel-making. Magnetic ceramics are made from ferrites such as $MnFe_2O_4$ and $BaFe_{12}O_{19}$.

Calcium oxide (lime = CaO) is usually obtained by heating limestone (calcite = $CaCO_3$) to drive off carbon dioxide. Apatite ($Ca_5(PO_4)_3OH$), dolomite ($CaMg(CO_3)_2$, gypsum ($CaSO_4 \cdot 2H_2O$), and anorthite feldspar are other common calcium minerals. Hydrated lime (portlandite = $Ca(OH)_2$), gypsum, and clay are key components of cement and concrete.

Sodium and sodium oxide (soda = Na_2O) are usually extracted from rocksalt (halite = NaCl). Trona ($Na_3(CO_3)(HCO_3) \cdot 2H_2O$ is another mineral source for the sodium used in glass-making. A more common occurrence in nature is the feldspar mineral albite = $NaAlSi_3O_8$.

Potassium is extracted from chloride minerals such as sylvite (KCl) and carnallite ($KMgCl_3 \cdot 6H_2O$). The oxide potash (K_2O)) is a major constituent of agricultural fertilizers and a minor additive in many glass compositions. Orthoclase ($KAlSi_3O_8$) is potash feldspar and muscovite ($KAl_2(AlSi_3O_{10})(OH)_2$ is common mica.

Magnesium oxide (magnesia = MgO) is also known by its mineral name periclase. Mg enters into many rock-forming minerals including olivine (mostly Mg_2SiO_4) and pyroxenes such as diopside ($CaMgSi_2O_6$). Of greater economic importance are the carbonates magnesite ($MgCO_3$) and dolomite ($CaMg(CO_3)_2$) used in making refractories.

Hydrogen is widely distributed on the Earth's crust, not only as water and ice (H_2O), but also as partially hydrated minerals such as micas, clays, and amphiboles (tremolite =

$Ca_2Mg_5(Si_4O_{11})_2(OH)_2$). Clean drinking water and abundant hydrogen fuel are major engineering objectives in the years ahead.

2.5   Ore Minerals

Metallic elements are often very specific in regard to the type of minerals they form. A few elements, such as platinum, palladium, and gold, occur primarily as *metallic alloys*, while others like copper and zinc are found primarily as *sulphides*. As expected, based on elemental abundances in the Earth's crust, *oxides* are the most common ore minerals. Silicon, aluminum, magnesium, and iron are generally recovered from oxygen compounds. A few elements like the inert gases form no minerals at all. Geochemists classify these four groups as lithophile, chalcophile, siderophile, and atmophile (see Table 2.3).

The geochemical character of an element is governed by the type of chemical bonding involved. *Lithophile* elements are readily ionized to form oxides or stable oxyanions such as silicates, carbonates, sulfates, or phosphates. The bonding has a strong ionic character. *Chalcophile* elements ionize less easily and tend to form covalent bonds with elements like S, Se, and Te. Chalcopyrite, an important ore of copper, is a typical chalcophile mineral (see Fig. 2.2), as is cinnabar, an important ore of mercury (Fig. 2.3). *Siderophile* elements are those for which metallic bonding is the normal condition.

2.6   The Oceans

Oceans cover about 70% of the Earth's surface, some 1.5×10^{21} liters of seawater. Table 2.4 shows the concentration of eight elements, 3.5% of which consist of dissolved substances. Water is an excellent solvent for many ionic compounds because the dipole moment of water effectively shields the charged cations and anions from each other, as shown in Fig. 2.4.

For several centuries, scientists and engineers have been seeking ways of removing salts from seawater to augment the supply of freshwater. The three principal technologies for desalination are distillation, freezing, and reverse osmosis. *Distillation* is the oldest and most widely used method. The process involves vaporizing seawater and then condensing the purified water vapor. To reduce costs, solar radiation is used to evaporate the seawater in large "solar stills." *Freezing* methods are also under development for desalination. When seawater is frozen, the solid ice is nearly pure H_2O. Much less energy is required to freeze seawater than to evaporate it, since the heat of fusion is only 6.01 kJ/mole, compared with 40.79 kJ/mole for vaporization. Even less energy is involved in the third method of desalination, *reverse osmosis*. No phase change is required for this method, making it potentially much less expensive. Reverse osmosis uses a high pressure of 30 atmospheres or more to purify seawater by forcing it through a semipermeable membrane. The main technical problem is development of inexpensive membranes that can be used for prolonged lifetimes under high-pressure conditions.

The ocean is also host to a large variety of raw materials in the continental shelf and ocean basins. These include mineral sands of titanium, manganese, and iron ores, phosphates, limestone, and diamonds. Bromine, magnesium, and sodium are recovered

Table 2.3 Typical ore minerals

Lithophile elements	Symbol	Typical ore
Aluminum	Al	Bauxite, $Al_2O_3 \cdot 2H_2O$
Barium	Ba	Barite, $BaSO_4$
Beryllium	Be	Beryl, $Be_3Al_2Si_6O_{18}$
Boron	B	Borax, $Na_2B_4O_7 \cdot 10H_2O$
Calcium	Ca	Calcite, $CaCO_3$
Cerium	Ce	Monazite, $CePO_4$
Chromium	Cr	Chromite, $FeCr_2O_4$
Iron	Fe	Hematite, Fe_2O_3
Lithium	Li	Spodumene $LiAlSi_2O_6$
Magnesium	Mg	Magnesite, $MgCO_3$
Manganese	Mn	Pyrolusite, MnO_2
Niobium	Nb	Columbite, $FeNb_2O_6$
Phosphorus	P	Apatite, $Ca_5(PO_4)_3OH$
Potassium	K	Sylvite, KCl
Silicon	Si	Quartz, SiO_2
Sodium	Na	Halite, $NaCl$
Strontium	Sr	Celestite, $SrSO_4$
Sulphur	S	Gypsum, $CaSO_4 \cdot 2H_2O$
Tantalum	Ta	Tantalite, $FeTa_2O_6$
Tin	Sn	Cassiterite, SnO_2
Titanium	Ti	Ilmenite, $FeTiO_3$
Tungsten	W	Scheelite, $CaWO_4$
Uranium	U	Uraninite, UO_2
Zirconium	Zr	Zircon, $ZrSiO_4$
Chalcophile elements	**Symbol**	**Typical ore**
Antimony	Sb	Stibnite, Sb_2S_3
Arsenic	As	Arsenopyrite, $FeAsS$
Cadmium	Cd	Greenockite, CdS
Cobalt	Co	Cobaltite, $CoAsS$
Copper	Cu	Chalcopyrite, $CuFeS_2$
Lead	Pb	Galena, PbS
Mercury	Hg	Cinnabar, HgS
Molybdenum	Mo	Molybdenite, MoS_2
Nickel	Ni	Pentlandite $(Ni,Fe)S$
Silver	Ag	Argentite, Ag_2S
Zinc	Zn	Zincblende, ZnS
Siderophile elements	**Symbol**	
Gold	Au	
Platinum	Pt	
Palladium	Pd	
Atmophile elements	**Symbol**	
Helium	He	
Neon	Ne	
Argon	Ar	
Krypton	Kr	
Xenon	Xe	
Nitrogen	N_2	

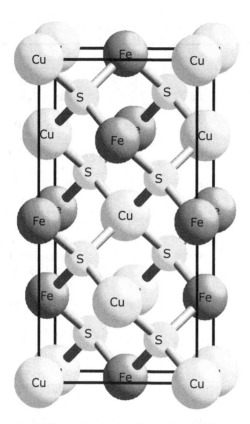

Crystal System: Tetragonal
Lattice parameters
    a = 5.24Å
    c = 10.30Å
Space Group: $I\overline{4}2d$

Fig. 2.2 The unit cell of chalcopyrite, $CuFeS_2$. Iron and copper are bonded to four sulfurs in tetrahedral configuration. Each S is bonded to two Cu and two Fe.

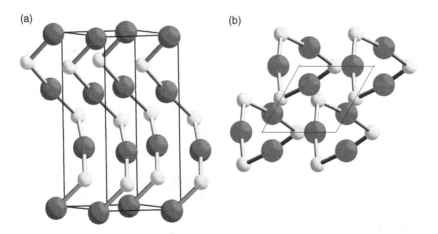

Fig. 2.3 Crystal structure of cinnabar, HgS. (a) Several chains of Hg and S atoms spiraling parallel to c. (b) Four spirals looking down the c axis. Crystal system: the crystal system is trigonal, with lattice parameters $a = 4.149$ Å, $c = 9.495$ Å, and space group: $P3_121$.

Table 2.4 Typical composition of seawater. Salinity ranges from less than 1% in arctic waters to around 4% in the Red Sea

Element	Weight %	Atomic %
Oxygen	85.77	32.92
Hydrogen	10.87	66.38
Chlorine	1.90	0.33
Sodium	1.15	0.28
Magnesium	0.14	0.03
Sulfur	0.09	0.02
Calcium	0.04	0.02
Potassium	0.04	0.01

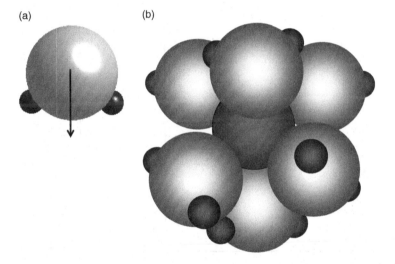

Fig. 2.4 (a) Image of a water molecule. The small H atoms are partially buried inside the oxygen, yielding a very short O–H bond length of ~0.96 Å. The arrow represents the dipole moment of the water molecule. (b) Representative configuration of water molecules around an Na^+ ion.

from seawater, and petroleum products and sulfur from the continental shelf. Calcite and aragonite (two forms of $CaCO_3$) are the main constituents of seashells.

2.7 Atmosphere

We live in an ocean of air composed mainly of nitrogen and oxygen (see Table 2.5 and Fig. 2.5), a few trace elements, water vapor, and various pollutants. Because of gravitational forces, air is much denser near the surface of the Earth. At sea level, atmospheric pressure is 1 atm = 1.01×10^5 Pa = 1.01×10^5 N/m^2 = 14.7 psi. 99% of the atmosphere is within 32 km = 20 miles of the Earth's surface.

Table 2.5 Average composition of the Earth's atmosphere with variable amounts of H_2O, CO_2, and other molecules

Nitrogen (N_2)	75.38 wt %	77.87 atomic %
Oxygen (O_2)	23.18	21.12
Argon (Ar)	1.39	0.47

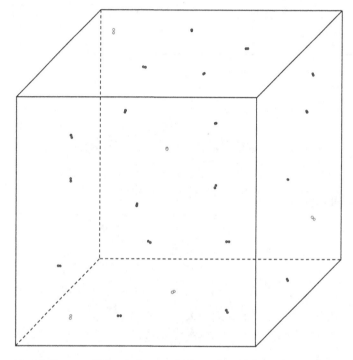

Fig. 2.5 Oxygen (O_2) and nitrogen (N_2) are the chief constitutes of the Earth's atmosphere. The nitrogen molecules are shown in filled circles, the oxygen in white. The interatomic bond length is 1.10 Å in N_2 and 1.21 Å for O_2. The figure shows a representation of a volume of air of about 100 Å x 100 Å x 100 Å at room temperature and pressure.

Only about 2% of the Earth's nitrogen is free in the atmosphere. Most of the remainder is trapped in rocks. Plants depend on the availability of nitrogen-containing compounds produced from the Earth's atmosphere either biologically or commercially. Crop productivity is generally limited by the availability of nitrogen fertilizers.

Molecular oxygen is essential to respiration and animal life, and is formed as a waste product by many forms of vegetation. One mature tree supports two human beings. Oxygen (O_2) takes part in the combustion of fuels that support heat, light, and power and reacts chemically with all but a few elements to form oxide minerals, ceramics, and water. Commercial O_2 is obtained by cryogenic distillation and by gas absorption methods.

Table 2.6 Composition of fossil fuels (weight %)

Element	Coal	Crude oil
Carbon	60–95	80–87
Hydrogen	2–6	10–14
Sulfur	0.3–13	0.05–6
Nitrogen	0.1–2	0.2–3
Oxygen	2–30	0.05–1

(a)   (b)

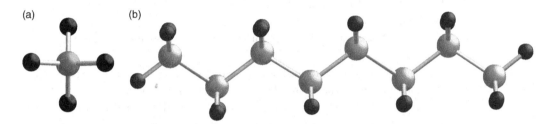

Fig. 2.6 Molecules of (a) methane and (b) *n*-octane (also called isooctane). The larger atoms are carbon, the smaller ones are hydrogen.

2.8   Fossil Fuels

Coal, crude oil, and natural gas are of organic origin. The organisms that provide the basic material for crude oil are believed to be the tiny plants and animals known as plankton. Over time, they sink to the bottom of the ocean and form deposits of considerable thickness. Eventually, this mass is converted into fuel. Wood (mainly cellulose) and other plant life are involved in the formation of coal.

Typical compositions of coal and crude oil are listed in Table 2.6. Note that coal possesses more oxygen and sulfur and less carbon and hydrogen than crude oil and natural gas. For many years, coal has been the principal source of inexpensive electrical power. Combustion of coal takes place in two stages, thermal decomposition to combustible gases and a residue of carbon, followed by combination of these products with oxygen. On combustion, the sulfur in the coal is released as sulfur dioxide, which is problematic from a pollution perspective. *Natural gas* is a mixture of naturally occurring hydrocarbon gases found in porous geologic formations beneath the earth's surface. Methane (Fig. 2.6) is the principal constituent with minor amounts of ethane (C_2H_6), propane (C_3H_8), and butane (C_4H_{10}). Natural gas is used as a primary fuel and chemical feedstock throughout the industrialized world. *Petroleum ethers* with five or six carbons are generally used as solvents for organic compounds. *Gasoline* and other automobile fuels are made up of liquids containing molecules with 6–12 carbon atoms. Isooctane (see Fig. 2.6) is a typical constituent. Jet engines and some home heating units use *kerosene* with 11–16 carbons per molecule. *Fuel oils* for domestic heating and for the production of electricity possess molecules with 14–18 carbons, while *lubricating oils* for automobiles and heavy machinery contain molecules with 15–24 carbons.

2.9   Problems

(1)   Learn the first 30 elements of the periodic table, their elemental symbols, and their common oxidation states.

(2)   Write the electron configurations for:
    (a)   Li
    (b) Li^{1+}
    (c) Fe^{2+}
    (d) Cu^{1+}.

(3)   What is the most common oxidation state of calcium? Justify this on the basis of the filling of the electron orbitals.

(4)   What would you expect the chemical formula of a compound of magnesium and sulfur to be?

(5)   Elemental analysis is performed on a sample of sodium silicate, and the Na:Si ratio is found to be 2:1. What is the chemical formula of the compound?

(6)   Suppose you have an ionic compound made up of silicon and fluorine. On the basis of your knowledge of common oxidation states, write the chemical formula of the compound.

(7)   What is the molecular weight of kaolinite ($Al_2Si_2O_5(OH)_4$)? Show your work.

(8)   What are the charges on each of the ions in apatite: $Ca_5(PO_4)_3OH$?

3 Chemical Bonding and Electronegativity

3.0   Introduction

The previous chapter discussed the elements as well as many of the compounds commonly available to scientists and engineers. In the solid phase, the atoms are held together by chemical bonds. Quantum mechanics provides valuable insights into the nature of these bonds, and the distribution of electrons around the atoms. However, in many cases, simplified models of the attractive forces responsible for bonding are of great benefit, and provide an intuitive understanding of critical factors. This chapter provides a brief review of the major types of chemical bonds and their influence on crystal structure and physical properties.

3.1   Chemical Bonds

Five primary types of chemical bonds have been identified as shown schematically in Fig. 3.1: covalent, ionic, metallic, hydrogen, and van der Waals bonding.

Covalent bonds involve a sharing of electrons between neighboring atoms. The electrons are supplied by one or both of the atoms and are generally located between the atoms in molecular orbitals. Because the molecular orbitals have well-defined geometries, covalent bonds tend to be highly directional. Covalent bonds are widely found between atoms in columns III, IV, and V in the periodic table.

Diamond is generally accepted as an ideal covalently bonded material. Each carbon atom has six electrons arranged in three shells: $1s^2 \, 2s^2 \, 2p^2$. The two inner electrons form a closed shell while the four 2s and 2p electrons participate in bonding. Together they form the tetrahedral sp^3 hybrid bonds found in diamond (Fig. 3.2). The structure and bonding in silicon and germanium are similar, but the bondlengths are somewhat longer.

Ionic bonds involve the transfer of electrons from one atom to another, creating positively charged cations and negatively charged anions. Ionic crystals are held together by the attractive Coulomb forces between cations and anions. Because the attractive force does not depend greatly on the geometric arrangement of neighbors, but only on their distance, ionic bonds are not directional in character. As a result, the coordination of atoms in ionically-bonded crystals (where coordination includes the number and type of near neighbors along with their geometric arrangement) is quite

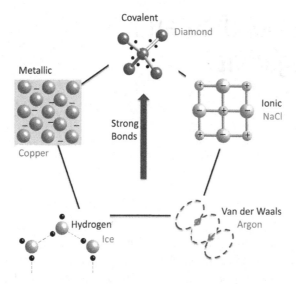

Fig. 3.1 Five major types of chemical bonds typified by diamond, rocksalt, copper, ice, and argon. Covalent bonds are generally the strongest, followed by metallic and ionic bonds, which are approximately equal in strength. Hydrogen bonds and van der Waals forces are the weakest. The dotted ellipses around the atoms in the van der Waals bonds are the electric field lines.

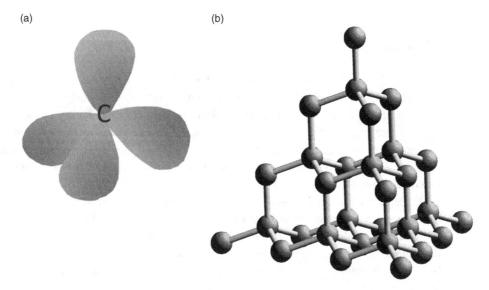

Fig. 3.2 (a) The sp^3 hybrid orbitals and (b) crystal structure of diamond.

varied. Some of the factors that govern coordination will be discussed in subsequent chapters (Chapters 8–13).

Rocksalt (= halite) is a good example of a largely ionic solid (Fig. 3.3). Before bonding, the electron configuration of neutral sodium is $1s^2\ 2s^2\ 2p^6\ 3s^1$, and chlorine

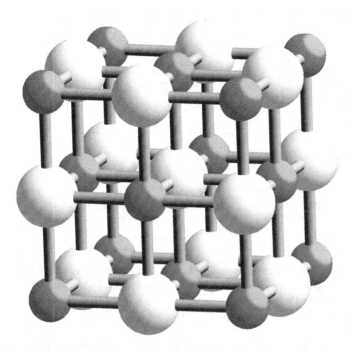

Fig. 3.3 Crystal structure of halite (NaCl). The larger spheres are Cl$^-$ and the smaller spheres are Na$^+$.

$1s^2\,2s^2\,2p^6\,3s^2\,3p^5$. Na has one electron outside the closed $2p^6$ shell, while Cl lacks one electron to complete the $3p^6$ shell. As a result, Na donates an electron to the neighboring Cl, forming spherically symmetric Na$^+$ cations and Cl$^-$ anions:

$$\text{Na}^+ \quad 1s^2\,2s^2\,2p^6$$

$$\text{Cl}^- \quad 1s^2\,2s^2\,2p^6\,3s^2\,3p^6$$

Metallic bonds arise from the attractive forces between nearly free electrons and positive ion cores. Copper has an electron configuration of $1s^2\,2s^2\,2p^6\,3s^2\,3p^6\,3d^{10}\,4s^1$. In metallic copper, the $4s^1$ electron is a mobile charge carrier which conducts heat and electricity. The attractive force between the free electron and the ionized cores can be thought of as a resonating covalent bond. From a band diagram perspective, metallic bonding results when there is overlap of orbitals between adjacent atoms that results in a partially filled conduction band. Here also, the bonds are not strongly directional.

Hydrogen bonds are very important in organic and biological chemistry. When a hydrogen atom loses its electron and becomes H$^+$, (i.e. a bare proton) it also loses most of its size. As a point positive charge, it readily forms a bridge between electronegative anions such as O^{2-} or F$^-$. In water, for example, the hydrogen bond is the bond *between* discrete H$_2$O molecules; this bond is substantially longer and weaker than the bonding within the molecule. In the inorganic world, hydrogen bonding is important in water and ice, as well as in partially hydrated minerals such as kaolinite (Al$_2$Si$_2$O$_5$(OH)$_4$) and gypsum (CaSO$_4 \cdot$ 2H$_2$O). It is also essential in many organic compounds and polymers.

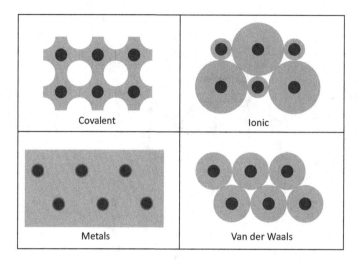

Fig. 3.4 Schematic illustration of electron distribution (light) around atom cores (dark) for different bond types. Adapted, by permission, from N. W. Ashcroft and N. D. Mermin, *Solid State Physics, 1E.* Saunders College, Philadelphia, PA 1976 Fig. 19.3, p. 378. Copyright Brooks/Cole, a part of Cengage Learning Inc. www.cengage.com/permissions

Hydrogen bonds are roughly an order of magnitude weaker than those of a typical covalent bond. As a result, they are much more readily broken either with increasing temperature, or with applied mechanical stresses.

Van der Waals bonding is present in all materials, even in the gaseous state, but the energies are very small. This type of bonding arises from the weak dipole or quadrupole forces between atoms or molecules. In many neutral atoms, the distribution of electrons around the nucleus is approximately spherical, averaged over time. However, instantaneously, the electrons can be distributed asymmetrically about the nucleus, so that the center of negative charge is not co-located with the center of positive charge in the atom. This produces an instantaneous dipole. This dipole can interact with the electron clouds of adjacent atoms, inducing a small attractive force between the atoms. In inert gases such as argon, it is the dominant type of bonding which causes the gas to liquefy and solidify at low temperatures. These bonds are typically on the order of one hundredth of the strength of a typical covalent bond.

For each of the bond types, the attractive forces holding the solid together are fundamentally electrostatic in origin. A schematic contrasting the electron distribution around atom cores for several bond types is shown in Fig. 3.4. Note that in covalent bonds there is significant electron density between atoms. In contrast, in ideal ionic bonds, the electron density drops to zero somewhere along the line connecting atom centers, since electrons are transferred from cations (smaller) to anions (larger). Bonds which have both covalent and ionic character are intermediate between the two extremes shown above, so that there are aspects both of electron transfer and electron sharing. In metallic bonding, the electron sea can be treated as belonging equally to all of the ion cores in the system. In van der Waals bonds, the electron clouds of the atoms undergo comparatively little perturbation on bonding. These differences produce bonds of significantly different strengths. A summary of the differences in

Table 3.1 Characteristic properties for different families of bonding

Property	Ionic bonding	Covalent bonding	Metallic bonding	Hydrogen bonding	Van der Waals bonding
Mechanical	Strong, hard materials	Strong, hard materials	Variable strength, plastic deformation more likely	Low strength, soft materials	Low strength, soft materials
Thermal	Medium to high melting points. Usually low thermal expansion	High melting points, usually low thermal expansion	Variable melting points, intermediate thermal expansion. High thermal conductivity	Low melting points, large thermal expansion coefficients	Very low melting points, larger thermal expansion coefficients
Electrical	Often wide band gap insulators	Often semiconductors	Conduction by free electron transport		Often insulators
Bond directionality	High coordination numbers result from low directivity	Spatially directed bonds produce low coordination numbers (often 3–4)	Coordination numbers often ≥ 8 due to low directivity in bonding		Coordination numbers often ~12

properties that are often observed in materials with these different bond types is given in Table 3.1.

3.2 Bond Length–Bond Strength Correlations

In understanding the factors that control the measurable material properties, it is useful to recognize that there is a strong correlation between bond length and bond strength within a given bond type.[1] *Short bonds are strong bonds*; longer separation distances between atoms are characteristic of weaker bonds. Since atom sizes increase as inner electron shells are added, this means that the strongest bonding is often found between atoms near the top of the periodic table. This correlation is illustrated in Fig. 3.5 for single bonds with mixed ionic–covalent character. The vertical axis in the figure is the average enthalpy per mole for bonds appearing in several different compounds. It is clear that the enthalpies rise for shorter bond lengths. As a result, more work would have to be done to break the shorter primary bonds.

As an example, consider the case of ice (H_2O). Ice has two different types of bonds: bonds with mixed ionic–covalent character that hold together the water molecules, and hydrogen bonding that provides the bonding between water molecules. The bond length within a molecule is approximately half the distance of the hydrogen bond. As a result, when ice melts, it is the weak hydrogen bonds that are broken. Water molecules retain their identity even in steam to much higher temperatures.

---

[1] L. Pauling, "The dependence of bond energy on bond length," *J. Phys. Chem.*, **58** (8), 662–666 (1954).

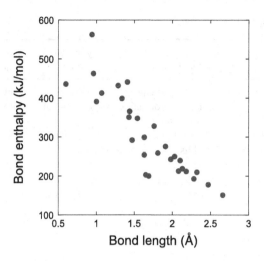

Fig. 3.5 Relationship between the average bond enthalpy and bond length in a series of single bonds with mixed ionic–covalent character.

A wide variety of material properties depend on the strength of some or all of the bonds that hold the solid together. As a result, many properties scale with the bond lengths, including the electronic band gaps, the elastic moduli, the thermal expansion coefficients, melting points, and others. Numerous examples of such correlations are given throughout this book.

It should also be noted that there are a number of factors that can lead to poor correlations between bond strength and bond length. For example, in metallically-bonded solids, the bond strength also depends on the number of electrons available to participate in the bonding per atom. As a result, there are anomalies in the bond strength–bond length trends associated with d-electron participation in bonding. In bonding between two atoms near the top of the periodic table, it is also worth recognizing that repulsion between the unshared pairs of electrons on neighboring atoms can reduce the bond strength.

3.3   Electronegativity

It is important to remember that the bond types discussed above represent limiting cases. In many materials, the bonding involves some mixture of these types of bonds. For example, the bonding in common ceramics and minerals is partly ionic and partly covalent. Electronegativity differences provide a way to estimate the ionicity of a bond. The larger the difference in electronegativity between two atoms, the more ionic the bond; the smaller the difference, the more covalent the bond.

Electronegativity is a measure of the ability to attract electrons. Atoms like fluorine and oxygen easily form anions and are the most electronegative of the elements. Alkali atoms such as Cs, Rb, K, and Na have the lowest electronegativity because of the ease with which they give up electrons. The alkali metals have low ionization potentials

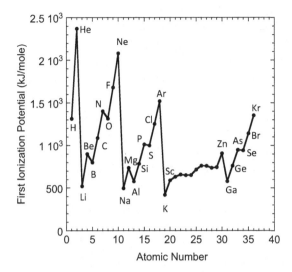

Fig. 3.6 Ionization potentials for the common elements. Alkali metals are easily ionized and readily form ionic bonds.

(Fig. 3.6). As a general trend, on moving left to right across the periodic table within a period, electronegativity values increase (though there are exceptions among the transition metals). Moving down the periodic table, the electronegativity values decrease.

Numerous methods for determining electronegativity values have been developed. Electronegativity values for the common elements based on the Pauling scale are shown in Fig. 3.7(a). A smooth increase takes place moving from column I of the periodic system to column VII. The dividing line between metals and non-metals is about 2 on the electronegativity scale. Figure 3.7(b) plots the percent ionicity versus the difference in electronegativity. This graph is commonly used to estimate the type of bonding in common oxides (Table 3.2). The percentage ionic character can be approximated as:

$$\% \ ionic \ character = 100^* \left[1 - e^{\frac{-1}{4}(\chi_A - \chi_B)^2} \right],$$

where χ_A and χ_B are the Pauling electronegativity of the anion and cation, respectively. For the main group elements there is a correlation between bond type and bond length. The stronger, more covalent bonds, like Si–O, are shorter than the weaker, more ionic, bonds such as Na–O. The chemical bonds in many ceramics are 50–90% ionic.

The electronegativity of elements is also related to their propensity for oxidation, as shown in Fig. 3.8. This is critical both in the corrosion of metals and in the development of new battery technologies. Elements with low reduction potentials are likely to oxidize when in the presence of an element with a higher reduction potential. The strong correlation between reduction potential and electronegativity means that electronegativity can be used as a guide in predicting corrosion behavior. In much the same way, electronegativity is an important concept in the processing of ceramic bodies. Much of that processing is done in aqueous solutions where some metals form bases, others form acids, and others form condensed oxides.

Table 3.2 Ionicity and bondlengths for common oxides

Bond	Percent ionicity	Bond length
Si–O	45%	1.6 Å
Al–O	57%	1.8 Å
Mg–O	68%	2.1 Å
Fe–O	47%	2.1 Å
Ca–O	77%	2.5 Å
Na–O	79%	2.5 Å
K–O	82%	2.8 Å

(a)

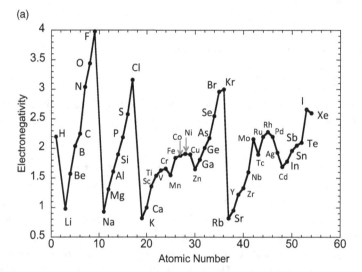

(b)

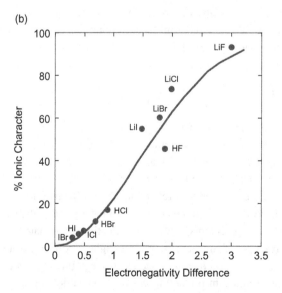

Fig. 3.7 (a) Electronegativity values for main group elements; (b) the percent ionicity of a chemical bond can be estimated from the difference in electronegativity. Adapted by permission from L. Pauling, *General Chemistry*, Dover Publications, New York (1970).

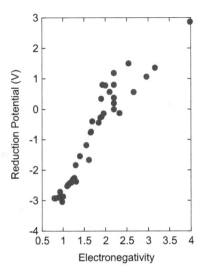

Fig. 3.8 Relation between reduction potential and electronegativity.

Electronegativity is a useful concept that helps systematize our understanding of a wide variety of phenomena. It serves, in many ways, as an empirical approach to describing the quantum mechanical properties of atoms. More sophisticated computational approaches which treat the quantum character directly are seeing increasing use, but are more challenging to apply in a general fashion in interpreting material properties.

3.4   Bonds with Mixed Character

The idea of electronegativity helps to quantify how bonds can have different characters; although many physical models treat different types of bonds with different formalisms, in reality bond types are not pure. As discussed above, mixtures of the strong bonding types are common in materials. Most ceramic materials and rocks have a mixture of ionic and covalent bonding characteristics. In the same way, metallically conducting oxides such as $SrRuO_3$ or ReO_3 have bonding that can be regarded as a mixture of predominantly ionic and metallic bonding, since some of the electrons reside in partially filled bands.

Finally, in all solid materials, once the atoms are close enough, there is a van der Waals contribution to the bonding. This is often a minor correction to the total bonding energy when other bond types are present, but it needs to be accounted for to make accurate calculations of the total bonding energies.

3.5   Problems

(1)   What is the percent ionic character for the following?
    (a)   Na–Cl
    (b)   Al–N

  (c) Ca–O

  (d) Si–N

  (e) Mg–O

  (f) Cu–Zn

  (g) Si–Ge

(2) Draw a picture of the sp^3 orbitals in diamond. Relate this to the observed crystal structure.

(3) Discuss (do not just list) the types of bonds that are observed in solids. Give an example of a material that shows each bond type. If there is more than one type of bond in any compound you choose as an example, be sure to specify which bond you are referring to.

(4) Discuss the types of bonds that are observed in materials. In each case, describe the origin of the attractive force. Give an example of a material that shows each bond type. If there is more than one type of bond in any compound you choose as an example, be sure to specify which bond you are referring to.

4 Hardness, Melting Points, and Boiling Points

4.0 Introduction

From an engineering perspective, the real importance of crystal chemistry is to understand how the crystal structure and bonding control the physical and chemical properties of matter. It is worthwhile considering first why there might be a relationship between bond strength or crystal structure and the boiling point, melting point, or the hardness. This can be thought of in the following manner; one means to measure the strength of the bond is to break it. Bonds can be broken mechanically in hardness tests or thermally on vaporizing a solid. Thus, an understanding of bond types and the way bonds are arranged in space (e.g. in a particular crystal structure) provides a useful way of rationalizing observed properties.

4.1 Hardness

The Mohs hardness scale (see Table 4.1) is a qualitative scale that illustrates key links between crystal structure, bond strength, and properties.[1] Mohs hardness measures how difficult it is to leave a scratch on a solid when two materials are rubbed together; the higher the hardness, the more difficult the material is to scratch. Although it is a qualitative test, it is a useful way to identify minerals. For engineering purposes, indentation hardnesses are conducted by pushing a tip with a well-defined shape into the sample surface. This permits more quantitative measurements; Mohs scale and more quantitative measurements correlate reasonably well.

The stronger a bond is, the more difficult it will be to break. If these strong bonds are arranged in three dimensions, the solid will be very hard. On the other hand, secondary bonds such as van der Waals or hydrogen bonds are easily broken. Materials connected together by these bonds have low hardnesses. An excellent example of this is talc; although the mineral has strong bonds such as Mg–O and Si–O within layers, the layers are bonded by van der Waals bonds. Therefore, it is the lowest hardness on Mohs scale. Thus, it matters not just how strong the bonds are, but how the bonds are arranged in space.

---

[1] Portions of this section are adapted from R. E. Newnham, *Structure—Property Relations*, Springer-Verlag, New York (1975) (by permission).

Table 4.1 Mohs hardness scale illustrating the bonds broken during the scratch test. Hardness is determined by the chemical bonds in the weakest direction. Generally speaking, covalent bonds are stronger than metallic and ionic bonds. Hydrogen bonds and van der Waals forces are weakest of all

Mohs hardness	Material	Bond broken	Comments	
10	Diamond C	C–C	3D array of strong bonds	Covalent
9	Corundum Al_2O_3	Al–O	Close-packed O network Produces a 3D array of strong Al–O bonds	
8	Topaz $Al_2SiO_4F_2$	Al–O Al–F Si–O	Close-packed O network	
7	Quartz SiO_2	Si–O	Low density of strong Si–O bonds	
6	Orthoclase $KAlSi_3O_8$	Al–O Si–O K–O	Cleavage plane	
5	Apatite $Ca_5(PO_4)_3F$	Ca–O	Primarily ionic bonds broken	
4	Fluorite CaF_2	Ca–F		
3	Calcite $CaCO_3$	Ca–O	Cleaves between CO_3 groups	Ionic
2	Gypsum $CaSO_4 \cdot 2H_2O$	OH—O		Hydrogen bonds
1	Talc $Mg_3Si_4O_{10}(OH)_2$	O—O	Cleaves between silicate layers	van der Waals bonds

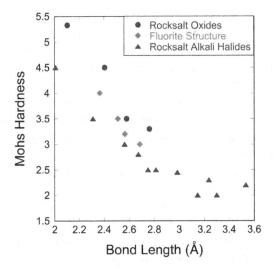

Fig. 4.1 Comparing solids with similar structures. Hardness generally decreases with bond length and increases with valence. Note that there is considerable scatter in hardness data for any particular mineral, which adds uncertainty to the absolute y axis values on the plot.

Another example of this can be seen in molecular solids. Molecular solids have strong bonds within the molecules, but the molecules are held together with secondary bonds. As a result, the material can be scratched by breaking only the weaker bonds between molecules. Thus, most molecular solids have low hardnesses (generally 3 or below). The hardness increases as the number of bonds that must be broken rises; this trend can be used to assess the relative hardnesses between hydrogen bonded solids.

Higher hardesses are obtained when the bonds that must be broken are stronger bonds such as metallic, ionic, or covalent bonds. The highest hardnesses are obtained in materials with covalent bonds that form a three-dimensional (3D) network. Diamond and SiC are excellent examples of this.

In comparing ionically bonded solids, the bond strength and hardness are directly proportional to valence and inversely proportional to interatomic distance. The correlations between hardness and valence, and between hardness and bond length, are illustrated in Fig. 4.1. Divalent alkaline-earth oxides with the rocksalt structure are much harder than the monovalent alkali halides. Fluorite-family alkaline-earth halides with divalent cations and univalent anions have intermediate hardnesses. In the same way, the hardness of $CaCO_3$ (3) exceeds that of $NaNO_3$ (2) because the $Ca^{2+}-O^{2-}$ bond is stronger than the $Na^{1+}-O^{2-}$ bond due to the higher valence. The two materials have the same structure and the Na–O and Ca–O bond lengths are nearly identical.

The trends with bond length are equally clear in Fig. 4.1. Longer bonds lead to lower Coulomb forces and lower hardnesses. Short bonds are strong bonds.

Hardness also rises as the number of bonds that must be broken increases. Thus, compounds with higher bond densities (number of bonds/unit volume) have higher hardnesses. This is easiest to confirm by comparing the relative hardnesses of polymorphs. Polymorphs have the same composition, but different crystal structures.

The crystal structure with the higher bond density will produce the higher hardness. Thus diamond is a 10 on Mohs scale, while graphite has a Mohs hardness of 1–2. In diamond, the density is 3.51 g/cm^3 and graphite has a density of 2.27 g/cm^3. The crystal structures of these materials are described in Chapter 8. Smaller differences in hardnesses are observed when the structures are more similar. Consider, as an example, silicate framework structures. The zeolites have hardnesses from 4 or 5, orthoclase feldspar has a hardness of 6, and quartz has a Mohs hardness of 7. The densities follow the same order, 2.2 for zeolites, 2.56 for orthoclase and 2.65 g/cc for quartz.

Similar trends between bond strength and hardness are observed in metallically bonded solids. As more d electrons participate in the bonding, the bond strength increases and the hardness rises. This can be seen by comparing the hardnesses of alkali metals (~0.5) to alkaline-earth metals (~2), and the transition metals (4–7). In general, the hardness of metals covers about the same range as ionic crystals, showing that the bonds are comparable in strength. Soft metals can be hardened by modifying the composition to make it more difficult to slip planes of atoms over each other (see Chapter 30). Thus, interstitial atoms such as carbon increase the hardness of steel to 6–8.

Crystal structures where covalent bonds connect small atoms in three-dimensional networks have the highest hardnesses. Thus, many of the hardest materials contain atoms near the top of the periodic table like carbon, boron and nitrogen. Diamond, cubic boron nitride (BN), boron carbide (B_4C), and silicon carbide (SiC) are all very hard.

Hardness decreases with increasing temperature, with soft materials showing a decrease of about 1 in hardness for a 20–100 °C temperature rise.

4.2   Melting Points, Boiling Points, and Bonding

Boiling points correlate closely with bond strength because when bonds break, molecules evaporate.[2] The relationship between bond strength and melting point is less obvious but some interesting generalizations apply. In structures containing both weak and strong bonds it is sometimes the weak bonds that determine melting, while strong bonds govern vaporization.

Substances with extended covalent bonding generally have high melting points. Crystalline diamond, graphite, silicon carbide, and boron nitride are stable to temperatures exceeding 3000 °C. Melting of such materials necessarily involves extensive disruption of strong covalent bonds, and high temperatures are required to achieve this. In molecular compounds, however, where only the molecular units are covalently bonded, the melting points are much lower. Titanium tetrachloride consists of tetrahedral $TiCl_4$ molecules weakly held together by van der Waals forces. $TiCl_4$ melts at –23 °C and boils at 137 °C. The weak intermolecular forces constitute the points of structural weakness in such a crystal, and fusion occurs at relatively low temperatures. Inert gases with only van der Waals bonding exhibit some of the lowest melting points.

---

[2] Portions of this section are adapted from R. E. Newnham, "Phase diagrams and crystal chemistry," in *Phase Diagrams: Materials Science and Technology, Volume V*, Academic Press, New York (1978) (by permission).

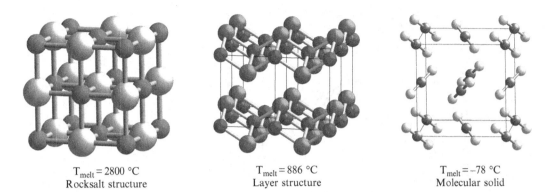

T$_{melt}$ = 2800 °C
Rocksalt structure

T$_{melt}$ = 886 °C
Layer structure

T$_{melt}$ = −78 °C
Molecular solid

Fig. 4.2 Three-dimensionally bonded crystals often have higher melting points than do molecular solids with weak intermolecular bonding. (Left) MgO is used as a refractory because of its high melting point while (middle) PbO with its weak interlayer bonding is often used as a flux to promote melting. Dry ice (CO_2) (right) has a molecular structure held together by weak dipole forces and sublimes below room temperature.

The melting points of ionic crystals are generally higher than molecular solids but lower than solids with extended covalent bonding. Sodium chloride melts at 801°C and boils at 1465 °C. The relationships between melting point and crystal structure are nicely illustrated with the oxides MgO, PbO, and CO_2 (Fig. 4.2). PbO, for example, has a layer structure made up of tetragonal pyramids of Pb^{2+} ions bonded to tetrahedral O^{2-} ions. The PbO bonds lengths *in the layer* are 2.2 Å, much shorter than the interatomic distance *between layers* (about 3.5 Å). The peculiar coordination of divalent lead, along with the low melting temperature, is attributed to the lone-pair electron configuration. Pb^{2+} has two electrons outside of a closed d shell. CO_2 has strongly bonded molecules held together by van der Waals forces and low melting and boiling points. MgO is a three-dimensional solid with strong bonds, and a higher melting temperature.

Among simple ionic crystals, the melting point is a measure of the strength of bonding. Comparing the melting points of RX compounds, for example, divalent substances have higher melting points than monovalent materials. Melting points in the alkali-halide family range from a low of 450 °C for LiI to a maximum of 988 °C for NaF, far less than the isostructural alkaline-earth oxides. Trends among the rocksalt family are shown in Fig. 4.3. The melting points of MgO, CaO, SrO, and BaO are 2800, 2580, 2430, and 1923 °C, respectively. The oxides also have high boiling points in the 2000–4000 °C range, whereas the halides all boil at less than 1700 °C. Divalent ions have twice the charge and approximately four times the Coulomb force of a monovalent ion, giving rise to larger heats of fusion and higher melting and boiling points. Among the most refractory oxides are thoria, urania, and zirconia, all of which possess the fluorite structure. The cubic fluorite structure resembles the arrangement that would be made if two PbO layers were connected back-to-back to make a three-dimensional structure.

Many of the best high-temperature materials are from the light elements in the first row of the periodic system. Graphite, boron nitride, beryllium oxide, and boron carbide have melting points in the 2300–3700 °C range, and boiling points above 3500 °C.

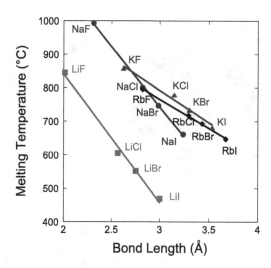

Fig. 4.3 Melting points for alkali halides generally decrease as bond lengths increase, although the lithium salts appear to be anomalously low. This has been attributed to the small ionic radius of Li^+, which brings the halogen ions into closer contact with one another. This increases the anion–anion repulsive forces and lowers the stability of the crystal.

Because nearly all the electrons in these compounds are involved in bonding, they have higher melting points than similar materials of higher atomic number. For the same reason, oxides have higher melting points than do sulfides, selenides, and tellurides. The heavier elements contain additional electrons arranged in closed shells which contribute little to the attractive forces between atoms and much to the repulsive forces. When the closed shells of neighboring atoms overlap, there is Coulomb repulsion, increasing the interatomic distance and lowering the melting point. These trends are clearly obvious among the various materials used in the semiconductor industry (Table 4.2), as well as in metals (Fig. 4.4).

Oxides, borides, carbides, and other ceramic materials can generally be used at higher temperatures than metals. However, metals are generally stronger than ceramics, so that ceramic-coated metals are a good compromise. The coatings not only improve the high-temperature characteristics, providing erosion, corrosion, and oxidation protection, but are also used to modify the electrical and optical properties of the surface.

Some of the coating materials employed in the aerospace industry are listed in Table 4.3. Coatings can be applied by flame-spraying molten or partially molten particles onto a cold metal surface. Impacting droplets flatten, interlock, and overlap to provide a dense coherent coating of the desired thickness. Surface materials are selected to have suitable characteristics for absorbing, reflecting, and emitting radiation in the required amounts. Coatings must also be sufficiently durable to withstand harsh environments.

Attractive forces associated with van der Waals bonds cause inert gases and many small molecules to condense into liquids and freeze into solids at low temperatures. The boiling points provide a measure of the molecular motion required to overcome these attractive forces. In general, the van der Waals attraction increases with the number of

Table 4.2 Semiconductor materials show a clear relationship between melting point and bond strength, with the longer, weaker bonds of heavy elements giving lower melting points

Group IV elements
Diamond structure

C	>3500 °C		
Si	1417		
Ge	937		

III–V compounds
Zincblende structure

GaP	1465 °C	InP	1070 °C
GaAs	1237	InAs	942
GaSb	712	InSb	525

II–VI compounds
Rocksalt structure

PbS	1077 °C		
PbSe	1062		
PbTe	904		

III–VI compounds
Layer structure

Bi_2S_3	850 °C		
Bi_2Se_3	706		
Bi_2Te_3	580		

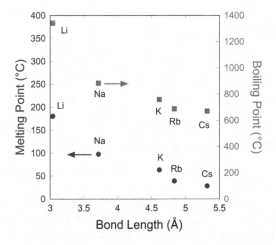

Fig. 4.4 Melting and boiling temperatures of the alkali metals decrease with bond length.

electrons per molecule, due to the larger dipoles that are generated. Therefore, large molecules with large molecular weight have higher boiling points than those of low molecular weight. The trends are illustrated in Fig. 4.5. Note that in this case the increased polarizability of the electron cloud on moving down the periodic table increases the bond strength enough that the melting point rises, even though the van der Waals radius (and hence the bond lengths) increase on moving from He to Rd.

Table 4.3 Approximate melting points of some refractory oxides, carbides, borides, and nitrides

Oxides		Carbides		Borides and nitrides	
ThO_2	3300 °C	HfC	3900 °C	HfB_2	3250 °C
UO_2	2880	TaC	3900	TaB_2	3100
MgO	2800	ZrC	3500	ZrB_2	3060
HfO_2	2780	NbC	3500	NbB_2	3000
CeO_2	2730	TiC	3250	TiB_2	2980
ZrO_2	2600	VC	2800	CrB_2	2760
Y_2O_3	2400	Al_4C_3	2800	SiB_6	1950
BeO	2350	WC	2770	BN	3000
Cr_2O_3	2260	MoC	2700	TiN	2940
$MgAl_2O_4$	2130	SiC	2700	AlN	2200
Al_2O_3	2070	B_4C_3	2450	Si_3N_4	1900

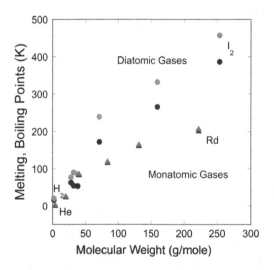

Fig. 4.5 Melting (darker symbol) and boiling points (lighter symbol) of monatomic and diatomic gases. The strength of the van der Waals bonding in the liquids and solids increases with molecular weight.

The melting and boiling points of hydrocarbons tend to increase with molecular weight and with symmetry. The melting point of the linear normal paraffins (Fig. 4.6) increases smoothly from *n*-butane (–138 °C) to higher molecular weights. In this case, all of the molecules consist of chained CH_2 groups terminated by CH_3 groups. Here, the van der Waals bond strength is not changing. Instead, as the number of carbon atoms in the chain rises, there are more species interacting that contribute to the overall cohesive strength. The energy holding the molecules together thus increases with molecular weight in a manner that depends on the atomic arrangement, as is discussed in Chapter 11.

The effect of symmetry is more subtle, and may offset the effect of molecular weight. When side groups are added to normal paraffins, the mirror symmetry is destroyed, and

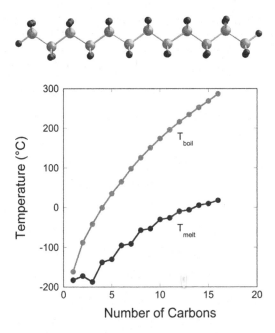

Fig. 4.6 Melting points and boiling points of long-chain hydrocarbon (top) plotted as a function of backbone length. Melting points increase with length because of the tendency of chains to entangle. Boiling points increase even faster because of the large number of bonds to be broken before vaporization is possible.

the branched-chain hydrocarbon melts far below the normal paraffins with the same molecular weight. The effect is illustrated with n-hexane and its isomers. n-Hexane, a straight-chain hydrocarbon containing six carbon atoms, has the highest melting point and the highest symmetry, $2/m$. Its isomer 2-methylpentane with a trimethyl side group belongs to point group 1 and has a much lower melting point. The other three isomers in Table 4.4 show a similar trend, with the low-symmetry isomers melting at lower temperatures. Note that the boiling points are nearly equal.

Most asymmetric molecules are difficult to crystallize. The molecules do not pack together in an efficient manner to meet the requirements of a densely packed crystal structure. Branched aliphatics and most substituted cyclics are used as lubricants because they crystallize slowly, if at all. On cooling, they merely increase in viscosity to glass-like solids.

Flat, high-symmetry molecules like benzene (point group $6/mmm$) pack together very efficiently in the crystalline state and have relatively high melting points (5.5 °C). The molecular weight of benzene is slightly lower than n-hexane, but the melting point is much higher because the more efficient packing allows more van der Waals interactions between molecules. The hydrocarbons illustrated in Table 4.4 all contain six carbons, but have very different molecular structures and symmetries. Their melting points differ by as much as 70 °C, but the boiling points by only 30 °C.

Boiling points are much less affected by molecular symmetry than are the melting points. Materials with anisotropic molecular units behave like tangled threads, increasing viscosity and the melting points. Methane, ethane, propane, butane, and the other

Table 4.4 Melting points, boiling points, and symmetry groups of C_6H_{14} isomers. High-symmetry molecules crystallize more easily than low-symmetry molecules and have higher melting points

Material	Symmetry	Melting point (°C)	Boiling point (°C)
n-hexane	2/*m*	−94	68.8
2-methylpentane	1	−153.7	60.2
3-methylpentane	*m*	−118	63.2
2,2-dimethylbutane	*m*	−98.2	49.7
2,3-dimethylbutane	1	−128.8	58

chain hydrocarbons illustrate the effect nicely (Fig. 4.6). Note that the separation between boiling and melting increases with chain length. This is because in the long hydrocarbons many more bonds must be ruptured in boiling the materials. Melting points increase less rapidly because the van der Waals bonding between molecules does not change much.

There are several relations between melting and structure, and between melting and other physical properties. Two of these are with thermal expansion and elasticity.

Perhaps the simplest concept of melting is Lindemann's observation that melting occurs when the amplitude of vibration exceeds a certain critical fraction of the interatomic distance. If x_m is the vibration amplitude at melting and we assume simple harmonic motion around the equilibrium site with a spring constant k, then the energy is

$$\frac{1}{2}kx_m^2 = k_B T_m,$$

where k_B is Boltzmann's constant and T_m is the melting point. The force constant k is related to elastic stiffness c and the interatomic distance d by $k = cd$ so that

$$T_m = cdx_m^2/2k_B = cd^3\beta^2/2k_B,$$

where $\beta = x_m/d$, the relative vibration amplitude at melting. Substituting measured values from T_m, c, and d for a wide range of materials, it is found that β is surprisingly constant, about 1/7. Thus, many materials melt when the thermal vibration amplitude is about 15% of the interatomic distance, the so-called 15% rule.

Because of the close relation with vibration amplitude, it is perhaps to be expected that melting and thermal expansion are correlated. Materials with low melting points have large thermal expansion coefficients (α) because all materials expand by about the same amount (3–10% in volume) before melting. Therefore αT_m is nearly constant. Fortunately there is enough scatter so that materials with different melting points can be bonded together – enamel and iron, for example.

4.3 Heat of Vaporization

The heat of atomization – the amount of heat required to vaporize one mole – is a good measure of bond strength.[3] The values quoted here are given in kilocalories per mole, and refer to solid elements at 300 K or at the melting point, whichever is lower.

The heats required to atomize rare-gas solids are small: He 0.5, Ar 1.8, Kr 2.6, and Xe 3.6 kcal/mole. The atoms have closed electron shells so that only van der Waals forces act between atoms. Close-packed structures are favored by the non-directional van der Waals forces. Solid He is hexagonally close-packed (see Chapter 10), and the other inert-gas solids are cubic close-packed.

Halogens have seven electrons per atom and bond together to form diatomic molecules. The energy of the electron pair bond can be estimated from the heats of atomization: F 20, Cl 32, Br 28, and I 26 kcal/mole. The bond energy is an order of magnitude larger than that of van der Waals solids. The halogens form molecular solids consisting of diatomic molecules. Melting points are low because the forces between molecules are weak.

Column VI elements have even higher heats of atomization: O 60, S 66, Se 49, Te 46, and Po 35 kcal/mole. The bonding energies are roughly twice those of the halogens,

---

[3] Portions of this section are adapted from R. E. Newnham, "Phase diagrams and crystal chemistry," in *Phase Diagrams: Materials Science and Technology, Volume V*, Academic Press, New York (1978) (by permission).

since two pairs of electrons are involved. The crystal structures of sulfur, selenium, and tellurium consist of rings or chains in which each atom is bonded to two others. Each bond is a single electron-pair bond. Solid oxygen contains O_2 molecules with double electron-pair bonds.

Heats of atomization for the fifth-group elements are as follows: N 114, P 80, As 69, Sb 62, and Bi 50 kcal/mole. These elements lack three electrons for a filled shell. Phosphorus, arsenic, antimony, and bismuth crystallize in puckered layers, with each atom forming three single bonds, each involving an electron pair. As expected, the heat of atomization is about three times that of column VII elements. Solid nitrogen contains N_2 molecules, with three electron pairs concentrated between nitrogen atoms. Multiple bonds are common in first row elements, but not elsewhere.

Carbon, silicon, and the other elements of column IV lack four electrons for a filled octet. The large heats of atomization (C 171, Si 108, Ge 90, Sn 72, and Pb 47 kcal/mole) reflect the increase in the number of bonding electrons. Several of the elements crystallize in the diamond structure in which tetrahedrally coordinated atoms form four single bonds. The tendency for first-row elements to form multiple bonds is again reflected in graphite, the second common polymorph of carbon.

Before discussing the structures of metals, we recapitulate the bonding in non-metallic elements. Most non-metals obey the *8 − N rule*: elements in column *N* of the periodic table form $8 − N$ covalent bonds. Many of the crystal structures can be explained by this rule. Each atom forms $8 − 4 = 4$ covalent bonds in diamond, $8 − 5 = 3$ in bismuth, $8 − 6 = 2$ in sulfur, $8 − 7 = 1$ in bromine, and $8 − 8 = 0$ in argon (see Fig. 4.7). The multiple bonds formed in first row elements are exceptions to the $8 − N$ rule since fewer (but stronger) bonds are formed. Heats of atomization among the non-metals are proportional to the number of bonding electrons. The heats increase steadily from column VIII elements in which there are no bonding electrons to column IV elements with four. There is also a marked dependence on the row of the periodic table, as well as the column. Within a column, the heats of atomization generally decrease with increasing atomic number. The value for carbon, for instance, is nearly four times that of lead. This behavior can be ascribed to the influence of the inner closed electron shells. Inner electrons contribute little to bonding, while increasing the interatomic distance, because of overlap repulsion. Note that the trend is reversed in rare-gas elements: xenon has a greater heat of atomization than argon. Inner electrons enhance the dipole interactions responsible for van der Waals attraction.

The energies required to atomize metals are comparable to those of non-metals, showing that the bonding energies are similar. For alkali metals in column I, the heats of atomization (Li 38, Na 26, K 22, Rb 20, and Cs 19 kcal/mole) span the same range as the halogens in column VII. In both cases, there is one electron available for bonding.

Similar correlations exist between column II and VI, and between III and V. Heats for the alkaline-earth elements (Be 78, Mg 35, Ca 42, Sr 39, Ba 43 kcal/mole) are comparable to those of the sulfur family, and about twice as large as corresponding alkali metals. Group IIB elements are slightly lower: Zn 31, Cd 27, Hg 15 kcal/mole, indicating fewer bonding electrons. Atomization energies for group IIIA elements

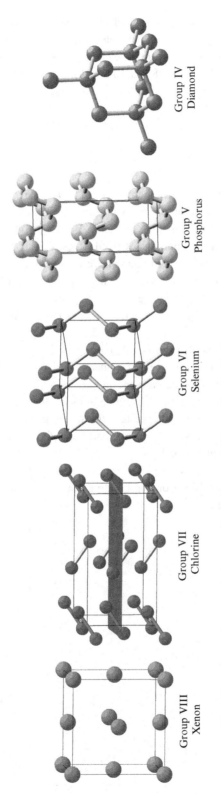

Fig. 4.7 Representative crystal structures from elements following the $8 - N$ rule.

(B 135, Al 78, Ga 69, In 58, and Tl 43 kcal/mole) are nearly the same as those for the nitrogen family. There are three electrons per atom in both groups, even though some are metals, and some are not. The elements of group IIIB have somewhat larger heats of atomization (Sc 88, Y 98, La 102 kcal/mole). The similarity in energy points out the similarity between covalent and metallic bonding. Metallic bonding can be visualized as resonating electron-pair bonds.

4.4   Problems

(1) Describe the relative hardnesses of SiO_2 (quartz), fluorite, and Al_2O_3. Explain why they have the order observed. Justify your reasoning.

(2) Rank order the melting points of Al, ZrO_2, MgO, and H_2O. Justify your reasoning based on the bonding involved. If more than one type of bonding is present, be specific about which you believe is broken on melting.

(3) Rank the following materials in terms of their hardness: CaF_2, gypsum, and Al_2O_3. Explain your reasoning.

(4) Cinnabar (HgS) has a space group of $P3_121$ with Hg atoms at the 0.72, 0, 0.3333 positions; and S atoms at 0.485, 0, 0.83333. It is a trigonal material with lattice parameters of $a = 4.149$ Å, $c = 9.495$ Å.

    (a) Draw a unit cell of cinnabar. This can be done using a number of crystal drawing software packages.

    (b) How many HgS formula units are there per unit cell?

    (c) Plot several unit cells of this material and show the bonds that connect atoms.

    (d) Would you expect the hardness of the material to be high or low? Explain your reasoning.

(5) Rank the following materials in terms of their melting point. Explain your reasoning based on the bonding and the crystal structures involved: Cu, C_6H_6, Ne, and ZrO_2.

5 Planes, Directions, and Morphology

5.0 Introduction

In this chapter the crystallographic notation used to label planes and directions in crystals is reviewed. This notation is widely used both to discuss crystals, and to describe the macroscopic characteristics of crystalline materials. For example, crystals often grow in characteristic shapes bounded by flat faces. The preferred shape is often called the *morphology* of the crystal. Some of the factors governing the crystal morphology are discussed.

5.1 Crystallographic Nomenclature

All crystals have translational symmetry with a fundamental unit, called the *unit cell*, repeated over and over many millions of times.[1] A generic unit cell is shown in Fig. 5.1; it is described in terms of three non-parallel vectors $\vec{a}$, $\vec{b}$, and $\vec{c}$, with lengths a, b, and c, and three interaxial angles, α, β, and γ. The cell lengths are called the *lattice parameters*. The angle α is the one between the b and c axes (sometimes called the angle "across from a"), while β is between a and c, and γ is between a and b. As will be described in detail in Chapter 6, the shape of the unit cell is linked to the symmetry of the crystal.

5.2 Zone Axes – Interaxial Angles

Taking any lattice point as the origin, the vector to any other lattice point may be expressed as $\vec{R} = u\vec{a} + v\vec{b} + w\vec{c}$, where u, v, and w are integers. Directions in crystals are denoted in a three-index notation $[uvw]$, with square brackets. Thus, vectors along the three principal axes of the unit cell are represented by [100], [010], and [001], as shown in Fig. 5.2. A vector in the $-\vec{a}$ direction is written as $[\bar{1}00]$. A family of equivalent directions related by symmetry is written in carat brackets $\langle uvw \rangle$. For a cubic crystal such as NaCl, the symmetry-related directions $\langle 100 \rangle$ are [100], [010], [001], $[\bar{1}00]$, $[0\bar{1}0]$, and $[00\bar{1}]$. In a triclinic crystal like kaolinite, with no other

---

[1] Excepting complications associated with quasi-crystals and incommensurate structures.

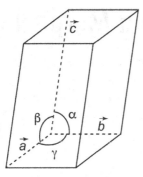

Fig. 5.1 Unit cell showing the cell lengths and cell angles. Note that, in general, the cell lengths may all be different from each other, as may the cell angles.

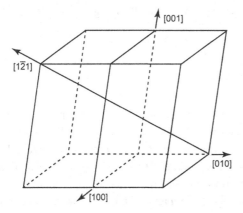

Fig. 5.2 Examples of some directions in crystals. A direction can be shown by picking an origin, and counting out u units of $\vec{a}$, v units of $\vec{b}$, and w units of $\vec{c}$, sequentially. Then the origin is connected to the final point.

symmetry elements than translational symmetry, there are no symmetry-related directions to [100], so $\langle 100 \rangle = [100]$.

It is useful to remember that since the directions are written in terms of the unit cell parameters, the absolute lengths of a unit in the directions of [100], [010], and [001] may be different. Likewise, the unit vectors are not necessarily orthogonal to each other. The angle made between any two vectors, $\vec{R}_1$ and $\vec{R}_2$ can be calculated from $\vec{R}_1 \cdot \vec{R}_2$, where:

$$\vec{R}_1 = u_1\vec{a} + v_1\vec{b} + w_1\vec{c}$$

$$\vec{R}_2 = u_2\vec{a} + v_2\vec{b} + w_2\vec{c}$$

$$\vec{R}_1 \cdot \vec{R}_2 = R_1 R_2 \cos \phi$$

$$= u_1 u_2 a^2 + v_1 v_2 b^2 + w_1 w_2 c^2 + (u_1 v_2 + u_2 v_1) ab \cos \gamma$$

$$+ (u_1 w_2 + u_2 w_1) ac \cos \beta + (v_1 w_2 + v_2 w_1) bc \cos \alpha$$

and

$$R_1^2 = u_1^2 a^2 + v_1^2 b^2 + w_1^2 c^2 + u_1 v_1 ab \cos \gamma + u_1 w_1 ac \cos \beta + v_1 w_1 bc \cos \alpha$$

$$R_2^2 = u_2^2 a^2 + v_2^2 b^2 + w_2^2 c^2 + u_2 v_2 ab \cos \gamma + u_2 w_2 ac \cos \beta + v_2 w_2 bc \cos \alpha.$$

Thus,

$$\cos \phi = \frac{\vec{R}_1 \cdot \vec{R}_2}{\sqrt{R_1^2} \sqrt{R_2^2}}.$$

In cubic crystals, $\cos \alpha = \cos \beta = \cos \gamma = 0$, and $a = b = c$. In this case:

$$\cos \phi = \frac{(u_1 u_2 + v_1 v_2 + w_1 w_2)}{\sqrt{u_1^2 + v_1^2 + w_1^2} \sqrt{u_2^2 + v_2^2 + w_2^2}}.$$

As an example, the cosine of the angle between directions [111] and [1$\bar{1}$1] is $\frac{1-1+1}{\sqrt{3}\sqrt{3}} = \frac{1}{3}$. Thus, the angle is 70° 32′.

5.3   Miller Indices and Planes

Miller indices are used to designate planes. If a plane intersects the unit cell axes at $\vec{a}/h$, $\vec{b}/k$, and $\vec{c}/l$, then the Miller indices are written in parentheses as (hkl). Several important planes are illustrated in Fig. 5.3.

For all symmetries above monoclinic, there can be planes which are symmetry-related to one another. These are called a family. A family of planes related by symmetry is denoted by the use of braces {hkl}. For a diamond crystal with cubic symmetry, there are eight symmetry-related planes in the {111} family, including: (111), ($\bar{1}$11), (1$\bar{1}$1), (11$\bar{1}$), (1$\bar{1}$ $\bar{1}$), ($\bar{1}$1$\bar{1}$), ($\bar{1}$ $\bar{1}$1), and ($\bar{1}$ $\bar{1}$ $\bar{1}$).

A fourth index is sometimes used in describing hexagonal and trigonal crystals. In the symbol ($hkil$), the first three numbers h, k, and i refer to the three axes at 120° to each other in the a–b plane. In this case the relationship holds that $i = -(h+k)$. The i index is redundant. A comparison between three- and four-index notation for directions is shown in Fig. 5.4. The utility of the four-index notation can be seen from a comparison of the symmetry-related planes in trigonal or hexagonal crystals. Consider, for example, the family of planes defining the boundaries of a hexagonal prism, as shown in Fig. 5.5. In three-index notation, {100} = (100), (010), ($\bar{1}$10), ($\bar{1}$00), (0$\bar{1}$0), and (1$\bar{1}$0). It is not obvious simply by looking at the indices that these planes belong in one family. However, in four-index notation, the symmetry-related planes {10$\bar{1}$0} = (10$\bar{1}$0), ($\bar{1}$010), (01$\bar{1}$0), ($\bar{1}$010), (1$\bar{1}$00), and ($\bar{1}$100). Here, it is easier to identify the symmetry-related planes by inspection.

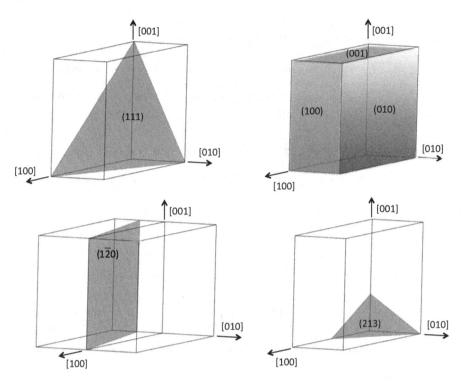

Fig. 5.3 Examples of planes in crystals. Miller indices are proportional to the reciprocals of the intercepts with the crystallographic axes.

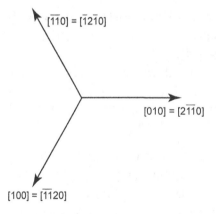

Fig. 5.4 Three- and four-index notation for directions looking directly down the c axis.

5.4   Relations between Zone Axes and Planes

Orthogonality: Since many crystals have inclined axes, one cannot assume that the direction $[hkl]$ is perpendicular to the plane (hkl) with the same indices. Only for cubic crystals is $[hkl]$ always $\perp$ (hkl). For the lower crystal systems, there are a few

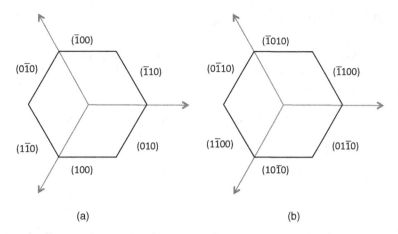

Fig. 5.5 Projection of a hexagonal prism on the a–b plane using (a) three-index and (b) four-index notation.

high-symmetry directions where this is true, but in general [hkl] is not perpendicular to (hkl). In hexagonal crystals, for example, [001] is perpendicular to (001) but [100] is not perpendicular to (100). In triclinic crystals, the orthogonality condition is never true.

Parallelism: The zone law is used to determine whether or not the direction [uvw] is parallel to, or contained in, the plane (hkl). If $hu + kv + lw = 0$, then [uvw] is parallel to (hkl). The zone law holds for all crystal systems regardless of symmetry. As examples, the [$5\bar{1}1$] direction is parallel to the ($11\bar{4}$) plane because $5 - 1 - 4 = 0$, while [210] is not parallel to ($\bar{1}35$) because $-2 + 3 + 0 \neq 0$.

The zone law also allows one to determine which directions should lie within a given plane. As an example, consider which directions lie in (120). The zone law shows that $u + 2v = 0$. Thus, $u = -2v$. Examples of directions that would satisfy this criterion are [$\bar{2}10$], [$\bar{2}10$], and [$2\bar{1}l$], where l can be anything.

Planar intersections: Two intersecting planes ($h_1\ k_1\ l_1$) and ($h_2\ k_2\ l_2$) form a line [uvw] given by the equations

$$u = k_1 l_2 - k_2 l_1$$

$$v = h_2 l_1 - h_1 l_2$$

$$w = h_1 k_2 - h_2 k_1.$$

As an example, the planes (123) and (210) intersect along the [$\bar{3}6\bar{3}$] = [$\bar{1}2\bar{1}$] direction.

This can be seen by taking the determinant $\begin{vmatrix} \vec{a} & \vec{b} & \vec{c} \\ 1 & 2 & 3 \\ 2 & 1 & 0 \end{vmatrix}$. In the same way, the *plane formed by two intersecting zone axes*, [$u_1\ v_1\ w_1$] and [$u_2\ v_2\ w_2$] is given by

$$h = v_1 w_2 - v_2 w_1$$
$$k = w_1 u_2 - w_2 u_1$$
$$l = u_1 v_2 - u_2 v_1.$$

5.5   Interplanar Spacings and Bragg's Law

The interplanar separation distance between two adjacent planes with indices (hkl), referred to as d_{hkl}, is calculated along a line perpendicular to the planes. For a material with triclinic symmetry, $a \neq b \neq c$, $\alpha \neq \beta \neq \gamma$:

$$\frac{1}{d^2} = \frac{1}{V^2} \left[h^2 b^2 c^2 \sin^2 \alpha + k^2 a^2 c^2 \sin^2 \beta + l^2 a^2 b^2 \sin^2 \gamma + 2hkabc^2 (\cos \alpha \cos \beta - \cos \gamma) \right.$$
$$\left. + 2kla^2 bc (\cos \beta \cos \gamma - \cos \alpha) + 2hlab^2 c (\cos \alpha \cos \gamma - \cos \beta) \right],$$

where the volume of the unit cell, V, is given by:

$$V = \vec{a} \times \vec{b} \cdot \vec{c} = abc \sqrt{1 - \cos^2 \alpha - \cos^2 \beta - \cos^2 \gamma + 2 \cos \alpha \cos \beta \cos \gamma}.$$

Thankfully, as the symmetry increases, the expression simplifies progressively. For example, in cubic crystals, the volume reduces to $V = a^3$, and $\frac{1}{d^2} = \frac{a^2}{h^2 + k^2 + l^2}$.

Crystal structures are often determined by careful measurements of the diffraction patterns of X-rays, neutrons, or electrons from the crystal. X-rays have a wavelength that is similar to the interatomic distances in solids (e.g. 1.540 56 Å for Cu Kα_1) and are therefore strongly scattered by periodic arrangements of atoms. The situation is illustrated graphically by Fig. 5.6. When the extra distance traveled by the radiation striking the second plane of atoms is an integral number of wavelengths, the two reflections will add, and there will be a measurable X-ray intensity at the angle 2θ. For all other angles, however, the reflections from the lower planes cancel the waves reflecting from the top. At those angles, there will be no reflected beam. The condition $m\lambda = 2d \sin \theta$ is referred to as Bragg's law, where m is the integer number of wavelengths.

In a polycrystalline material, many small crystallites are oriented randomly with respect to the X-ray beam, so several distinct peaks are observed at different angles. Information on the symmetry and internal atomic arrangements in the material is

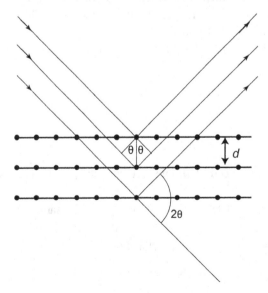

Fig. 5.6 Illustration of Bragg's law.

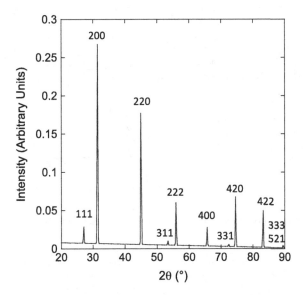

Fig. 5.7 Calculated X-ray diffraction pattern for polycrystalline NaCl with Cu Kα radiation.

contained in the placement of the peaks (corresponding to different d values) and in the relative intensity of each peak. Because these are different for different materials, the X-ray diffraction pattern acts like a material fingerprint. An example diffraction pattern is shown in Fig. 5.7 for NaCl.

In a glass, on the other hand, no long-range order is present. Because there is no periodic arrangement of atoms, there are no sharp diffraction peaks. Instead, there appears a diffuse X-ray pattern. If this is carefully measured, the data can be Fourier-transformed to obtain a radial distribution function which describes the short-range order in the material. The peak positions give the distance between neighbors, next-nearest neighbors, etc., while the area under the peak measures how many neighbors there are. This is how we know that Si also adopts tetrahedral coordination in oxide glasses with the same bond length as it shows in crystals. The breadth of the peaks in the radial distribution function relates to the degree of structural variability in the glass – the radial distribution function for a crystal would have much sharper lines that would continue out indefinitely on the distance axis.

5.6 Crystal Morphology

Crystal morphology describes the natural shape of a crystal. Before the discovery of X-ray diffraction, crystallographers determined the symmetry of crystals by measuring the interfacial angles between different faces. Crystals with different chemical composition, but the same shape and the same symmetry were called *isomorphs*. Those with the same chemical composition, but different morphology were referred to as *polymorphs*. As examples, diamond and silicon are isomorphs, while diamond and graphite are polymorphs.

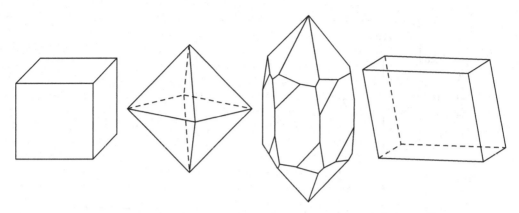

Fig. 5.8 Crystal morphologies preferred by (from left to right) rocksalt, diamond, quartz, and calcite crystals.

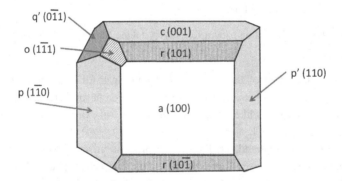

Fig. 5.9 Morphology of a sucrose crystal.

Under favorable conditions, crystal growth takes place in such a way that the external surface is bounded by a set of plane faces. Common crystal habits for rocksalt, diamond, quartz, and rutile are illustrated in Fig. 5.8. The preferred shape of rocksalt family crystals is a cube bounded by six symmetry-related {100} faces. For diamond, an octahedral shape with eight {111} faces often appears. Quartz and calcite belong to lower symmetry crystal systems with more anisotropic morphologies. Quartz crystals often elongated along the c axis with a hexagonal cross section bounded by six {100} faces while the ends are terminated by {101} and six {011} faces. Calcite is a mineral with many different morphologies. It is often formed in rhombs.

Sweetness is a fascinating property involving receptors on the human tongue. It is believed that these receptors operate through a lock-and-key mechanism. Many sweet-tasting molecules have common geometric and chemical characteristics. These are some-times referred to as the triangle of sweetness, or Kier's rule: among organic molecules, sweetness is associated with a triangle made up of two H-bonding groups and one H-repelling group separated by distances of 3–3.5 Å. Presumably, this triangle corresponds to active receptor sites in the human tongue. In sucrose, the constituent groups of the triangle are found on the glucose and fructose rings that make up the molecule (see Fig. 5.9).

5.7   Crystal Growth

Perhaps the simplest notion regarding the relationship between crystal morphology and crystal structure is the concept of "dangling bonds." Consider the coordination of the Na^+ and Cl^- atoms in rocksalt. Inside the crystal each ion is bonded to six neighbors forming an octahedron. But on the outer surface of a crystal the situation is different. On the (100) face of the crystal each sodium atom is bonded to four surface chlorines and to a fifth Cl neighbor inside the crystal. This leaves one "dangling" bond protruding from the surface. The unsatisfied bond becomes a point of attachment for an anion in the surrounding medium.

Other faces on the rocksalt crystal will have different numbers of dangling bonds per unit area, and therefore different growth rates. For the {110} dodecahedral faces each surface atom has four neighbors and two dangling bonds (Fig. 5.10), and for the {111} octahedral faces the surface atoms have three neighbors and three dangling bonds.

Because of these large numbers of dangling bonds, the atoms on {110} and {111} surfaces are less well bonded than those on the cubic {100} faces. It also means that outside atoms or molecules will attach to the {110} and {111} faces more easily than to {100} faces. Therefore $\langle 111 \rangle$ directions are fast-growing directions and $\langle 100 \rangle$ axes are slow-growing directions with $\langle 110 \rangle$ in between. Figure 5.11 illustrates the relationship between fast- and slow-growing faces and crystal morphology for rocksalt. Imagine a cubic crystal of rocksalt that has been cut in such a way that the fast-growing {110} faces are exposed. During crystal growth Na^+ and Cl^- ions are rapidly attached to these dangling bonds, causing fast growth. As growth proceeds, fast-growing faces tend to disappear, leaving the slow-growing faces behind.

Crystals growing from a melt or a solution begin as small nuclei. Figure 5.12 shows a schematic of morphology development as a function of growth rate. In

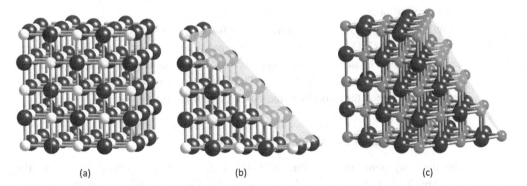

(a)   (b)   (c)

Fig. 5.10 Rocksalt crystal structure with one of the (a) {100}, (b) {110}, and (c) {111} planes marked. There is one dangling bond per surface atom on the {100} faces, two dangling bonds per surface atom on the {110} faces, and three dangling bonds per surface atom on the {111} faces. The larger sphere is the Cl, and the smaller sphere is Na. The large number of dangling bonds on the {110} and {111} faces of rocksalt leads to faster growth in $\langle 110 \rangle$ and $\langle 111 \rangle$ directions, relative to the $\langle 100 \rangle$ directions.

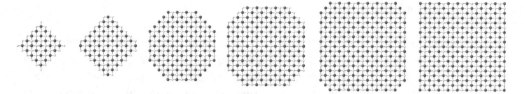

Fig. 5.11 Schematic showing the evolution of the morphology of rocksalt as a function of time during crystal growth. Time is increasing to the right in the figure. It is clear that the morphology evolves towards that of a cube.

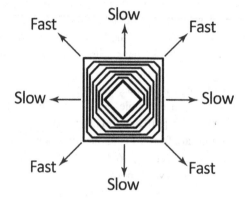

Fig. 5.12 Fast growing faces disappear while slow-growing faces remain. The surfaces mark the crystal dimensions at different times. It starts small (innermost figure). As growth proceeds at different rates in different directions, the characteristic morphology develops.

general, because the slowest-growing faces are the best bonded, they control the final crystal shape. In general, the morphological faces are often parallel to low-index crystal faces.

Figure 5.13 shows how the morphology of the rutile crystal results from the crystal structure. Rutile (TiO_2) also tends to be elongated along c sometimes forming long slender needles. The needles have a square cross section consisting of four {110} faces terminated by eight small {111} faces. The faces that are the best bonded are slow-growing, and hence they dominate the observed morphology.

Surface Modification and Morphology Control

The dangling-bond model gives a rough idea of crystal growth but is vastly oversimplified. Other factors that influence growth rate and crystal morphology include surface rearrangements, crystal defects, surface poisons, and pseudomorphic structures.

Not all surfaces resemble the full structure. Silicon is isostructural with diamond, with each silicon atom bonded to four Si neighbors arranged in a tetrahedron. Each Si–Si bond is a covalent electron-pair bond, but the surface atoms with their dangling

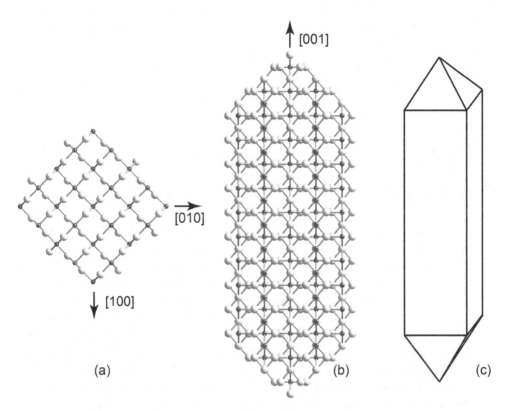

Fig. 5.13 The morphology of rutile TiO_2 crystals. (a) Looking down the c axis, (b) along the a axis, and (c) schematic of needle shape. The {110} family of planes have a high density of bonding, and hence are slow growing. This, in conjunction with a fast-growth direction along the c axis, produces needle-like crystals. In the drawings the Ti atoms are the small filled spheres, the O are the larger spheres.

bonds sometimes rearrange themselves to lower the symmetry of silicon as shown in Fig. 5.14. Normally this surface would have one dangling bond per surface atom, but under certain circumstances the surface silicons form paired dangling bonds by slightly altering their positions.

Surface structures, dangling bonds, growth rates, and crystal morphology are also strongly influenced by contaminants. Foreign atoms or molecules sometimes attach to certain faces preferentially. This can have the effect of changing a fast-growing face into a slow-growing face, and altering the shape of a crystal. In the crystal-growth literature these shape-changing additives are known as "poisons." Ammonium chloride poisons the fast-growing faces of NaCl and changes the morphology from a cube to an octahedron. A listing of surface poisons can be found in the classic book *Crystal Growth*.[2]

[2] H. E. Buckley, *Crystal Growth*, John Wiley and Sons, New York (1951).

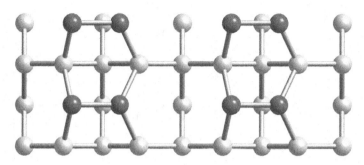

Fig. 5.14 A portion of a reconstructed surface of silicon showing pairing of atoms to reduce the number of dangling bonds. The atoms shown as darker gray are the ones participating in the pairing.

Another way of altering crystal shape is to make use of intermediate pseudomorph phases. A pseudomorph is a crystal that resembles a chemically and structurally different crystal. Pseudomorphs are very common in nature where one mineral is often converted to another without changing shape. This is often an atom-by-atom replacement process in which for example, a cubic iron sulfide such as pyrite (FeS_2) is converted to orthorhombic goethite (FeOOH) without losing the cubic morphology. This same process is used in the reverse direction in making magnetic tape. To obtain the desired needle-like morphology, the iron oxide is first prepared as orthorhombic goethite and then oxidized to cubic γ-Fe_2O_3 while retaining the fibrous morphology.

Crystal morphology depends on growth conditions as well as on thermodynamic factors. Snowflakes are an important and interesting example. Contrary to popular opinion, snowflakes are seldom symmetric, but are highly modified by complicated growth patterns. Laboratory and meteorological studies have determined the temperature and humidity conditions under which certain forms are stable; these are summarized in Fig 5.15.[3] Snowflakes are formed when water molecules crystallize around a nucleus, usually a dust particle or a pollen grain. Twinning is common, giving rise to delicate dendritic forms in seemingly infinite varieties. It is said that no two snowflakes are ever exactly alike, which is probably true for any crystals. The shapes of precipitated snowflakes can be correlated with the temperature and height from which snow has fallen. Hexagonal prisms and prism clusters with pyramid terminations form in high wispy cirrus clouds at 30 000–50 000 feet, where temperatures are usually low. These tiny flakes are common in the arctic. Hexagonal columns and plates terminated by two parallel faces occur in alto-cumulus and alto-stratus clouds at 10 000–30 000 feet, and are somewhat larger than those in cirrus clouds. Nimbo-stratus, stratus, and strato-cumulus clouds are found at lower altitudes, where the temperatures are higher. Stellar snowflakes grow under these conditions, providing the classic shapes with elaborate fern-like extensions.

There are several types of snow. *Precipitated snow* falls from the sky, often in forms with very high surface to volume ratios. Surface free energy of fallen snow is reduced by vaporization and recondensation into the simpler forms of *metamorphosed snow.*

---

[3] Reproduced by permission from: http://www.its.caltech.edu/~atomic/snowcrystals/primer/primer.htm

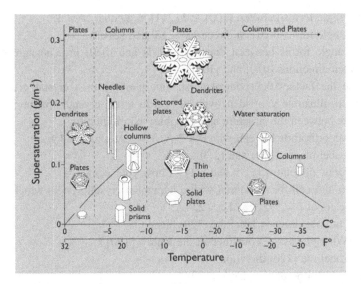

Fig. 5.15 Morphology of snowflake crystals as a function of the formation conditions. Reproduced with permission from Kenneth G. Libbrecht.

Two other types are *rime* and *hoarfrost*. If growing snow crystals fall through a cloud of supercooled water droplets, they collect a layer of *rime*. When the droplets contact the snow crystals, they freeze as globules. Similar to dew, *hoarfrost* results from the sublimation of ice crystals from water vapor in the air. It can form a thick coating on snow, usually a layered structure, which is an excellent lubricant for fast skiing and slab avalanches.

5.8 Polymorphs, Polytypes, Isomorphs, and Isostructural Compounds

Since the advent of X-ray crystallography and the determination of crystal structures, the words isomorph and polymorph now refer to atomic structures rather than morphology. Most scientists use the words *isomorph* and *isostructural* interchangeably. The words *polymorphism* and *polytypism* have similar but not identical meanings. *Polymorphs* refer to two or more different crystal structures of a given chemical compound. Diamond and graphite are polymorphs, but they are not polytypes. The atomic coordination in diamond and graphite are quite different. *Polytypism* refers to two polymorphs which differ only in the stacking sequence of identical units. Wurtzite and zincblende can be described as polytypes (or polymorphs) of ZnS. Silicon carbide is famous for its many stacking sequences. About a hundred different polytypes have been reported.

5.9 Problems

(1) What direction is formed by the intersection of (112) and (110)?

(2) What directions lie in (120)?

(3) What angle forms between [100] and [$\bar{2}$11] in a cubic crystal?

(4) What angle forms between directions [2$\bar{1}$3] and [2$\bar{1}$0]?

(5) What angle forms between directions [001] and [010] in a hexagonal crystal? Between directions [010] and [100]?

(6) Define the factors that control crystal morphology in a crystal growing from solution. Illustrate your ideas with respect to a real crystal structure, describing the link between the structure and the observed morphology.

(7) Draw an orthorhombic cell with $a = 2$ Å, $b = 3$ Å, and $c = 4$ Å.

    (a) Label planes (111) and (2$\bar{1}$0).

    (b) Draw directions [$\bar{1}\bar{1}$0] and [211]. Show your work.

(8) Draw a unit cell for a material with a tetragonal crystal system which has lattice parameters of $a = 5.0$ Å and $c = 2.0$ Å.

    (a) On your drawing label the (120) plane.

    (b) What planes are contained in the family {120}?

    (c) Label the [120] direction.

    (d) Is [120] perpendicular to (120)?

    (e) What direction is formed by the intersection of (12$\bar{1}$) with (203)?

(9) Scapolite crystals belong to the tetragonal crystal class and have the morphology shown with a {100}, m {110}, r {101}, and z {211}.

    (a) Label all visible faces.

    (b) What are the indices of the line at the intersection of the planes marked z and m?

    (c) Is (211) perpendicular to [211]?

    (d) Is [$\bar{1}$02] parallel to (312)?

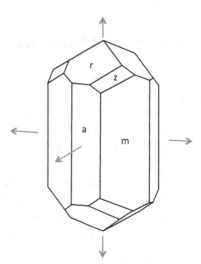

(10) Zircon crystals are in the tetragonal crystal system and have the morphology shown with x {100}, y {101}, and z {301}.

    (a) Label all visible faces.

(b) What are the indices of the line at the intersection of the planes marked *x* and *y*?

(c) What are the indices of the line at the intersection of the plane *y* and the face to its right?

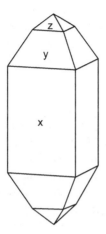

(11) Give an example of a material with a needle or fibrous morphology. Explain on the basis of the crystal structure and bonding why this shape is adopted.

6 Crystal Systems and Theoretical Density

6.0 Introduction

Crystals, by definition, can be built up by repeating a motif of atoms in a fashion that is consistent with translational symmetry. A brief description of unit cells was given in Chapter 5. In this chapter, unit cells are classified based on their shape and symmetry into seven crystal systems. Relationships for calculating volume and interatomic separation distances are given for the different crystals systems. Finally, the structure–property relationships for density are reviewed.

6.1 Crystal Systems and Unit Cells

As shown in Fig. 6.1, the unit cells for crystals are described with three cell dimensions, (a, b, and c) and three interaxial angles (α, β, and γ). Alpha (α) is the angle between b and c; beta (β) is the angle between edges a and c; and gamma (γ) is the angle between edges a and b. The angles are usually expressed in degrees and minutes, and the cell dimensions in ångström units (1 Å = 0.1 nm = 10^{-10} m).

The seven crystal systems are listed in Table 6.1. Rhombohedral crystals, a subset of the trigonal system, are included as well. Also shown are the volume for each type of unit cell, the minimum symmetry and examples of materials from the mineral kingdom.

While symmetry will be discussed in detail in Chapter 7, it is worth mentioning a little here. An n-fold rotation axis is a symmetry operation in which a structure is rotated around a line in increments of $360°/n$. After each rotation, it is impossible to distinguish the structure from its starting configuration. Some of the minimum symmetry values are self-evident. For example, in a tetragonal material, the unit cell is square in the a–b plane, and c is either longer or shorter. It is apparent that if you looked straight down the c axis, through the middle of the square, that you should be able to rotate the square by 90° and not be able to tell the difference when you were done. Hence the minimum symmetry is a four-fold rotation axis parallel to [001]. In much the same way, an orthorhombic system has the minimum symmetry of a brick. With a brick, there is a two-fold rotation axis normal from the middle of each face of the brick. Since a brick is typically different lengths along the three principal axes, there are no symmetry operations relating the a to the b axis, for example.

Somewhat less obvious may be the fact that the minimum symmetry of a cube includes four three-fold rotation axes around each of the cube body diagonals. This is

Table 6.1 Unit cells for the seven crystal systems

Crystal system	Axial lengths and angles	Minimum symmetry	Volume formula[*]	Name
Triclinic	$a \neq b \neq c$ $\alpha \neq \beta \neq \gamma$	None	$V = abc(1 + 2\cos\alpha\cos\beta\cos\gamma -$ $\cos^2\alpha - \cos^2\beta - \cos^2\gamma)^{1/2}$	Kaolinite, $Al_2Si_2O_5(OH)_4$
Monoclinic	$a \neq b \neq c$ $\alpha = \gamma = 90°$ $\beta \neq 90°$	Two-fold axis along [010]	$V = abc(\sin\beta)$	Orthoclase, $KAlSi_3O_8$
Orthorhombic	$a \neq b \neq c$ $\alpha = \beta = \gamma = 90°$	Three $\perp$ two-fold axes along [100], [010], [001]	$V = abc$	Forsterite, Mg_2SiO_4
Trigonal	$a = b \neq c$ $\alpha = \beta = 90°$ $\gamma = 120°$	Three-fold axis along [001]	$V = a^2c(\sqrt{3}/2)$	Quartz, SiO_2
Rhombohedral[+]	$a = b = c$ $\alpha = \beta = \gamma \neq 90°$	Three-fold axis along [111]	$V = a^3(1 - 3\cos^2\alpha + 2\cos^3\alpha)^{1/2}$	Calcite, $CaCO_3$
Tetragonal	$a = b \neq c$ $\alpha = \beta = \gamma = 90°$	Four-fold axis along [001]	$V = a^2c$	Rutile, TiO_2
Hexagonal	$a = b \neq c$ $\alpha = \beta = 90°$ $\gamma = 120°$	Six-fold axis along [001]	$V = a^2c(\sqrt{3}/2)$	Beryl, $Be_3Al_2Si_6O_{18}$
Cubic	$a = b = c$ $\alpha = \beta = \gamma = 90°$	Four three-fold axes along $\langle 111 \rangle$	$V = a^3$	Halite, NaCl

[*] All volume formulae can be derived from that of the triclinic cell.
[+] Subclass of trigonal.

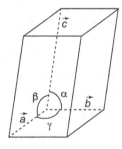

Fig. 6.1 Unit cell showing the cell lengths and cell angles.

illustrated in Fig. 6.2. If you consider the three-fold rotation axis that connects the lower back left corner of the cube to the upper front right corner, then it should be apparent that this rotation axis relates the front face of the cube to the top of the cube and the right face of the cube.

The shape of a rhombohedral crystal can be envisaged by either elongating or compressing the cube along one of its body diagonals. When this is done, one three-fold axis will be retained (the stretching or shrinking axis); the others will be destroyed by the deformation. One three-fold rotation axis is also the characteristic minimum symmetry of a material in the trigonal crystal system. Thus, all rhombohedral crystals are a subset of the trigonal crystal class, and can be written in trigonal coordinates with

Table 6.2 Distribution of crystals amongst the crystal classes

Crystal system	Inorganic	Organic
Triclinic	14.54%	20.42%
Monoclinic	38.14%	52.97%
Orthorhombic	18.95%	21.28%
Trigonal	6.03%	1.59%
Tetragonal	7.66%	2.70%
Hexagonal	5.22%	0.55%
Cubic	9.49%	0.48%

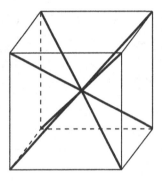

Fig. 6.2 A cubic unit cell showing the three-fold rotation axes along the body diagonals.

an appropriate transformation of the unit cell. The minimum symmetry will be discussed in more detail in Chapter 23.

Population statistics for the seven crystal systems from more than 280 000 chemical compounds are shown in Table 6.2. Inorganic and organic crystals are distributed differently among the two families of materials; it is clear that the inorganic crystals populate the higher symmetry crystal classes more frequently.

6.2   Transformations Between Cells

Unit cells are somewhat arbitrary constructions, in that the choice of origin and shape is not unique. There are many possible non-parallel vectors $\vec{a}_i, \vec{b}_i$, and $\vec{c}_i$ that could describe the repeating unit of a crystal. By convention, the choice is usually made of the smallest possible unit cell that readily shows the symmetry in the material.

However, because more than one choice of unit cell is possible, it is useful to be able to transform from one unit cell to another. If the two unit cells are the same shape and orientation, but different origins are chosen, then one unit cell can be transformed to the other by adding the fractional coordinates x, y, and z to all of the coordinates of the atoms in the cell, where $x\vec{a} + y\vec{b} + c\vec{z}$ is the vector that connects the origin of the second cell to the first.

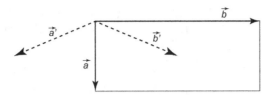

Fig. 6.3 Two different possible sets of unit cell vectors shown in a plane.

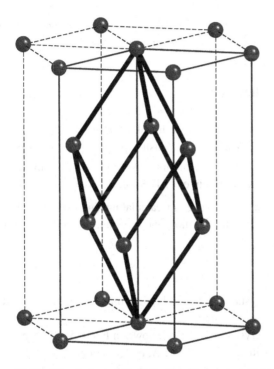

Fig. 6.4 Rhombohedral (bold line) and hexagonal (lighter line) lattices.

If the unit cells are chosen to have different shapes, then the transformation between the two cells becomes more complicated. In this case, the new unit vectors, $\vec{a}'$, $\vec{b}'$, and $\vec{c}'$ are written in terms of the original vectors, $\vec{a}$, $\vec{b}$, and $\vec{c}$. For example, consider the figure shown in Fig. 6.3. In this case, $\vec{a}' = \frac{1}{2}(\vec{a} - \vec{b})$, and $\vec{b}' = \frac{1}{2}(\vec{a} + \vec{b})$. If it is assumed that old and new $\vec{c}$ axes are the same, then $\left(\vec{a}'\vec{b}'\vec{c}'\right) = \left(\vec{a}\,\vec{b}\,\vec{c}\right)\begin{pmatrix} 1/2 & 1/2 & 0 \\ -1/2 & 1/2 & 0 \\ 0 & 0 & 1 \end{pmatrix}$.

One common example is the transformation required to move between rhombohedral and hexagonal coordinates, as shown in Fig. 6.4. In this case, the [111] axis of the rhombohedral unit cell corresponds to the [001] axis of the hexagonal unit cell. As shown, the volume of the rhombohedral cell is 1/3 that of the trigonal cell. Here the transformation of the lattice parameters is given by

$$\left(\vec{a}'\vec{b}'\vec{c}'\right)_{Rhomb} = \left(\vec{a}\,\vec{b}\,\vec{c}\right)_{Hex} \begin{pmatrix} 2/3 & -1/3 & -1/3 \\ 1/3 & 1/3 & -2/3 \\ 1/3 & 1/3 & 1/3 \end{pmatrix}. \text{ The indices of a plane (HKL)}$$

in the hexagonal cell is related to those of plane (*hkl*) in the rhombohedral cell via:

$$H = h - k$$

$$K = k - l$$

$$L = h + k + l.$$

The *International Tables for Crystallography* is a valuable resource describing the conventions, along with the location of the symmetry elements.

6.3   Unit Cell Volumes and Densities

The density of a crystal depends on the chemical composition and crystal structure through the relation $\rho = \frac{MZ}{N_A V}$. That is, the density, ρ (g/cm^3), is determined by the molecular weight M, the number of formula units per unit cell Z, Avogadro's number N_A, and the unit cell volume measured in cm^3. The volume, V, of a unit cell is given by $V = \vec{a} \times \vec{b} \cdot \vec{c}$. It is critical to remember in calculating this that the unit cell is not necessarily orthogonal, and that there is no guarantee that the cell lengths are equal. Thus, for a triclinic crystal:

$$V = abc\left(1 + 2\cos\alpha\cos\beta\cos\gamma - \cos^2\alpha - \cos^2\beta - \cos^2\gamma\right)^{1/2}.$$

Unit-cell volumes for any crystal system can be determined from the triclinic formula by inserting values for a, b, c, α, β, and γ.

Consider, for example, a crystal of NiAs, as shown in Fig. 6.5. NiAs belongs to the hexagonal crystal system, with $a = b = 3.4392$ Å, $c = 5.3484$ Å, $\alpha = \beta = 90°$, $\gamma = 120°$. Thus, $V = a^2 c \sqrt{3}/_2 = 54.79$ Å^3. For NiAs, the molecular weight, $M = 133.61$ g/mole:

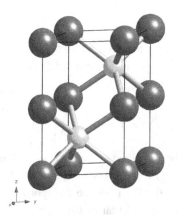

Fig. 6.5 The unit cell of NiAs. Ni is shown as the darker atom and As as the lighter one.

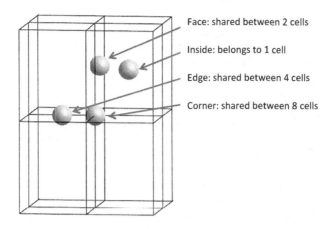

Fig. 6.6 The outlines of eight unit cells with selected atoms shown.

$N_A = 6.023 \times 10^{23}$ NiAs/mole. Examining the unit cell drawing, it can be seen that there are eight Ni atoms at the corners of the unit cell and four on the cell edges. Since each corner atom is shared between eight unit cells and each edge between four unit cells, the number of Ni atoms belonging to the unit cell shown $= \frac{8}{8} + \frac{4}{4} = 2$. The two As atoms shown belong completely inside the unit cell, and so are counted completely. Thus, $Z = 2$ NiAs formula units per unit cell. Collecting all of the numbers, the theoretical density for NiAs is

$$\rho = \frac{MZ}{N_A V} = \frac{\left(133.61 \ \frac{g}{mole}\right)\left(2 \ \frac{\text{NiAs formula units}}{\text{unit cell}}\right)}{\left(6.023 \times 10^{23} \ \frac{\text{NiAs formula units}}{mole}\right)\left(54.79 \ \frac{\mathring{A}^3}{\text{unit cell}}\right)\left(\frac{cm}{10^8 \ \mathring{A}}\right)^3} = 8.098 \ \frac{g}{cm^3}.$$

For more symmetric crystal systems, the calculation is considerably simpler. Gold is cubic with $a = 4.0786$ Å and $Z = 4$ in a face-centered cubic arrangement. The cell volume and density are $V = a^3 = 67.85$ Å^3 and $\rho = 19.3$ g/cm^3.

Before moving on, here are a few notes about correctly calculating Z. First, as shown in Fig. 6.6, unit cells tile together to build up the three-dimensional structure. Atoms at the unit cell corners are shared between eight unit cells, and so belong 1/8 to any given cell. Atoms on a face belong 1/2 to a given cell; the other 1/2 belongs to the neighboring cell with which it is shared. Atoms on a unit cell edge belong 1/4 to each of the four adjoining cells, while atoms inside the cell are counted as belonging to only that cell. Secondly, when you count up atoms in a compound AX_2, within one unit cell, there should be twice as many X atoms as A atoms. This can be used as a double check on one's calculations. Thirdly, Z is the number of *formula units per unit cell*, not the number of atoms per unit cell. In the case of the NiAs structure discussed above, $Z = 2$ NiAs formula units per unit cell. Thus, there are two Ni atoms and two As atoms per unit cell.

In comparing the densities of related materials, the density usually increases with molecular weight. For example, among the isomorphs of the fluorite structure: ρ (CaF$_2$) = 3.18 g/cm^3; ρ (SrF$_2$) = 4.24 g/cm^3; ρ (BaF$_2$) = 4.89 g/cm^3; and ρ (PbF$_2$) = 8.24 g/cm^3.

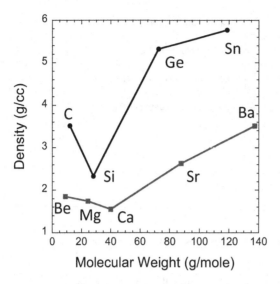

Fig. 6.7 Density vs. atomic weight for column IV and column II elements. Lighter elements are often denser than expected because of their small atomic radii.

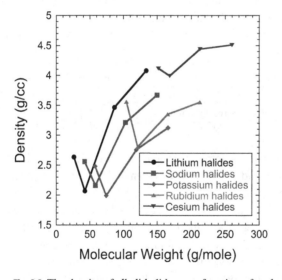

Fig. 6.8 The density of alkali halides as a function of molecular weight. Density usually increases with weight, but size changes cause anomalies.

Among the lighter elements, however, anomalies in this relationship often occur. Figures 6.7 and 6.8 plot the density of several inorganic families against atomic or molecular weight. Metallic and covalently bonded elements are shown in Fig. 6.7, and ionic alkali halides are shown in Fig. 6.8. In all of the cases shown, there is a minimum in the density vs. molecular weight relationship. Despite its higher weight, silicon has a lower density than does diamond, calcium is less dense than beryllium and magnesium, and among the alkali halides, the fluorides are denser than the heavier chlorides.

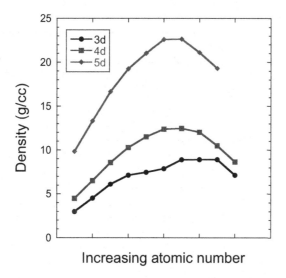

Increasing atomic number

Fig. 6.9 Density of transition metals as a function of increasing atomic number. The densest naturally occurring elements are those in the 5d transition metal series. Densities correlate well with melting points for these refractory metals.

This unexpected behavior is caused by the rapid increase in the unit cell volume on moving down the periodic table. In fact, the fluorides are more dense than the chlorides because of the large increase in ionic radius going from F^- to Cl^-. The low density of Mg-based metals is of particular interest for aerospace applications.

For metallically bonded solids, transition metal elements in the 4d and 5d series have the highest densities. As shown in Fig. 6.9, the maximum densities occur near the middle of the 4d and 5d series, as in these materials, the d electrons participate strongly in interatomic bonding. The short, strong bonds in these elements also lead to very high melting and boiling points. In the 3d transition metals, the d electrons do not extend to as large a fraction of the atomic radius, and so contribute less in bonding. Transition metals have densities several times larger than those of the more weakly bonded alkali- and alkaline-earth metals.

The densest ceramics are nitrides and carbides like WC, with densities exceeding 15 g/cm^3. The densities of several important oxides are listed in Table 6.3. It is not surprising that oxides with a close-packed oxygen network have larger densities than those with open structures. Thus, close-packed structures like those of corundum, spinel, chrysoberyl, and olivine give densities of ~3.3–4.0 g/cm^3. In contrast, even though they contain many of the same elements, more open structures such as quartz, beryl, and feldspar have densities near 2.7 g/cm^3. The differences in densities of these minerals means that they can be separated by flotation methods.

Note the low densities of solid CO_2 and H_2O. Weak van der Waals and hydrogen bonds give rise to long intermolecular distances and, combined with low molecular weights, result in low densities. The carbon, nitrogen, and oxygen atoms in organic crystals all have roughly the same size, and the hydrogen atoms are much smaller. Moreover, the intermolecular contact distances are also all about the same. This means

Table 6.3 Specific gravities for common oxides

Material	Molecular weight (g/mole)	Density (g/cm^3)	Material	Molecular weight (g/mole)	Density (g/cm^3)
Rocksalt structures			Corundum structures		
MgO	40	3.7	Al_2O_3	102	4.0
CaO	56	3.4	Cr_2O_3	152	5.2
SrO	104	4.7	Fe_2O_3	160	5.2
BaO	153	5.7	Ga_2O_3	187	6.4
MnO	71	5.4			
FeO	72	5.7	Fluorite structures		
CoO	75	6.5	ZrO_2	123	5.5
NiO	75	7.5	CeO_2	172	7.3
			ThO_2	264	9.7
Wurtzite structures			UO_2	270	10.9
BeO	25	3.0			
ZnO	81	5.6	Quartz structures		
			SiO_2	60	2.6
Molecular structures			GeO_2	105	4.7
H_2O	18	1.0			
CO_2	44	1.6			

the number of molecules in the unit cell can be estimated from the chemical formula and the unit cell volume. The Kempster–Lipson rule states that the number of C, N, and O atoms in the unit cell of an organic crystal is approximately equal to the volume of the unit cell (in $Å^3$), divided by 18.[1] As an example, consider urea $OC(NH_2)_2$. Crystalline urea is tetragonal with $a = 5.67$ Å and $c = 4.73$ Å. The unit cell volume is $a^2c = 152.1$ $Å^3$; dividing by 18 gives 8.4. Rounding off, there are 8/4 = 2 molecules/unit cell.

For a constant composition, high pressures favor higher-density polymorphs. As a result, cristobalite ($\rho = 4.35$ g/cm^3) is more stable at elevated pressures than is α-quartz ($\rho = 2.65$ g/cm^3). The thermodynamic basis for this will be discussed in detail in Chapter 20. In contrast, elevated temperatures tend to favor lower-density polymorphs. As an example, quartz, $\rho = 2.65$ g/cm^3 is denser than its high-temperature polymorph tridymite ($\rho = 2.17$ g/cm^3).

Glasses and other amorphous solids are usually less dense than crystalline materials of the same composition. For example, many silicate glasses are 0–20 % less dense than their crystalline counterparts.

6.4   Close-Packing

In metals, and in atoms or ions with closed outer electron shells, the electron cloud for each atom is approximately spherical. As a result, many crystal structures are based on the close-packing of spherical atoms (Table 6.4). Close-packing allows many bonds to be made, and enables high-density crystal structures. The common close-packed metal

---

[1] C. J. E. Kempster and H. Lipson, *Acta Cryst.*, B28, 3674 (1972).

Table 6.4 Structures based on close-packing: f_o and f_t denote the fractional filling of octahedral and tetrahedral interstices in the close-packed lattice

Structure	Close-packed atoms	Stacking sequence	Octahedral atoms	f_o	Tetrahedral atoms	f_t
Mg (HCP)	Mg	AB	–	0	–	0
K_2GeF_6	KF_3	AB	Ge	1/8	–	0
UCl_6	Cl	AB	U	1/6	–	0
$Cs_3Tl_2Cl_9$	$CsCl_3$	AB	Tl	1/6	–	0
$CsNiCl_3$	$CsCl_3$	AB	Ni	1/4	–	0
PdF_3	F	AB	Pd	1/4	–	0
BiI_3	I	AB	Bi	1/3	–	0
CdI_2	I	AB	Cd	1/2	–	0
TiO_2	O	AB	Ti	1/2	–	0
α-Al_2O_3	O	AB	Al	2/3	–	0
NiAs	As	AB	Ni	1	–	0
$AlBr_3$	Br	AB	–	0	Al	1/6
Al_2Se_3	Se	AB	–	0	Al	1/3
Al_2ZnS_4	S	AB	–	0	Al_2Zn	3/8
ZnS (wurtzite)	S	AB	–	0	Zn	1/2
Mg_2SiO_4	O	AB	Mg	½	Si	1/8
Cu	Cu	ABC	–	0	–	0
K_2PtCl_6	KCl_3	ABC	Pt	1/8	–	0
$Cs_3As_2Cl_9$	$CsCl_3$	ABC	As	1/6	–	0
$SrTiO_3$	SrO_3	ABC	Ti	1/4	–	0
ReO_3	O	ABC	Re	1/4	–	0
$CrCl_3$	Cl	ABC	Cr	1/3	–	0
$CdCl_2$	Cl	ABC	Cd	1/2	–	0
$Cu_2(OH)_3Cl$	$(OH)_3Cl$	ABC	Cu	1/2	–	0
NaCl	Cl	ABC	Na	1	–	0
SnI_4	I	ABC	–	0	Sn	1/8
HgI_2	I	ABC	–	0	Hg	1/4
α-Ga_2S_3	S	ABC	–	0	Ga	1/3
Ag_2HgI_4	I	ABC	–	0	Ag_2Hg	3/8
ZnS (zincblende)	S	ABC	–	0	Zn	1/2
Bi_2O_3	Bi	ABC	–	0	O	3/4
CaF_2	Ca	ABC	–	0	F	1
Co_9S_8	S	ABC	Co	1/8	Co	1/2
$MgAl_2O_4$	O	ABC	Al	1/2	Mg	1/8
$BiLi_3$	Bi	ABC	Li	1	Li	1
α-Nd	Nd	ABAC	–	0	–	0
K_2MnF_6	K_1F_3	ABAC	Mn	1/8	–	0
CdI_2	I	ABAC	Cd	1/2	–	0
TiO_2 (brookite)	O	ABAC	Ti	1/2	–	0
SiC (4H)	Si	ABAC	–	0	C	1/2
$Al_2SiO_4F_2$	O_2F	ABAC	Al	1/3	Si	1/12
$K_3W_2Cl_9$	KCl_3	ABCACB	W	1/6	–	0
$BaTiO_3$ (hex)	BaO_3	ABCACB	Ti	1/4	–	0
α-Sm	Sm	ABABCBCAC	–	0	–	0

From A. F. Wells, *Structural Inorganic Chemistry*, 5th Edition, Clarendon Press, Oxford (1984).

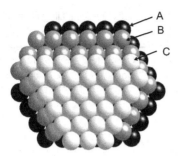

Fig. 6.10 Cubic close-packing of identical spheres in which each sphere contacts 12 others, showing the layering sequence. Tetrahedral and octahedral interstices are common in all structures based on close-packing of identical spheres. For each close-packed sphere, there are two tetrahedral and one octahedral interstitial sites.

structures, where all of the atoms in the crystal have the same size, are described in detail in Chapter 10. Here, the basic ideas behind close-packing are reviewed. In brief, in this type of packing, the atoms have 12 coordination, with six neighbors in the close-packed layer, three neighbors above, and three below. This is illustrated in Fig. 6.10. Each atom in layer C touches six other atoms in the layer, arranged in a hexagon. Atoms in the adjacent layer occupy the B or A positions, but not both. Three B atoms contact each C atom in the layer above.

Three stacking sequences are common amongst the close-packed structures: cubic close-packed (which has the stacking sequence ABCABC...), hexagonal close-packed (stacked as ABAB...), and "double" hexagonal close-packed (stacked as ABACABAC...). Thus, in cubic close-packing, the fourth layer repeats the position of the first layer. In hexagonal close-packing (hcp), the third layer is superimposed directly on top of the first, and in double hcp, the repeat pattern requires four different close-packed layers. In principle many complex patterns could develop, but in nature the shorter repeat units are more common.

It is also worth noting that many crystal structures can be thought of as deriving from a close-packed lattice, even if none of the atomic sublattices in the material is truly "close-packed." That is, if the atoms were regarded as hard spheres, the geometric arrangement of some atoms may correspond to close-packed positions, but they may not in fact be touching 12 other atoms of the same kind. For example, in rocksalt, the Cl^- ions are arranged as in a close-packed array, but they do not, in fact, touch each other. The close-packed plane in this case is the $\{111\}$ plane.

Higher-density structures are often more most stable, since they allow more bonds to be made. However, it is challenging to calculate packing densities for some compounds. Calculation of the relative packing efficiency of ionically bonded solids depends extremely strongly on the values used for the ionic radii, and so is not of tremendous practical importance.

In Table 6.4 it is clear that in ionically bonded compounds, anions are more likely to form close-packed layers than cations. This is reasonable since anions are generally larger than cations. In close-packed oxides, there are thus three common roles for cations.

(1) large cations like K^+, Ba^{2+}, or Pb^{2+} participate in the close-packing, and are 12 coordinated.

(2) Cations with sizes of 0.5–0.8 Å, which includes Mg^{2+}, many of the transition metal cations, and sometimes Al^{3+}, are often found in octahedral interstices of close-packed structures.

(3) Small cations such as Si^{4+} and sometimes P^{5+} appear in tetrahedral interstices of close-packed lattices.

In more complex structures based on close-packed arrays, some of the interstitial sites have three- and five-fold coordinations. Kotoite, $Mg_3B_2O_6$, contains interlinked MgO_6 octahedra and BO_3 triangles. In kotoite, the oxygen sublattice is hexagonally close-packed; each oxygen is coordinated to 3 Mg and 1 B neighbor in a distorted tetrahedral coordination. $BaFe_{12}O_{19}$ is industrially important as a permanent magnet with a large magnetic anisotropy. This material is an isomorph of magnetoplumbite in which the O^{2-} and Ba^{2+} ions form a close-packed array with the complex stacking sequence, ABACBCBCAB. In this sequence, most of the close-packed layers have only O^{2-} ions; in one of every five layers, the Ba^{2+} ions occupy 1/4 of the positions in the close-packed layer. Three different Fe^{3+} coordinations are observed: octahedral, tetrahedral, and a 5-coordinated site with a trigonal bipyramid geometry.

Close-packing can also develop with atoms of more than one element, if the atoms are approximately the same size. Several compositions for close-packed layers are known, including XY, XY_2, XY_3, XY_4, and XY_6. For example, in WAl_5, WAl_2 close-packed layers are stacked alternately with Al_3 layers. If the compound has a significant amount of ionicity in the bonding, it is imperative that the stacking of close-packed layers not bring ions of like charge together. XY_2 layers cannot be stacked without bringing X atoms into contact, hence few examples are observed among ionic compounds. Figure 6.11 shows examples of the ordered close-packed (or nearly so) layers from a variety of compounds. This figure also makes it clear that, while some of the structures can be visualized based on close-packing as a template, distortions often occur, so that the close-packed layers may not be flat, as in the ideal situation.

Chapter 11 describes the packing of organic crystals in detail. Kitaigorosdskii[2] has applied close-packing concepts to organic crystals, demonstrating that many molecules pack together in a way that fills space efficiently. Thus, long rod-like molecules pack like cylinders, while molecules with complicated shapes achieve high packing densities by arranging the molecules in interlocking patterns.

6.5   Bond Length Calculations, Atomic Coordinates, and Bond Angle Calculations

Knowledge of the sizes of atoms is very useful in predicting and understanding the behavior of materials. The size of an atom is predominantly controlled by the valence electron shell. In practice, radii are difficult to determine directly, at least in part due to

---

[2] A. I. Kitaigorodskii, *Organic Chemical Crystallography*, Consultants Bureau, New York (1961).

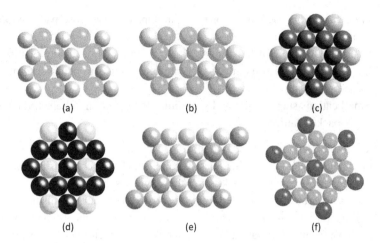

Fig. 6.11 Two elements of comparable size can crystallize in ordered close-packed arrays. The superstructures shown here are for (a) Cu(OH)Cl with 1:1 ordering of OH and Cl, (b) $Al_2Si_2O_4F_2$ with 1:2 ordering of F:O, (c) Cu_3Au with 1:3 ordering of Au:Cu, (d) $TiAl_3$ with 1:3 ordering of Ti: Al, (e) $MoNi_4$ with 1:4 ordering of Mo:Ni, and (f) $KOsF_6$ with 1:6 ordering of K:F.

the "fuzzy," probabilistic nature of the electron cloud. However, interatomic distances can be measured precisely via diffraction techniques. Radii can be estimated from these distances. It is important to note that the radii for a given element vary as a function of the valence, coordination number and bond type.

Bond Length

The position of any atom in a unit cell can be expressed in terms of fractions of the unit cell vectors. Thus the vector from the origin to the first point (x_1, y_1, z_1) shown in Fig. 6.12, is given by $\vec{R_1} = x_1\vec{a} + y_1\vec{b} + z_1\vec{c}$. Likewise, for point 2 with coordinates (x_2, y_2, z_2) is $\vec{R_2} = x_2\vec{a} + y_2\vec{b} + z_2\vec{c}$. The distance between these two vectors is:

$$\vec{R_1} - \vec{R_2} = (x_1 - x_2)\vec{a} + (y_1 - y_2)\vec{b} + (z_1 - z_2)\vec{c} \ .$$

The distance between the two points, $d_{12} = |\vec{R_1} - \vec{R_2}| = \sqrt{(\vec{R_1} - \vec{R_2}) \cdot (\vec{R_1} - \vec{R_2})}$. For a triclinic unit cell, this leads to:

$$d_{12}^2 = a^2(x_1 - x_2)^2 + b^2(y_1 - y_2)^2 + c^2(z_1 - z_2)^2 + 2ab(x_1 - x_2)(y_1 - y_2)\cos\gamma$$
$$+ 2bc(y_1 - y_2)(z_1 - z_2)\cos\alpha + 2ac(x_1 - x_2)(z_1 - z_2)\cos\beta.$$

For higher symmetries, the formula can be simplified by inserting the correct values for the unit cell parameters and angles.

Bond Angle

The angle θ made between two bonds can be calculated from $\cos\theta = \frac{R_{12}^2 + R_{23}^2 - R_{13}^2}{2R_{12}R_{23}}$, as shown in Fig. 6.13. Consider, for example, the NaCl structure shown in Fig. 3.3.

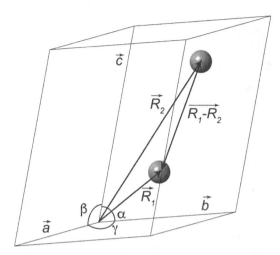

Fig. 6.12 A triclinic unit cell with two atoms shown at arbitrary locations.

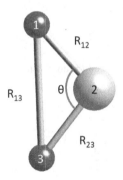

Fig. 6.13 Calculation of bond angles.

The Na–Cl bond distance is $a/2$, while the Cl–Cl distance is $\frac{\sqrt{2}a}{2}$. Inserting these into the formula yields a Cl–Na–Cl bond angle

$$\cos\theta = \frac{\left(\frac{a}{2}\right)^2 + \left(\frac{a}{2}\right)^2 - \left(\frac{\sqrt{2}a}{2}\right)^2}{2\left(\frac{a}{2}\right)\left(\frac{a}{2}\right)} = \frac{\frac{1}{4} + \frac{1}{4} - \frac{2}{4}}{\frac{1}{2}} = 0.$$

Thus, $\theta = 90°$.

To illustrate the interatomic distance formula, the bond lengths and bond angles are calculated from the two forms of tin: gray tin and white tin. Gray tin has the cubic diamond structure (Fig. 6.14) with neighboring atoms at $(0, 0, 0)$ and $(1/4, 1/4, 1/4)$. The lattice parameter is $a = 6.4912$ Å. For a cubic crystal, $a = b = c$, and $\alpha = \beta = \gamma = 90°$. The bond length is

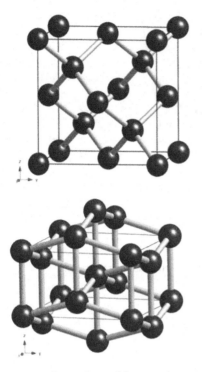

Fig. 6.14 Comparison of the crystal structures of (top) gray tin and (bottom) white tin. The solid lines mark the unit cell.

$$d^2 = \left(\frac{1}{4} - 0\right)^2 a^2 + \left(\frac{1}{4} - 0\right)^2 a^2 + \left(\frac{1}{4} - 0\right)^2 a^2$$

$$= 3a^2/16.$$

Hence, $d = \sqrt{3}a/4 = 2.8108$ Å. There are three other near neighbors for the tin atom located at $(0, 0, 0)$; they are located at $(-1/4, -1/4, 1/4)$, $(-1/4, 1/4, -1/4)$, and $(1/4, -1/4, -1/4)$. All four near neighbors have the same bond length.

The twelve next-near neighbors are located at the face-centered positions such as $(1/2, 1/2, 0)$. The interatomic distance is

$$d^2 = \left(\frac{1}{2} - 0\right)^2 a^2 + \left(\frac{1}{2} - 0\right)^2 a^2 = \frac{a^2}{2}.$$

Hence, $d = a/\sqrt{2} = 4.5900$ Å. Using the near neighbor and next-near neighbor distances and the law of cosines, the bond angle is given by

$$\cos\theta = \frac{\left(\frac{\sqrt{3}a}{4}\right)^2 + \left(\frac{\sqrt{3}a}{4}\right)^2 - \frac{a^2}{2}}{2\left(\frac{\sqrt{3}a}{4}\right)\left(\frac{\sqrt{3}a}{4}\right)} = -\frac{1}{3}.$$

This is the tetrahedral angle of $109.5°$ for sp^3 covalent bonds.

White tin, the stable form at room temperature, has a tetragonal unit cell with $a =$ 5.8197 Å and $c = 3.1749$ Å. There are four atoms in the unit cell located at $(0, 0, 0)$, $(0, 1/2, 1/4)$, $(1/2, 0, 3/4)$, and $(1/2, 1/2, 1/2)$. Each atom is surrounded by a distorted tetrahedron of four nearest neighbors, and two more slightly further away along the c axis. The atomic arrangement is illustrated in Fig. 6.14.

The shortest bond length is between atoms located at $(0, 0, 0)$ and $(0, \frac{1}{2}, \frac{1}{4})$. For a tetragonal unit cell, $\alpha = \beta = \gamma = 90°$: $d^2 = \left(0 - \frac{1}{2}\right)^2 a^2 + \left(0 - \frac{1}{4}\right)^2 c^2 = \frac{a^2}{4} + \frac{c^2}{16}$.

Hence, $d = 3.0162$Å for the four nearest neighbors located in the deformed tetrahedron. The distance between atoms at $(0, 1/2, 1/4)$ and $(1/2, 0, 3/4)$ is $d^2 = \frac{a^2}{4} + \frac{a^2}{4} + \frac{c^2}{4} = \frac{2a^2 + c^2}{4} = 19.454$ Å^2.

Hence, $d = 4.4107$ Å. The angle between bonds in the deformed tetrahedron is given by $\cos\theta = \frac{2(3.0162^2 - 19.454)}{2(3.0162)^2}$, so $\theta = 93.071°$.

Gray and white tin are also good examples for density calculations. The atomic weight of tin is 116.69 g/mole. Gray tin has the cubic diamond structure with $Z = 8$ atoms/unit cell. The

density is $\rho = \dfrac{\left(8\,\dfrac{\text{Sn atoms}}{\text{unit cell}}\right)\left(118.69\,\dfrac{\text{g}}{\text{mole Sn}}\right)}{\left(6.023\times10^{23}\,\dfrac{\text{Sn atoms}}{\text{mole Sn}}\right)(6.4912)^3\,\dfrac{\text{Å}^3}{\text{unit cell}}\left(\dfrac{\text{cm}}{10^8\text{Å}}\right)^3} = 5.765$ g/cm^3.

Gray tin, the metallic form, has $Z = 4$ atoms/unit cell.

$$\rho = \frac{\left(4\,\dfrac{\text{Sn atoms}}{\text{unit cell}}\right)\left(118.69\,\dfrac{\text{g}}{\text{mole Sn}}\right)}{\left(6.023\times10^{23}\,\dfrac{\text{Sn atoms}}{\text{mole Sn}}\right)(5.8197\text{Å})^2(3.17488\text{Å})\,\dfrac{\text{Å}^3}{\text{unit cell}}\left(\dfrac{\text{cm}}{10^8\text{Å}}\right)^3}.$$

So $\rho = 7.332$ g/cm^3.

6.6   Trends in Radii

In elemental metals, the radius can be obtained directly from half of the interatomic distance of adjacent atoms. Measured radii vary from 0.83 Å for boron to 2.65 Å for cesium. In general, metallic radii drop on moving across the periodic table from left to right within a row; this can be visualized as a consequence of the increasing nuclear charge exerting a larger pull on the electrons in the outermost shell. Metallic radii increase on moving down a column in the periodic table as inner electron shells are added.

Covalent radii are harder to define because the electron distributions around the atom core are not spherical. This is a consequence of the fact that the electrons are shared between adjacent atoms, so that there are lobes of charge directly along the bonds. Here again, the covalent radii are often taken as half the interatomic distance between covalently bonded atoms. In general, the covalent radii increase on going down the periodic table within a column. The radii also decrease as the number of electrons participating in a bond increase. Thus, the radius for a carbon atom in a double bond is smaller than that for singly bonded carbon atoms. Typical bond lengths are discussed in detail in Chapter 8.

Many ions are approximately spherical as a result of achieving a closed electron shell. Ideally, in a purely ionic bond, the electron density drops to zero between atoms, so that the ion size can be determined directly from electron-density maps. In this case, it is found that the radii cannot be defined as half of the interatomic distance. Instead, cations tend to be significantly smaller than anions. Details on ionic radii will be given in Chapter 9.

Van der Waals radii can be determined from the separation distances in molecular solids.

6.7   Problems

(1) *Drilling fluids* perform a number of functions that influence the drilling rate of wells, and control blow-outs. They are generally water-based muds with dissolved additives to control pH and corrosion, and finely divided particles of clay and polymers to control viscosity. Density is an important factor that counterbalances the pressures associated with underground fluids and gases. Barite ($BaSO_4$) is a dense, insoluble powder widely used in drilling muds. Drilling mud densities as high as 2 g/cm^3 are sometimes required for oilfields. What percent barite must be suspended in water to achieve these densities? Barite is orthorhombic with a = 7.154 Å, b = 8.879 Å, and c = 5.454 Å. The arrangement of atoms in the unit cell is shown in Fig. 6.15.

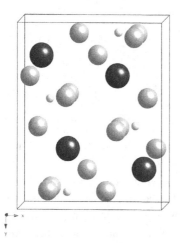

Fig. 6.15 Crystal structure of barite. The largest spheres are Ba, the intermediate size is oxygen, and the small atoms are S. All atoms shown are inside the unit cell.

(2) For rutile TiO_2, the Ti atoms are at positions (0, 0, 0) and (1/2, 1/2, 1/2), while the O are at $\pm$ (u, u, 0) and (u + 1/2, 1/2 – u, 1/2), where u = 0.3053. Rutile is tetragonal, with a = 4.593 73 Å and c = 2.958 12 Å.

   (a) Using Crystal Maker or some comparable software package, draw a unit cell of rutile.

   (b) What is the coordination of Ti? What is the coordination of O?

   (c) How many TiO_2 formula units are there per unit cell?

(d) What is the theoretical density of rutile? Show your work.

(e) Calculate the two Ti–O bond lengths in rutile.

(f) What atoms are on the (100) and (110) planes of rutile, in their standard setting? Show the atomic arrangement.

(3) Molybdenite (MoS_2) is the chief ore of molybdenum. It has a hexagonal layer structure which cleaves readily into flat platelets. The unit cell contains two Mo atoms at $\pm(1/3, 2/3, 1/4)$ and four S atoms at $\pm(1/3, 2/3, u)$ and $\pm(2/3, 1/3, u+1/2)$. X-ray measurements give $u = 0.629$, and lattice parameters $a = 3.1604$ Å and $c = 12.295$ Å.

(a) Make drawings of the molybdenite structure as viewed parallel and perpendicular to the hexagonal c axis.

(b) Calculate the Mo–S bond length.

(c) Describe the coordination of molybdenum and sulfur atoms.

(d) Compare the S–S distances in the MoS_2 layers and between layers.

(e) Determine the theoretical density of molybdenite.

(4) Cubic ZrO_2 has the fluorite structure with $a = 5.07$ Å. Tetragonal ZrO_2 has $a = 5.07$ Å, $c = 5.16$ Å. Monoclinic ZrO_2 has $a = 5.1454$ Å, $b = 5.2075$ Å, $c = 5.3107$ Å, and b $= 99°14'$. For all of these crystal structures, $Z = 4$ ZrO_2/unit cell. Calculate the densities of each of the phases of zirconia.

(5) Elemental Se is trigonal, with long zig-zag chains arranged in helices about the c axis. The helices can be either right-handed (space group $P3_121$) or left-handed (space group $P3_2$) as shown in Fig. 6.16. There are three Se atoms in the unit cell ($Z = 3$). In right-handed Se, the atoms are located in positions $(x, 0, 1/3)$; $(0, x, 2/3)$; $(-x, -x, 0)$, with $x = 0.217$. Atom locations in the left-handed enantiomorph are $(x, 0, 2/3)$; $(0, x, 1/3)$, again with $x = 0.217$. The coordinates are the same for tellurium, except that $x = 0.269$. For Se, the room temperature lattice parameters are $a = 4.355\ 17$ Å and $c = 4.949\ 45$ Å. For Te, the room temperature lattice parameters are $a = 4.446\ 98$ Å and $c = 5.914\ 92$ Å. Calculate the Se–Se and Te–Te bond lengths.

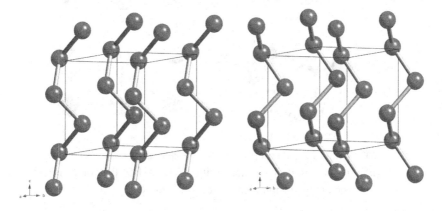

Fig. 6.16 Left- and right-handed forms of Se (on left and right respectively).

(6) Consider the mineral beryl, $Be_3Al_2Si_6O_{18}$. It is hexagonal with space group P6/ *mcc*. The a and c axes are 921 pm, the b axis is 920 pm. The atoms are located at:

Al   0.33333, 0.66667, 0.25000

Be   0.50000, 0.00000, 0.25000

O   0.48000, 0.15000, 0.15000

Si   0.39000, 0.12000, 0.00000

O   0.30000, 0.24000, 0.00000.

(a)   Plot a unit cell of the crystal structure.

(b)   On your diagram show (312) and [121].

(c)   What are the interatomic distances between the following atoms?

    (i)   Al at (1/3, 2/3, 1.25) and O at (1/3, 0.48, 1.15),

    (ii)   Be at (1/2, 0, 1.25) and O at (0.48, 0.15, 1.15),

    (iii)   Si at (0.12, 0.39, 0.5) and O at (0.24, 0.3, 0.5).

(7)   Monazite ($CePO_4$) is monoclinic, with $a = 6.76$ Å, $b = 7.00$ Å, $c = 6.44$ Å, and $\beta = 103.63°$. One of the P atoms is at position (0.467, 0.43, 0.265). One of the O is at (0.292, 0.344, 0.081). What is the interatomic distance between these two atoms?

(8) $Ca_5(PO_4)_3F$ is hexagonal, with $a = 9.3684$ Å, and $c = 6.8841$ Å. One of the Ca atoms is at position 2/3, 1/3, 0.501, and one of the O is at 0.877, 0.411, 0.25. What is the interatomic distance between these two atoms?

(9)   Two views of a unit cell of the anatase (a polymorph of TiO_2) crystal structure are shown below. All of the bonds are between Ti and O atoms. It belongs to a tetragonal crystal system (space group $I4_1/amd$) with $a = 3.785$Å, and $c = 9.515$. The atomic weight of Ti is 47.867 g /mole. O is 16 g/mole. Avagadro's number is 6.02×10^{23}/mole.

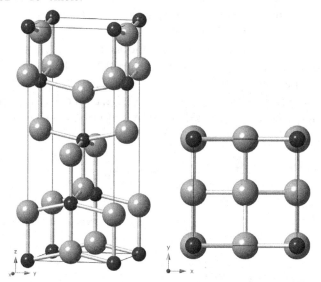

(a)   What is the theoretical density of anatase?

(b)   The Pauling electronegativity of Ti is 1.5; that of O is 3.5. What is the percentage of ionic character of the bonds?

7 Symmetry, Point Groups, and Stereographic Projections

7.0 Introduction – Why Symmetry?

Symmetry can be beautiful. It also governs many of the physical and chemical properties of crystals. As examples, consider the following:

- Cubic crystals, like copper, conduct electricity equally along [100], [111], and [110] directions. Non-cubic crystals do not.
- Crystals having a center of symmetry are not piezoelectric or pyroelectric.
- Monoclinic crystals have two optic axes which change orientation with wavelength. Trigonal crystals have only one optic axis which is the same for all wavelengths.
- Triclinic crystals have 21 different elastic constants. Cubic crystals have only three.
- Cubic rocksalt crystals dissolve at the same rate along [100], [010], and [001] directions. Polar hexagonal crystals like ZnO dissolve at different rates along [001] and [00$\bar{1}$].

Thus, it is important to have a good understanding of symmetry and its consequences. This chapter describes the various symmetry operators, and discusses how they can be combined in crystals. Stereographic projections are discussed as a short-hand means of illustrating symmetry.

7.1 Symmetry in Crystals

All crystals have translational periodicity.[1] Translational periodicity means that the object appears periodically some repeat distance away, without any mirrors or rotations, as shown in Fig. 7.1. In crystals, the periodicity develops in three dimensions, with repeat units $\vec{a}$, $\vec{b}$, and $\vec{c}$.

There are 14 essentially different kinds of translational periodicity. By definition, a *lattice is a set of points, each of which has identical surroundings*. The fourteen Bravais lattices are shown in Fig. 7.2. It can be seen that the different lattices can be differentiated based on both the shape of the unit cell, and whether or not the lattice is primitive.

---

[1] This statement describes conventional crystals. Exceptions occur in crystals with incommensurate modulations.

Repeat distance

Fig. 7.1 Professor Robert Newnham in translational symmetry. The repeat distance is shown by the arrow.

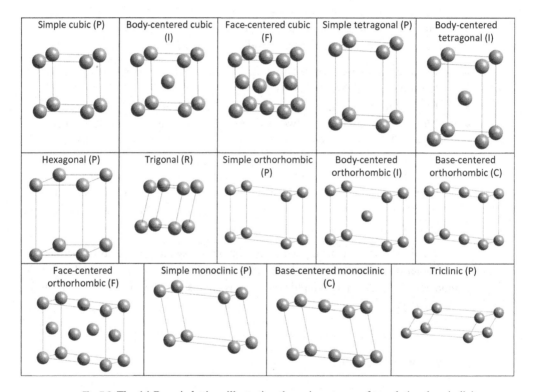

Fig. 7.2 The 14 Bravais lattices illustrating the unique types of translational periodicity.

For example, amongst the cubic Bravais lattices, simple cubic, body-centered cubic, and face-centered cubic versions all exist.

It is instructive to consider why no two face-centered lattices appear. Figure 7.3 shows that, if you try and construct such a lattice, the environment of points C and B differ. Thus, this cannot be considered to be a lattice. That is, around point B, there are four near neighbors at the top corners of the unit cell. For point C, there are no atoms at equivalent positions. They would have been located at the centers of the front and back faces of the unit cell as shown. This would have produced a face-centered orthorhombic Bravais lattice.

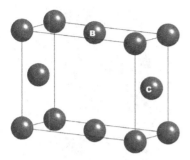

Fig. 7.3 Attempt to construct an orthorhombic lattice centered on two bases. It is clear that the geometric arrangement of neighbors is different for atoms B and C. Thus, this is not a lattice.

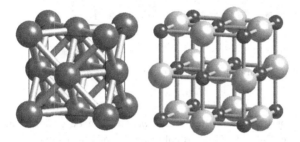

Fig. 7.4 Comparison of the crystal structure of Ni and NiO. Both are based on face centered cubic lattices. For Ni (left) the basis is one atom; for NiO (right) there are two atoms in the basis.

To describe a crystal structure, a *basis* of atoms is operated on by the lattice periodicity. As an example, consider a comparison of the crystal structures of Ni and NiO. Ni has a face-centered cubic lattice with a basis of one Ni atom on the corners of the unit cell. For NiO, the basis is two atoms, one Ni at (0, 0, 0) and O at (1/2, 0, 0). When this pair of atoms is operated on with the FCC lattice, the structure of rocksalt results, as shown in Fig. 7.4.

It is important to note that if the lattice is body-centered, then each atom in the basis also has body-centered translational periodicity. For example, in rutile, there are Ti atoms at positions (0, 0, 0) and (1/2, 1/2, 1/2) in the unit cell. However, since the O do *not* possess body-centered translational symmetry, the lattice is a primitive one.

Most crystals possess other symmetry elements in addition to translational periodicity. These symmetry elements include reflection across a mirror plane, rotation around a line (also called a rotation axis) in space, or inversion through a point. After the symmetry element has been applied, the object looks the same as it did before the operation. Thus, the symmetry operator can be thought of as relating one part of an object to another.

A mirror plane operates in much the same way as a bathroom mirror; mirror symmetry is present if any object on one side of the mirror appears an equal and opposite distance on the other side of the mirror. The object should appear identical before and after the symmetry operator is applied. That is, for a set of orthogonal axes, X, Y, and Z, a mirror plane at $Y = 0$ takes any point (x, y, z) and symmetry relates it to

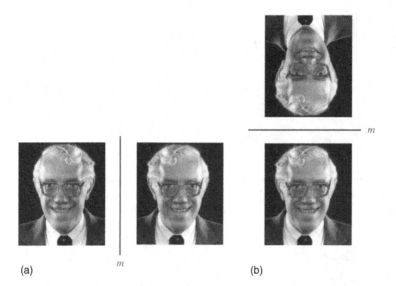

Fig. 7.5 Professor Newnham illustrating mirror symmetry: (a) vertical mirror and (b) horizontal mirror.

one at $(x, -y, z)$. Mirrors are given the symbol m, and can be arranged in different orientations, as shown in Fig. 7.5. The vertical mirror symmetry shows that the parting in Professor Newnham's hair, which was on the left side of the picture, flips to the right side of the picture with the symmetry operation. In the same way, mirror planes symmetry-relate a right hand to a left hand, a point that will become important in considering properties like optical activity that depend on a unique handedness for the material. The same change in handedness also develops for horizontal mirror planes.

An n-fold rotation axis means that the object should appear to be the same when it is rotated about the axis in increments of $360°/n$. By convention, the rotation is counter-clockwise. Examples are shown in Fig. 7.6. At first blush, the figure for the two-fold rotation axis appears to be similar to that of the horizontal mirror. However, close inspection of the figures makes it clear that the parting in Professor Newnham's hair is in different locations in the two instances. Likewise, application of a two-fold rotation axis symmetry-relates a right hand to a right hand, while the mirror symmetry relates the right hand to a left hand. Thus, rotation axes do not destroy chirality (handedness) in crystals or molecules. For a two-fold rotation axis coincident with the Z axis, a point at (x, y, z) is rotated 180° about Z to $(-x, -y, z)$, changing the sign of two of the coordinates. Rotation axes can be oriented along different directions of the crystals.

An inversion center, written as $\bar{1}$, is also called a center of symmetry. This symmetry-relates an object through operation around a point. That is, a line can be drawn from any point on an object through the center of symmetry. The object should re-appear an equal and opposite distance on the other side. This is illustrated in Fig. 7.7, where the center of symmetry is located on Professor Newnham's nose. Thus, an inversion center at the origin of the coordinate system transforms (x, y, z) to $(-x, -y, -z)$. This insures that a center of symmetry symmetry-relates a right hand into a left hand, and so will not appear in chiral crystals.

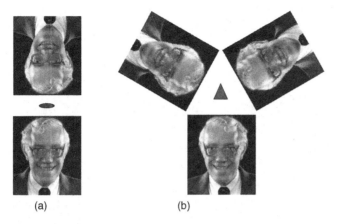

(a)   (b)

Fig. 7.6 Professor Newnham illustrating (a) a two-fold rotation axis normal to the page and (b) a three-fold axis normal to the page.

Fig. 7.7 Inversion symmetry-relates a right hand to a left hand.

In addition to these simple symmetry elements, there are also combination symmetry elements, which entail application of two symmetry elements *without stopping*. Examples of these include screw axes, glide planes, and roto-inversion axes. Let's consider each one of these independently.

A *screw axis* combines rotation and translation in a fashion that operates in much the same way as a spiral staircase. That is, one applies the rotation and translation axes concurrently. Screw axes have a two-part symbol; the main part of the symbol denotes the amount of rotation, while the magnitude of the translation is $\frac{subscript}{main\ symbol}$ of a repeat unit parallel to the rotation. For example, the screw axis 2_1 denotes a rotation of $360°/2 = 180°$ with a translation of 1/2 of the unit cell parallel to the screw axis. It is important to recognize that you continue to apply this element until the original position (or one translationally equivalent to it) is achieved. Figure 7.8 illustrates 3_1 and 3_2 screw axes. The construction is the same in both cases – the end result corresponds to screws that rotate with opposite handedness.

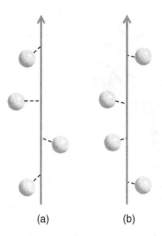

(a)   (b)

Fig. 7.8 (a) 3_1 and (b) 3_2 screw axes.

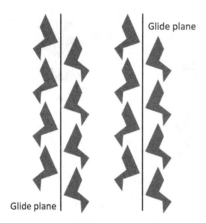

Glide plane

Glide plane

Fig. 7.9 Illustration of a glide plane.

A *glide plane* is a mirror + 1 translation, while a *net plane* is a mirror + 2 translations. In both cases, the translation is parallel to the mirror plane, often a distance of 1/2 of a repeat unit. The symbols a, b, and c are used to denote glide planes with translation components parallel to the $\vec{a}$, $\vec{b}$, and $\vec{c}$ axes respectively. Thus, for a, the glide plane would be perpendicular to either [010] or [001], and the glide vector is $\vec{a}/_2$. It is somewhat unfortunate that a, b, and c are also the symbols used for the lengths of the cell edges; the meaning needs to be ascertained from context. The symbol n is used for a net plane (sometimes call a diagonal glide plane); for the net plane there are translations along both of the allowed directions. For a net plane perpendicular to [001], the glide vector is $\frac{\vec{a}+\vec{b}}{2}$. A diamond glide entails a diagonal glide, with a translation of a quarter of the repeat unit, rather than half the repeat unit. Figure 7.9 gives an example of a drawing with a glide plane.

A *roto-inversion axis* combines rotation with inversion through a point on the rotation axes. This is given the symbol $\overline{2}$, $\overline{3}$, $\overline{4}$, or $\overline{6}$. An example is shown in Fig. 7.10.

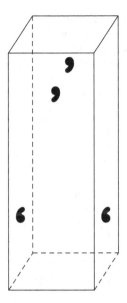

Fig. 7.10 A $\overline{4}$ roto-inversion axis.

7.2   Crystal Classes (Point Groups)

The crystal classes (also called the crystallographic point groups) are important because they govern the symmetry characteristic of the physical properties of crystals.[2] A point group is a self-consistent set of symmetry elements operating around a point. Thirty-two point groups are consistent with the translational periodicity found in crystals. A listing of the classes with their symmetry elements is in Table 7.1.

There are two different sets of symmetry notations that are in wide use to describe the crystal classes: Hermann–Mauguin notation and Schoenflies notation. This book utilizes the former. In Hermann–Mauguin notation, rotation axes are described using a number. As described above, for an n–fold rotation axis, a rotation of $\frac{360°}{n}$ leaves the object unchanged. Thus an object that has a three-fold symmetry axis can be rotated by $360°/3 = 120°$ (or $240°$ or $360°$) and will look the same; the symbol used to describe this is 3. Mirror planes are denoted by m; the relative orientation of a mirror with respect to other symmetry elements depends on the crystal system. For all crystal systems, if a slash appears between a rotation axis and the mirror, the mirror is perpendicular to the rotation axis. Thus, $4/m$ means that the four-fold axis is perpendicular to the mirror plane. In class $3m$, the three-fold axis lies *in* the mirror plane. Except for some triclinic crystals, the inversion symmetry operator does not appear in the notation used for the crystal class, even if a $\overline{1}$ is present. This is because the symbol is chosen to list only the independent symmetry elements. For example, the monoclinic class $2/m$ has a two-fold rotation axis along the high symmetry b axis, a

[2] Portions of this section were reproduced by permission from R. E. Newnham, *Structure–Property Relations*, Springer-Verlag (1975).

Table 7.1 The 32 crystal classes. Hermann–Mauguin and Schonflies symbols are given in the first column. The number of rotation axes and rotoinversion axes are listed in subsequent columns. The center of symmetry is $\bar{1}$, and $m = \bar{2}$ represents mirror planes. The last two columns give the populations (in percent) for inorganic and organic crystals. The statistics are based on structure analyses for over 280 000 compounds. Data from G. Johnson

Crystal class	2	3	4	6	$\bar{1}$	m	$\bar{3}$	$\bar{4}$	$\bar{6}$	Inorganic	Organic
$1 = C_1$	0	0	0	0	0	0	0	0	0	0.67	1.24
$\bar{1} = C_i$	0	0	0	0	1	0	0	0	0	13.87	19.18
$2 = C_2$	1	0	0	0	0	0	0	0	0	2.21	6.70
$m = C_s$	0	0	0	0	0	1	0	0	0	1.30	1.46
$2/m = C_{2h}$	1	0	0	0	1	1	0	0	0	34.63	44.81
$222 = D_2$	3	0	0	0	0	0	0	0	0	3.56	10.13
$mm2 = C_{2v}$	1	0	0	0	0	2	0	0	0	3.32	3.31
$mmm = D_{2h}$	3	0	0	0	1	3	0	0	0	12.07	7.84
$3 = C_3$	0	1	0	0	0	0	0	0	0	0.36	0.32
$\bar{3} = C_{3i}$	0	1	0	0	1	0	1	0	0	1.21	0.58
$32 = D_3$	3	1	0	0	0	0	0	0	0	0.54	0.22
$3m = C_{3v}$	0	1	0	0	0	3	0	0	0	0.74	0.22
$\bar{3}m = D_{3d}$	3	1	0	0	1	3	1	0	0	3.18	0.25
$4 = C_4$	1	0	1	0	0	0	0	0	0	0.19	0.25
$\bar{4} = S_4$	1	0	0	0	0	0	0	1	0	0.25	0.18
$4/m = C_{4h}$	1	0	1	0	1	1	0	1	0	1.17	0.67
$422 = D_4$	5	0	1	0	0	0	0	0	0	0.40	0.48
$4mm = C_{4v}$	1	0	1	0	0	4	0	0	0	0.30	0.09
$\bar{4}2m = D_{2d}$	3	0	0	0	0	2	0	1	0	0.82	0.34
$4/mmm = D_{4h}$	5	0	1	0	1	5	0	1	0	4.53	0.69
$6 = C_6$	1	1	0	1	0	0	0	0	0	0.41	0.22
$\bar{6} = C_{3h}$	0	1	0	0	0	1	0	0	1	0.07	0.01
$6/m = C_{6h}$	1	1	0	1	1	1	1	0	1	0.82	0.17
$622 = D_6$	7	1	0	1	0	0	0	0	0	0.24	0.05
$6mm = C_{6v}$	1	1	0	1	0	6	0	0	0	0.45	0.03
$\bar{6}m2 = D_{3h}$	3	1	0	0	0	4	0	0	1	0.41	0.02
$6/mmm = D_{6h}$	7	1	0	1	1	7	1	0	1	2.82	0.05
$23 = T$	3	4	0	0	0	0	0	0	0	0.44	0.09
$m3 = T_h$	3	4	0	0	1	3	4	0	0	0.84	0.15
$432 = O$	9	4	3	0	0	0	0	0	0	0.13	0.01
$\bar{4}3m = T_d$	3	4	0	0	0	6	0	3	0	1.42	0.11
$m3m = O_h$	9	4	3	0	1	9	4	3	0	6.66	0.12

mirror plane perpendicular to b, and a center of symmetry at the origin. The two-fold rotation axis means that any point (x, y, z) is symmetry equivalent to $(-x, y, -z)$. The mirror plane symmetry relates $(-x, y, -z)$ to $(-x, -y, -z)$. Thus, the combination of 2 and the perpendicular mirror m relates (x, y, z) to $(-x, -y, -z)$; this is the same symmetry relation provided by the $\bar{1}$.

The 32 crystal classes are derived by assessing how symmetry elements can be combined without symmetry-relating one point in space to an infinite number of other points. This imposes some restrictions: only some rotational axes are consistent with translational

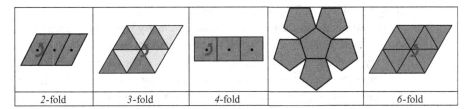

| 2-fold | 3-fold | 4-fold | | 6-fold |

Fig. 7.11 Illustrations of units with two-, three-, four-, and six-fold rotation axes. Pentagons, with five-fold symmetry cannot be used to fill space, unless they are coupled with another unit of a different shape. Thus, five-fold symmetry does not exist in crystals, although apparent five-fold symmetry appears in quasi-crystals.

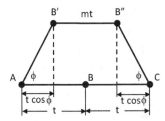

Fig. 7.12 One-dimensional chain with rotation axes normal to the paper. The only allowed values of ϕ are those corresponding to one-, two-, three-, four-, and six-fold symmetry axes. Redrawn from Newnham, "Phase Diagrams and Crystal Chemistry", in *Phase Diagrams: Volume V*, Ed. Allen M. Alper, Academic Press, NY 1978, Fig. 2 (chapter 1), redrawn with permission from Elsevier.

symmetry, rotation axes can meet only at some particular angles, there are only a finite number of ways in which rotation, mirror, and inversion symmetries can be combined.

The only rotation axes which are compatible with translation are one-, two-, three-, four-, and six-fold rotation axes. The one-fold axis corresponds to rotation of an object by 360°/1, bringing any object back into coincidence with itself. A pictorial representation of why these rotation axes are the important ones in crystals is shown for two dimensions in Fig. 7.11. It is clear there that one can fill space with parallelograms, triangles, squares, or hexagons. However, space cannot be filled only by tiling pentagons with a five-fold rotation axis. In this case, two geometric units (for example pentagons and diamonds) would be required to fill space, though the result would not be compatible with translational symmetry.

A more complete geometric proof of the allowed rotation axes in crystals is illustrated in Fig. 7.12. Consider starting with a one-dimensional set of lattice points separated by a distance t. By definition, lattice points are required to have identical surroundings. Suppose an n-fold rotation axis perpendicular to the plane passes through each lattice point. This means that the object can be rotated by $\phi = \frac{360°}{n}$. When the rotation axis through A acts on point B it makes points B and B' symmetry-equivalent. Likewise, acting on B with the rotation axis through C takes B to B''. Points B' and B'' define a line parallel to A – B – C which has the same repeat distance as ABC. The distance between B' and B'' is thus mt, where m is an integer. Inspection of the geometry points to the observation that $mt = 2t - 2t\cos\phi$. Various solutions can then be tried for different integer values of m. The outcome is that there are five permissible values for ϕ: $60°$, $90°$, $120°$, $180°$, and $360°$. Thus, translational symmetry is compatible with rotation axes, 6, 4, 3, 2, and 1. Thus, while other types of rotation axes (5, 7, 10, etc.) are found in biology, molecules, art, and quasi-crystals, they do not occur in crystals.

Symmetry axes cannot be oriented at arbitrary angles with respect to each other. The key point to recognize is that two intersecting rotation axes always generate a third. As a result, point groups have either *one* rotation axis or *more than two*. The five crystal classes with one rotation axis are 1, 2, 3, 4, and 6. It is possible to use spherical geometry to prove that when multiple rotation axes exist, that they meet at specific angles. If axes OA and OB intersect at point O, then the angle between the two axes, AOB is:

$$\cos(AOB) = \frac{\cos(\phi_3/2) + \cos(\phi_1/2)\cos(\phi_2/2)}{\sin(\phi_1/2)\sin(\phi_2/2)},$$

where ϕ_1, ϕ_2, and ϕ_3 are the rotation angles of the three intersecting axes. As an example, consider the case where there are two three-fold axes and one two-fold axis. The cosine of the angle between the two three-fold axes will be given by

$$\cos(AOB) = \frac{\cos(180°/2) + \cos(120°/2)\cos(120°/2)}{\sin(120°/2)\sin(120°/2)} = \frac{1}{3}.$$

Thus, angle AOB is $70.5°$. In the same way, the angle between the two-fold axis and either of the three-fold axes is $54.7°$. These angles correspond to angles between two body diagonals of a cube, and between the body-diagonal and a cube edge. Hence, the combination of the symmetry elements 332 can appear in a cubic crystal system. Not all combinations of symmetry elements are permitted. For example, calculations of cos(AOB) for the case of 643, produces a value >1, an impossible situation. Six- and four-fold axes never occur together in crystals. The combinations of rotation axes that are possible include 222, 322, 422, 322, and 432, as shown in Fig. 7.13. All other combinations

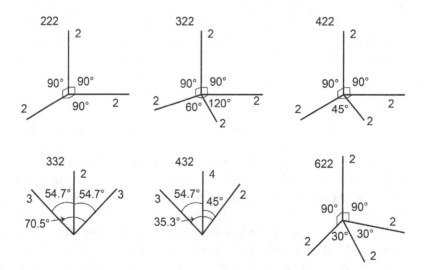

Fig. 7.13 Drawing of axial combinations with interaxial angles.

of the one, two, three, four, and six-fold rotation axes that satisfy Euler's equations are equivalent to one of these. These six permissible combinations correspond to the point groups 222, 32, 422, 622, 23, and 432, respectively.

Recall that a rotation axis does not change the handedness of an object. Thus, the point groups that have only rotation elements (e.g., 1, 2, 3, 4, and 6, 222, 32, 422, 622, 23, and 432) are the eleven enantiomorphic crystal classes. Right- and left-handed crystals can occur in these classes. These are the crystal classes in which properties that require handedness (e.g. optical activity), can appear.

As was shown in Chapter 6, all crystal structures belong to one of the following crystal systems: cubic, hexagonal, tetragonal, trigonal, rhombohedral, orthorhombic, monoclinic, and triclinic. Each of the crystal systems is characterized by a minimum symmetry. The minimum axial symmetry for the cubic system is four three-fold axes that appear along the body diagonals of the cube. The hexagonal, tetragonal, and trigonal crystal systems have a single six-, four-, and three-fold axis, respectively; these rotation axes occur along the crystallographic c axis. For the rhombohedral crystal system, there is also a single three-fold rotation axis which appears along the [111] axis. In some texts, rhombohedral crystals are not listed, since they are a subset of the trigonal system (e.g. they have the same minimum symmetry, and hence all rhombohedral crystals can be expressed in a trigonal unit cell, although not all trigonal systems have rhombohedral symmetry). The orthorhombic system has three mutually perpendicular two-fold rotation axes. Monoclinic crystals have a single two-fold axis along the b axis. Triclinic crystals possess only a one-fold rotation axis.

The minimum symmetry imposes restrictions on the unit cell dimensions. To illustrate, an orthorhombic unit cell has three perpendicular two-fold rotation axes parallel to a, b, and c; this follows from the minimum symmetry 222 and the allowed angles between these rotation axes. These rotation axes symmetry-relate a to $-a$, b to $-b$, and c to $-c$. However, there is nothing that symmetry-relates the cell edges, a, b, and c. That is, a is never transformed to b, etc. As a result, in an orthorhombic crystal system, $a \neq b \neq c$, and $\alpha = \beta = \gamma = 90°$. However, for higher symmetries, the rotation axes make some of the lattice parameters equivalent to each other. In the trigonal system, the minimum symmetry is a three-fold axis parallel to c. This rotation symmetry-relates a to b, and since a unit cell is left unchanged by its symmetry elements, a must be equal in b in length. The minimum symmetry for a trigonal system does not impose any restrictions on c. Thus for a trigonal system, $a = b \neq c$, and $\alpha = \beta = 90°$, and $\gamma = 120°$.

So far, we have considered only the 11 axial point groups: 1, 2, 3, 4, and 6, 222, 32, 422, 622, 23, and 432. The remaining 21 crystal classes are obtained by combining the eleven axial classes with inversion and mirror symmetry.

Consider first the addition of a center of symmetry. Adding a center to the identity operator 1 gives $\bar{1}$, a centric triclinic crystal class.

There are two ways of adding a center of symmetry to the classes with a single rotation axis. Consider the combination of a center of symmetry with a single two-fold rotation axis. The inversion center is constrained to lie on the axis, in order to avoid generating more lines and more points. The inversion and rotation operations may be applied *consecutively* or *concurrently*; these would correspond to the cases where there is both a 2 and a $\bar{1}$, or the case where the symmetry operation is a $\bar{2}$, respectively.

Four equivalent points are generated under *consecutive* application of the two symmetry operations. A two-fold axis parallel to Y takes a point at (x, y, z) to $(-x, y, -z)$. Operation on these two points with the inversion operator at the origin takes them to $(-x, -y, -z)$ and $(x, -y, z)$, respectively, giving a total of four equivalent points. Note also that a mirror plane at $Y = 0$ relates (x, y, z) to $(x, -y, z)$ and $(-x, -y, -z)$ to $(-x, y, -z)$. Thus, the point group contains an inversion center, a two-fold rotation axis along Y, and a mirror plane perpendicular to the two-fold axis. This is the monoclinic point group $2/m$. Applying the inversion center and the two-fold axis along Y *concurrently* gives a different result, taking (x, y, z) to $(x, -y, z)$. This is point group $\bar{2} = m$. This crystal system is also monoclinic, but with fewer equivalent points. Addition of inversion symmetry to the three-, four-, or six-fold axis gives point groups $\bar{3}, \frac{4}{m}, \frac{6}{m}$ when applied consecutively and $\bar{3}, \bar{4}, \bar{6} = \frac{3}{m}$ when applied concurrently. Thus, eight new point groups are obtained by combining an inversion center with a single rotation axis.

Consider next the addition of mirror symmetry to a single rotation axis. To avoid generating more rotation axes, the mirror plane must be either perpendicular to the rotation axis or contain the rotation axis. When the mirror plane is perpendicular to the rotation axis the following crystal classes result: $\frac{1}{m} = m, \frac{2}{m}, \frac{3}{m} = \bar{6}, \frac{4}{m}, \frac{6}{m}$; all of these were generated by rotation–inversion combinations. When the mirror contains the rotation axis (e.g. the mirror and the rotation axis are parallel) the following classes are generated: $1m = m, 2m = mm, 3m, 4m = 4mm$, and $6m = 6mm$. The latter four are new, giving a total of 23 point groups.

The remaining nine point groups arise from combining mirror planes with 222, 32, 422, 622, 23, and 432. Six of these result from placing a mirror perpendicular to the leading symmetry axis: $\frac{2}{m}22 = \frac{2}{m}\frac{2}{m}\frac{2}{m} = mmm, \frac{3}{m}2 = \frac{3}{m}m2 = \bar{6}m2, \frac{4}{m}22 = \frac{4}{m}\frac{2}{m}\frac{2}{m} = \frac{4}{m}mm, \frac{6}{m}22 = \frac{6}{m}\frac{2}{m}\frac{2}{m} = \frac{6}{m}mm, \frac{2}{m}3 = \frac{2}{m}\bar{3} = m3, \frac{4}{m}32 = \frac{4}{m}\bar{3}\frac{2}{m} = m3m$. The final three classes are derived by adding parallel planes to the combinations 222, 322, and 23, yielding $\bar{4}2m, \bar{3}\frac{2}{m} = \bar{3}m$, and $\bar{4}3m$, respectively.

7.3　　Space Groups

The full microscopic symmetry of a crystal structure is given by the space group; all crystal structures belong to one of the 230 space groups.[3] A space group can be thought of as a self-consistent set of symmetry operations acting around a lattice. Thus, space groups necessarily contain translational symmetry, where point groups do not. A complete listing of the space groups, together with the equipoint positions and symmetry drawings, is given in the *International Tables for Crystallography*.[4] A few examples, and an explanation of the Hermann–Mauguin symbols are presented here.

---

[3] Portions of this section were reproduced by permission from R. E. Newnham, *Structure–Property Relations*, Springer-Verlag (1975).

[4] *International Table for Crystallography, Volume A: Space Group Symmetry*, ed. Theo Hahn.

Table 7.2 Symbols and directions used to designate space groups

Lattice type	Lattice point positions				Symbol
Primitive	0, 0, 0				P
(100)-face-centered	0, 0, 0;	0, ½, ½			A
(010)-face-centered	0, 0, 0;	½, 0, ½			B
(001)-face-centered	0, 0, 0;	½, ½, 0			C
Body-centered	0, 0, 0;	½, ½, ½			I
Face-centered	0, 0, 0;	½, ½, 0;	½, 0, ½;	½, ½, 0	F
Rhombohedral	0, 0, 0;	$^1/_3, {}^2/_3, {}^2/_3$;	$^2/_3, {}^1/_3, {}^1/_3$		R

Symbol	Symmetry operation
m	reflection (mirror plane)
a, b, c	axial glide planes
n	diagonal glide plane, or net plane
d	diamond glide plane
1	identity (monad)
2, 3, 4, 6	rotation axes (diad, triad, tetrad, and hexad)
$\bar{2}, \bar{3}, \bar{4}, \bar{6}$	inversion axes
$2_1, 3_1, 3_2, 4_1, 4_2, 4_3,$ $6_1, 6_2, 6_3, 6_4, 6_5$	screw axes

Order of position of symbols in point groups and space groups

Crystal system	Primary position	Secondary position	Tertiary position
Triclinic	–	–	–
Monoclinic	[010]	–	–
Orthorhombic	[100]	[010]	[001]
Trigonal	[001]	$\langle 100 \rangle^{\dagger}$	$\langle 210 \rangle^{\dagger}$
Tetragonal	[001]	$\langle 100 \rangle^{\dagger}$	$\langle 110 \rangle^{\dagger}$
Hexagonal	[001]	$\langle 100 \rangle^{\dagger}$	$\langle 210 \rangle^{\dagger}$
Cubic	$\langle 100 \rangle$	$\langle 111 \rangle$	$\langle 110 \rangle^{\dagger}$

† These directions are not symmetry axes in all classes within the crystal system.

All space group symbols begin with a capital letter describing the translational symmetry of the lattice. The symbols for this, along with the equivalent lattice points, are given in Table 7.2. The second symbol corresponds to the primary symmetry direction. If secondary and tertiary symmetry elements are present, they appear in the third and fourth positions. The relative orientation of the different symmetry elements depends on the crystal system; the rules for the orientation are shown in the table. While the crystal system is not stated explicitly in the space group symbol, it is easily deduced from the symmetry elements. For example, if the space group is converted to a point group that shows three symbols which are either 2 or m, this corresponds to the orthorhombic crystal class. Cubic crystals can be recognized by a 3 in the third position. Hexagonal, tetragonal, and trigonal systems can be identified by a 6, 4, and 3, respectively in the secondary position showing the primary symmetry direction. Rhombohedral systems are identified by use of the R translational symmetry element.

To illustrate, consider the space group of fluoroapatite, $P6_3/m$. The lattice has primitive translational symmetry, as noted by the P. It is apparent that the crystal

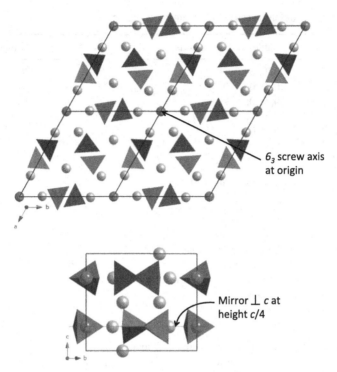

Fig. 7.14 Apatite crystal structure showing $(PO_4)^{3-}$ tetrahedra. The smaller filled circles are Ca^{2+}, the larger are the F^- ions. (Top) View of crystal structure down the c axis, showing the location of the screw axis ∥ c. (Bottom) View of crystal structure down a axis, showing the mirror plane perpendicular to c.

belongs to the hexagonal crystal system since a six-fold axis is present. The complex symbol $6_3/m$ counts as a single symbol that refers to the [001] direction, the c (or Z) axis. The 6_3 is a screw axis parallel to [001], and consists of a 60° rotation accompanied by a translation parallel to [001] of $\frac{3}{6}c$, half of the unit cell. The /m means that there is a mirror plane perpendicular to the 6_3 axis (the c axis). Fluoroapatite does not have either secondary or tertiary symmetry symbols in the space group. The structure and symmetry elements in apatite are shown in Fig. 7.14.

As a second example, rutile has the space group $P\frac{4_2}{m}\frac{2_1}{n}\frac{2}{m}$. The P symbol describes the fact that the lattice is primitive. The crystal system is tetragonal because of the four-fold axis. (Note that while cubic crystals sometimes have four-fold symmetry, it is apparent that rutile is not cubic because there is no three-fold rotation axis in the symbol. All cubic crystals contain four three-fold axes along the $\langle 111 \rangle$ directions.) Referring back to Table 7.2, the first symmetry element, $4_2/m$, refers to the primary symmetry direction [001]. That is, there is a 4_2 screw axis parallel to c, and a mirror plane m perpendicular to c. Rutile possesses secondary and tertiary symmetry elements along the $\langle 100 \rangle$ and $\langle 110 \rangle$ directions, respectively. There is a 2_1 screw axis parallel to [100] that denotes a 180° rotation with a translation of $\frac{\vec{a}}{2}$ without stopping. The net plane n is perpendicular to the 2_1 screw axis. A net plane consists of reflection – across (100) in this case – followed by a diagonal translation parallel to the mirror. The translation vector is $\frac{(\vec{b}+\vec{c})}{2}$

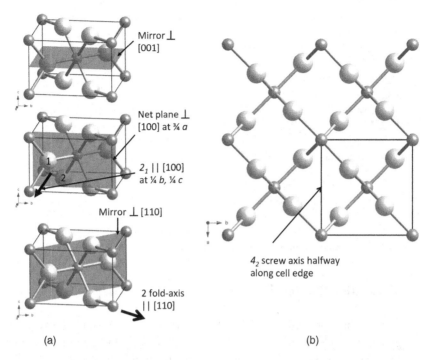

Fig. 7.15 Location of symmetry elements on rutile. (a) Schematic showing the mirror and net planes. The net plane in the middle panel relates atom 1 to atom 2. (b) Location of screw axis. Equivalent screw axes are located at comparable positions on each edge of the unit cell, and are parallel to the c axis.

for a net plane perpendicular to [100]. The tertiary direction in rutile is [110], which is parallel to a two-fold axis and perpendicular to a mirror plane. In tetragonal crystals, the four-fold rotation axis symmetry relates a and b. Thus the [100] and [010] directions are equivalent, as are [110] and [$\bar{1}$10]. For simplicity, the symmetry properties of [010] and [$\bar{1}$10] are not explicitly written out in the space group symbol. The location of each of these symmetry elements is shown on the structure in Fig. 7.15.

Space groups are used to describe the full symmetry present in crystals. However, the full symmetry is not needed to describe either morphology or physical properties; only the macroscopic symmetry defined by the point groups is needed for this. Property measurements are insensitive to the unit-cell scale translations which differentiate space groups and point groups. Thus, a crystal with a two-fold axis has the same tensor symmetry as one with a two-fold screw axis.

Space groups can be converted to point groups by dropping all reference to translation. The lattice symbol describes the translation symmetry of the lattice. Thus, the lattice symbol does not appear in the point group. In the same way, since a screw axis is a rotation plus a translation, when the translation is eliminated, the screw axis becomes a rotation axis. Both glide and net planes are changed to mirrors. For example, the space group $Fm3m$ corresponds to point group $m3m$. For rutile, the space group symbol $P\frac{4_2}{m}\frac{2_1}{n}\frac{2}{m}$ converts to point group $\frac{4}{m\,m\,m}\frac{2}{m}\frac{2}{m} = \frac{4}{m}mm$. For apatite, the point group is $6/m$.

North

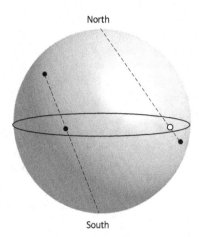

South

Fig. 7.16 Construction of a stereographic projection.

7.4   Stereographic Projections

Stereographic projections offer a compact means of illustrating the symmetry of point groups. Consider a sphere with a point in an arbitrary location on its surface. One can apply all of the symmetry operations in the point group progressively to the sphere to illustrate the symmetry. Drawing a proper representation of this in three dimensions, is, however, rather a painful exercise. To avoid needing to perform miracles of three-dimensional rendering, stereographic projections can be utilized instead. In a stereographic projection, all of the points on the sphere are projected onto the equatorial plane. This is illustrated in Fig. 7.16. An object in the Northern hemisphere is connected to the South pole. The point at which it intersects the equatorial plane is shown as a solid dot on the stereographic projection. Objects in the Southern hemisphere are connected to the North pole; their position is marked as an open circle on the equatorial plane. The equatorial plane, with all of the symmetry-related points marked, is the stereographic projection of the point group.

To draw such a stereographic projection, start with a point in an arbitrary position in the Northern hemisphere. The point is operated on with each symmetry element. Consider, for example a three-fold rotation axis (shown projecting out of the plane of the page from the center of the circle in Fig. 7.17). The starting point in the lower right is spun 120° around the three-fold axis, where it reappears. This symmetry element is then applied as often as necessary until you arrive back at the starting point. The result is the stereographic projection for point group 3.

Symmetry-related points resulting from a mirror plane are drawn by imagining a line from the point perpendicular to the mirror plane, then going an equal and opposite distance to the other side. Mirrors that are parallel to the plane of the page symmetry-relate points in the Northern hemisphere to points in the Southern hemisphere. Mirrors that are perpendicular to the plane of the page take a point in the Northern hemisphere to a symmetry-related position in the Northern hemisphere. Both possibilities are illustrated in Fig. 7.18.

Shown in Fig. 7.18 is the effect of a center of symmetry on a stereographic projection. The center relates points in the Northern hemisphere to ones on the "opposite side of the

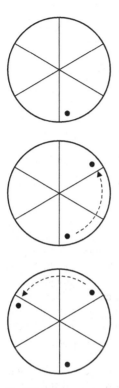

Fig. 7.17 Construction of stereographic projection for point group 3. The arrows show the 120° rotations.

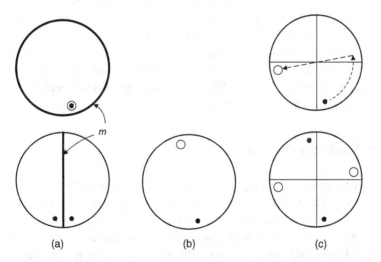

(a)   (b)   (c)

Fig. 7.18 (a) Stereographic projections for mirror planes in different orientations. (b) Stereographic projection for a center of symmetry. (c) Construction of a stereographic projection of a $\overline{4}$ roto-inversion axis projecting out from the middle of the diagram; (top) the point is rotated by 90° and inverted without stopping. The symmetry element is then applied repeatedly until the original point is reached to produce (bottom) the full stereographic projection.

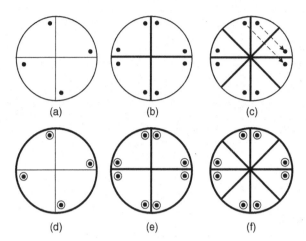

Fig. 7.19 (a–c) Construction of the stereographic projection for point group 4*mm*. (a) Four-fold rotation axis. (b) Mirror planes, shown as bold lines, perpendicular to [100]. Note that the four-fold axis also acts on the symmetry elements, so that there is also a mirror perpendicular to [010]. (c) The mirror planes perpendicular to [110] do not introduce any new points to the stereographic projection. (d–f) Construction of the stereographic projection for point group 4/*mmm*. (d) 4/*m*. (e) Mirror planes, shown as bold lines, perpendicular to [100]. (f) The mirror planes perpendicular to [110] do not introduce any new points to the stereographic projection.

sphere" in the Southern hemisphere. Other roto-inversion axes are constructed by taking the starting point, rotating through the appropriate amount, then inverting through the center without stopping. This sequence is repeated until arriving back at the initial point.

Stereographic projections for point groups with more than one symmetry element can be drawn by applying each of the symmetry elements in sequence. For example, point group 4*mm* has a four-fold rotation axis parallel to *c*, and mirrors perpendicular to [100] and [110]. Figure 7.19 shows how the stereographic projection is built up. In point group 4/*mmm*, the first symmetry element is 4/*m*; the mirror perpendicular to the four-fold rotation axis symmetry-relates the points in the Northern hemisphere to points in the Southern hemisphere.

In a comparable way, stereographic projections for all 32 point groups can be elucidated, as shown in Fig. 7.20. The rules governing the placement of the symmetry elements are those shown in Table 7.2.

7.5   Group–Subgroup Relations

One space group or point group is considered a subgroup of another if it contains fewer symmetry elements than the original group, without the addition of any new symmetry. For example, the point group 23 is one of the cubic point groups. If the cube is distorted by elongating it along one body diagonal, then the two-fold axis is lost, and only one of the original three-fold axes is retained. That is, the point group 3 is a subgroup of the point group 23. The point group 6 is *not* a subgroup of point group 23, since there is no six-fold rotation axes in point group 23. Table 7.3 shows a full listing of the point groups, along with their subgroups.

Group–subgroup relations in point groups can be determined from the stereographic projections. A subgroup should have fewer points than the group from which it comes,

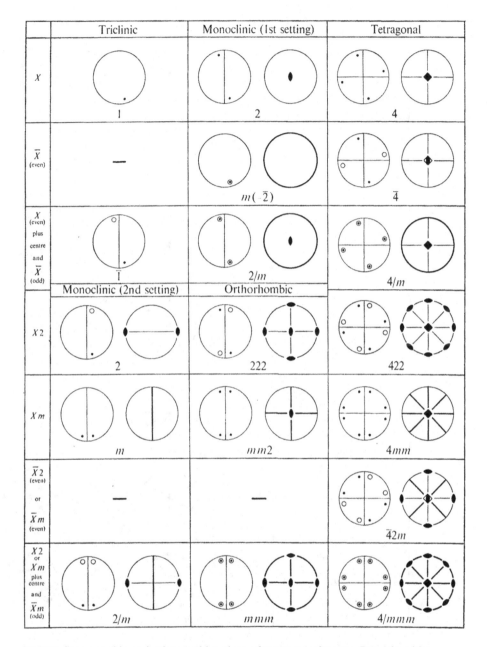

Fig. 7.20 Stereographic projections and locations of symmetry elements. Reproduced by permission from *International Tables for X-Ray Crystallography, Vol. I Symmetry Groups*, ed. N. F. Henry and K. Lonsdale, Kynoch Press, Birmingham, England (1952). Fig. 3.3.1, pages 26 and 27. http://it.iucr.org/

with no new points being added. Referring back to Fig. 7.19, it is clear that point group 4*mm* is a sub-group of point group 4/*mmm*.

Group–subgroup relations are often useful in describing phase transitions, as well as in describing the relationship between a parent and a derivative structure.

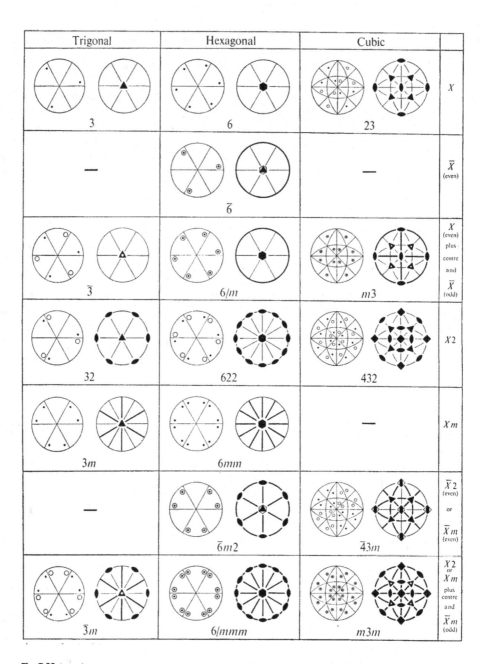

Fig. 7.20 (*cont.*)

7.6   Curie Groups

The net symmetry of liquids, gases, non-crystalline solids, and polycrystalline materials with a random arrangement of crystallites cannot be described using space or point groups. In this case, the Curie groups shown in Fig. 7.21 are useful.

Table 7.3 Point groups (†) and their subgroups (•)

$\frac{4}{m}\bar{3}\frac{2}{m}$	†																															
$\bar{4}3m$	•	†																														
432	•		†																													
$\frac{2}{m}\bar{3}$	•			†																												
23	•	•	•	•	†																											
$\frac{6}{m}\frac{2}{m}\frac{2}{m}$						†																										
$\bar{6}m2$						•	†																									
$6mm$						•		†																								
622						•			†																							
$\frac{6}{m}$						•				†																						
$\bar{6}$						•	•			•	†																					
6						•	•	•	•	•	•	†																				
$\bar{3}\frac{2}{m}$	•					•							†																			
$3m$	•	•				•	•	•					•	†																		
32	•		•			•	•		•				•		†																	
$\bar{3}$	•			•		•				•			•			†																
3	•	•	•	•	•	•	•	•	•	•	•	•	•	•	•	•	†															
$\frac{4}{m}\frac{2}{m}\frac{2}{m}$	•																	†														
$\bar{4}2m$	•	•																•	†													
$4mm$	•																	•		†												
422	•		•															•			†											
$\frac{4}{m}$	•																	•				†										
$\bar{4}$	•	•																•	•			•	†									
4	•		•															•		•	•	•		†								
$\frac{2}{m}\frac{2}{m}\frac{2}{m}$	•			•		•												•							†							
$2mm$	•	•		•		•	•	•										•	•	•					•	†						
222	•	•	•	•	•	•			•									•	•		•				•		†					
$\frac{2}{m}$	•			•		•				•			•					•				•			•			†				
m	•	•		•		•	•	•		•	•		•	•				•	•	•		•			•	•		•	†			
2	•	•	•	•	•	•	•	•	•	•		•	•		•			•	•	•	•	•	•	•	•	•	•	•		†		
$\bar{1}$	•			•		•				•		•		•		•		•				•			•			•			†	
1	•	•	•	•	•	•	•	•	•	•	•	•	•	•	•	•	•	•	•	•	•	•	•	•	•	•	•	•	•	•	•	†

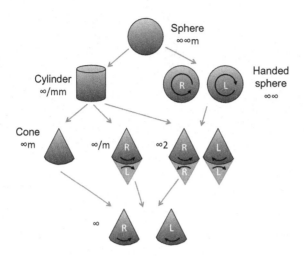

Fig. 7.21 Curie groups. The R and L denote right- and left-handedness, respectively.

The Curie groups are sometimes called the limiting groups or the continuous groups. All Curie groups possess at least one infinite-fold rotation axis. Consider, for example, molten metals such as gold. On average, the structure cannot be differentiated when it is rotated by any amount, with the net result that there is in infinite-fold rotation axis. A second infinite-fold rotation axis is perpendicular to the first. Since there is no sense of handedness in this melt, there are also an infinite number of mirror planes. In short, the melt has the symmetry of a sphere, denoted as $\infty\infty m$. In a cylinder, the symmetry drops to ∞m. In the case of solutions with dissolved handed molecules of one type, the solution also has a handedness, which eliminates mirror planes.

7.7   Problems

(**1**)   Show the arrows that would be symmetry related to the first by a $\bar{6}$ axis normal to the page. Show the top and bottom of the arrow in two different colors.

(**2**)   High quartz has space group $P6_222$, and can be described on a hexagonal cell with $a = 5.01$ Å and $c = 5.47$ Å. Si atoms are at 0.5, 0, 0 and O atoms are at 0.2, –0.2, 0.8333.

    (**a**)   Using a crystal drawing program, show a unit cell.

    (**b**)   Mark the screw axis and show which Si atoms are related. (Hint: it is easier to see if you show several unit cells.)

(3) If the line represents a mirror coming out of the page, where is the symmetry related arrow?

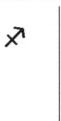

(4) Consider the space group $I4_1/a$.
  (a) What does each of the symbols mean?
  (b) What is the corresponding point group?
  (c) Draw the stereographic projection for the point group.

(5) Draw a figure showing the symmetry $\bar{4}$.

(6) What point group would result if you added a center of symmetry to point group $4mm$? Be sure to detail your reasoning (graphically is fine).

(7) Prove graphically that a $\bar{3}$ axis possesses a center of symmetry.

(8) Prove that a $\bar{6}$ axis is identical to a three-fold axis with a perpendicular mirror plane.

(9) Suppose you have a four-fold rotation axis parallel to the c axis of a unit cell intersecting the position $(0.5, 0.5, 0)$. If you have an atom at position $(0.75, 0.5, 0.3)$, what are the atomic coordinates of the atoms that are related to that atom by the four-fold axis?

(10) An atom in a tetragonal unit cell has fractional coordinates of $(0.1, 0.13, 0.2)$. Give the coordinates of the second atom in the unit cell that is related by the following (separately applied).
  (a) Body centering.
  (b) A center of symmetry at the origin.
  (c) A mirror plane containing the x and z axes.
  (d) A 4_2 axis parallel to z and passing through the origin.
  (e) An a glide plane intersecting point $(0.5, 0.5, 0.5)$.

(11) Identify the symmetry characteristic of the macroscopic shape of phosgenite crystal as shown below. The faces pointing to the left and towards you have the same dimensions.

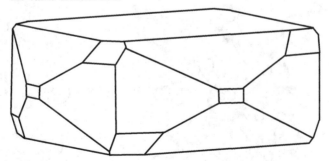

(12) Find on-line a rectilinear image of M.C. Escher's "Angels and Demons". Identify the symmetry elements in the figure and locate the smallest unit cell.

(13) BaTiO$_3$ is tetragonal with the lattice parameters $a = 3.99$Å, $c = 4.03$ Å. Ba has an atomic weight of 137.33 g /mole and an electronegativity of 0.89. Ti has an

atomic weight of 47.90 g/mole and an electronegativity of 1.54. O has an atomic weight of 16.00 g /mole and an electronegativity of 3.44. The unit cell is shown below. Avogadro's number is 6.02×10^{23}.

(a)   What is the theoretical density of $BaTiO_3$?

(b)   What is the percent ionic character of the Ba–O bond?

(c)   The space group is $P4mm$. What does each of the symbols mean?

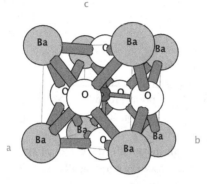

(14)   Shown below is a unit cell of scheelite ($CaWO_4$). Its space group is $I4_1/a$. The atomic weight of Ca = 40.08 g/mole, W = 183.9 g/mole, and O = 16 g/mole. The unit cell is tetragonal with $a = 5.24$Å and $c = 11.38$Å. Avogadro's number is 6.02×10^{23}. The Ca are at positions $(0, 0, 0.5)$ and symmetry-related positions, W is at $(0, 0, 0)$ and symmetry related positions, and O is at $(0.25, 0.15, 0.075)$ and symmetry related positions.

(a)   Calculate the theoretical density of scheelite.

(b)   What is the W–O bond length?

(c)   Define each of the symbols in the space group.

(d)   What is the corresponding point group?

Looking down c Looking perpendicular to c

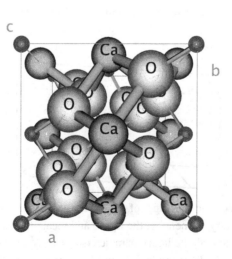
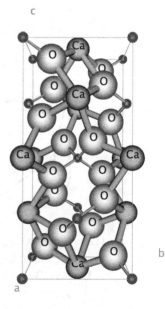

(15) Log on to: http://escher.epfl.ch/escher/ and read the accompanying documentation. Play around with the drawing program enough that you are comfortable with the various symmetry elements in two-dimensional space.

(a) Create figures showing the following symmetries: *pm*, *pg*, and *p2mm*.

(b) On one of the figures, illustrate the differences between placing a point on a high-symmetry position versus a low-symmetry position.

8   Covalent Crystals and Semiconductors

8.0   Introduction

Unlike ionic bonding, there is no good classical description of the covalent bond. Instead, within covalently bonded molecules and solids, quantum mechanical approaches are used to describe the system. Covalent single bonds are the result of sharing electrons between adjacent atoms. Thus, in electron-density maps of covalently bonded materials, there is a significant electron density between the atoms, as shown in Fig. 8.1. This differs from ionic bonding, where the charge density goes through a minimum along a line through the atoms, and the electron distribution around an atom core is often nearly spherical. Double and triple bonds can also form as more electrons are contributed per atom.

8.1   Covalent Bonds

Covalent bonding occurs between atoms of similar electronegativity on the right-hand side of the periodic table, while metallic bonding tends to be favored by the elements on the left-hand side of the periodic table. Metallic character also tends to increase on moving down a column in the periodic table (for example, the group IV elements Sn and Pb). Thus, excellent examples of covalently bonded solids include the crystal structures of the light column IV elements.

Because the covalent bond entails electron sharing between atoms, the bonds are highly directional, and bond angles are well-defined. This is a result of the observation that an *atom can form an electron pair bond for each stable orbital*; the orbital shapes are a consequence of quantum mechanics. As an example, consider the several common near-neighbor geometries adopted by covalently bonded carbon, as illustrated in Fig. 8.2. Carbon has the electron configuration $1s^2 2s^2 2p^2$. Different coordination geometries are produced depending on how the four valence electrons of carbon are partitioned in bonds to neighboring atoms. In diamond, sp^3 hybrid orbitals result when the 2s and 2p electrons are promoted to form four equivalent orbitals, with one electron in each orbital. This produces regular tetrahedral coordination characterized by four equivalent bond lengths and C–C–C bond angles of $109.5°$. This angle keeps the orbitals as far apart from each other as possible, and so minimizes overlap between the electron clouds on a single atom. Triangular coordination results when the four valence electrons are shared equally in bonds with three near neighbors. In this case, the

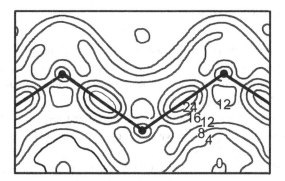

Fig. 8.1 Relative values for density for valence electrons of silicon measured in the plane of the bonds for Si. Note the concentration of electrons between the Si atoms. The atom cores themselves are shown as dark circles. Redrawn, by permission, from J. R. Chelikowsky and M. L. Cohen, "Electronic charge densities and the temperature dependence of the forbidden (222) reflection in silicon and germanium," *Phys. Rev. Lett.*, 33 (22), 1339–1342 (1974).

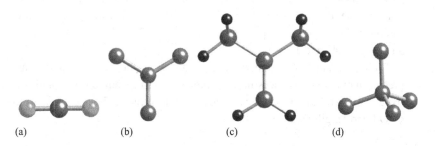

(a)   (b)   (c)   (d)

Fig. 8.2 Typical C coordinations: (a) sp coordination resulting from the formation of two double bonds, from the CO_2 molecule; (b) sp^2 C with three equivalent bond lengths, resulting from the formation of a resonating double bond, as in graphite. Here the C–C bond length is 1.42 Å at room temperature; (c) sp^2 C with one single bond and two equivalent shorter bonds, as in isobutene (the small atoms here are hydrogen); (d) sp^3 C with four single bonds to neighboring atoms, as in diamond. The C–C bond length in diamond is 1.54 Å.

characteristic C–C–C bond angle is 120°, as is observed in graphite. Asymmetric triangular coordination develops when the four electrons are distributed unequally between the neighbors. This can occur if a double bond is formed to one neighbor, and single bonds to two neighbors, or if there is one single-bonded neighbor, and two neighbors with a resonating double bond. The characteristic geometry for carbon forming two double bonds from sp hybrid orbitals (as is shown in CO_2) is a dumbbell shape, where the O=C=O bond angle is 180°.

Because covalent bonding develops with electron concentrations between adjacent atoms, the bonds are highly directional. The orientation of the bonds depends on the orbital used in the bonding. For diamond, the bonds are strongest, and the system is most stable, when the four bonds from each carbon atom are arranged as a regular tetrahedron. The resulting bonds resist deformation, producing stiff materials.

8.2   Covalent Radii and Bond Lengths

Interatomic distances between covalently bonded atoms have been measured for thousands of molecules and crystals. Based on these measurements, it is possible to deduce atomic radii for various types of chemical bonds. Relatively few elements are involved in organic chemistry, and the geometrical parameters vary little between related compounds. Table 8.1 lists typical bond lengths. These data illustrate the fact that the covalent bond length increases on going down columns and decreases going across a row on the periodic table.

As examples, the C–C distance in diamond is 1.54 Å, and the Cl–Cl distance in Cl_2 is 1.98 Å, suggesting single-bond covalent radii of 0.77 Å for carbon, and 0.99 Å for chlorine. Adding these numbers gives a predicted bond length of 1.76 Å for the C–Cl bond. Diffraction measurements on CCl_4 confirm this prediction. Other single-bond covalent radii are listed in Table 8.1.

The bonds in double and triple bonded atoms are shorter and stronger than single bonds. Carbon–carbon bond lengths range from 1.54 Å in diamond to the 1.2 Å length bonds in ethylene. Bonds of intermediate order lie between these extremes, as shown in Fig. 8.3. For example, the C=C double bond in ethylene is 1.33 Å long. The bond orders of the carbon–carbon bonds of graphite and benzene are 1.33 and 1.5, respectively, and the bond lengths are 1.42 and 1.39 Å.

The C, N, and O covalent radii calculated from these bond lengths can be used to predict interatomic distances in other molecules. The double-bond radii for carbon and oxygen are 0.67 and 0.60 Å, respectively. The sum (1.27 Å) is close to the observed C=O bond length (1.25 Å) in urea.

8.3   Diamond, Zincblende, and Wurtzite Structures

Diamond

Diamond is the high-pressure form of carbon. Graphite, the low-pressure form, will be discussed later. In the cubic crystal structure of diamond, every carbon atom is tetrahedrally bonded to four other carbons, as shown in Fig. 8.4. The key structural

Table 8.1 Covalent bond lengths (Å)

C–C 1.54	N–N 1.40	O–O 1.32	F–F 1.44	C–N 1.47	C–O 1.47	C–H 1.08
C=C 1.33	N=N 1.20	O=O 1.10	F=F 1.08	C=N 1.32	C=O 1.23	N–H 1.02
C≡C 1.20	N≡N 1.10	O≡O 1.00		C≡N 1.16		O–H 0.97
					N–O 1.36	H–H 0.6
Si–Si 2.34	P–P 2.20	S–S 2.08	Cl–Cl 1.98		N=O 1.24	
Si=Si 2.14	P=P 2.00	S=S 1.88	Cl=Cl 1.78			
Ge–Ge 2.44	As–As 2.42	Se–Se 2.34	Br–Br 2.28			
Ge=Ge 2.24	As=As 2.22	Se=Se 2.14	Br=Br 2.08			
Sn–Sn 2.80	Sb–Sb 2.82	Te–Te 2.74	I–I 2.66			
Sn=Sn 2.60	Sb=Sb 2.62	Te=Te 2.54	I=I 2.46			

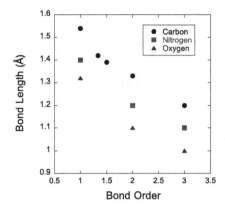

Fig. 8.3 Relationship between bond length and bond order for selected covalent bonds.

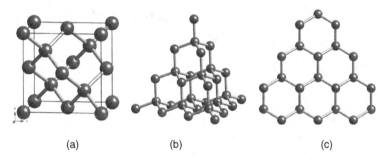

(a)   (b)   (c)

Fig 8.4 Crystal structure of diamond showing (a) unit cell, (b) stacking of puckered hexagonal rings and, (c) a (111) plane.

feature is a set of puckered hexagonal rings parallel to the {111} planes. This is illustrated in Fig. 8.5, where it is apparent that subsequent layers of the puckered rings are stacked offset from each other. There are eight carbon atoms per unit cell in diamond.

It should be clear from the figures that if the crystal is terminated parallel to the puckered hexagonal rings to make a surface, then each surface carbon atom would have three out of its four neighbors. The "missing" bond is the one that would connect to the next plane of puckered rings. This is the minimum number of dangling bonds for the surface atoms on a diamond crystal. Thus, the {111} planes are the best-bonded planes. As described in Chapter 4, the morphologically stable faces are often the best bonded faces. The octahedral morphology of diamond is a result of terminating the crystal with the {111} family of planes. Likewise, cleavage will take place parallel to these planes.

Cell parameters and bond lengths for diamond and the isostructural elements silicon, germanium, and gray tin are given in Table 8.2. The extremely short C–C separation distance in diamond leads to the high mechanical stiffness and scratch resistance of diamond. The high hardness makes diamond useful for gemstones, drills, saws, and polishing abrasives.

Table 8.2 Lattice parameters (*a*) and bond lengths (*d*) for column IV elements with the diamond structure

	a	*d*
C (diamond)	3.57 Å	1.54 Å
Si	5.43	2.35
Ge	5.66	2.45
Sn	6.49	2.81

(a)   (b)

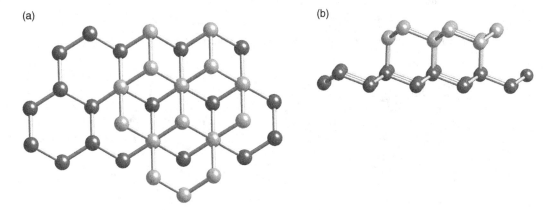

Fig. 8.5 Sections of the diamond structure showing the stacking of the puckered hexagonal rings: (a) top view and (b) side view. The lighter atoms show the second layer.

Zincblende and Derivative Structures

The zincblende structure (Fig 8.6) is closely related to the diamond structure. Its crystal structure is easily visualized from that of diamond by replacing alternate carbon atoms with zinc and sulfur. Both Zn and S are tetrahedrally coordinated with a bondlength of 2.34 Å. As was the case for diamond, the puckered hexagonal rings in zincblende are offset from one another, producing a cubic crystal system. The mineral zincblende (ZnS) is the chief ore of zinc. Zincblende is often called a II–VI compound in reference to the positions of Zn and S in the periodic system. Other II–VI, I–VII, III–V, and IV–IV compounds with the cubic zincblende structure are listed in Table 8.3.

Zincblende is one of a number of structures that are easily visualized from the diamond structure. These are called *derivative structures* in the crystal chemistry literature. Formally, successive suppression of symmetry elements leads to a series of lower symmetry structures which are "derived" from a prototype structures. This can be done by changing the geometry, the composition, or both. For example, symmetry is lowered geometrically when an atom is displaced from a high-symmetry position in the lattice at a phase transition. Alternatively, derivatives can be made when two or more atoms are placed in an ordered way on the sites occupied by one atom in the parent

Table 8.3 Lattice parameters and bond lengths for
zincblende structures

I–VII	a	d
CuF	4.26 Å	1.85 Å
CuCl	5.41	2.34
CuBr	5.69	2.56
CuI	6.04	2.62
IV–IV		
SiC	4.35	1.88
II–VI		
ZnS	5.41	2.34
ZnSe	5.67	2.46
ZnTe	6.09	2.64
BeS	4.85	2.10
BeSe	5.07	2.20
BeTe	5.54	2.40
III–V		
AlP	5.45	2.36
AlAs	5.62	2.44
AlSb	6.13	2.66
BN	3.62	1.57
BP	4.54	1.97
BAs	4.78	2.07
GaP	5.45	2.36
GaAs	5.65	2.45
GaSb	6.12	2.65
InP	5.87	2.54
InAs	6.04	2.62
InSb	6.48	2.80

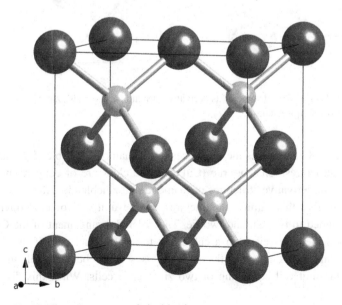

Fig. 8.6 Crystal structure of zincblende.

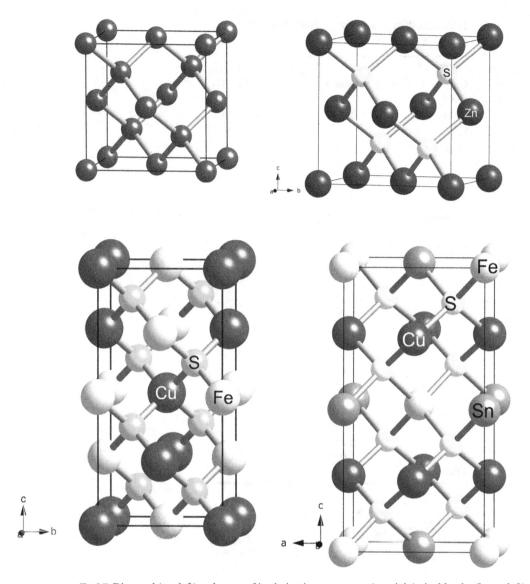

Fig. 8.7 Diamond (top left) and some of its derivative structures: (top right) zincblende, (lower left) chalcopyrite, (lower right) stannite.

structure. Figure 8.7 compares the sequences in the diamond, zincblende, chalcopyrite, and stannite structures. Chalcopyrite ($CuFeS_2$), the chief ore of copper, and stannite (Cu_2FeSnS_4) are derivative structures of diamond and zincblende. All atoms are tetra-hedrally bonded in this family of structures. In the unit cell of chalcopyrite, each S atoms is bonded to two Cu and two Fe. The ordered arrangement of the Cu and Fe atoms changes the unit cell from a cubic one to a tetragonal one. If the distinction between Cu and Fe is ignored, the structure is seen to be identical with that of zincblende, the unit cell consisting of two zincblende cells. When this distinction is

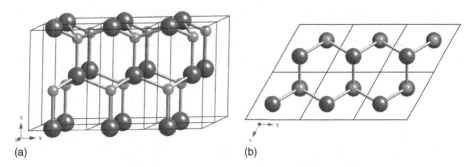

Fig. 8.8 Crystal structure of wurtzite looking (a) approximately down the *a* axis, and (b) directly down the *c* axis.

taken into account, however, the atoms at the corners of the sub-cell are no longer identical, and the larger cell becomes the true repeat unit.

The space groups of these materials show the symmetry suppression. Diamond crystallizes in $Fd\bar{3}m$, zincblende in $F4\bar{3}m$, chalcopyrite in $I\bar{4}2d$ and stannite in $I\bar{4}2m$. The space group $F4\bar{3}m$ is a symmetry subgroup of $Fd\bar{3}m$, verifying that zincblende is a derivative of diamond. Chalcopyrite and stannite are derivatives of both diamond and zincblende, but not of one another. $I\bar{4}2m$ cannot be derived from $I\bar{4}2d$ simply by removal of symmetry elements, hence it is not a subgroup of $I\bar{4}2m$.

A similar relationship exists between the structures of rutile, TiO_2, and $FeNb_2O_6$, the latter being identical with that of rutile if the Fe and Nb atoms are not distinguished. When this distinction is made, the true unit cell is three times as large and corresponds to three cells of rutile. Owing to the close relationship of cell dimensions, derivative structures often form oriented overgrowths. Solid solution between them, however, appears to be rare.

Wurtzite

Wurtzite is another form of zinc sulfide which crystallizes in a hexagonal pattern rather than in a cubic unit cell like zincblende. The chemical bonding is basically the same, with every zinc bonded tetrahedrally to four sulfurs, and vice versa. The two structures differ only in the stacking sequence of the tetrahedra (Fig. 8.8); in wurtzite, the puckered hexagonal rings are superimposed. Zinc atoms in zincblende are in a face-centered cubic lattice, whereas in wurtzite they are nearly in the positions of hexagonal close packing. If the structure of zincblende is pictured standing on a tetrahedron base, the relation between zincblende and wurtzite becomes clear. The hexagonal axis of wurtzite is polar. The sheets of atoms parallel to the basal plane all have Zn above and S below in each pair. The rare mineral greenockite, CdS, has a similar structure as also do MgTe, CdSe and AlN. Zincite (ZnO) and beryllia (BeO), two important oxides with the wurtzite structure, are listed in Table 8.4, along with a number of other isomorphic compounds.

Table 8.4 Lattice parameters and approximate bond lengths for wurzite family compounds

	a	c	d
AlN	3.11 Å	4.98 Å	1.87 Å
BeO	2.70	4.38	1.64
CdS	4.13	6.75	2.53
GaN	3.18	5.17	1.94
InN	3.53	5.69	2.13
NbN	3.02	5.58	2.09
SiC	3.08	5.05	1.90
ZnTe	4.27	6.99	2.62
TaN	3.05	4.94	1.85
ZnO	3.25	5.21	1.96
ZnS	3.81	6.23	2.34
ZnSe	3.98	6.53	2.45

8.4   Band Gaps in Semiconductors

The energy levels for electrons in solids are quantized. For a crystal containing N atoms, a set of $2N$ closely spaced energy levels develops from each atomic orbital. These closely spaced energy levels form allowed *energy bands* for the electrons. The width of the band depends on the overlap between atomic wave functions. Tightly bound inner orbitals give rise to narrow bands, whereas the bands formed by valence orbitals are relatively wide. Owing to the confinement of electrons (either near the atom cores or in bonds), the allowed energy levels are separated by energy levels that the electron cannot have, called *band gaps*, E_g, or forbidden energy levels. The energy that describes the highest occupied quantum state at 0 K is the Fermi energy, E_F. In semiconductors such as Si, the Fermi energy lies at a forbidden energy level, resulting in a filled valence band and an empty conduction band at 0 K.

Most of the commercially important semiconductors crystallize into the diamond, zincblende, or wurtzite crystal structures. For example, carbon, silicon, germanium, and gray tin all crystallize with the diamond structure. For this series, the bonds that hold the material together are strongest for diamond, and weakest in tin. Short, strong bonds lead to large forbidden energy gaps because a large amount of energy is required to liberate an electron from its bond. This trend is apparent in Table 8.5. Similar trends are observed in III–V and II–VI compounds. The boundary between semiconductors and insulators is somewhat of an arbitrary one. Here, we will treat materials as insulators if $E_g \geq 3$ eV, and as semiconductors if the band gap is smaller.

There are multiple bands per crystal. The inner electron shells tend to be very narrow, usually with wide spacings in energy between the bands, because those electrons are buried deep in the atom, and interact little with neighboring atoms. Inner electron shells are used in a variety of core level spectroscopies as fingerprints of particular atoms. Outer electrons are the ones that participate in bonding, and so must necessarily interact

Table 8.5 Band gaps (at 300 K) and bond lengths for group IV elements

Element	E_g (eV)	Bond length (Å)	Dielectric constant
C	5.47	1.54	5.5
Si	1.12	2.34	11.8
Ge	0.66	2.44	15.8
Sn	0.08	2.80	

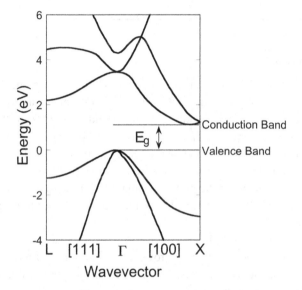

Fig. 8.9 Band structure for Si showing the band gap, E_g, between the valence band and the conduction band.

with other atoms. The outer bands govern the transport properties such as the electrical conductivity of materials. An example of the outer electron band structure is shown in Fig. 8.9 for Si. At 0 K, Si has a full valence band and an empty conduction band. Because only partially filled bands are able to contribute to electrical transport, the electrical conductivity is low at 0 K. As the temperature rises, electrons can be promoted from the valence band to the conduction band, leaving behind a hole in the valence band. Both the promoted electrons and the holes can contribute to the electrical conductivity, σ. If n is the concentration of electrons (e⁻) in the conduction band, and p is the concentration of holes (h⁺) in the valence band, then the total conductivity will be $\sigma = nq\mu_e + pq\mu_h$, where q is the charge per carrier, and μ_e and μ_h are the mobilities of the electrons and holes, respectively.

Group IV elements shift from covalent to metallic bonding with increasing atomic number and increasing size. Thus, gray tin (β-Sn) is a semiconductor with the diamond structure, while white tin (α-Sn) is a metal. Lead, the heaviest of the column IV elements, has a face-centered cubic structure and metallic conductivity.

Table 8.6 A number of compound semiconductors and the difference in electronegativity

III–V	$\Delta\chi$	II–VI	$\Delta\chi$	Others	$\Delta\chi$
AlSb	0.3	CdS	1.0	SiC	0.7
GaP	0.5	CdSe	0.8	AgI	0.8
GaAs	0.4	CdTe	0.6	CuI	0.8
GaSb	0.2	ZnS	1.0	Mg_2Ge	0.5
InP	0.5	ZnSe	0.8	Mg_3Sb_2	0.6
InAs	0.4	HgSe	0.4	ZnSb	0.3
InSb	0.2	HgTe	0.2	CdSb	0.3

Compound Semiconductors

The valence bond treatment of semiconductors proposed by Mooser and Pearson[1] provides a simple model for predicting new semiconductors and classifying their properties. The atoms in silicon, germanium, gallium arsenide, and most of the useful semiconductors are in tetrahedral coordination with sp^3 hybrid bonds. The covalent bonds for three-dimensional network structures run throughout the crystal.

As with other covalently bonded solids, silicon follows the $8 - N$ rule, which states that each atom forms $8 - N$ bonds, N being the column number in the periodic system. Silicon and germanium are in column IV, and form four covalent bonds to neighboring atoms in the structure.

The nature of bonds in compound semiconductors is somewhat more difficult to judge, but the difference in electronegativity provides a useful guide. Table 8.6 lists a number of compound semiconductors and their electronegativity difference $\Delta\chi$. Compound semiconductors generally have $\Delta\chi < 1$. This corresponds to 75% covalent and 25% ionic character. The percentage covalency ranges from 100% in Si and Ge ($\Delta\chi = 0$) to 75% for ZnS and CdS.

Comparing the band gaps and bond lengths of the compounds shown in Fig. 8.10, it is apparent that the band gap increases with ionicity. This is associated with a change in the details of the band structure. Covalent bonding is conducive to narrow band gaps and semiconducting behavior because of the electron bridges between bonded atoms. In ionically bonded materials, there is a difference in potential between the cations and anions; the asymmetric part of the potential energy flattens the bands and opens up the band gap. As the bonding in most oxides is predominantly ionic, the band gaps are generally very large. MgO, for instance, has a band gap of 8.3 eV, making it a very good electrical insulator, when pure. This also implies that strongly ionic materials, including the fluorides, will be transparent deep into the ultraviolet (for example, for mask blanks in lithography).

Metal-to-semiconductor transitions are usually accompanied by a change in crystal structure, as in the gray tin (semiconductor) to white tin (metallic) transition. The cases

---

[1] E. Mooser and W. B. Pearson, "The chemical bond in semiconductors," *Journal of Electronics,* 1 (6), 629–694 (1956).

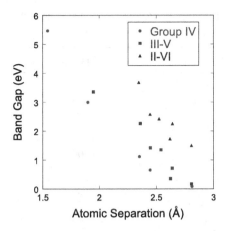

Fig. 8.10 Relationship between band gap and near neighbor distance for tetrahedral semiconductors. The iconicity increases from IV–IV to III–V to II–VI semiconductors.

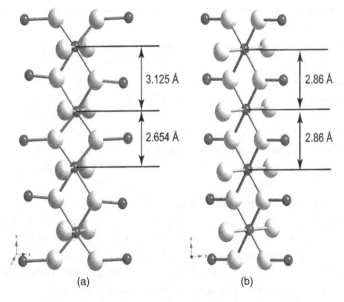

Fig. 8.11 VO_2 crystal structures: (a) semiconducting monoclinic room temperature phase, and (b) metallic tetragonal phase stable above 68 °C.

in which a change in electrical properties is not accompanied by a major change in structure are rare, and will occur only if the atomic positions are sufficiently variable to accommodate a major change in bonding. The lower oxides of vanadium, V_2O_3 and VO_2, are examples. At room temperature, VO_2 has a monoclinic crystal structure, which can be regarded as a distorted rutile structure, as shown in Fig. 8.11. The distortion leads to the d electrons being confined on the V atoms, producing semiconducting characteristics. At 68 °C, a higher symmetry tetragonal phase is recovered. As this occurs, the V–V distances along the diamond chain become uniform, and are short enough that band conduction becomes possible. This produces metallic conductivity.

Table 8.7 Interatomic distances (d), band gaps (E_g), electron mobility (μ_e) and hole mobility (μ_h) for commonly used semiconductors at 300 K

Material	d (Å)	E_g (eV)	$\mu_e \left(\frac{cm^2}{V \cdot s}\right)$	$\mu_h \left(\frac{cm^2}{V \cdot s}\right)$
C	1.54	5.33	1800	1200
Si	2.35	1.12	1500	450
Ge	2.45	0.66	3900	1900
GaP	2.36	2.26	110	75
GaAs	2.45	1.42	8500	400
GaSb	2.65	0.72	5000	850
InP	2.54	1.35	4600	150
InAs	2.62	0.36	33,000	460
InSb	2.80	0.17	80,000	1250
CdS	2.52	2.42	340	50
CdSe	2.62	1.70	800	—
CdTe	2.80	1.56	1050	100

Mobilities for several important semiconductors are listed in Table 8.7. Carrier scattering due to perturbations in the periodic potential, including scattering from defects in the materials and atoms displaced off their equilibrium sites by thermal vibrations, governs the mobility. It is observed that the mobilities for electrons are generally higher than those for holes. The largest mobilities are for electrons in heavy metal semiconductors with narrow band gaps and long chemical bonds. Heavy atoms reduce the amplitude of atomic vibrations, and so lead to less scattering of the carriers, and higher mobilities.

It should also be noted that it is possible to obtain semiconducting characteristics through electron hopping mechanisms in materials with multivalent ions. The $Ti^{4+} + e^- \leftrightarrow Ti^{3+}$ transfer operates in n-type $BaTiO_3$. In electron hopping, the charge carrier is trapped on ions; this differs from band conduction, where the carrier is free to move. In the electron-hopping mechanism, semiconducting behavior results from the thermally activated carrier mobility. For the $BaTiO_3$ example, this can be thought of as resulting from the fact that the Ti^{3+} ion is larger than the Ti^{4+} ion. Thus, as the electron moves between trapping sites, a lattice distortion must move with it.

8.5   Silicon and Applications

Silicon is the most common cation in the Earth's crust, occurring mainly as SiO_2 (quartz) and various silicates. From the electronic viewpoint, Si forms sp^3 hybrid bonds. It is tetrahedrally connected in SiO_2, SiH_4, and in elemental Si. The interatomic distances are 1.61 Å in quartz, 1.48 Å in silane and 2.35 Å in Si. Unlike carbon, silicon has very little tendency to form π bonds.

Elemental silicon is obtained by reducing SiO_2 with carbon at about 1700 °C:

$$SiO_2 + C \rightarrow Si + CO_2 \uparrow .$$

After reduction, the Si is in the form of fine-grained powder and contains many impurities. It is transformed to a gas to remove the impurities. A reaction with hydrochloric acid

Table 8.8 Segregation coefficients, K, for various elements in solid and liquid Si

Element	K
Li	1×10^{-2}
Cu	4×10^{-4}
Ag	1×10^{-6}
B	8×10^{-1}
Al	2×10^{-3}
C	6×10^{-2}
P	4×10^{-1}
As	3×10^{-1}
Sb	2×10^{-2}
O	1.4
Ti	9×10^{-6}
Fe	8×10^{-6}

at 300 °C gives trichlorosilane gas $Si + HCl \leftrightarrow SiHCl_3 \uparrow + H_2 \uparrow$. The low boiling point of $SiHCl_3$ (32 °C) allows purification by fractional distillation. Electronic grade Si is obtained by reducing $SiHCl_3$ in a hydrogen atmosphere at 900 °C. Single crystals of silicon are grown by the Czochralski technique and then further purified by the float-zone method. The Si specimen is locally melted by an induction coil and the molten zone is forced to travel from one end of the specimen to the other. Most of the remaining impurities segregate in the melt and are removed. Segregation coefficients between solid and molten silicon are listed in Table 8.8. Elements with K values <1 prefer the molten state, those with $K > 1$ prefer the diamond structure of solid Si. It is interesting to speculate why. Most elements are easily accommodated in the irregular structure of the melt, but find it much more difficult to fit into the rigid structure of the crystal. Oxygen bonds strongly to silicon, just as it does in most silicate minerals. In quartz, tridymite, and cristobalite, each silicon atom is bonded to four oxygens, and each oxygen is bonded to two Si. Note the similarity between the structures of silicon and cristobalite (Fig. 8.12).

The oxygen in silicon comes from the reduction of crucible SiO_2 by molten Si:

$$SiO_2 + Si \rightarrow 2SiO \uparrow.$$

SiO, in turn, is incorporated into the molten silicon, and then into the crystal ingot. In solid silicon, oxygen occupies an interstitial site, bridging the two nearest Si atoms, just as it does in cristobalite. The oxygen does not diffuse rapidly in the solid because of the strong bonds with the Si.

The first transistors were made from germanium, but rapid developments in microelectronics took place when Ge was replaced by Si. Integrated circuits made of silicon are the heart of every computer made today. The excellent insulating properties and interface characteristics of SiO_2 are far superior to the oxides of competing semiconductors. CO_2 is a gas, while GeO_2 is water soluble. Planar technology and integrated circuits became possible with Si and SiO_2. In integrated circuit processing, amorphous SiO_2 is utilized. In the interface region between Si and SiO_2, Si is bonded to between

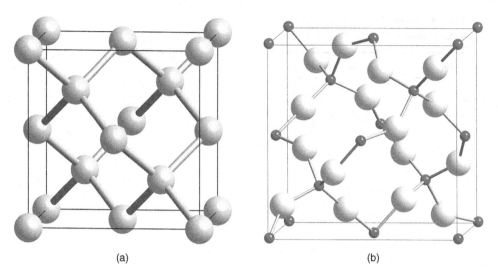

(a)   (b)

Fig. 8.12 Comparison of the structures of (a) Si and (b) an idealized version of cristobalite.

zero and four oxygens. The width of the interface region is about 25 Å, or about four unit cells. There is a resurgence of interest in Si-Ge alloys for high speed computing.

Semiconductor Lasers

Light-emitting diodes (LEDs) are semiconductor p–n junctions capable of emitting radiation in the ultraviolet, visible, or infrared regions of the electromagnetic spectrum. A quantized packet of light, called a photon, can be emitted when an excited electron drops to a lower energy level. The energy of the photon emitted is given by $E_g \leq h\nu = \frac{hc}{\lambda}$, where h is Planck's constant ($h = 6.6262 \times 10^{-34}$ J s), ν is the frequency, λ is the wavelength of light, and c is the speed of light in a vacuum ($c = 2.9979 \times 10^8$ m/s). Alternatively, the extra energy can be lost to heat, through emission of one or more phonons (lattice vibrations). The probability of light emission depends on the band gap. Photons carry energy, but very little momentum. Carrier momentum changes require changes in wavevector. *Thus, photons produce nearly vertical transitions on energy–wavevector diagrams.*

This brings up the important distinction between *direct* and *indirect* gap semiconductors. In a direct band gap semiconductor, the minimum in the conduction band occurs at the same wavevector as the maximum in the valence band. For an indirect gap semiconductor, the maximum in the valence band is at a different wavevector than the minimum in the conduction band. This distinction is illustrated in Fig. 8.13. Direct gap semiconductors are much more likely to produce a photon when an excited electron drops down to the valence band, and so are useful in low power light emitting devices. In contrast, for an indirect gap semiconductor, millions of electrons may need to be excited and relax back to the ground state releasing heat, before one drops back to the valence band emitting a photon. Thus, GaN, which is a direct gap semiconductor, has supplanted SiC, an indirect gap

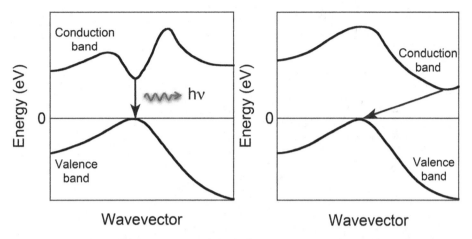

Fig. 8.13 Comparison of (left) direct and (right) indirect band gap semiconductors. The upper curves are for the conduction band, the lower curves are for the valence band. For direct gap semiconductors, there is a higher probability that electron–hole recombination is accompanied by emission of a photon. Adapted, by permission, from B. E. A. Saleh and M. C. Teich, *Fundamentals of Photonics*, John Wiley & Sons, New York (1991). Fig. 18.3–6.

material, for blue light-emitting diodes. The nature of the band gap also influences the shape of the optical absorption spectrum of semiconductors, as is described in Chapter 27.

Since the human eye is sensitive to light with wavelengths between around 0.4 and 0.7 µm, the energy gap of the semiconductor must lie between about 1.8 and 3 eV for lighting applications. Wavelengths of interest in optical communications are 0.9 µm, 1.3 µm, and 1.55 µm, where the losses in SiO_2 fibers are very low. Direct band gap semiconductors are especially important for electroluminescent devices, because there is no phonon involved in the transition from the conduction band to the valence band. As a result, the quantum efficiency is much higher than in indirect band gap semiconductors like Si and Ge.

GaAs and GaN, along with some of their alloys, are two of the most important families of direct band gap semiconductors. The laser wavelengths, energy band gaps and lattice constants for several semiconductor solid solutions are plotted in Fig. 8.14. All have the zincblende, wurtzite, or chalcopyrite structures. To achieve thin-film junctions with negligible interface traps, the lattice parameters of the film and the substrate are carefully matched. Using GaAs crystals as the substrate, $Al_xGa_{1-x}As$ layers have a mismatch of less than 0.5%. In a similar way, substrates can be matched with $Ga_xIn_{1-x}As_ySb_{1-y}$ compositions that generate laser light in the near infrared range.

Solar Cells

Solar cells convert sunlight directly to electricity with reasonably high conversion efficiency. They provide a permanent power source at low operating cost.

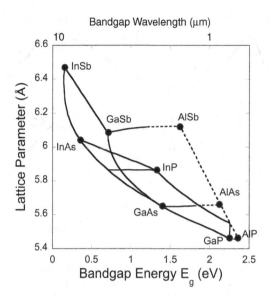

Fig. 8.14 Relationship between lattice parameter and band gap for several III–V semiconductors. Indirect band gaps are noted with dashed lines. Direct band gaps are connected by solid lines. Adapted, by permission, from B. E. A. Saleh and M. C. Teich, *Fundamentals of Photonics*, John Wiley & Sons, New York (1991). Fig. 15.1–5, p. 550.

The broad optical spectrum of the sun corresponds to a blackbody radiator of 5800 K. Useful wavelengths range from 0.3 μm in the near ultraviolet to 3 μm in the near infrared. The ideal solar-cell efficiency is plotted as a function of semiconductor band gap in Fig. 8.15.[2] Efficiency has a broad maximum and does not depend critically on E_g. As a result, many semiconductors with band gaps between 1 and 2 eV are potential solar cell materials. Substantial increases in efficiency can be achieved by using two or more band gaps in multilayer solar cell designs. For two semiconductors with band gaps of 1.56 and 0.94 eV in series, the ideal efficiency is 50%. There are many factors that degrade the efficiency of solar cells down into the 10% range.

For terrestrial flat-plate systems, important materials include Si single crystals, polycrystalline Si, amorphous Si, and thin-film Cd chalcogenides. The main advantage of thin film solar cells is low-cost processing, but long-term stability can be a problem. There are many ternary compounds that are potential candidates for solar cell applications. Figure 8.16 shows some of the I–III–VI$_2$ and II–IV–V$_2$ semiconductors which have energy gaps in the range of interest.

8.6   Graphite and Related Structures

The structure of graphite consists of flat interconnected hexagonal rings of carbon atoms (Fig. 8.17). Within these strongly bonded layers each carbon is covalently linked to three others but the bonding between layers is very weak. The difference in bonding is

[2] S. M. Sze, *Physics of Semiconductor Devices*, John Wiley and Sons, New York (1981).

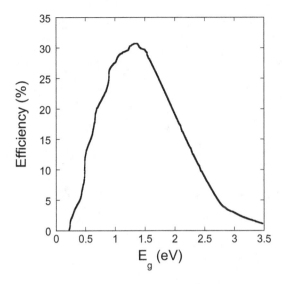

Fig. 8.15 Ideal solar cell efficiency at 300 K plotted as a function of semiconductor band gap. Minor irregularities are caused by atmospheric absorption.

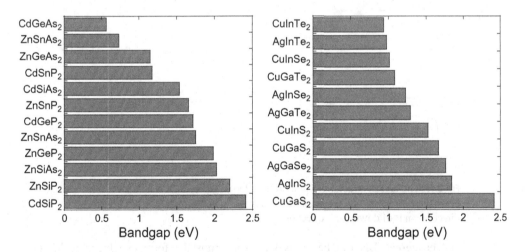

Fig. 8.16 Energy gaps of two families of compound semiconductors for possible use in photovoltaic energy conversion.

quite apparent from the C–C interatomic distances: 1.42 Å within the layer and 3.35 Å between layers. The extremely easy cleavage of graphite is between the strongly bonded layers, where the bonding is very weak.

A closely related structure is that of hexagonal boron nitride. In graphite, the layers are staggered in position with only half the carbons over carbons in the layer below, while in BN the layers are directly superposed but with B over N, and N over B. Again, there is a large difference between the covalent bond length within the BN layer (1.45 Å) and the van der Waals bonds holding layers together (3.33 Å). Both graphite

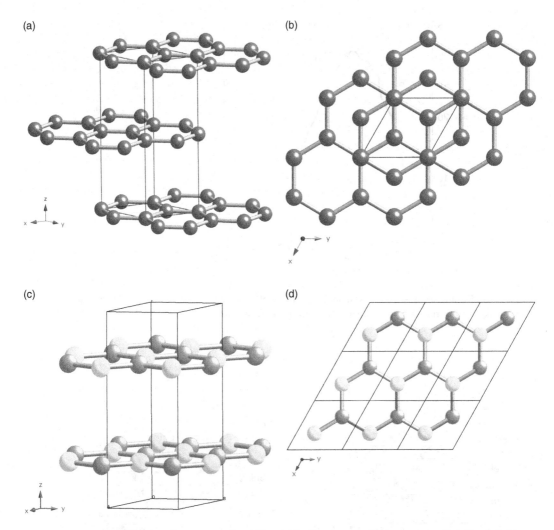

Fig. 8.17 The crystal structure of graphite: (a) looking perpendicular to the c direction and (b) down the c direction, and the graphite-like form of boron nitride (c) looking perpendicular to the c direction and (d) down the c direction.

and hexagonal BN can be converted to a turbostatic form in which the layers are randomly arranged to give two-dimensional disorder. Graphite and hexagonal boron nitride both are very refractory with melting points exceeding 3000 °C but they are very different optically and electrically. Graphite is black (the "lead" in a lead pencil) and is sufficiently conducting to be used as electrodes, while BN is a very good insulator and is often referred to as "white graphite."

It is possible for a graphene sheet (one layer of graphite) to curl up on itself to make either carbon nanotubes or a variety of large molecules such C_{60}, also known as a "buckeyball." Illustrations of the atomic arrangements of such species are shown in Fig. 8.18. Flat layers of graphene are an excellent example of a two-dimensional crystal, and so offer a good opportunity to study fundamental physics. This resulted in the

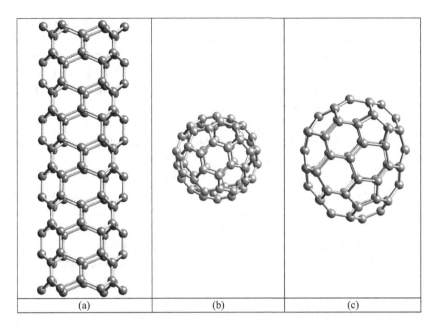

Fig. 8.18 Other forms of carbon: (a) a single-walled nanotube, (b) a C_{60} buckeyball, and (c) a C_{70} molecule. One can imagine the nanotube as being a rolled layer of graphite where the sides have been attached to each other. Metallic and semiconducting versions of nanotubes occur depending on how much the edges of the graphene sheet are displaced parallel to the nanotube length prior to bonding.

2010 Nobel Prize in Physics being awarded for work in graphene. It has been demonstrated that graphene has extremely high electron mobilities and thermal conductivities in the plane.

8.7   Problems

(1)   The stable form of carbon at room temperature and pressure is graphite (space group 186: $P6_3mc$ with $a = 2.456$ Å, $c = 6.696$ Å) and C at $(0, 0, 0)$ and $(1/3, 2/3, 0)$ and all symmetry-related positions.

   (a)   Draw a few unit cells of this structure showing the bonds that connect the C ions.

   (b)   Show the coordination (i.e. the number and geometric arrangement of near neighbors) of C atoms.

   (c)   Determine the closest C–C distances parallel and perpendicular to the c axis. Based on your knowledge of bonding, describe what you believe to be the predominant bond type(s) present in graphite.

   (d)   What would you expect the equilibrium morphology of graphite to be and why?

(2)   At high pressures, graphite undergoes a phase transformation to the diamond crystal structure. Diamond is cubic and has space group 227: $Fd\overline{3}m$ with $a = 3.56$ Å and

C at (0, 0, 0) and all symmetry-related positions. Si and Ge, the two most important elemental semiconductors have this same crystal structure.

(a)   Draw a unit cell, showing the bonds that connect the C atoms.

(b)   What is the coordination of the C atoms? What is the C–C bond length? Compare the bond length to those determined in Problem (1) above. What does this suggest to you?

(c)   Diamond is phenomenally hard, while graphite is soft enough to use as a solid-state lubricant. Discuss why this should be, based on the structures and bonding. Recognize that hardness (scratch resistance) is one measure of bond strength.

(3)   Wurtzite (one of the forms of ZnS) has the space group $P6_3mc$. Thus, it is from the hexagonal crystal system with $a = 3.81$ Å and $c = 6.23$ Å. Several unit cells are shown below. The large atoms correspond to Zn, while the small atoms are S. In describing crystal structures, the atom positions are frequently given in terms of fractions of the unit-cell parameters, a, b, and c. As drawn, there is a Zn at (2/3, 1/3, 0), and S at (2/3, 1/3, 0.375). That is, one Zn atom is positioned at $\frac{2}{3}\vec{a}$, $\frac{1}{3}\vec{b}$, 0, etc. The 6_3 screw axis is directed along the c axis (vertical in the drawing).

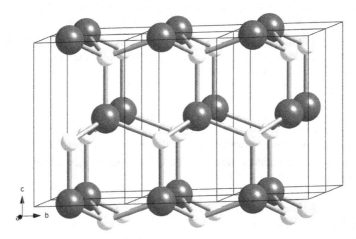

(a)   Give the positions of the Zn and S atoms symmetry-related to the ones given by the screw axis.

(b)   What is the volume of the unit cell?

(c)   Calculate the distances between Zn and S near neighbors, and between S and S next near neighbors.

(4)   ZnS is also a polymorphic (same composition, different structures) compound, with two known structures. Sphalerite (or zincblende) is cubic (space group 216: $F\overline{4}3m$, with $a = 5.41$ Å) and Zn at (0, 0, 0), S at (0.25, 0.25, 0.25) and all symmetry-related positions). Wurtzite is described in Problem (3) above. These two structures are adopted by many semiconducting materials. The differences in the structures profoundly impact the properties.

(a)   Draw a unit cell of sphalerite. Describe the relation between this and the diamond structure.

(b)   Contrast the crystal structures of zincblende and wurtzite, comparing coordinations of atoms, bond lengths, and assembly of the polyhedra into the crystal structure.

(5) BN is polymorphic; a cubic form comparable to sphalerite exists; in addition a hexagonal form is characterized by (space group 186: $P6_3mc$ with $a = 2.504$ Å, $c = 6.6612$ Å) and B at (1/3, 2/3, 1/4) and N at (1/3, 2/3, 3/4) and all symmetry-related positions.

    (a) Draw a unit cell of this structure showing the bonds that connect the B and N ions.

    (b) What is the coordination number for the B and N atoms?

    (c) Contrast this structure to that of graphite.

    (d) Hexagonal BN is an electrical insulator, while graphite is an electrical conductor. Based on what you know about the bonding, explain this.

(6) Draw a picture of the sp^3 orbitals in diamond. Relate this to the observed crystal structure.

(7) The electron-density map for silicon is shown in Fig. 8.1.

    (a) Taking numbers (in electrons/ Å^3) from this graph, plot the electron density as a function of distance for a line trace between the Si atoms (shown as black dots). Use an interatomic separation distance of 2.33 Å for scaling. The electron densities are high on the Si atom cores.

    (b) Compare and contrast your results with those for a more strongly ionic bond.

(8) Explain the trends that underlie the three sets of data on the following graph.

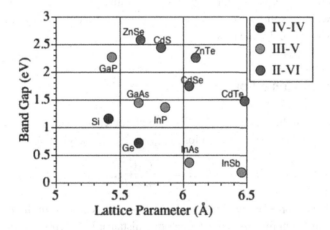

(9) Draw the crystal structures characteristic of diamond and zincblende (projections are fine). Compare the structures to that of wurtzite.

(10) What is the atomic coordination for a Si atom in crystalline Si? Justify this based on the bonding type and the electronic configuration.

9 Ionic Crystals

9.0 Introduction

Many elements ionize to reach closed electron shells, with nearly spherical electron distributions around the ion. This is apparent in Fig. 9.1, which shows the electron-density map of LiF.[1] In contrast to the case of covalent bonding, the electron concentration drops to a low value between the constituent ions.

It is possible to develop a reasonable picture of an ionic bond without resorting to quantum mechanics. Instead, we can envision the ionic bond as resulting from the transfer of electrons from one atom to another, with subsequent electrostatic attraction between the ions that are formed. This chapter discusses the origin of the electrostatic attraction between a pair of ions, and then generalizes this picture to describe the ionic bonding within a solid. Also discussed are a classification of ion types, typical coordination polyhedra, and ionic radii.

9.1 Nature of the Ionic Bond

Ionic crystals are held together by the Coulomb forces between cations and anions. The attractive force between two such ions is given by the expression:

$$F = -\frac{Q_A Q_C}{4\pi\varepsilon_0 d^2},$$

where F is the force in Newtons, Q_A and Q_C are the charges on the anion and cation expressed in Coulombs, d is the interatomic distance between the cation and anion in meters, and $4\pi\varepsilon_0$ is a constant ($\varepsilon_0 = 9.85 \times 10^{-12}$ F/m).

The cohesive energy for an ionic crystal can be calculated by summing the Coulomb energy for all ion pairs. Because the Coulomb potential is quite long-ranged, it is not possible to make an accurate calculation of this energy if only the first few sets of near (and next-near...) neighbors are considered. Consider, for example, the rocksalt structure shown in Fig. 9.2. For simplicity, all ions are treated as point charges. Around any given cation, there are six anions at a distance d_0 away. There are 12 other cations at a distance $\sqrt{2}d_0$, and eight anions at a distance $\sqrt{3}d_0$. Summing these terms for the

---

[1] H. Witte and E. Wolfel, "Electron distributions in NaCl, LiF, CaF_2, and Al," *Rev. Mod. Phys.*, **30** (1) 51–55 (1958).

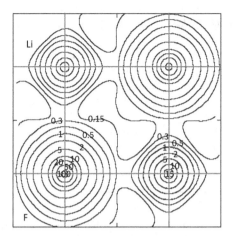

Fig. 9.1 Electron-density map of LiF showing nearly spherical ions. The numbers are in electrons/Å^3. Adapted, by permission, from A. R. West, *Basic Solid State Chemistry, 2nd Ed.* John Wiley and Sons, NY 1999, Fig. 2.1 p. 68.

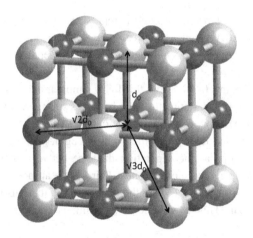

Fig. 9.2 A section of the rocksalt structure, showing the distances to the three sets of nearest neighbors. Adapted from A. R. West, *Basic Solid State Chemistry*, 2nd Ed. John Wiley and Sons, NY 1999, Fig. 2.1, p. 68.

attractive potential yields $V_{attractive} = \frac{Q_{Na}Q_{Cl}}{4\pi\varepsilon_0 d_0}\left(6 - \frac{12}{\sqrt{2}} + \frac{8}{\sqrt{3}} + \cdots\right)$. The term in the parentheses is the same for any material with the rocksalt crystal structure, and is called the Madelung constant, A, for the structure. In practice, care needs to be taken in setting up the series so that it converges. Table 9.1 shows the Madelung constants for a number of important ionic crystal structures. It should be noted that when the charges of the cations and anions differ, the Madelung constant determined using a reference cation will differ by the ratio of the charges from those where the Madelung constant is calculated using the anion as the reference ion. The overall potential is determined by repeating the attractive energy calculation for all ions in the structure.

A good elementary description of the potential energy for an ionic crystal contains this attractive term, and a repulsive term due to the Pauli exclusion principle. Accurate

Table 9.1 Madelung constants for some ionic crystals calculated for the cation site

Structure type	Madelung constant
Rocksalt	1.747 565
CsCl	1.762 675
Wurtzite	1.641
Sphalerite	1.6381
Fluorite	5.0388
Rutile	4.816
Corundum	25.031

calculations of the lattice energy, U, of an ionic crystal also require that the van der Waals and zero point terms be added to the potential

$$U = \frac{Q_{anion}Q_{cation}\,AN_A}{4\pi\varepsilon_0 r} + BN_0 e^{-\frac{r}{\rho}} - \frac{CN_0}{r^6} + zero\ point\ energy.$$

Here, the first term is the attractive term, the second is repulsive due to electron cloud overlaps, and the third term is a (typically) small correction due to van der Waals attractions. In the equation, N_A is Avogadro's number, while r is the separation distance between the atoms, and B, C, and ρ are constants.

There are a number of useful points to take from this fairly simple model. First, as the valence of the ions increases, the electrostatic attraction becomes stronger. Roughly, as the valence doubles, the cohesive energy rises by a factor of four. Thus, properties such as melting point or hardness, which depend on the bond strength, will rise as well. Secondly, the attractive force drops as the separation distance between atoms rises. Thus, the bonding tends to be stronger for shorter bonds. Thirdly, the attraction between cations and anions depends on distance, rather than the orientation of the ions when the ions can be treated as spherical. This is why ionic bonding is non-directional in character.

In many materials, the bonds aren't fully ionic. Even in MgO, the real charges appear to be somewhat closer to Mg^+O^- than $Mg^{2+}O^{2-}$. Nevertheless, the fully ionic description is extremely useful as a starting point in many cases.

9.2   Classification of Ions

Ions can be classified into several families, depending on their location on the periodic table.

(1)   *Noble gas core ions* have completely closed electron shells. For example, the electron configuration of Li^+ is $1s^2$. Such ions are approximately spherical in shape, and tend to participate in strongly ionic bonding.

(2)   The *lanthanide ions* shown in Table 9.2 have 4f orbitals buried inside the 5s and 5p electron shells. These ions often act as spherical ions, although exceptions have been noted in chalcogenides and pnictides.

Table 9.2 Examples of lanthanide ions

f^0	f^1	f^2	f^3	f^4	f^5	f^6	f^7	f^8	f^9	f^{10}	f^{11}	f^{12}	f^{13}	f^{14}
Y^{3+}	Ce^{3+}	Pr^{3+}	Nd^{3+}	Pm^{3+}	Sm^{3+}	Eu^{3+}	Gd^{3+}	Tb^{3+}	Dy^{3+}	Ho^{3+}	Er^{3+}	Tm^{3+}	Yb^{3+}	Lu^{3+}
La^{3+}	Pr^{4+}					Sm^{2+}	Eu^{2+}							
Ce^{4+}							Tb^{4+}							

(3) *Transition metal ions* are the source of many discrepancies in classical crystal chemistry. In the 3d electron series, the d electrons are drawn in close to the noble gas core. As such, they are fairly well localized, and do not participate as much in the bonding. In contrast, for the 4d and 5d rows, the d orbitals extend to a larger fraction of the atom size, and so are more likely to interact with anions. This has a tendency to introduce more covalent or metallic characteristics into the bonding.

Transition metals and rare earths often have multiple oxidation states. The possible valence states and their relative stability are determined experimentally. A good rule of thumb is that atoms ionize by losing electrons from partially filled electron shells from the outside in. Thus, when Fe^0, with an electron configuration of $1s^2 2s^2 2p^6 3s^2 3p^6 3d^6 4s^2$, is ionized to Fe^{2+}, the electron configuration becomes $1s^2 2s^2 2p^6 3s^2 3p^6 3d^6 (4s^0)$. That is, the 4s electrons are lost before those in the 3d shell. For Fe^{3+}, the electron configuration is $1s^2 2s^2 2p^6 3s^2 3p^6 3d^5$. In the lanthanides, electrons are removed first from the 6s, then from the 4f shells, rather than breaking into the filled 5s and 5p shells. Examples of the transition metal ions are given in Table 9.3.

(4) The d^{10} *ions* have filled shells and, as a result, have a tendency towards covalent bonding. This is a result of the fact that sd^3 hybrid orbitals can mix with the sp^3 to stabilize the tetrahedral coordination. As a result, Zn^{2+} is often found in 4-coordination, though size alone would suggest 6-coordination.

(5) *Rydberg ions* (also called lone pair ions) have a lone pair of electrons from the ns^2 shell. This produces non-spherical ions and strongly asymmetrical coordination polyhedra.

9.3   Ionic Radii

Ionic radii are determined from measurements of interatomic distances, and are found to be reasonably constant from one material to another. For example, the Si^{4+}–O^{2-} bond length is found to vary between ~1.60 and 1.64 Å in a large number of compounds. This approximation of constant radii is best for the smaller ions. For large ions, the electron cloud tends to be more deformable, and the effective size will change depending on the compound in which the ion is found. Indeed, it is often preferable to think of ions not as hard spheres with well-defined radii, but as stiff ion cores embedded in a more compliant valence electron cloud.

Table 9.3 Examples of transition metal ions

d^0	d^1	d^2	d^3	d^4	d^5	d^6	d^7	d^8	d^9	d^{10}
3d										
Ti^{4+}	Ti^{3+}	V^{3+}	Cr^{3+}	Cr^{2+}	Mn^{2+}	Fe^{2+}	Co^{2+}	Ni^{2+}	Cu^{2+}	Cu^+
V^{5+}	V^{4+}	Cr^{4+}	Mn^{4+}	Mn^{3+}	Fe^{3+}	Co^{3+}	Ni^{3+}			Zn^{2+}
Cr^{6+}				Fe^{4+}						Ga^{3+}
Mn^{7+}										
4d										
Zr^{4+}	Nb^{4+}	Mo^{4+}	Tc^{4+}	Ru^{4+}	Rh^{4+}	Rh^{3+}		Pd^{2+}	Ag^{2+}	Ag^+
Nb^{5+}	Tc^{6+}				Ru^{3+}	Pd^{4+}				Cd^{2+}
Mo^{6+}										In^{3+}
Tc^{7+}										Sn^{4+}
Ru^{8+}										Sb^{5+}
5d										
Hf^{4+}	Ta^{4+}	W^{4+}	Re^{4+}	Os^{4+}	Ir^{4+}	Pt^{4+}		Pt^{2+}	Hg^+	Hg^{2+}
Ta^{5+}	Re^{6+}					Ir^{3+}		Au^{3+}		Au^+
W^{6+}										Tl^{3+}
Re^{7+}										Pb^{4+}
Os^{8+}										Bi^{5+}

Ionic radii are harder to determine than are metallic and covalent radii, since it is not possible to divide the interatomic distance by two. The usual approach is to establish a baseline number for a single ion, and then calculate a series of self-consistent radii from a large database of compounds. Numerous approaches have been taken to accomplish this; tables of ionic radii are available from Pauling, Goldschmidt, Shannon–Prewitt, and others. Since these different tables often assign a different fraction of the interatomic distance to the cation, it is quite important not to mix radii from different tables.

At present, the Shannon–Prewitt radii are in widest use. Alas, there are two sets of these. The Shannon–Prewitt *crystal radii* come from electron density maps. Here, a typical radius for O^{2-} is 1.26 Å, while that for F^- is 1.19 Å. The Shannon–Prewitt *ionic radii* for the same two ions are 1.40 Å and 1.33 Å, respectively. These numbers are somewhat closer to the values from the Pauling and Goldschmidt tables. A comparison of the different radii is shown for the case of LiF in Fig. 9.3. Shannon–Prewitt radii for different ions are given in Appendix A.

The following points are useful in understanding the trends in ionic radii.

(1) Cations are smaller than neutral atoms of the same element. Thus, the radii for metallic Li, $r_{Li} = 1.5$ Å, while $r_{Li^+} = 0.74$ Å. As electrons are given away, the electron cloud shrinks.

(2) Anions are larger than neutral atoms. The $r_{Cl} = 1.0$ Å, while $r_{Cl^-} = 1.8$ Å.

(3) Radii shrink as the cation valence increases. As an example, the radius of Ni^{2+} in six-coordination (shown in Roman numerals) is $r_{Ni^{2+}(VI)} = 0.69$ Å, while $r_{Ni^{3+}(VI)} = 0.4$ to 0.56 Å, depending on the spin state (more on spin in a moment). This is illustrated for a number of cations in Fig. 9.4.

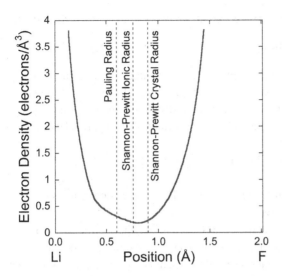

Fig. 9.3 Electron density as a function of position in LiF. Several measures of the ionic radii are also shown for comparison. Redrawn, by permission, from A. R. West, *Basic Solid State Chemistry, 2nd Ed.* John Wiley and Sons, NY 1999, Fig. 2.2, p. 69.

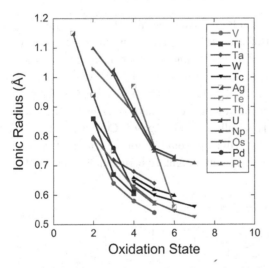

Fig. 9.4 Relationship between ionic radius and oxidation state. All of the data shown are for 6-coordination.

(4) On moving down a column in the periodic table, the ionic radii increase. For typical coordination numbers:

$$r_{Li^+} = 0.74\,\text{Å}$$

$$r_{Na^+} = 1.02\ \text{Å}$$

$$r_{K^+} = 1.38\ \text{Å}$$

$$r_{Rb^+} = 1.49\ \text{Å}$$

$$r_{Cs^+} = 1.7\ \text{Å}.$$

(5)   Going left to right across a row in the periodic table, the cation radii drop:

$$r_{Na^+} = 1.0 \text{ Å}, r_{Mg^{2+}} = 0.7 \text{ Å}, r_{Al^{3+}} = 0.5 \text{ Å}, r_{Si^{4+}} = 0.3 \text{ Å}.$$

This is a consequence of several factors acting together. First, the higher cation charge produces stronger Coulombic attractive forces, which pulls the anions in more closely, lowering the separation distances. Secondly, the nuclear attraction is increasing left to right on the periodic table. Thirdly, the degree of covalency is also rising.

(6)   There is a contraction in radii in the lanthanides. That is, the ionic radii fall as the 4f shell is populated. As an example, the La^{3+}–O^{2-} bond length is 2.44 Å, while the Lu^{3+}–O^{2-} bond length is 2.33 Å. This contraction is a result of the fact that the f electrons are pretty well buried in the atom. The f electrons do not shield the nuclear charge efficiently. As a result, filling the f orbitals doesn't increase the size much, while the increasing nuclear charge tends to decrease the radii. One of the important consequences of the lanthanide contraction is that ions immediately following the rare-earth elements have approximately the same size as the ions in the row above on the periodic table. Thus, the Zr:Hf, Nb:Ta, and Mo:W pairs have very similar radii, and tend to substitute extensively for each other in solid solutions.

(7)   Negative radii are possible. For example, the H^+ ion has lost its only e^-, and hence has lost most of its size. The resulting proton tends to bury itself in the electron cloud of an adjacent O^{2-} ion. This is evident from the fact that typical ionic radii for O^{2-} range from 1.3 to 1.4Å, while the O–H separation distance is usually ~1 Å. As a result, the Shannon–Prewitt ionic radius for $r_{H^+} = -0.38$ Å.

(8)   Across a transition metal row, there are anomalies in the radii associated with crystal field theory. This is illustrated in Fig. 9.5. Considering the 2+ cations from Ca^{2+} to Zn^{2+}, a simple prediction would be for the radii to decrease in a straight line across the periodic table left to right. Experimentally, this is not observed; the

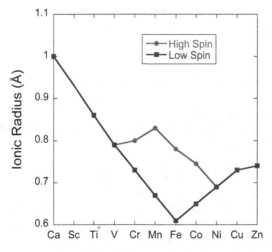

Fig. 9.5 Ionic radii for divalent cations in 6-coordination for the 3d transition metal series.

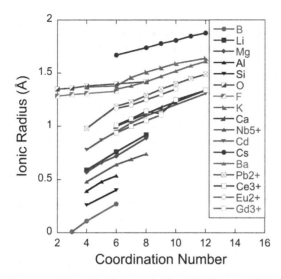

Fig. 9.6 Relationship between ionic radius and coordination number.

d^0, d^5, and d^{10} ions match this prediction fairly well, while the others fall below the line. These anomalies can be explained by the fact that the anions can be drawn in closer to the cation when the d electrons are concentrated in orbitals whose spatial extent minimized overlap (and hence repulsion) with the anion electron cloud. The crystal field theory that explains this will be discussed in more detail in Chapter 14. That said, the radii aren't vastly different, with the result that extensive substitutional solid solution between transition metal cations is possible.

(9)   The spin state affects the radius. A high spin state would be one in which as many as possible of the spins of the partially filled inner electron shells (like the d shells) are unpaired. The low spin state pairs all possible spins for these electrons. Because the low spin state tends to concentrate the electrons in orbitals which reduce overlap with adjacent anions, the radii tend to be smaller, as shown in Fig. 9.5.

(10) Ionic radii increase as the coordination number rises for a given atom as illustrated in Fig. 9.6. This can be thought of as a result of deformation of the electron cloud due to the electric field of the neighboring atom. This deformation is largest in structures of low or irregular coordination. As shown in the figure, anions show the same trend as cations, although the radius tends to rise more slowly with the coordination number.

9.4   Coordination Polyhedra

The number, type, and geometric arrangement of atoms bonded to a particular atom are together referred to as the coordination of that atom. Table 9.4 shows that the cation size (and hence its bond length with a given anion) is closely correlated with its coordination

Table 9.4 Typical coordination numbers with oxygen for common cations

Ion	Coordination	Bond length with oxygen (Å)
H^+	1, 2	1.0
C^{4+}	3	1.5
B^{3+}	3, 4	1.5
Si^{4+}	4	1.6
Be^{2+}	4	1.7
Zn^{2+}	4	1.8
Al^{3+}	4, 5, 6	1.8–1.9
Ti^{4+}	6,4	2.0
Cr^{3+}	6	2.0
Fe^{3+}	6, 4	2.0
Ni^{2+}	6	2.1
Mg^{2+}	6, 4, 8	2.1
Fe^{2+}	6, 4, 8	2.2
Mn^{2+}	6, 4, 8	2.2
Zr^{4+}	6, 7, 8	2.2
Na^+	6–10	2.3–2.6
Ca^{2+}	6–10	2.3–2.6
K^+	8–12	2.5–3.0

number. That is, the larger the cation, the more anions can be fit around the cation, while maintaining cation–anion contact. Generally, the more highly charged the cation, the shorter the interatomic distance, and the lower the coordination number.

The stability of ionically bonded materials is increased the more bonds that can be made. As a result, the cation coordination tends to be as high as possible, consistent with geometric packing of spheres of different radii. This differs from the case for covalently bonded materials, where the number of bonds is governed by the electron distribution around the atoms, and low coordination numbers are often favored.

Table 9.5 shows typical regular coordination polyhedra. The polyhedra are named for the shape of the geometric figure made by connecting the near neighbors around a given atom. Many crystallography books sort structures based on the arrangement and type of cation-centered polyhedra. It is apparent that the number of neighbors that can be accommodated around any given cation is a function of the relative sizes of the cation and the anion. This is the basis of Pauling's rules, as will be described in Chapter 13.

In addition to the common polyhedra shown in the table, numerous other geometries are observed in real materials. For example, a trigonal bipyramid would have three neighbors arranged in a plane around the cation, with the fourth and fifth neighbors forming pyramids at the top and bottom of the figure, respectively. This produces a 5-coordination for the cation.

The coordination polyhedra for small, highly charged cations are often very similar from one structure to another. As a result, Si^{4+} is found in a tetrahedral coordination with O^{2-} in thousands of crystals. Boron, which is slightly smaller, is found either in 3- or 4-coordination with oxygen in a large number of materials. In contrast, as the cation size increases, its electron cloud tends to become more distortable. Thus, the larger cations have a wider range of coordinations with a given anion, when surveyed

Table 9.5 High-symmetry coordination polyhedra

Linear (1 or 2)   H in H_2O

Triangular (3)   C in $CaCO_3$

Tetrahedral (4) Si^{4+} in SiO_2

Octahedral (6) Mg^{2+} in MgO

Cubic (8) Ca^{2+} in CaF_2

Close-packed (12) Sr^{2+} in $SrTiO_3$
(Icosohedron)

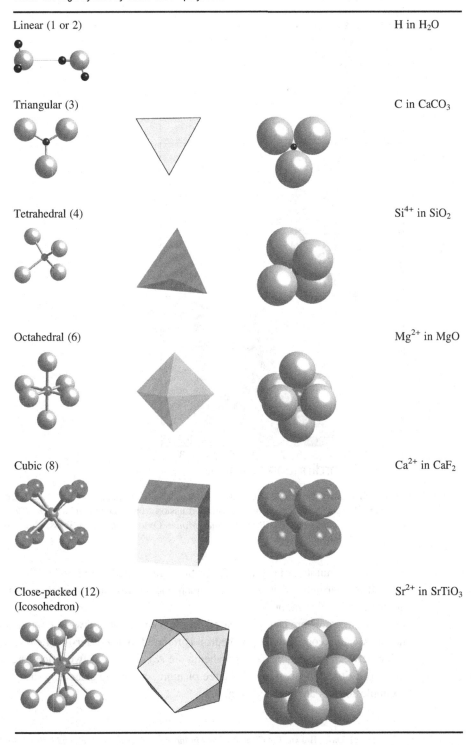

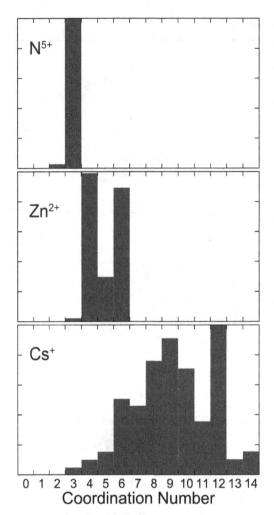

Fig. 9.7 Histogram showing coordinations for a small (N^{5+}), medium (Zn^{2+}), and large cation (Cs^+) in different structures. Redrawn with permission from I. David Brown, *The Chemical Bond in Inorganic Chemistry: The Bond Valence Model*, Oxford University Press, New York (2002) American Chemical Society.

across a large number of crystals. This trend is illustrated in Fig. 9.7 for N^{5+}, Zn^{2+}, and Cs^+.[2] It is clear that Cs^+ is found in a much wider array of coordination numbers than are the two smaller cations.

Many structures are classified based on the connectivity between coordination polyhedra. A cation coordination polyhedron is referred to as *corner-shared* if it shares one anion between two cations. The polyhedra are *edge-shared* if two anions are shared by the two cations, or *face-shared* if three or more anions are shared by the two cations. Examples of this are shown in Fig. 9.8.

---

[2] Data from I. David Brown, *The Chemical Bond in Inorganic Chemistry: The Bond Valence Model*, Oxford University Press, New York (2002).

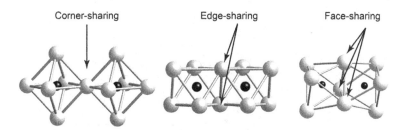

Fig. 9.8 Examples of corner-, edge-, and face-sharing of polyhedra (in this case octahedra).

9.5   Simple Ionic Crystal Structures

There are an extremely large number of structures observed in ionically bonded materials. This chapter will provide the framework for some of the simple ionic structures, others are sprinkled throughout the book. Separate chapters describe the crystal structures of silicate phases (Chapter 19).

One of the most important simple ionic crystal structures is *rocksalt* (also known as halite). The unit cell is a cube, with anions and cations alternating along the three principal directions. In the rocksalt structure, every Na^+ ion is octahedrally bonded to six Cl^-, and conversely with six Na^+ bonded to each Cl^-, as shown in Fig. 9.9. The interatomic bond length is 2.82 Å. The lattice parameter is twice this bond length, or 5.64 Å.

There are several ways of thinking about the structure of rocksalt. First, in terms of cation polyhedra, Fig. 9.9 shows that the octahedra share edges. Secondly, if you consider the Cl and Na sublattices separately, it is clear that both are face-centered, so that the same atom appears at the corners of the unit cell and the center of each face. This is illustrated in Fig. 9.10. The space group (a short-hand means of identifying the symmetry in the material) is *Fm3m*. Here, the *F* identifies that the lattice is face-centered; the first *m* denotes the mirror plane which is perpendicular to each cell edge. The 3 is a three-fold rotation axis along the cube body diagonal. That is, if you rotate the cube 120°, 240°, or 360° around the body diagonal, it will look the same after the rotation. The final *m* is for the mirror plane perpendicular to the face diagonals. A longer description of space groups and symmetry appears in Chapters 6 and 7.

Among the dozens of oxides with the rocksalt structure are magnesia, also called periclase (MgO), lime (CaO), and wustite (FeO). Most of the alkali halides are isomorphs with NaCl, as well, including sylvite (KCl). Galena (PbS), the chief ore of lead, is another. Lattice parameters for these isomorphs are listed in Table 9.6. Note that the cell size increases dramatically with atomic number except for the transition metal oxides where a small decrease is observed, consistent with the decrease in ionic radii on moving left to right across the 3d transition metal series.

The rocksalt structure is also adopted by some materials that are less ionic in character, including nitrides such as TiN, ScN, as well as carbides such as VC and UC.

There are also numerous derivatives of the rocksalt structure, many of which result from substituting groups of atoms on either the cation or anion sublattice. Several of these are shown in Fig. 9.11. For example, in the case of pyrite, FeS_2, S_2 dumbbells

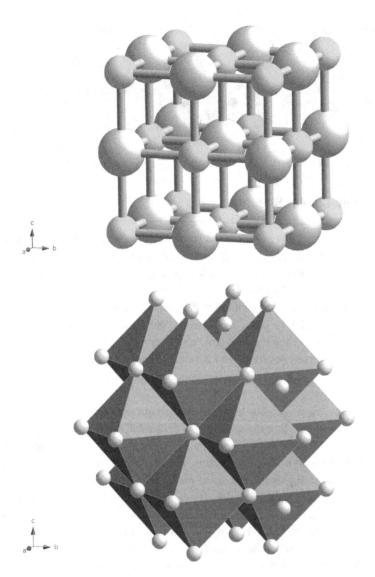

Fig. 9.9 The rocksalt (NaCl) structure. (Top) Ball and stick model. The large spheres are the Cl⁻ ions, the smaller spheres are the Na⁺. (Bottom) Drawing showing the connection of the Na octahedra. The Cl are shown as small spheres.

occupy the positions of the Cl ions. Cubic symmetry is retained because the dumbbells cant with respect to the cubic axes. In contrast, in CaC_2, the C_2 dumbbells are all oriented parallel to the c axis, distorting the unit cell from cubic to tetragonal. In calcite $CaCO_3$, the CO_3^{2-} molecules form planar triangles, all of which are oriented perpendicular to one of the original three-fold axes. The resulting unit cell is rhombohedral. In contrast, in Cu_2AlMn, the Al and Mn atoms are arranged in a rocksalt configuration. The Cu atoms occupy the 1/4, 1/4, 1/4 positions of the unit cell.

The structure of *fluorite* (CaF_2), the chief ore of fluorine, is shown in Fig. 9.12. The Ca atoms lie on a face-centered cubic lattice. Each F is at the center of one of the smaller

Table 9.6 Lattice parameters of some oxides and alkali halides with the rocksalt structure. Interatomic distances are equal to a/2

Alkaline-earth oxides

MgO	4.21 Å
CaO	4.81
SrO	5.16
BaO	5.52

Transition metal oxides

MnO	4.44 Å
FeO	4.31
CoO	4.27
NiO	4.17

Alkali halides

	Li	Na	K	Rb	Cs
F	4.02	4.62	5.35	5.64	6.01 Å
Cl	5.13	5.64	6.79	6.58	7.02
Br	5.50	5.97	6.60	6.85	
I	6.00	6.47	7.07	7.34	

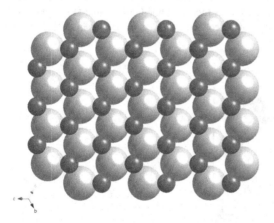

Fig. 9.10 Section of an NaCl crystal cut parallel to the (111) plane. Atoms are shown at their full Shannon–Prewitt ionic radii. Both the Na and Cl sublattices are in a geometric arrangement comparable to that of cubic close packing, although there is no anion–anion or cation–cation contact.

cubes obtained by dividing the unit cube into eight parts. The F are surrounded by four Ca atoms, and each calcium atom is surrounded by eight F atoms (8:4 coordination). The Ca–F bond length is 2.36 Å. The space group is $Fm3m$, like that of rocksalt, even though the atomic arrangement is completely different.

One of the most notable structural features of the fluorite structure is the large cavity at the center of the unit cell. Indeed, the atoms form a large cage-like structure around the center of the cell, producing a lower density material with large interstices. This is one of the reasons why many materials with the fluorite structure have unusually high ionic conductivity.

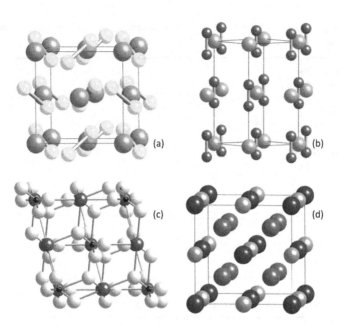

Fig. 9.11 Derivatives of the rocksalt structure: (a) FeS_2 pyrite, (b) CaC_2, (c) $CaCO_3$ calcite with Ca shown as the large filled circles, C as the small filled circles, and O in light gray, and (d) Cu_2AlMn with Mn on the corners of the cell, Al on the edges, and Cu in the 1/4, 1/4, 1/4 positions.

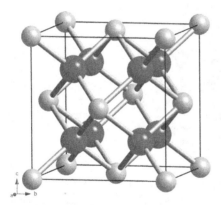

Fig. 9.12 The fluorite (CaF_2) structure. The smaller ions are Ca^{2+}; the larger are F^-.

Among oxides, zirconia (ZrO_2) together with hafnia (HfO_2), uraninite (UO_2), and thoria (ThO_2) possess the fluorite structure. Potash (K_2O), soda (Na_2O), and lithia (Li_2O) have the antifluorite structure in which cation and anion positions are interchanged. Lattice parameters are listed in Table 9.7.

The fluorite structure is also adopted by several intermetallic compounds, including Mg_2Si, Mg_2Sn, Mg_2Pb, $AuAl_2$, $AuGa_2$, $GeMg_2$, AgAsMg, CoMnSb, NiMnSb, and NiMgBi.

As was the case for rocksalt, there are a number of derivatives of the fluorite structure. Two examples of these are shown in Fig. 9.13. In MgAgAs, the Ag and

Table 9.7 Lattice parameters of oxides with the fluorite and antifluorite structure. Bond lengths are equal to $\sqrt{3}a/4$

Fluorites		Antifluorite	
AmO_2	5.38 Å	Li_2O	4.62 Å
CeO_2	5.41	Na_2O	5.55
CmO_2	5.37	K_2O	6.44
HfO_2	5.12	Rb_2O	6.74
NpO_2	5.43		
PaO_2	5.51		
PrO_2	5.47		
ThO_2	5.60		
UO_2	5.47		
ZrO_2	5.07		

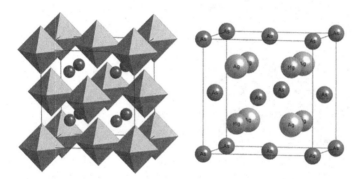

Fig. 9.13 Two derivatives of the fluorite structure: (left) K_2PtCl_6, with the $PtCl_6$ octahedra shown; and (right) MgAgAs.

Mg atoms alternate on the F sites, while the As atoms occupy the Ca positions. In K_2PtCl_6, the $PtCl_6$ octahedra adopt the Ca positions, while the K sit on the F sites. The pyrochlore crystal structure can be treated as a fluorite derivative in which 1/8 of the anion sites are vacant, in an ordered array.

The *rutile* crystal structure of TiO_2 is shown in Fig. 9.14. The unit cell is a tetragonal one, with Ti atoms at the corners and center of the cell. The Ti are octahedrally coordinated in a somewhat distorted octahedron, while the O is triangularly coordinated. The TiO_6 octahedra are edge-shared along the c axis. The columns of edge-shared octahedra are corner shared with adjacent columns. Cations fill only half the available octahedral sites, and the closer packing of oxygen ions around the filled cation sites leads to the distortion of nearly hexagonally close packed anion lattice. The rutile structure is characteristic of MgF_2, ZnF_2, MnF_2, FeF_2, CoF_2, NiF_2, and of TiO_2, SnO_2, PbO_2, VO_2, NbO_2, TeO_2, MoO_2, WO_2, MnO_2, RuO_2, OsO_2, and IrO_2.

Rutile is an important material industrially for a number of applications. The high refractive index makes the material useful as an additive to paints. It is also, along with ilmenite, a key ore of Ti metal. As was shown in Chapter 4, growth of rutile crystals is faster along the c direction than in the a–b plane, resulting in needle-like morphologies. Aligned rutile needles in doped Al_2O_3 crystals produce the "star" effect in star sapphires.

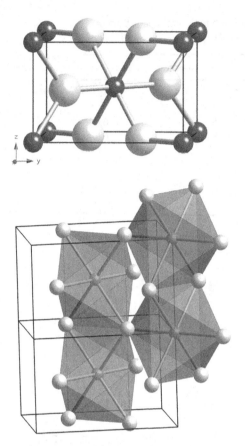

Fig. 9.14 Two views of the structure of rutile: (top) a unit cell, and (bottom) two unit cells, showing the connectivity of the TiO_6 octahedra.

Alumina, Al_2O_3, is a material for which many polymorphs have been reported. The crystal structure of α-Al_2O_3, also known as *corundum*, is the form stable under ambient conditions. It belongs to the rhombohedral crystal system. In corundum, the Al^{3+} ions are 6-coordinated with oxygen in an octahedral arrangement, while the O^{2-} ions are tetrahedrally coordinated, with an Al–O bond length of ~1.9 Å. The oxygens are hexagonally close-packed, and the Al fit into 2/3 of the octahedral interstices. The result is a very dense crystal structure in which the alumina octahedra face-, edge-, and corner-share. The theoretical density is just under 4 g/cm^3, much higher than that of α-quartz (SiO_2), which has a theoretical density of ~2.65 g/cm^3, even though Si and Al are next-door neighbors on the periodic table. The high density of strong bonds also produces the high hardness and high melting point of corundum. These, in turn, lead to the use of corundum as an abrasive and as a refractory material.

This structure is illustrated in Fig. 9.15. Along the c axis, the cations are in an ABCABC stacking arrangement, while the anions are in an ABAB stacking sequence. Therefore, the c axis is rather long, ~13 Å. Isomorphs of corundum include α-Ga_2O_3, Ti_2O_3, Cr_2O_3, Fe_2O_3, Sc_2S_3, Lu_2S_3, and some of the smaller rare-earth sulfides. The brilliant red color of ruby is associated with substitution of a small amount of Cr on the

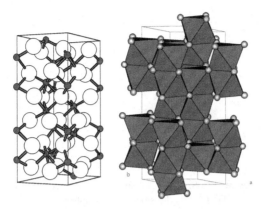

Fig. 9.15 Two views of the unit cell of corundum, showing (left) the atom coordinations and (right) the connectivity of the AlO_6 octahedra.

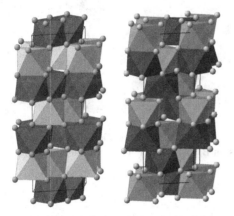

Fig. 9.16 Two derivatives of corundum (left) $FeTiO_3$ and (right) $LiNbO_3$. The c axis is vertical.

Al sites in corundum, while the deep blue of sapphire is due to a combination of Fe and Ti doping.

The ilmenite, $FeTiO_3$, and lithium niobate, $LiNbO_3$, structures can be visualized as derivatives of the corundum structure. The key differences lie in the stacking of the cations. In ilmenite, Fe-centered and Ti-centered octahedra alternate in layers. In contrast, in $LiNbO_3$, the stacking along the c axis is Li–Nb–vacancy, Li–Nb–vacancy.

The ABO_3 *perovskite* crystal structure is important in both geology and electro-ceramics. It is believed to be the most common crystal structure in the Earth. Figure 9.17 shows a unit cell for the cubic form of $SrTiO_3$. The Ti atom on the B site is octahedrally coordinated to oxygen, and forms a corner-linked three-dimensional framework for the structure. The large A-site ion adopts 12-coordination to the surrounding O; each O thus has two Ti and 4 Sr neighbors. The Sr and O together make a cubic close-packed array. It is believed that deep in the Earth many silicates adopt the perovskite structure.

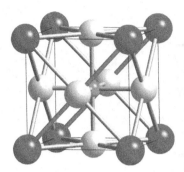

Fig. 9:17 Prototype perovskite structure. The Sr atoms are shows as the large filled spheres, the O are the large open spheres, and the Ti atoms is the smaller filled sphere.

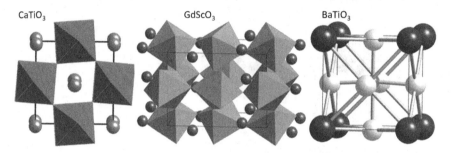

Fig. 9.18 Three examples of distorted perovskites. $CaTiO_3$ and $GdScO_3$ show two different types of octahedral rotations. In $BaTiO_3$, the Ti atom off-centers from the high-symmetry position.

Two principal families of distorted perovskites structure exist: those involving rotations of the oxygen octahedra around one or more axes, and those with off-centering of the B-site ions. Examples of both of these are shown in Fig. 9.18. When the A-site ion is large, or when the temperature is high, the A-site ion adopts a symmetrical 12-coordination, and the angle from B-site ion–O–B-site ion is 180°. However, in the mineral perovskite, $CaTiO_3$, the Ca^{2+} ion is slightly too small for 12-coordination with oxygen at room temperature. In order to reduce the size of the cavity in which is sits (and so reduce its effective coordination number) the octahedral framework rotates (or tilts) so that the B-site ion–O–B-site ion angle is no longer 180°. Within a plane of octahedra, if one rotates clockwise, the adjacent octahedra must rotate counterclockwise in order to avoid breaking any bonds. Many different distortions can arise depending on the number of axes about which rotations develop, the magnitude of the rotations, and the relative orientations of the rotations in one layer with respect to the next.

Other distortions of the perovskite structure involve the displacement of one or more ions in the structure from their high-symmetry positions. Shown in Fig. 9.18 is the example of tetragonal $BaTiO_3$, in which the Ti ion is shown shifted slightly up from the

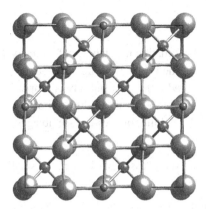

Fig. 9.19 Spinel crystal structure viewed looking down the x axis. The smallest ions are the octahedrally coordinated Al; the Mg are the intermediate size sphere (tetrahedral coordination), and the anions are the largest ion shown.

center of the unit cell. Distortions of this type are particularly important in inducing a spontaneous polarization in the material.

The perovskite structure is a garbage-bag of a crystal structure, in that most of the elements on the periodic table can be accommodated. Thus, there is a wide range of stoichiometries with this structure, providing charge balance is maintained. The structure is also unusually tolerant of non-stoichiometry. For example, WO_3 can be regarded as a derivative of the perovskite structure in which the A site is completely vacant. When two or more cations occupy one of the sites, ordering is observed if the ions are sufficiently different in size or valence.

Because of this, a wide variety of functional properties are observed in perovskites, ranging from high-permittivity insulators, to semiconductors, to metallic conductors to superconductors. Magnetism and ferroelectricity are also possible.

Spinel is the mineral name for $MgAl_2O_4$. This structure is based on oxygen atoms in a nearly cubic close-packed arrangement, with one of the cations in tetrahedral coordination, with two others occupying octahedral sites. The oxygen has a distorted tetrahedral coordination, with one tetrahedral cation neighbor, and three octahedral cation neighbors. This is illustrated in Fig. 9.19.

In $A^{2+}B^{3+}_2O_4$ spinels, several different arrangements for the cations are known. *Normal spinels*, such as $MgAl_2O_4$ place the 2+ cation on the tetrahedral site, and the two 3+ cations on octahedral sites. *Inverse spinels*, such as Fe_3O_4 have one of the 3+ cations on the smaller tetrahedral site. The other 3+ cation and the 2+ cation then fill the octahedral sites. *Disordered spinels* have random occupancy of the cation sites. *Defect spinels* such as γ-$Fe_2O_3 \equiv Fe_{8/3}\square_{1/3}O_4$ maintain charge balance with only 3+ cations by introducing vacancies, $\square$, on the cation sites. The site occupancy in different spinels offers a rich field for exploration of crystal field theory. It also offers practically a means of controlling the electrical conductivity as well as the magnetic behavior of a number of spinels.

9.6   Ramberg's Rules

Coulomb's law provides a simple way of estimating the stability of ionic crystals. If we assume that the charges are proportional to the valences of the ions and the interatomic distance is equal to the sum of their ionic radii, then the stability is proportional to the product of the two valences, and inversely proportional to the sum of the radii. This leads to Ramberg's rules[3] for predicting the stability of various assemblages of ionically bonded materials.

Rule 1: For ions of similar valence, the stable assemblage pairs the smallest and largest ions:

$$LiCl + NaF \rightarrow LiF + NaCl.$$

Rule 2: For ions of similar size, highly charged cations pair with highly charged anions:

$$MgF_2 + Li_2O \rightarrow MgO + 2LiF.$$

Rule 3: Small cations combine with highly charged anions, and large cations with lower charged anions:

$$2LiF + Na_2O \rightarrow Li_2O + 2NaF.$$

Rule 4: Small anions pair with highly charged cations, and large anions with lower charged cations:

$$Li_2O + MgS \rightarrow MgO + Li_2S.$$

All four rules are based on Coulomb energy. As an illustration, consider a melt containing equimolar concentrations of K^+, Li^+, Cl^- and F^-. The KCl + LiF combination is more stable than KF + LiCl because the more efficient packing of the two smallest and two largest ions leads to lower Coulomb energies.

9.7   Problems

(1) Halite = rocksalt = NaCl (space group 225: $Fm\bar{3}m$ with $a = 5.68$ Å) and Na at (0, 0, 0) and Cl at (0.5, 0, 0) and all symmetry-related positions.

    (a) Draw a unit cell of this structure showing the bonds that connect the Na and Cl ions.

    (b) Show the coordination (i.e. the number and geometric arrangement of near neighbors) of Cl and Na atoms.

    (c) One of the factors that controls the morphology (i.e. shape) of crystals is the number of dangling bonds per atom at the surface. The fewer the dangling bonds there are, the more nearly satisfied each surface atom is, the more stable the surface. Plot the {100}, {110}, and {111} planes. Determine the

[3] H. Ramberg, *Amer. Min.*, **39**, 79 (1954).

number of dangling bonds per surface atom on each of these planes. (Hint: it helps if you show the plane with a finite thickness below the plane. You can then count the number of bonds present and so determine the number missing, based on the coordination number. It may also help if you plot a big enough piece of the plane that you can look at atoms near the center of your drawing, so you are not confused by edge effects.)

(d)   Which face should be lowest in free energy?

(e)   Calculate the theoretical density (in g/cc) NaCl, showing all steps.

(f)   The surface termination of a crystal can also be important in a variety of cases (for example in growth of thin films on surfaces). Compare a {111} surface with a {222} surface. How do these surfaces compare in terms of the number of dangling bonds per atom? How do they compare in terms of surface composition?

(2)   The electron density map for LiF is shown in Fig. 9.1.

(a)   Taking numbers from this graph, plot the electron density as a function of distance for a line trace along the [100] direction. Use an interatomic separation distance of 2.0315 Å for scaling.

(b)   Estimate the percent ionicity of the Li–F bond.

(3)   There are many derivatives of the rocksalt structure. For example, it is often observed that you can replace either the cation or the anion with complex ions. The following problems can be viewed as replacements of this type. Please note that the resulting structures sometimes bear only a passing resemblance to NaCl itself in terms of their properties and morphology.

Pyrite is the mineral name for FeS_2, and is colloquially known as fool's gold. It is cubic and has space group 205: $Pa\overline{3}$ with $a = 5.408$Å) and Fe at $(0, 0, 0)$ and S at $(0.389, 0.389, 0.389)$ and all symmetry-related positions.

(a)   Draw a unit cell, showing the bonds that connect the Fe–S and S–S atoms.

(b)   What are the coordinations of the Fe and S atoms?

(c)   Explain how this crystal structure is related to that of rocksalt.

(d)   Describe the type of bonds that appear in this structure.

(4) Cu_2AlMn is a Heusler alloy. It is cubic (space group: $Fm3m$, with $a = 5.947$ Å). Mn is at $(0, 0, 0)$ Al at $(1/2, 1/2, 1/2)$, Cu at $(1/4, 1/4, 1/4)$ and all symmetry-related positions.

(a)   Draw a unit cell.

(b)   How does this structure relate to rocksalt?

(5) $CaCO_3$ is a rhombohedral crystal (space group 167: $R\overline{3}c$, with $a = 6.36$ Å and $\alpha = 46.08°$) and Ca at $(0, 0, 0)$, C at $(0.25, 0.25, 0.25)$, O at $(0.506, -0.007, 0.25)$ and all symmetry-related positions).

(a)   Draw a unit cell, showing the C–O and Ca–O bonds.

(b)   Show the coordinations of a C, Ca, and O.

(c)   How does this structure relate to rocksalt?

(6)   Suppose you had a mixture of Ca, K, Cl, and F. What compounds would you expect to form and why?

(7)   Cristobalite is a cubic polymorph of SiO_2 (space group 198: $P2_13$, with $a = 7.20$ Å) and Si at $(-0.008, -0.008, -0.008)$ and $(0.255, 0.255, 0.255)$, and O at

(0.125, 0.125, 0.125) and (0.44, 0.34, 0.16). This structure is slightly less symmetrical than the usual idealized cristobalite that is pictured in many textbooks, but is closer to the real structure observed at high temperatures.

(a) Draw a unit cell, showing the Si–O bonds.

(b) Describe how this structure relates to that of diamond. (Hint: try making the O atoms invisible, and plotting more than one unit cell.)

(c) Calculate the first three terms in the Madelung constant for cristobalite.

(8) CaC_2 is a tetragonal material (space group 139: $I4/mmm$, with a = 3.633 Å and c = 6.036 Å) and Ca at (0, 0, 0), C at (0, 0, 0.395) and all symmetry-related positions).

(a) Draw a unit cell, showing the bonds that connect the C atoms.

(b) How does the unit cell relate to that of rocksalt?

(9) Radii and bond distances.

(a) Compare the covalent and ionic radii of an atom like Fe. Explain which you expect to be larger and why.

(b) Describe the trends in ionic radii as you go across the periodic table in a row, or down within a column. Be clear about what happens in the transition metal series as well.

(c) $YMnO_3$ is hexagonal, with a = 6.1387 Å, and c = 11.4071 Å. One of the Mn atoms is at position (0.3352, 0, 0.24738), and one of the O atoms is at (0.6667, 0.3333, 0.2660). What is the interatomic distance between these two atoms?

(10) NiAs structure (space group 186: $P6_3mc$ with a = 3.4392 Å, c = 5.3484 Å) and Ni at (0, 0, 0) and As at (2/3, 1/3, 1/4) and all symmetry related positions.

(a) Draw a unit cell.

(b) What are the nearest-neighbor coordinations of the Ni and As atoms? Describe both the coordination number and the coordination geometry.

(c) Describe the connectivity of the Ni-centered polyhedra in terms of corner-, edge-, and face-sharing.

(d) What is the closest Ni–Ni distance? This is so close to the distance between two metallically bonded Ni that the coordination of Ni often includes the two nearest Ni neighbors. As a result, the NiAs structure often has a metallic component to its bonding. This contrasts strongly with materials with the NaCl structure (a competitor), which are typically electrical insulators.

(11) CsCl has a very simple structure with space group 221: $Pm3m$ with a = 4.11 Å, and Cs at (0, 0, 0) and Cl at (1/2, 1/2, 1/2) and all symmetry-related positions. Draw a unit cell of CsCl. Describe the relation between this and one of the common metal structures.

(12) For the NaCl crystal structure, sketch a conventional unit cell. On your diagram show a primitive unit cell. How many atoms are in the conventional cell and the primitive cell? (Please recall that in counting the number of atoms that occur in a unit cell, any atom completely inside the unit cell is counted. Any atom on a unit cell corner is shared by eight unit cells, and so can only be counted 1/8 into the cell in question. Any atom on a cell edge is shared by four unit cells, and counts only 1/4 towards a given cell; any atom on a face is shared by two unit cells, and counts only 1/2 towards a given cell.)

(13) For each pair, indicate which atom or ion is larger. If no oxidation state is given, treat the material as being unionized.

(a) C   Si

(b) Li^+ Na^+

(c) Mg^{2+} Al^{3+}

(d) Ca   Fe

(e) Ni^{2+} Ni^{3+}

(14) Describe the trends in ionic radii across the periodic table. Be sure to include the transition metals and lanthanides in your description. Why are these trends observed?

(15) Draw a schematic electron-density map for an ionically bonded solid.

(a) Explain how you would expect the map to change as the degree of covalency in the bonding was increased.

(b) Explain the origin of ionic bonding.

(c) How would you go about calculating the lattice energy of such a material?

(d) How would you expect the melting temperatures of materials with this type of bonding to change with the sizes of the constituent atoms? Would this be the same as, or different to, the trend with atom size for a materials with van der Waals bonding? Describe your reasoning in detail.

10 Metals and Steel

10.0 Introduction

In the free-electron model, the interatomic bonding in metals is viewed as cation cores embedded in a sea of free electrons. The electrostatic attraction between the negative electrons and the positively charged ions holds the structure together, and the free electrons provide an explanation of the large electrical and thermal conductivity of metals. In metallic sodium, for example, the outer 3s electron is ionized away to form a cloud of nearly free electrons, leaving Na^+ ion cores behind.

A chemical approach to the metallic bond is to assign the bonding electrons to bands formed by the outermost orbitals. This becomes difficult when the atoms come close together, and the electronic states broaden and mix. Iron, nickel, and other elements in the first transition metal series form bands from a complex mixture of 4s, 4p, and 3d orbitals, requiring band theory calculations to quantify the electronic structure. As a result, it is difficult to explain the structures of metals and their properties in a simple way.

Fortunately, there are only three important structures for elemental metals, and a few simple rules to explain why. The subsequent sections describe these structures and the nature of their bonds. The remainder of the chapter describes the Fe–C alloys known as steel, by far the most useful of metals, as well as some important intermetallic compounds.

10.1 Metal Structures

The crystal chemistry of the metallic elements is relatively simple, with only three common crystal structures: body-centered cubic (BCC), face-centered cubic (FCC), and hexagonal close-packed (HCP). Face-centered cubic is also referred to as cubic close-packing (CCP).

Many metals crystallize in either cubic close-packing or hexagonal close-packing arrangements. In these structures each atom is surrounded by twelve others: six neighbors in the close-packed plane. Three additional neighbors are in the layer above, nestling over the interstices of the first layer. Three more neighbors in the layer below complete the 12-fold coordination of each metal atom. Many different stacking sequences are possible but only two are common: hexagonal close-packing with a two-layer ABABAB sequence and cubic close-packing with a three-layer ABCABC

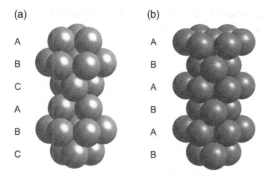

Fig. 10.1 Stacking sequences for (a) cubic close-packing and (b) hexagonal close-packing.

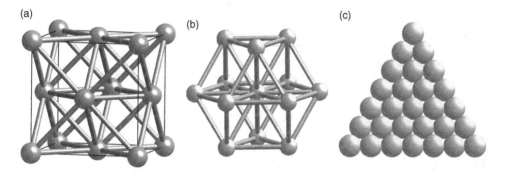

Fig. 10.2 Face-centered cubic metal structure (a) one unit cell, (b) schematic showing the 12-coordination of the central atom and (c) the close-packed {111} plane showing each atom surrounded by six other atoms in the plane.

sequence. The nomenclature here means that in cubic close-packing, the first and third layers are staggered. In the hexagonal close-packed structure, however, layers one and three are superposed, as shown in Fig. 10.1.

As shown in Fig. 10.2, cubic close-packed structures can be described by a face-centered cubic (FCC) unit cell. There are four atoms in the unit cell: one at the corners (0, 0 ,0) and three in face-centered positions (1/2, 1/2, 0), (1/2, 0, 1/2), and (0, 1/2, 1/2). A number of important metals (Cu, Ag, Au, Pt, Pd, Ni, Al) possess this structure, as do the inert gases when they solidify at low temperatures. The close-packed plane in FCC metals belong to the {111} family. Unit-cell parameters and bond lengths for FCC metals are listed in Table 10.1. There are two principal types of interstices (or holes) in the FCC lattice. At the center of the unit cell in Fig. 10.2 is an octahedrally coordinated interstice, with six adjacent metal atoms; comparable interstices also appear half-way along each unit-cell edge. There are also tetrahedrally coordinated interstices at (3/4, 3/4, 1/4) and equivalent positions. Looking perpendicular to the close-packed planes, octahedral interstices are located on top of tetrahedral interstices and vice versa.

Copper is a reddish-orange metal with the FCC crystal structure. It is used extensively for delivering electrical power for heating and home appliances, as well as the

Table 10.1 Lattice parameters and bond lengths for metals with the FCC structure

Material	Lattice parameter, a (Å)	Nearest-neighbor distance (Å)
Al	4.05	2.86
Cu	3.61	2.56
Ag	4.09	2.89
Au	4.08	2.88
Ni	3.52	2.49
Pd	3.89	2.75
Pt	3.92	2.77
Ca	5.57	3.94
Sr	6.08	4.30
Pb	4.95	3.50
Ne*	4.51	3.21
Ar*	5.43	3.84
Kr*	5.69	4.03
Xe*	6.25	4.42

* At low temperatures.

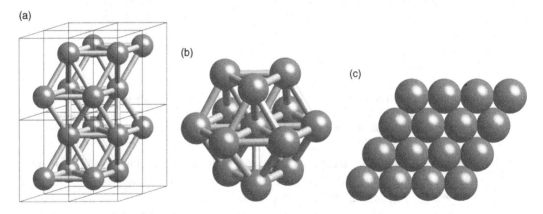

Fig. 10.3 Crystal structure of magnesium and other metals with the hexagonally close-paced metal structure: (a) eight unit cells, (b) schematic showing the 12-coordination of one atom, and (c) the close-packed {001} plane showing each atom surrounded by six other atoms in the plane.

interconnect metal in most silicon integrated circuits. Four characteristics that make copper attractive as an engineering material are high electrical conductivity, high thermal conductivity, ease of fabrication, and excellent corrosion resistance. Its high ductility and corrosion resistance make Cu an excellent choice for domestic plumbing.

The crystal structure of hexagonally close-packed metals is shown in Fig. 10.3. The close-packed plane in HCP metals are the {001} family. As in the FCC structure, each metal atom is bonded to 12 others, but the bond lengths may not be exactly the same. Atoms in the same close-packed plane are separated by a distance equal to the a lattice parameter while the others, three above and three below, are at a distance $\sqrt{\frac{a^2}{3} + \frac{c^2}{4}}$ away.

Table 10.2 Lattice parameters, a and c, along with the interatomic separation distance for HCP metals. Most rare-earth metals possess this structure as well

	a (Å)	c (Å)	d (Å)
Be	2.2866	3.5833	2.23
Cr	2.722	4.427	2.72
Cd	2.978 87	5.617 65	3.14
Co	2.5057	4.0686	2.50
Hf	3.1967	5.0578	3.16
La	3.75	6.07	3.74
Mg	3.209 27	5.210 33	3.20
Os	2.7352	4.3190	2.71
Ru	2.703 89	4.281 68	2.68
Ti	2.950	4.686	2.81
Y	3.6474	5.7306	3.60
Zn	2.6648	4.9467	2.79
Zr	3.232	5.147	3.21

All 12 bonds are equal if the c/a ratio is 1.63. Lattice parameters for several hexagonal close-packed (HCP) metals are given in Table 10.2. For many metals, however, c/a is closer to 1.60, making the atoms above and below slightly closer. Cd and Zn are exceptions, with large c/a ratios of 1.886 and 1.856, respectively. As in FCC metals, there are two principal types of interstices: octahedral and tetrahedral. The layer stacking arrangement in HCP metals leads to tetrahedral interstices being on top of tetrahedral interstices, while octahedral interstices are stacked on top of octahedral interstices.

In both ideal hexagonally close-packed and cubic close-packed materials, 74% of the volume of the unit cell is occupied by atoms. This represents the upper limit for packing efficiency with spheres of one size.

The structure of a body-centered cubic metal is shown in Fig. 10.4. The unit cell is a simple one in which the same atom appears at the corners of the unit cell and at the center of the unit cell. Each atom in BCC metals has eight near neighbors. However, it is worth noting that the distance from the center of one unit cell to the center of the next is only 15% further away than the near-neighbor distance. This is sufficiently close that many people regard the coordination of each atom to be 14 (i.e. 8 + 6), as shown in Fig. 10.4(b). This geometry is not close-packed, but is nearly so, with a packing fraction of 68%. The almost close-packed (110) plane has four neighbors that are touching, and two next-near neighbors arranged to complete a distorted hexagon.

Lattice parameters and bondlengths for body-centered cubic metals are given in Table 10.3. The eight near-neighbor bondlength is $\frac{\sqrt{3}a}{2}$, while the six next-nearest atoms are a distance a away.

In BCC metals, there are two families of interstices that are important: tetrahedral interstices like those at (1/2, 1/4, 0) (which could accommodate an atom of size 0.36 Å in α-Fe without significant distortion) and distorted octahedral sites at positions like (0, 0, 1/2). If one were to treat the ion cores in α-Fe as being incompressible spheres, the octahedral interstice could accommodate only atoms of size 0.19 Å. In reality,

Table 10.3 Body-centered cubic metals with cell edges, *a*, and bond lengths, *d*

	Alkali metals			Transition metals	
	a	*d*		*a*	*d*
Li	3.51Å	3.04Å	V	3.02 Å	2.62 Å
Na	4.29	3.72	Cr	2.88	2.50
K	5.34	4.63	Fe	2.87	2.48
Rb*	5.63	4.88	Nb	3.30	2.86
Cs*	6.06	5.25	Ta	3.31	2.863
			Mo	3.15	2.73
			W	3.16	2.74

* Measured at −173 °C.

(a)   (b)   (c)

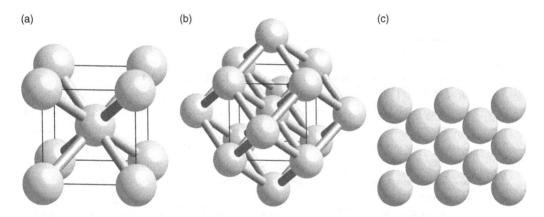

Fig. 10.4 Structure of body centered cubic metals such as Fe: (a) unit cell, showing eight near neighbors, (b) effective 14-coordination of central atom, and (c) almost close-packed (110) plane.

these octahedral sites can be occupied by larger atoms – the structure will distort in response to this.

The free energies for the three common metal structures are often very close to one another, so phase transformations are very common. Iron, for example, exists in all three phases (Fig. 10.5). BCC is stable at low temperatures and low pressure, FCC is stable near 1000 °C and low pressure, and HCP is stable at low temperatures and high pressure.

Other metals crystallize into close-packed arrangements with more complicated stacking sequences. Examples of this are shown in Fig. 10.6. It can be seen there that α-Nd crystallizes into a more complicated stacking sequence, ABAC. For α-Sm, a nine-layer stacking sequence: ABABCBCAC appears in the unit cell.

10.2   Bonding and Crystal Structure

Predicting crystal structures in metals is more difficult than in ionic and covalent crystals, but the number of bonding electrons is an important clue. Figure 10.7 shows

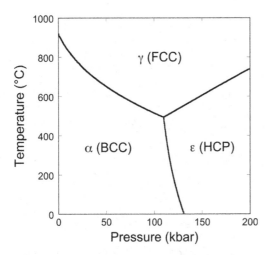

Fig. 10.5 Pressure–temperature phase diagram for elemental iron. Redrawn from F. P. Bundy, Pressure—Temperature Phase Diagram of Iron to 200 kbar, 900 °C, *J. Appl. Phys.* **36** (2) 616–620 (1965), with the permission of AIP Publishing.

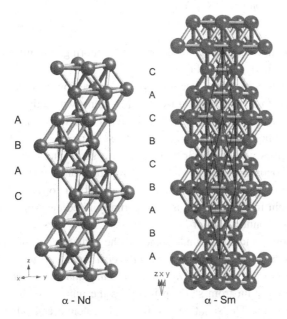

Fig. 10.6 Other close-packed metal stacking sequences.

the crystal structures of the metals on the periodic table. Each element has a distinctive electron configuration, such as outer electron shell configurations of $3s^1$ for sodium to $3s^2 3p^5$ in chlorine. The characteristic crystal structure is determined by this configuration, and in particular by the number of bonding electrons per atom (e/a).

Electronic configurations of the various chemical elements are based on extensive spectroscopic data. In metals, the bonding electrons are in partly filled shells, such as the

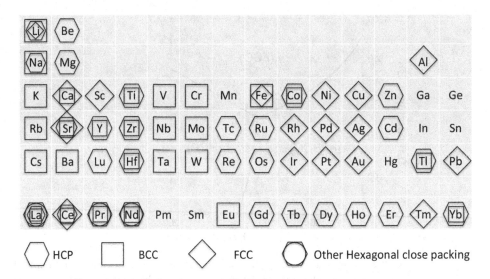

Fig. 10.7 Crystal structures adopted by elemental metals. The largest symbol denotes the room temperature structure. "Other hexagonal close packing" refers to stacking sequences such as ABAC, as was shown in Fig. 10.6. From data in: C. Barrett and T. B. Massalski, *Structure of Metals*, 3rd Rev. Edn, Pergamon Press, New York (1980).

outer 3s shell of sodium. Overlapping 3s shells of neighboring sodium atoms form resonating electron-pair bonds in metallic sodium.

One might imagine that in Mg, the s electrons would form a closed $3s^2$ shell. The $3s^2$ ground state in Mg leads to a vapor state of isolated Mg atoms with only weak van der Waals forces, similar to solid Ar. Such closed shells repel near neighbors, and do not promote strong chemical bonding, as is observed in the solid phase. Thus, the $3s^2$ state of isolated Mg atoms gives way to a 3s3p group state for short interatomic distances. Resonating sp hybrid bonds control the electronic properties and crystal structure of magnesium metal. The two unpaired electrons in the sp orbital form bonds with sp electrons of opposite spin in adjacent atoms. The partially filled electronic states also give rise to the electrical conductivity of Mg.

Thus, the first important idea for understanding the structure of metals is that *promotion of electrons from the ground state to an excited state can unpair the non-bonding valence electrons in the atom, and increase the number of electrons for bonding*. In Mg, the promotion from s^2 to sp increases the number of bonding electrons/atom from zero to two. The situation is different in calcium and other alkaline-earth metals, where d electrons begin to play a role. Here, the ground state, bond orbitals, and crystal structures are different.

The second main idea has to do with *bond order, the number of electrons per pair bond*. The effectiveness of bonding electrons depends on how crowded they are in the bond. In metallic magnesium, the two sp electrons per atom are spread among the 12 near neighbors in the HCP structure, and to more-distant neighbors as well. The resulting *low-order bonding* in solid Mg is *effective in offsetting the promotion energy required to produce the sp valence state*. If the sp bond produced an Mg_2 molecule,

Table 10.4 Ground states for isolated atoms and solids in one row of the periodic table

Element	Electron configuration of atom	e/a	Solid	Electron configuration in solid	e/a
Na	$3s^1$	1	BCC metal	s	1
Mg	$3s^2$	0	HCP metal	sp	2
Al	$3s^2 3p^1$	1	FCC metal	sp^2	3
Si	$3s^2 3p^2$	2	Diamond structure	sp^3	4
P	$3s^2 3p^3$	3	Molecular groups	$s^2 p^3$	3
S	$3s^2 3p^4$	2	Molecular groups	$s^2 p^4$	2
Cl	$3s^2 3p^5$	1	Diatomic molecules	$s^2 p^5$	1
Ar	$3s^2 3p^6$	0	Van der Waals solid	$s^2 p^6$	0

rather than Mg metal, the two electron pairs would be restricted to the bond between a pair of atoms, rather than distributed among many neighbors. With such a high concentration of bonding electrons (high-order bond) there is not sufficient energy to offset promotion.

The same argument holds for metallic Al and covalent Si. In both solids, an s electron is promoted to a p orbital to make all valence electrons available for bonding: sp^2 hybrid orbitals form in metallic aluminum, and sp^3 hybrid orbitals yield tetrahedral bonds in silicon. Note that promotion from sp^3 to sp^4 does not help phosphorus, because the number of bonding electrons remains the same. Table 10.4 summarizes the bonding in the row of the periodic table. To determine which valence states are likely to control the bonding, one must examine the spectroscopic evidence.

In the 1930s, Hume-Rothery[1] extended the correlation between electronic configuration and crystal structure to metals and their alloys. He noted that many of the intermetallic compounds of the non-transition metals showed a correlation between the observed crystal structure and the electron-to-atom (e/a) ratio. For alloys with less than 1.5 outer-shell s and p electrons, the BCC structure is observed, as in the alkali metals. Hexagonal close-packing is common for e/a ratios in the 1.7 to 2.1 range. Metallic magnesium is an example. Yet larger e/a values between 2.5 and 3 favor the cubic close-packed FCC structure. This works particularly well for structures near the melting temperature of the metal.

It is of interest to ask why. Alkali metals make use of the spherical s orbital. The BCC structure (Fig. 10.4) has eight nearest neighbors and six more next-nearest neighbors only 15% further away. The result is 14 neighbors arranged in a very symmetrical pattern, closely approximating a sphere. The rival HCP and FCC structures have slightly less regular coordination polyhedra. Here there are 12 near neighbors, but the next-near neighbors are >40% further away. Thus, BCC is slightly better for an e/a ratio near 1.

For e/a ratios near 2, the HCP structure is favored. This is the case for Mg and other metals making use of the sp hybrid orbital. Figure 10.8 illustrates the geometry of the sp bond. It has cylindrical symmetry with anisotropy extending along the p lobe of the

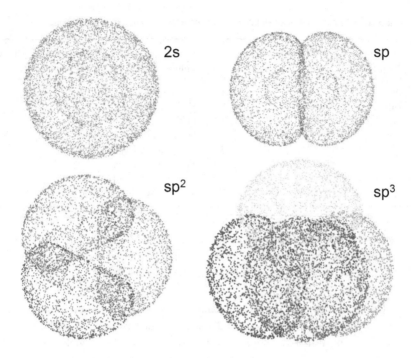

Fig. 10.8 Hybrid orbitals found in many engineering materials. The s orbitals are spherically symmetric (point group $\infty\infty m$), sp orbitals are cylindrically symmetric (∞/mm), sp^2 have triangular symmetry ($\bar{3}m$), and sp^3 are tetrahedral ($\bar{4}3m$).

wave function. The symmetry is closest to the $6/mmm$ point group of hexagonal close-packing. The BCC and FCC structures lack a special axis such as this, with an adjustable geometry.

The FCC structure dominates for e/a ratios near 3. Here the bonding makes use of sp^2 hybrid orbitals with triangular symmetry (Fig. 10.8). There are multiple three-fold rotation axes in the FCC structure (along each of the body diagonals). Perhaps this is related to the preference for the FCC structure at higher e/a ratios.

The sp^3 hybrid orbital favors a diamond crystal structure, as illustrated in Chapter 8.

Transition Metals

Engel[2] combined the ideas of Pauling and Hume-Rothery, and applied them to transition metals. He recognized that d orbitals are more localized than s and p orbitals, and do not play as important a role in intermetallic bonding. Vacant d orbitals often drain off electrons from the s and p orbitals, and delay the build-up of bonding electrons. As a result, the BCC structure is more common than might be expected. As shown in Fig. 10.7, all 18 metals of the first six transition metal groups possess the BCC structure at high temperatures. Proceeding across these groups, the d shell begins to fill with just

---

[2] N. Engel, *Acta Met.*, 15, 557 (1967).

one outer electron in a $d^{n-1}s$ hybrid. Adding one more electron converts the electron configuration to a d^5sp state and an HCP structure in the 4d and 5d series. The 3d transition metals behave somewhat differently as the additional electrons participate in magnetism. For example, as pointed out in Chapter 29, *ferromagnetism* is observed in iron, cobalt, and nickel alloys. Adding still more electrons leads to $d^{n-3}sp^2$ bonds and FCC structures, as in Cu, Ag, and Au.

Because the d electrons can participate in the bonding, they can increase the bond strength. As a result, the alkali metals that have one valence electron per atom have lower melting temperatures compared with transition metal alloys with the same structure. The contribution of d electrons to bonding increases with atomic number. Thus, 5d transition metals have higher melting points than 4d and 3d elements.

Among transition metals, the poor overlap of d orbitals compared with s and p orbitals makes it possible to predict structural changes at high pressures. *Compression improves the d overlap and favors the bonding orbitals with the most d electrons.* Consider the structural changes in elemental titanium from BCC to HCP. The BCC phase utilizes a d^3s orbital with three bonding d electrons. HCP has d^2sp bonds with only two bonding d states. As a result, the close-packed HCP structure is destabilized under pressure. Normally, one thinks of close-packed structures being favored at high pressures.

Note that the d electrons must be *unpaired* to participate in the bonding. Iron makes use of d^7s orbitals in the BCC structure and d^6sp in the HCP form: d^7s has only two bonding d electrons, whereas d^6sp has three. Therefore, the HCP structure is favored at high pressures, just the reverse of the titanium case (Fig. 10.5). In a similar way, pressure stabilizes HCP (d^4sp) over FCC (d^3sp^2) for d orbitals that are less than half full and FCC over HCP for those more than half full.

Summary of Chemical Bonding in Metals

(1) *Promotion to excited states* is often required to increase the number of bonding electrons: s, sp, sp^2, and sp^3 hybrid bonds are among the most important electron configurations involved.

(2) *Outer s and p electron shells overlap neighboring atoms* and are dominant in determining the crystal structure; d and f electrons are generally less important because they are located nearer the nucleus.

(3) *Hume-Rothery* points out the *correlation between bonding electron to atom (e/a) ratios and crystal structure.* For metal alloys with <1.5 outer shell s and p electrons/atom (e/a), the BCC structure is favored. HCP structures are formed for e/a ratios from 1.7 to 2.1, and CCP metals for e/a ratios of 2.5 to 3 s and p bonding electrons/atom. The diamond structure is stabilized for 4 e/a. Geometric arguments for these preferences were put forward earlier in this section.

(4) *Bond order* – the crowding of bonding electrons in the chemical bonds – is another important factor. In metals, the number of electron pairs per bond is generally small, being spread out among eight or 12 near neighbors, and even beyond.

(5) *Vacant d orbitals drain off electrons from the outer shell orbitals,* delaying the build-up of e/a ratios.

(6)   In 3d transition metals, and 4f rare earths, *magnetic phenomena* arise from unpaired localized electrons not used in bonding.

(7)   *Interstitial atoms* such as carbon, nitrogen, and oxygen are not directly involved in the bonding, but they *add to the e/a totals.*

(8)   *Inner shell electrons participate* much more *strongly in 4d and 5d transition metals,* because these electron shells extend to a larger fraction of the atom size.

(9)   *Substitutional non-transition metal neighbors destabilize such inner shell bonding.*

(10) *Substitutional transition metals favor inner-shell bonding.*

(11) *High pressures promote inner-shell bonding* by moving atoms closer together.

(12) *Phase changes with temperatures* are partly controlled by vibrational energy levels, heat capacity, and entropy. This leads to numerous phase transformations between the BCC, HCP, and FCC phases.

10.3   Slip in Metals

Metals deform more easily than ceramics because the close-packed planes glide past one another easily. In a simple metal, all atoms are alike; this allows bonds to break and reform with the next neighbor. Close-packed planes are flat, rather than corrugated, and have relatively few bonds to the next plane (three per atom). As a result, the energy cost to slip along close-packed planes is lower than along other planes. Slip leads to plastic deformation, so that the material does not recover its original shape when the stress is removed. In practice, slip typically occurs via motion of dislocations, as is discussed in Chapter 16.

The *slip system* describes both the plane along which slip occurs, as well as the slip direction. As shown in Fig. 10.9, slip tends to occur along the close-packed directions within a close-packed plane. For the close-packed (001) plane of the HCP metal shown, slip can occur along $\langle 100 \rangle$, resulting in three slip systems. Of these, two are independent, since slip in the $\langle \bar{1}\bar{1}0 \rangle$ direction can be described in terms of slip along $\langle 100 \rangle$ and $\langle 010 \rangle$. Likewise, in FCC metals, there are 12 slip systems given by the four $\{111\}$ planes, with slip along the $\langle 1\bar{1}0 \rangle$ directions. In BCC metals, the situation is somewhat complicated by the lack of a true close-packed plane. Thus, depending on temperature, different slip systems may be active. For example, in α-Fe, there are 12 slip systems in the $\{110\} \langle \bar{1}11 \rangle$ family, 12 more in the $\{211\} \langle \bar{1}11 \rangle$ family, and 24 in the $\{321\} \langle \bar{1}11 \rangle$ family.

The *Von Mises condition* states that good malleability will be observed in polycrystalline materials that have five or more independent slip systems. Thus, the abundance of slip planes in copper, silver, and gold account for the malleability of these metals.

10.4   Steels

Iron is by far the most abundant transition metal in the Earth's crust. The low cost, mechanical strength, and numerous applications make it the mainstay of structural engineering.

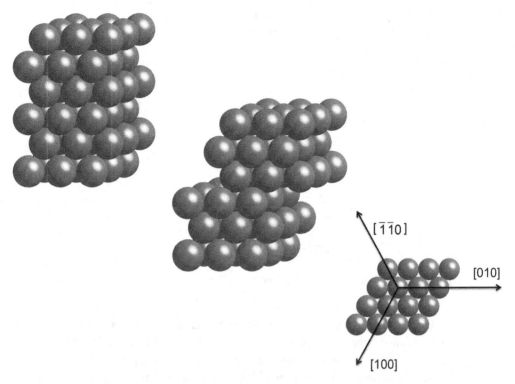

Fig. 10.9 (Top) unslipped crystal, (middle) slipped crystal, and (bottom) schematic showing the slip directions for an HCP metal on the close-packed (001) plane.

The principal ores of iron are the minerals magnetite (Fe_3O_4) and hematite (Fe_2O_3), usually contaminated by a number of silicate minerals. To remove the contaminants and reduce the oxides to molten iron metal, the ore is co-fired with limestone ($CaCO_3$) and coke (C) in a blast furnace (Fig. 10.10). All three are loaded from the top of the furnace and fill most of the stack. Air or oxygen heated to 1500 °C is pumped into the bottom. As the starting materials descend in the furnace, they are converted into flue gases, molten slag, and molten iron that is tapped off at the base of the furnace.

Four chemical reactions take place in the furnace. Coke and oxygen combine to form carbon monoxide:

$$2C + O \rightarrow 2CO.$$

Carbon monoxide reduces hematite to iron:

$$3\,CO + Fe_2O_3 \rightarrow 2\,Fe + 3CO_2.$$

Limestone decomposes to lime and carbon dioxide:

$$CaCO_3 \rightarrow CaO + CO_2.$$

Lime reacts with the silicate impurities to form molten slag:

$$CaO + SiO_2 \rightarrow CaSiO_3.$$

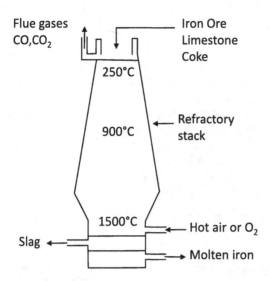

Fig. 10.10 A blast furnace for smelting iron ore.

The slag is actually a complex mixture of silicates, idealized here as $CaSiO_3$. Because of its lower density, the slag separates from molten iron and floats to the top. The heat given off by the blast furnace is used to preheat the incoming gases.

The pig iron castings prepared in a blast furnace are relatively impure, with significant amounts of C, Si, P, S, and Mn. To form steel, the castings are re-melted at higher temperatures and subjected to a stream of pressurized oxygen. Carbon and other impurities are oxidized and removed as gases or molten slag.

In pure iron, the room-temperature body-centered cubic form (α-Fe) changes to the face-centered cubic form (γ-Fe) at 910 °C. In the steel-making literature, BCC α-Fe is called *ferrite* and FCC γ-Fe is referred to as *austenite*. At 1400 °C, the structure reverts back to body-centered cubic (δ-Fe) before melting at 1536 °C. Iron atoms are coordinated to eight near neighbors in α-Fe and to 12 in γ-Fe. The interatomic distances and volume per atom are plotted in Fig. 10.11.

Steel is a billion dollar example of the importance of interstitial sites. Hagg's rule governs the formation of interstitial alloys: below a critical radius ratio of 0.59, non-metallic elements such as carbon, nitrogen and boron can occupy interstitial sites between metal atoms. Because of their small size (~0.8Å radius), carbon atoms enter interstitial sites in both BCC and FCC Fe. Octahedral sites are preferred in austenite (Fig. 10.12). The octahedral interstice in austenite is larger than the interstice in ferrite, and therefore it accommodates more carbon, as shown in the Fe–C phase diagram (Fig. 10.13). Nearly 2 wt% (~9 at%) enters the γ-Fe structure, about ten times more than α-Fe. This means that when austenite is cooled below 700 °C, the iron and carbon atoms must find new arrangements. Phase transformations such as this do not occur instantaneously, but are very time dependent. Many such transformations rely on atomic diffusion, which is often a slow process.

The separation of ferrite from austenite is a nucleation-and-growth process, which often takes place at grain boundaries. For compositions near the eutectoid (0.7 wt%

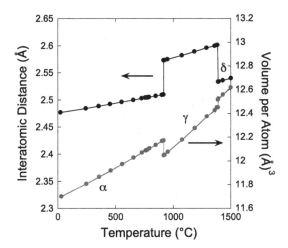

Fig. 10.11 Temperature dependence of the bond lengths and atomic volumes in pure iron. The interatomic distances are longer in FCC γ-Fe and the density is slightly higher than the BCC α and δ phases. Adapted, with permission, from W. Hume-Rothery, Z. S. Basinski, and A. L. Sutton, *Proc. R. Soc. Lond.*, A**229**, 459 (1955).

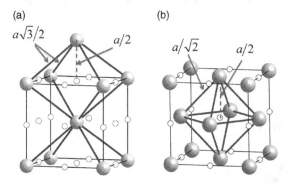

Fig. 10.12 Octahedral voids in the (a) ferrite BCC phase and (b) the austenite FCC phases of iron. The interstitial sites are substantially larger in austenite, allowing more carbon to enter the structure.

carbon), a lamellar microstructure called *pearlite* is formed. Pearlite is made up of ferrite (α-Fe) and *cementite* (Fe_3C) layers intimately bonded together. Cementite has a complex orthorhombic unit cell containing four carbon atoms and 12 iron atoms (Fig. 10.14). The cementite layers impart a high hardness to pearlite.

When austenite is cooled rapidly below 200 °C, the driving force transforming the FCC phase to BCC becomes so large that it occurs *before* carbon has time to form the cementite phase. Consequently, the BCC structure is supersaturated with carbon. The excess C distorts the BCC structure to a tetragonal form known as *martensite*. Carbon atoms occupy the (0, 0, $c/2$) interstitial sites, causing the tetragonality (Fig. 10.15). The c/a ratio is proportional to the amount of excess carbon. The large distortions in martensite make mechanical slip virtually impossible, leading to superb hardness for steels quenched in cold water. Disk-like platelets of twinned martensite

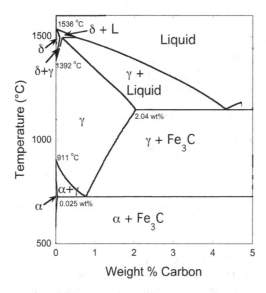

Fig. 10.13 The phase diagram of Fe–C alloys containing less than 5 wt% carbon. Note that γ-Fe is FCC, α- and δ-Fe are BCC. Cementite (Fe₃C) has a complex orthorhombic crystal structure. Adapted, with permission, from C. Barrett and T. B. Massalski, *Structure of Metals*, 3rd Rev. Edn, Pergamon Press, New York (1980).

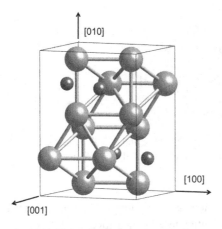

Fig. 10.14 Unit cell of cementite. The larger atoms are iron, the smaller ones are carbon.

contain up to 2 wt% carbon. Figure 10.16 shows the regions where annealing produces the metastable martensite phase and the important pearlite and bainite microstructures. Low carbon phases containing less than 2 wt% carbon are the most important.

Yet another microstructure appears in steels that are heat-treated between 250 and 500 °C. These temperatures are below those for pearlite but above the martensitic range (see Fig. 10.16). Bainite is a two-phase ferrite–carbide microstructure that differs from pearlite. Both are formed from supercooled austenite. In pearlite, cementite and ferrite crystallites form simultaneously within the FCC austenite phase. In bainite, the reverse

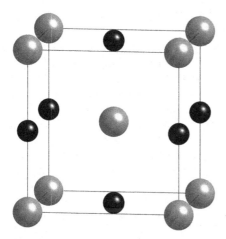

Fig. 10.15 Tetragonal unit cell of martensite showing the favored carbon interstitial positions (dark spheres) and the body centered tetragonal Fe lattice. The tetragonality increases with the carbon content.

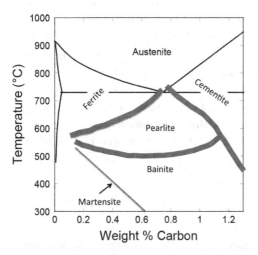

Fig. 10.16 Diagram showing processing regimes for production of desirable microstructures in Fe–C steels. Adapted, with permission, from C. Barrett and T. B. Massalski, *Structure of Metals*, 3rd Rev. Edn, Pergamon Press, New York (1980). Fig. 18–13b. Copyright Elsevier.

is true, with ferrite formed first, leaving a carbide behind. Elements like Mn are added to form carbides with acicular shape. Fibrous microstructure leads to mechanical toughening through crack shunting and fiber pullout mechanisms. The microstructure of bainite depends on temperature. Feather-like features are formed in upper bainite near 500 °C. A finer-grained microstructure is found in lower bainite.

Thus, a variety of microstructures are formed as Fe–C alloys are cooled through the eutectoid (0.7 wt% C, 730 °C). The sequence is austenite → coarse pearlite → fine pearlite → upper bainite → lower bainite → martensite. Careful control of times, temperatures, and composition are required to optimize the properties of steel. This is

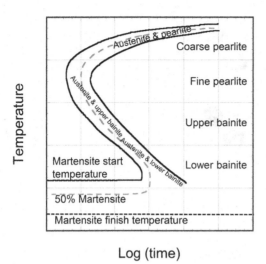

Fig. 10.17 Schematic time–temperature–transformation curve for a eutectoid steel.

illustrated in Fig. 10.17, where a time–temperature–transformation curve is shown for steel. Quantitative diagrams are available in other texts.[3]

The various Fe–C alloys encountered in steel and their crystal structures are of immense importance in industry. Figure 10.18 shows how the mechanical properties of hot-rolled steels depend on carbon content. Applications that require high ductility, such as forming automobile chasses, generally use *mild steel*, containing less than 0.3 wt% carbon. *Constructional steels* for girders or train rails use 0.4 to 0.6 wt% C steels with greater tensile strength. Harness is critical in *tool steels*, leading to the use of high C steels with 0.6 to 1.5 wt% C. Low-carbon steels possess greater malleability and impact strength, but high-carbon steels are superior in their tensile strength and hardness.

Habit Plane

When a second phase precipitates from a supersaturated solid solution, it frequently forms thin plates parallel to certain planes in the matrix. These so-called *habit planes* are designated by the orientation relationship between the precipitates (p) and the matrix (m). For example, when a FCC phase precipitates from an HCP matrix phase, the habit planes are usually given by $\{111\}_p \parallel \{0001\}_m$. This ensures that the close-packed planes are parallel to one another, minimizing strain mismatch at the interface.

The transformation between FCC and HCP involves a change in stacking sequence from ABCABC... to ABABAB... by a shearing motion parallel to the close-packed planes. The nucleus for such a phase change is generally a stacking fault. Cobalt undergoes a transformation from HCP to FCC on heating to 400 °C. This type of transformation is also observed in crystals with van der Waals bonding, such as Ar.

---

[3] *Atlas of Time–Temperature Diagrams for Irons and Steels*, ed. G. F. Van der Voort, ASM International, Materials Park, OH (1991).

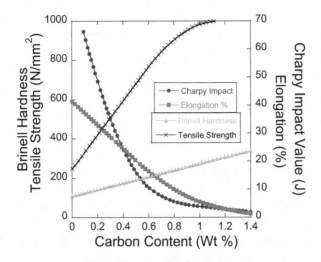

Fig. 10.18 Plot showing the relation between several functional properties and the carbon content of carbon steels. Replotted, by permission, using data from *ASM Specialty Handbook: Carbon and Alloy Steels*, ed. by J. R. Davis, ASM International, Materials Park, OH, 1996, p. 33, Fig. 13.

Iron transforms from BCC to FCC in the α–γ transition at 910 °C with the $(110)_\alpha$ plates parallel to the $(111)_\gamma$. Intricate lamellae patterns known as Widmanstatten structures are observed in Fe–Ni meteorites as a result of this. The (110) planes of BCC iron are similar to the close-packed {111} planes of FCC iron, as is seen by comparison of Figs. 10.2 and 10.4. A similar relationship is found in titanium where the BCC phase precipitates from the HCP phase. The habit relationship is $(110)_{BCC}$ ‖ $(001)_{HCP}$.

10.5   Common Alloys

There are numerous factors that control the crystal structures of alloys, including differences in size, electronegativity, and e/a ratio. Of these, differences in electronegativity are particularly important. The tendency of more electronegative atoms to complete an electron shell (either the octet from s and p electrons, or the d shell) decreases the extent of solid solution, and favor the formation of intermediate compounds. Differences in atom sizes drive ordered arrangements of atoms, rather than solid solutions. Finally, structures with hybrid orbital bonding with a significant component of d electron character tend to mix less well with some other elements.

Space-filling is important in alloy phases that favor high densities. Coordination numbers of 12, 14, 15, and 16 are found in bimetallic compounds in which n B atoms surround larger A atoms. Examples of some of these phases are given below.

The common *brass* used in hardware, electrical fixtures, and jewelry is an alloy of copper and zinc. Useful compositions range from 55 to 95% Cu. Copper-rich alloys have the rich reddish-gold colors of copper and are soft and malleable. Brass alloys are

Table 10.5 Typical compositions of austenitic specialty steels (atomic%)

Material	Al	Co	Cr	Fe	Mo	Ni
Nichrome heating elements	0	0	15	20	0	65
Heat-resistant alloys	0	0	16	53	6	25
Corrosion-resistant alloy (Hastelloy A)	0	0	0	20	20	60
Alnico magnetic alloys	11	5	0	57	0	27
High-permeability permalloy	0	0	0	21.5	0	78.5
Non-magnetic steel	0	0	0	83	5	12
Hardened steel	0	0	1	96	<1	2

highly formable, either hot or cold, and provide moderate strength and conductivity. Moreover, the alloys have a pleasing brass color and are amenable to polishing, plating, and soldering. The most widely used composition, near 30% Zn (cartridge brass), is used to make ammunition, doorknobs, and musical instruments.

Cu–Sn-based alloys are often referred to as *bronzes*. Because Sn is much larger in size than Cu, second phases begin to form if too much Sn is added to the alloy. The addition of ~8 mol% Sn to Cu significantly increases the hardness of the metal, as well as its strength. Many commercial bronzes are modified with Zn or P.

Since γ face-centered cubic iron forms a complete FCC solid solution with nickel and cobalt, there is no sharp break between austenite steels and the Ni–Co–Fe alloys used in a number of important specialty products. Typical compositions are given in Table 10.5.

Nichromes are widely used electrical heating elements. In addition to high electrical resistance, electrical heating elements must have good high-temperature toughness and high "scaling" temperatures. Resistance to oxidation is often expressed as a *scaling temperature* where thick oxide films (or scale) begin to form rapidly. For most mild steels the scaling temperature is about 550 °C. Nichromes have scaling temperatures of 1200 °C. As in stainless steels, the oxidation resistance of chromium is responsible for this dramatic improvement.

Heat-resistant alloys are also of great interest for gas turbines, jet propulsion, and rocket motors. Some of these alloys are austenitic steels modified with molybdenum or tungsten. Ductility is important to enhance machinability and toughness. Carbide-forming elements are avoided because of their brittleness. A typical composition is given in Table 10.5. Mo and W have much higher melting points than do the 3d transition metals.

The *Hastelloys* are a group of corrosion-resistant Ni–Mo alloys. Hastelloy is resistant to attack by sulfuric and hydrochloric acids. Other members of the family have been developed for use in environments rich in chlorine and nitric acid.

10.6   Intermetallic Compounds

Many metals have structures that can be regarded as *derivatives* of one of the three principal metal crystal structures, in which ordered substitutions are made. For example,

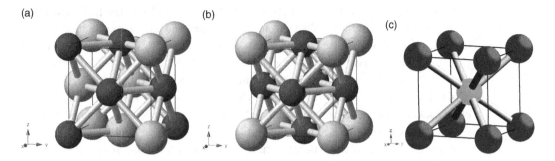

Fig. 10.19 (a) High-temperature disordered Cu_3Au phase; (b) low-temperature ordered phase Cu_3Au, a derivative of the FCC structure; and (c) CsCl, a derivative of the BCC structure.

ordered Cu_3Au is related to the FCC structure; the Cu atoms are placed on the face centers, while Au atoms are located at the corners of the cubic unit cell (see Fig. 10.19). Because the atoms at the face centers and the cell corners differ, the material no longer has a face-centered cubic lattice. This ordered arrangement of atoms is favored at lower temperatures. At higher temperatures, a more random distribution of the Cu and Au atoms on the available sites is favored, and the FCC lattice is recovered. A commercially important material with the ordered Cu_3Au structure is Ni_3Al, a precipitate in nickel-based superalloys that increases the hardness.

Likewise, the CsCl structure (see Fig. 10.19) is a derivative of the BCC structure, in which Cs is located at the body center, with Cl atoms at the cube corners. This is the same crystal structure as is observed in β-brass, (β′-CuZn). This type of structure appears in many other ordered alloys as well, including β-AlNi and β-NiZn.

Other intermetallics have structures similar to those of covalent and ionic crystals. In many cases, this is a result of chemical bonds with mixed metallic–covalent or metallic–ionic character. The rocksalt crystal structure (Chapter 9) is adopted by many alkaline-earth chalcogenides like MgSe or CaTe. The fluorite structure is adopted by other intermetallics including Mg_2Si, Cu_2S, and Na_2S. These compounds follow the $8 - N$ rule governing the covalent semiconductors like GaAs (III–V), and ZnS (II–VI), with the tetrahedral zincblende or wurtzite structures (Chapter 8). The nickel arsenide structures (Chapter 9) are another important member of this family; Ni, Co, Fe, Mn, and Cr frequently bond with group B metalloids such as S, Se, Te, As, Sb, and Bi to form this structure, which can be visualized as alternating layers of Ni and As.

Two generalizations apply to these normal valence compounds: many metals form normal valence compounds with group B elements of the periodic system, especially with columns IVB, VB, and VIB, and occasionally with IIIB elements such as In and Ga as well. The second generalization concerns electronegativity. The greater the difference in electronegativity between the metal and the group B element, the greater the stability of the normal valence compound.

The relationship between e/a ratios and crystal structures have been identified for intermetallic alloys, such as the C–Zn and Cu–Sn systems known as brass and bronze. There are three intermetallic compounds on the Cu–Zn phase diagram, the β, γ, and ϵ

electron phases with electron-to-atom ratios of 3/2, 21/13, and 7/4. The β phase is BCC with equal amounts of monovalent Cu and divalent zinc and an electron-to-atom ratio of (1+2)/2. A transition to the ordered β′ state takes place near 450 °C. The Cu_5Zn_8 (γ), and $CuZn_3$ (ε) phases have electron-to-atom ratios of 21/13 and 7/4, respectively. The γ-brass structure has a distorted BCC structure made up of 27 β-brass cells containing 52 Cu and Zn atoms and two vacancies. The ε electron phase is based on hexagonal close-packing. The electron compounds also turn up in the Cu–Sn bronzes. Cu_5Sn, Cu_3Sn_8, and Cu_3Sn have electron: atom ratios of 3/2, 21/13, and 7/4, respectively.

When two or more elements are combined, the smaller the differences there are in size and electronic structure of the constituent elements, the more likely the elements are to form a solid solution, where atoms randomly substitute for each other on the positions of the lattice (as will be described in Chapter 15). For example, Cu and Au both have the FCC crystal structure. Above 400 °C, the Cu and Au atoms randomly occupy the sites in the FCC lattice. As the difference in electronic structure rises, new alloy structures are adopted, typically with wide levels of solubility. Cn and Zn, for example, form an ordered CsCl structure, as shown in Fig. 10.19. However, larger difference in the electronic structure leads to decreases in the extent of substitutional solid solution, and the appearance of line compounds. Hundreds of intermetallic compounds involve close-packed arrangements of larger and smaller spheres.

The Laves phases are distinct, near-stoichiometric compositions, with the general chemical formula AB_2. Illustrations of three Laves phases are given in Fig. 10.20. $MgCu_2$ has a cubic crystal structure in which the Mg are in a diamond-like arrangement, and tetrahedral clusters of Cu are interspersed. Each Mg atom has four Mg neighbors and 12 Cu neighbors. This is the most common structure of the Laves phases. Ideally the atoms are all in contact when the radii of the metal atoms have the ratio 1.223. For $MgCu_2$, the ratio is 1.25, slightly larger. Figure 10.21 shows a section of the structure on the (110) plane. The dense packing of atoms with two different sizes can be seen. $MgZn_2$ forms a related structure based on a lonsdalite framework (the hexagonal version of C where the six rings are superimposed). In $MgNi_2$, there is a mixture of the cubic and hexagonal stacking; this complex arrangement is the least common of the crystal structures for Laves phases. Laves phases often occur when atom A has a strong tendency towards metallic bonding, and B is a transition metal with an incompletely filled inner electron shell, which has an atomic size at least 20% smaller than the A atom. Most Laves phases have a relatively small composition range over which the composition can vary. A number of important magnetostrictive materials such as $Tb_{1-x}Dy_xFe_2$ (often called Terfenol-D) adopt the crystal structure of one of the Laves phases.

The crystal structures of alloy phases are difficult to predict because the differences in the cohesive energies are small and a number of different factors are involved. Three of these factors are electron concentration, atomic size, and electronegativity. The electron concentration (bonding electrons per atom or e/a) involves the various Hume-Rothery rules described in Section 10.2. The atomic sizes and radius ratios are determined from interatomic distances and are related to space-filling concepts. Electronegativities (Chapter 3) govern the formation of intermetallic compounds containing electropositive and electronegative elements.

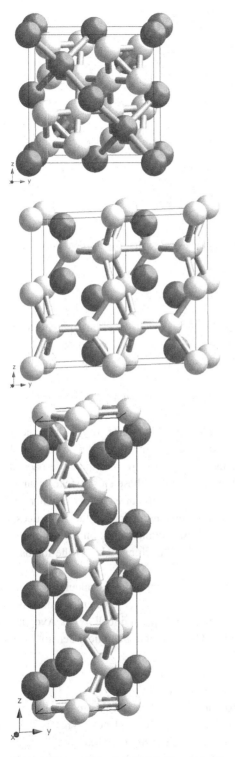

Fig. 10.20 Laves phases: (top) $MgCu_2$, with Mg as the darker atoms, and the Cu as the light atoms; (middle) $MgZn_2$; and (bottom) $MgNi_2$. Some of the bonds have been left out for clarity.

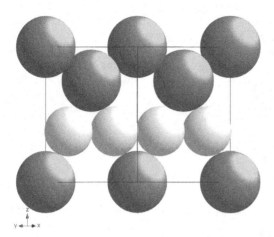

Fig. 10.21 The (110) plane of the $MgCu_2$ Laves phase showing the dense packing of atoms.

10.7   Problems

(1) Cu has the face centered cubic (FCC) metal structure (space group 225: $Fm\bar{3}m$ with a = 3.61 Å) and Cu at (0,0,0) and all symmetry related positions.

   (a) Draw a unit cell of this structure showing the bonds that connect the Cu atoms.

   (b) Show the coordination (i.e. the number and geometric arrangement of near neighbors) of a Cu atom.

   (c) Plot the atom arrangement on the (111) plane. If you draw the figure so that the atoms are shown at 100% of their metallic radius, you should see that the atoms are touching. This is often referred to as the close-packed plane.

   (d) Show the stacking of at least three of these close-packed planes.

   (e) Interstices are "holes" in the crystal structure (i.e. places where no atoms are). There are two main coordination numbers for the interstices in the FCC structure. What are these? Show the interstices on your drawing.

   (f) If you assume that the atoms are touching, what is the radius of a metallic Cu atom?

   (g) The theoretical density of a crystal can be calculated once the crystal structure is known. The formula is: $\rho = \frac{MZ}{N_A V}$, where ρ = density, M = atomic mass, Z = number of formula units per unit cell, N_A is Avogadro's number, and V = unit cell volume. Calculate the theoretical density (in g/cc) for Cu, showing all steps.

(2) Au has the FCC structure, with a = 4.08 Å.

   (a) Calculate the theoretical density of Au.

   (b) Sketch the (100), (110), and (111) planes of this material.

   (c) Suppose one made surfaces using these planes. What is the coordination number of a surface atom on each of these planes?

   (d) Which would you expect to be the equilibrium face of the material?

(3) Consider the crystal structure of Mg, a hexagonally close-packed metal. The space group is $P6_3/mmc$, $a = 3.2$ Å, and $c = 5.2$ Å. The atoms are at (0.667, 0.333, 0.25) and symmetry-related positions.

    (a) Plot several unit cells, showing the bonds that connect the atoms together (use metallic radii).

    (b) Compare the Mg coordination with that of Cu. Specify both number of neighbors and their geometric arrangement.

    (c) How does the stacking sequence of close-packed planes differ for HCP and FCC metals?

    (d) Identify the close-packed plane.

    (e) Identify the three slip directions on this plane, showing them on a diagram.

    (f) Is $[\bar{2}11]$ parallel to or contained in (315)?

    (g) Prove that if a crystal has an HCP structure, and all of the atomic sites are occupied by hard spheres of the same size which are in contact with each other, that the c/a ratio will be 1.633.

(4) Fe is a prototypical example of the body-centered cubic (BCC) crystal structure, with space group 229: $Im\bar{3}m$ with $a = 2.8664$ Å and Fe at (0, 0, 0) and all symmetry-related positions.

    (a) Draw a unit cell, showing the bonds that connect the Fe atoms.

    (b) What is the coordination number of the Fe? Now consider both near neighbors and next-nearest neighbors. What is the coordination number if both of these are taken into account?

    (c) Plot several unit cells worth of the (110) plane. Compare this to your drawing in Problem (1c).

(5) Cementite, Fe_3C, is an important phase in the Fe–C system. The crystal structure is orthorhombic, space group Pnma, with $a = 5.081$ Å, $b = 6.753$ Å, and $c = 4.515$ Å. The Fe are located at (0.1846, 0.0594, 0.3340) and (0.0346, 0.25, 0.8377) and all symmetry-related positions, while the C are at (0.8984, 0.25, 0.4467) and all symmetry-related positions.

    (a) Plot a unit cell, showing the Fe–Fe bonds.

    (b) Draw the stereographic projection for the point group for cementite.

    (c) Determine the distance between Fe atoms at (0.6847, 0.4406, 0.166) and the one at (0.8154, 0.5594, 0.666). Show your work.

    (d) Is your calculation in part (c) consistent with the metallic radii of Fe?

    (e) Explain the difference between this material and an interstitial solid solution of Fe and C.

(6) How does dislocation motion affect the mechanical properties of metals?

(7) Explain close-packing in metals. Be specific about the number of near neighbors, their geometric arrangement and the stacking sequences observed.

(8) How do close-packing and structure influence slip?

(9) Describe and contrast the crystal structures of MgO and Si. Describe the main features that you would expect in the electron density maps of these two materials. Justify your reasoning.

(10) Do the following for a material that undergoes slip:

  (a) Draw the crystal structure.

  (b) Indicate a slip plane and slip direction on the crystal structure, labeling (*hkl*) and [*uvw*].

  (c) Discuss the factors that make these the most likely slip system for the material you chose.

  (d) Give an example of a material with this crystal structure.

11   Molecular Crystals

In some materials, the bonding within a group of atoms is strong, while the bonding between these groups is weaker.[1] In *molecular crystals*, these groups of atoms (i.e. molecules) can be readily identified as discrete units. This is in contrast to crystal structures like rocksalt, in which each Na atom is shared equally with *six* Cl neighbors. Thus, one cannot identify individual NaCl "molecules" in the rocksalt structure. The molecules in molecular crystals are typically held together with either hydrogen or van der Waals bonding. This chapter will describe the intermolecular bond types, as well as the crystal structures of several molecular crystals and small molecules.

11.1   Hydrogen and van der Waals Bonding

Hydrogen forms strong, largely covalent bonds with many elements. Indeed, since hydrogen is quite small as an atom, one can often picture it as partially burying the atom core in the electron cloud of the atom to which it makes a primary bond. In addition to this, there are also weaker secondary bonds (the hydrogen bond) with nearby electronegative atoms like O, F, and N. These hydrogen bonds are strongly directional. That is, the H acts as though it is partially positive, and it weakly attracts adjacent electronegative atoms. Typically, the H-bond strength is on the order of 5 to 40 kJ/mole, roughly ten times stronger than a van der Waals bond, and approximately 1/10th the strength of a typical covalent bond. The strength of the hydrogen bond rises as the anion–anion separation distance decreases. In the case of an oxide, hydrogen bonds are observed when the separation distance between oxygens drops to ~3.1–3.2 Å. Hydrogen bonds are important in water, many organics and polymers, in biology, and in many hydrated minerals.

The second main type of secondary bond is given the name van der Waals bonding; this type of attraction occurs whenever any atoms, molecules, or surfaces approach closely. As a result, van der Waals attractions also occur whenever other types of bonding (e.g. ionic, covalent, metallic) are also present.

There are several potential sources to the attraction. First, *London forces* are momentary dipoles which arise from the instantaneous positions of electrons around a nucleus.

---

[1] Portions of this chapter are adapted from R. E. Newnham, "Phase diagrams and crystal chemistry," in *Phase Diagrams: Materials Science and Technology, Volume V*, Academic Press, New York (1978) (by permission).

Very often, the time-averaged position of the electron will produce an approximately spherical electron cloud around an atom. However, at any instant in time, the center of positive charge on the atom (the nucleus), may not be at exactly the same position as the center of negative charge for the atom). This will produce a net dipole moment, $\vec{p_1}$. Over time, the dipole averages to zero, but an instantaneous dipole is possible. This momentary dipole produces an electric field which is proportional to $\frac{p_1}{r^3}$, at a distance r. This electric field can induce a dipole moment on an adjacent atom 2, $p_2 = \alpha E$, where α is the polarizability of the electron cloud of atom 2, and E is the electric field. The resulting energy of the interaction $U_{London} = -\frac{3}{4}\frac{h\nu\alpha^2}{(4\pi\varepsilon_o)^2 r^6}$. In this equation, ν is the vibrational frequency of the bond (which typically on the order of 3×10^{15} Hz), and h is Planck's constant. Because the polarizability of the electron cloud increases as the atom size increases, larger atoms often have stronger van der Waals attractions. Typically, this produces an interaction energy on the order of 0.5 to 2 kJ/mole.

Debye interactions are the induced force that arises from the interaction of a polar molecule with the electrons of surrounding materials. For a molecule with a permanent dipole moment u, the energy of the interaction is given by $U_{Debye} = -\frac{u^2\alpha}{(4\pi\varepsilon_o)^2 r^6}$.

Keesom forces describe the attraction between two molecules with permanent dipole moments u_1 and u_2. Typical examples of molecules with permanent dipole moments include CO, NO, HCl, NH_3, and HF. Here, the attractive energy is given by: $U_{Keesom} = -\frac{2u_1^2 u_2^2}{3k_B T(4\pi\varepsilon_o)^2 r^6}$. Some authors describe hydrogen bonds as an example of Keesom forces.

Collectively, these three sources of attraction produce van der Waals bonding. It is important to note that for all of these interactions, the attraction falls off very rapidly with increasing separation distance.

In molecularly bonded solids, it is important to account for all of the atoms/molecules which interact. In many cases, there is a driving force to maximize the number of van der Waals bonds, in the absence of stronger bonding. This, in turn, will tend to favor close packing of molecules.

11.2   An Example Molecular Crystal: Sucrose

Sucrose – ordinary table sugar – is an excellent example of a molecular crystal. A typical American eats his or her own weight in sugar every 18 months, about three times the world average. As a result, sugar refining has developed into an enormous industry producing millions of tons of sucrose per year. Synthetic sucrose is rather expensive, so most table sugar comes from processed sugar cane and sugar beet. No other crystalline organic compound of comparable purity (better than 99.9%) has been offered to the market in such huge volumes at such a modest price. The sugar industry is the leading practitioner of crystallization from solution. Technically, the industry pioneered the development of the vacuum crystallizer as well as many other unit operations and unit processes of modern chemical engineering.

Chemically, sucrose is classified as a disaccharide made up of glucose and fructose rings linked by oxygen (see Fig. 11.1). The crystal structure of table sugar is illustrated

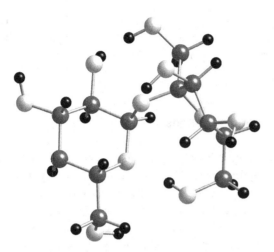

Fig. 11.1 The molecular structure of sucrose, ordinary table sugar. The chemical formula is $C_{12}H_{22}O_{11}$. Hydrogen is shown in small black circles, carbon in larger darker circles, and oxygen in large lighter circles.

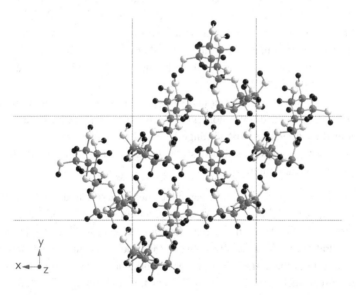

Fig. 11.2 The crystal structure of sucrose (viewed along [001]). Hydrogen is shown in black, carbon in larger dark circles, and oxygen in larger light circles.

in Fig. 11.2. Disaccharides are sweet soluble substances that crystallize readily from solution. At 20 °C, sucrose dissolves in about half its weight of water.

Sweetness is a fascinating property involving receptors on the human tongue. It is believe that these receptors operate through a lock-and-key mechanism. Many sweet-tasting molecules have common geometric and chemical characteristics. These are sometimes referred to as the triangle of sweetness, or Kier's rule: Among organic molecules, sweetness is associated with a triangle made up of two H-bonding groups and one H-repelling groups separated by distances of 3 to 3.5 Å. Presumably, this triangle corresponds to active receptor sites in the human tongue. In sucrose, the constituent groups of the triangle are found on the glucose and fructose rings that make up the molecule (see Fig. 11.1).

11.3   Molecular Packing in Organic Crystals

The bonding within organic crystals is predominantly covalent in character. These bonds satisfy the bonding requirements of the atoms involved, so only weak secondary bonding: hydrogen or van der Waals, holds the molecules together. The covalent bonds within the molecules are a result of the hybrid orbitals and are highly directional. This constrains the covalent bond lengths, and bond angles are very similar from compound to compound. Tetrahedral coordination typically is a result of the sp^3 hybrid orbital; the characteristic bond angles are always close to 109.5°. Triangular coordination develops from sp^2 hybrid orbitals; bond angles are usually within a few degrees of 120°. Typical covalent bond lengths were given in Chapter 8. Because covalent bonds are strong, they also tend to be short. Bonds between molecules are typically controlled by *van der Waals radii*, as shown in Table 11.1. These radii are often roughly double the radii for the same atoms in covalent bonds. In organic crystals, it is observed that the carbon atoms in different molecules come no closer than 3.4 Å, so that an effective inter-molecular radius for carbon is 1.7 Å. Accurate molecular drawings can be constructed using covalent and van der Waals radii.

Hydrogen bonds tend to be shorter and stronger than van der Waals bonds due to the existence of a permanent dipole moment. Characteristic separation distances between O atoms joined by a hydrogen bond range from 2.4 to 3.2 Å.

There are a number of trends that describe the systematics of how small molecules tend to align in crystal structures.

(1)   In cases where the bonding between molecules is largely hydrogen bonding, where OH^-, H_2O, or NH_3^+ are involved, steric conditions usually produce approximately a tetrahedral coordination for the electronegative atom. The angles around the tetrahedral atom are generally between 90° and 120°; in cases where the angle falls outside of this range, the hydrogen bond tends to be long and weak.

(2)   The closest packing of molecules lowers the free energy. This often produces configurations where each molecule is in contact with 10 to 14 other molecules. An excellent example of this is the crystal structure of adamantine, $C_{10}H_{16}$.

Table 11.1 Van der Waals radii for molecular crystals, after Pauling

H	1.2 Å	N	1.5 Å	O	1.4 Å	F	1.35 Å
C	1.7	P	1.9	S	1.85	Cl	1.80
		As	2.0	Se	2.00	Br	1.95
		Sb	2.2	Te	2.20	I	2.15

Source: L. Pauling, *General Chemistry*, New York, Dover Publications (1988).

(b)

(a)

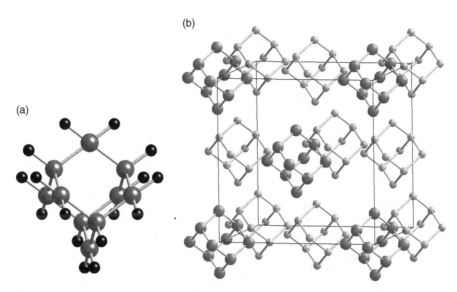

Fig. 11.3 Adamantane: (a) molecule, with the C shown as the larger tetrahedrally coordinated atoms and the H as the smaller atom in black; and (b) face-centered cubic structure of adamantine molecules. Hydrogens are not shown for clarity.

Figure 11.3 shows both a single molecule as well as the packing in the molecular crystal. It is clear that the carbon arrangements in the molecule look like a small piece out of the crystal structure of diamond, with the molecule terminated by hydrogen atoms. These roughly spherical units pack together in a face centered cubic crystal structure in which each molecule has 12 other molecule neighbors.

Close packing arrangements like this are common in molecular crystals, even when the molecules are not spherical. Figure 11.4 shows a close-packed plane in aspirin; the central molecule (and indeed all molecules) is surrounded by six others.

(3)   Organic molecules are generally constructed so that the projections on one molecule fit into indentations in adjacent molecules. This helps to maximize the packing density, and hence the number of van der Waals bonds that can be made.

(4)   Straight-chain hydrocarbons have zig-zig backbones and pack with the chains aligned along a common axis, often in hexagonal arrangements (Fig. 11.5). Planes where the molecules terminate are typically cleavage planes. It is important to note that most of the bonding energy is tied up in the covalent bonds within the molecules. Thus, changes in the molecule stacking cost little free energy (only the hydrogen or van der Waals bonding is changed), and so polymorphism is common. The number of carbons in the backbone of the chain also influences the packing: odd-numbered molecules from $C_{19}H_{40}$ to $C_{29}H_{60}$ have orthorhombic crystal systems, while even-numbered molecules from $C_{18}H_{38}$ to $C_{36}H_{74}$ are triclinic or monoclinic. Chains with –COOH or –NH$_2$ end groups typically pack together head to head.

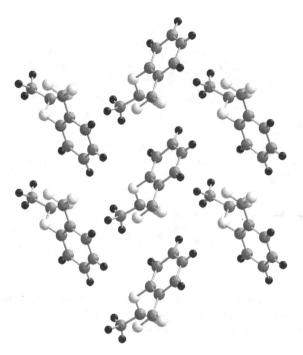

Fig. 11.4 Close packing in aspirin, $(HOOC)C_6H_4$-$OC(O)CH_3$. Carbons are shown as large darker circles, oxygen in large lighter circles, and hydrogen in black. The arrangement of six molecules around the central molecule within one plane is clear.

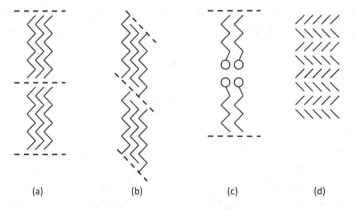

Fig. 11.5 Crystalline long-chain hydrocarbons pack together so as to maximize density. Multiple packing arrangements are often possible (a) and (b) since there is little free energy difference between polymorphs. When the end group of the molecule is polar (c) head-to-head arrangements may occur, while herringbone patterns (d) are observed in aromatic hydrocarbons. Adapted, with permission, from R. E. Newnham, "Phase diagrams and crystal chemistry," in *Phase Diagrams: Volume V*, ed. Allen M. Alper, Academic Press, New York (1978). Fig. 14, p. 34 & Fig. 16, p. 36. With permission by Elsevier.

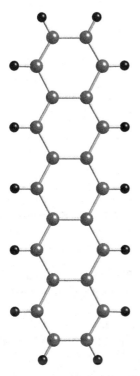

Fig. 11.6 Molecule of pentacene: large circles = C; small circles = H. The C–C bond length = 1.35–1.48 Å; C–H bond length = 1.1 Å; C coordination = 3; and H coordination = 1.

(5) Molecules with aromatic rings (e.g. benzene, anthracene, naphthalene) are generally planar. Higher densities and higher numbers of bonds can be achieved if these molecules crystallize face-to-face in staggered herringbone arrays. An excellent example of this is the molecular crystal pentacene: $C_{22}H_{14}$. One molecule has the structure shown in Fig. 11.6. Here, the C atoms are linked by resonating covalent bonds. The molecule structure looks like five benzene rings that are co-joined along a common line. As such, it belongs to the benzologue series:

> benzene: one ring
> naphthalene: two rings
> anthracene: three rings
> tetracene: four rings
> pentacene: five rings.

Because of the relatively small size of the molecules, as well as the weak intermolecular bonding, pentacene crystallizes easily. A projection of the unit cell is shown in Fig. 11.7. It is clear that the long axis of the $C_{22}H_{14}$ molecules line up parallel to each other in the structure with a herringbone arrangement. The unit cell of one of the polymorphs of pentacene has been reported to be triclinic with: $a = 7.93$ Å, $b = 6.14$ Å, $c = 16.03$ Å, $\alpha = 101.9°$, $\beta = 112.6°$, and $\gamma = 85.8°$. Its theoretical density is 1.303 g/cm^3.

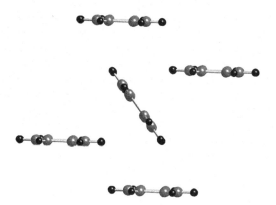

Fig. 11.7 Five pentacene molecules viewed end-on to show the stacking arrangement in the crystal.

Pentacene has recently received considerable interest for electronic applications. Thin films of pentacene have useful semiconducting properties, and have been utilized for flexible thin-film transistors.

The packing densities of molecules can be determined from the molecular volume, V_m, calculated using van der Waals radii. V_m is the sum of the volumes associated with the atoms in the molecule, where to a first approximation, all of the atoms are treated as spheres. Since van der Waals radii are longer than the covalent radii appropriate for bonds within the molecule, it is important to correct for "overlap" of the spheres.

For the molecule AB shown in Fig. 11.8: $V_m = V_A + V_B$ and $V_A = \frac{4}{3}\pi R_A^3 - \frac{\pi}{3}h_{AB}^2(3R_A - h_{AB})$, where $h_{AB} = R_A - \frac{(R_A^2 + d_{AB}^2 - R_B^2)}{2d_{AB}}$. A comparable calculation can be done for V_B. In aromatics with C–C = 1.40 Å and C–H = 1.08 Å, V_C is about 8 Å^3, and V_{C-H} about 14 Å^3. The molecular volume calculated in this way for anthracene, $C_{14}H_{10}$, is $4(8) + 10(14) = 172$ Å^3.

The packing efficiency, k, of the unit cell is the sum of the volume occupied by molecules over the volume of the unit cell, V. Thus, $k = ZV_m/V$, where Z is the number of molecules per unit cell. For anthracene, $k = 73\%$. This is only slightly less efficient than close-packing of spheres (as in the FCC and HCP metals), where $k = 74$ %. The ideal packing efficiency for parallel cylinders is over 90%; calculated efficiencies for aromatic compounds range from 60 to 80%. Higher packing densities are observed when the shapes of the molecules are more spherical or ellipsoidal in character. Thus, anthracene has a higher k value than phenanthrene, and hence a higher density (see Fig. 11.9). The more irregularly shaped a molecule is, the more challenging to pack the molecules together efficiently. One consequence of this is that irregularly shaped molecules are more likely to form amorphous, rather than crystalline solids.

The symmetry of a molecule in a crystal is often lowered relative to the same molecule in the gas state. This is a consequence of maximizing the packing density in the crystal. Close-packing of molecules is easier to achieve when there are more degrees of freedom for the molecule placement. The more symmetry the crystal has, the fewer

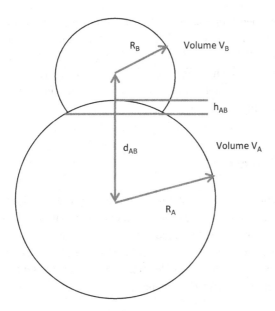

Fig. 11.8 Schematic drawing of a molecule AB showing overlapping spheres from use of van der Waals radii. Adapted, with permission, from R. E. Newnham, "Phase diagrams and crystal chemistry," in *Phase Diagrams: Volume V*, Ed. Allen M. Alper, Academic Press, New York (1978). Fig. 16, p. 36. With permission by Elsevier.

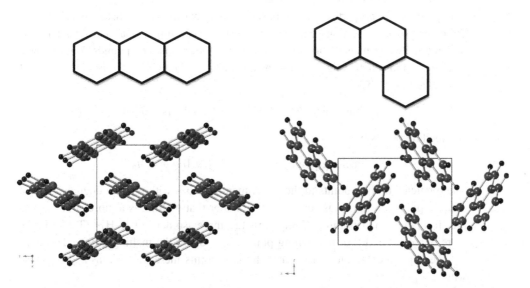

Fig. 11.9 Schematic showing the arrangement of carbon six rings in (left) anthracene and (right) phenanthrene. In the first, the carbon six rings are bonded in a line, in the second, they are not. The consequences in terms of molecule packing are shown in terms of the crystal structures underneath. The theoretical densities are 1.25 and 1.22 g/cm^3 for anthracene and phenanthrene, respectively.

degrees of freedom are retained. For example, a molecule in a general position has six degrees of freedom; for $\bar{1}$, three rotational degrees; m has one rotational and two translational; 2, one of each; and $2/m$, just one rotational degree of freedom. As a result, 1 and $\bar{1}$ are the preferred symmetries for close packing. Molecules tend to retain inversion symmetry in the solid state, but other symmetry elements are often lost unless the molecule had unusually high symmetry.

11.4   Structure–Property Relations

Organic chemists have long recognized that "function follows form," which means that the properties and behavior of organic molecules can be correlated with molecular geometry. The simplest approach is to represent molecular structure by a set of numbers, termed geometric indices. Two of the simplest indices for hydrocarbons are n_C, the number of carbon atoms in the molecule, and t_m the number of terminal methyl groups. As an example, let us consider isopentane.

$$CH_3$$
$$|$$
$$H_3C - CH - CH_2 - CH_3$$

Isopentane contains five carbons ($n_C = 5$) and three methyl terminations ($t_m = 3$). Elongated chain molecules have only two terminal methyl groups, while compact globular hydrocarbons may have more.

Representative properties of several saturated hydrocarbons are listed in Table 11.2. Structure–property relations can be quantified by carrying out a statistical analysis utilizing the experimental data for large numbers of organic compounds. With n_C as a measure of molecular weight, and t_m as a measure of branching, the boiling points of alkanes are given by

$$\text{boiling point (}°C\text{)} \cong -126.19 + 33.42n_C - 6.286t_m$$

and melting points by

$$\text{melting point (}°C\text{)} = 190.0 + 9.1n_C + 4.3t_m.$$

The predicted boiling points fit the experimental values within a few degrees, but the melting points do not. This serves as a reminder that melting is a more subtle phase transformation than boiling. Comparing the three isomers of C_5H_{12} (Table 11.2), there is a clear trend for the boiling points, but not for the melting points. Throughout the experimental data for alkanes, boiling points increase with n_C and decrease with t_m.

Heats of combustion (ΔH_c) and heats of vaporization (ΔH_v) can also be modeled successfully using a simple statistical analysis of the experimental data:

$$\Delta H_c(\text{kcal/mol}) = 48.67 + 147.12n_C - 1.29t_m,$$

$$\Delta H_v(\text{kcal/mol}) = 759.60 + 1246.09n_C - 390.04t_m.$$

Table 11.2 Selected properties of several saturate butanes (C_4H_{10}) and pentanes (C_5H_{12})

Compound	n-butane	i-butane	n-pentane	i-pentane	neo-pentane
Structure					
n_C	4	4	5	5	5
t_m	2	3	2	3	4
Melting point (°C)	−138.3	−159.4	−129.67	−159.77	−16.4
Boiling point (°C)	−0.50	−11.73	36.07	27.85	9.50
Molar volume (ml)	100.42	104.31	115.21	116.43	122.07
Critical temperature (°C)	152.01	134.98	196.62	187.8	160.60
Heat of vaporization (kcal/mol)	5035	4570	6316	5878	5205
Heat of combustion (kcal/mol)	635.05	633.05	782.04	780.12	777.37

Combustion of an alkane containing n carbons proceeds through a reaction with oxygen molecules to form water and carbon dioxide

$$C_nH_{2n+2} + \frac{1}{2}(3n+1)O_2 \rightarrow (n+1)H_2O + nCO_2.$$

This reaction involves breaking $(n-1)$C–C bonds, $(2n+2)$C–H bonds and $\frac{3n+1}{2}$O=O bonds, followed by formation of $2(n+1)$O–H bonds and $(2n)$C=O bonds. These are all strong covalent bonds: the weaker van der Waals forces and hydrogen bonds between molecules contribute little to the heat of combustion. As a result, the values of ΔH_c change linearly with n_C and are almost independent of t_m. The data in Table 11.2 bear this out.

This is not true for boiling points, molar volumes, critical temperatures, and the heat of vaporization. Here we see that there are clear trends with structure. More spherical isomers like neopentane have lower boiling points, lower critical temperatures, lower heats of vaporization, and higher molar volumes (lower densities). It is interesting to speculate why this is true. The shape of n-butane, n-pentane, and other chain hydrocarbons resemble closely packed cylinders while the other isomers are more globular in shape (resembling spheres). Close-packed cylinders fill 90.8% of space compared with 74.1% for close-packed spheres (Fig. 11.10). Thus, we expect cylindrically shaped molecules like n-butane and n-pentane to have smaller molar volumes (higher densities) than their isomers. The data in Table 11.2 confirm this expectation.

Packing effectiveness also helps explain the trends in boiling points, critical temperatures, and heats of vaporization. These properties depend on the chemical bonds between molecules rather than those within molecules. Van der Waals bonds, the dominant forces between molecules, depend on the sixth power of separation, so that

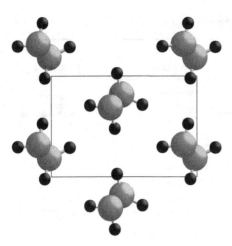

Fig. 11.10 End-on view of the hydrocarbon $C_{36}H_{74}$ looking down the c axis. Straight-chain hydrocarbons pack together like tiny gas cylinders. Carbon: large filled circles, H: small circles. Outline of unit cell shown.

densely packed molecules are held together more strongly than those which are not. Therefore, the cylindrically shaped isomers like *n*-pentane have higher boiling points, critical temperatures and heats of vaporization than the more globular isomers.

More complex shape indices have been introduced by organic chemists to describe molecular branching. The Wiener index and the Randic Topological index provide more detailed structure–property relationships. These indices have been applied to a wide variety of physical, chemical, and biological properties. Among these are solubilities, molar refractions, heats of formation, and a number of medical applications, including anesthetics, narcotics, enzyme inhibitors, and structure–activity relationships of interest in drug design.[2]

As will be discussed in Chapter 15, substitutional solid solution is less common in molecular crystals than in extended network crystals. This is not surprising, since a differently shaped molecule cannot be readily slotted into a space that is designed for another shape. Indeed, this poor solid solution can be used to purify molecular solids. The drug industry, for example, often utilizes crystallization for purification, since other molecules don't incorporate into the crystalline solid.

11.5   Hydrogen in Solids

The hydrogen atom has only one electron and plays a very special role in crystal chemistry. If it gains a second electron, completing the $1s^2$ shell, an H^- anion is formed. Lithium hydride (LiH) with the rocksalt structure, is an example.

[2] P. G. Seybold, M. May, and V. A. Bogal, "Molecular structure–property relationships," *J. Chem. Educ.*, 64, 575–581 (July 1987).

More commonly, however, hydrogen loses its electron to become an H^+ cation. At the same time, it loses its size, becoming a bare proton, and burying itself inside the electron shell of neighboring atoms. The C–H bond in methane and the O–H bond in the hydroxyl ion are two important examples. It should be noted, however, that it is difficult to completely remove the electron from the hydrogen, so there is often a strong degree of covalency in these bonds.

Several examples of hydrogen-bearing compounds are illustrated in Fig. 11.11. NaH and other *salt-like hydrides* are formed by heating the metals in hydrogen, forming colorless crystals. The crystals have the rocksalt structure and resemble alkali halides in many ways. The ionic radius of the H^- anion is about 1.4 Å, similar to those of O^{2-} and F^-. The hydrides react readily with water, evolving hydrogen gas in the process. The compounds of hydrogen with transition metals are very different from those with alkali or alkaline-earth metals. The alloys of transition metals behave like interstitial solid solutions with hydrogens occupying small interstitial cavities between the close-packed metal atoms. In the Ti–H system, for instance, titaniums form a face-centered cubic lattice with hydrogens in tetrahedral interstices. The structures of TiH and TiH_2 are isostructural with zincblende (Chapter 8) and fluorite (Chapter 9), respectively.

If a proton is removed from a water molecule, a *hydroxide ion* (OH^-) is formed. The hydroxide ion is about the same size as an oxygen ion and is a very common constituent of minerals. Portlandite, $Ca(OH)_2$, is hydrated lime (CaO), and an important component of cement (see Chapter 21).

If a proton is added to a water molecule, a *hydronium ion* (OH_3^+) is formed. In the hydronium ion, all three protons are attached to the oxygen atom by identical covalent bonds. Strong acids such as HCl dissociate in water to form hydronium ions with a characteristic sharp taste. Weak acids like acetic acid produce a smaller number of ions, whereas bases contain hydroxide groups that dissociate in aqueous solutions to form hydroxide ions. The OH^- ions impart a brackish taste to the solution.

Methane (CH_4) is a colorless, orderless gas consisting of a carbon atom linked to four hydrogen atoms by covalent electron pair bonds. In the United States of America, the annual consumption of methane natural gas exceeds two tons per person.

11.6   Ice and Water

Ice I is the crystalline form of H_2O obtained by freezing water at atmospheric pressure. A number of other forms are stable at high pressure. In Ice I, the O atoms occupy the Zn and S positions in wurtzite (see Chapter 8). Every oxygen is therefore surrounded tetrahedrally by four others with an O–O distance of 2.76 Å. The oxygens are bound together by hydrogen bonds. The hydrogen bonds in Ice I yield an asymmetric coordination for the hydrogen atom, in order to maintain the integrity of the water molecule. The O–H distance within a water molecule is 0.96 Å, while the O–H distance between molecules (the hydrogen bond) is 1.8 Å, as shown in Fig. 11.12. One of the key structural features of ice is the randomness in the placement of the hydrogen bonds. Around each oxygen, there are two near hydrogens (making up the water molecule) and two more distant hydrogens (the hydrogen bonds between water molecules).

Salt-like hydride
LiH showing lithium cation (small) and hydride anion (large) in ionic crystal

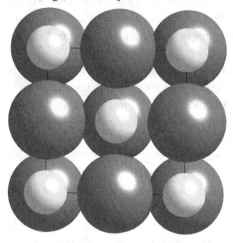

Interstitial hydride
TiH$_2$ showing Ti in a close-packed array (large) with hydrogens in the tetrahedral interstices (atoms show at 50% of their radii for clarity)

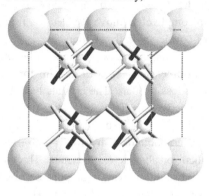

Hydroxyl ions
Ca(OH)$_2$: Ca is shown in intermediate size, medium gray, oxygen in white, and H is the smallest ion

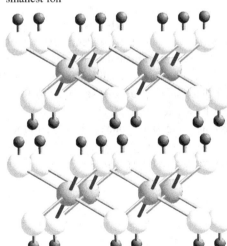

Methane molecules

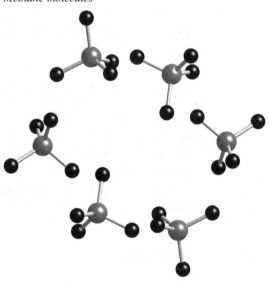

Fig. 11.11 Different examples of hydrogen in materials.

The hydrogens lie along lines connecting the oxygen atoms. There is no more than one hydrogen atom between any one pair of oxygen atoms. However, as shown in the figure, the orientation of the water molecules is otherwise random. This leads to an unusually high entropy in ice.

As pointed out previously, the intermolecular forces in molecular crystals (Section 11.2) are either weak van der Waals or hydrogen bonds. Melting points and boiling points provide evidence for hydrogen bonds in water (Fig. 11.13). The transitions for hydrogen

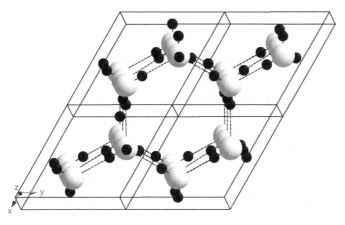

H₂O Molecule
     H – O bond length = 0.96Å
     H - - - O bond length ~ 1.8Å

Crystal System: Hexagonal
Lattice parameters (0°C)
     a = 4.523Å
     c = 7.367Å
Space Group: P6₃/mmc

Fig. 11.12 Crystal structure of ice looking nearly parallel to c, and the crystallographic data. The hydrogens are substantially smaller than the O.

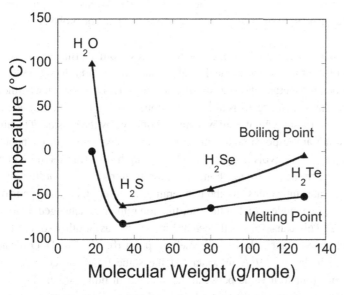

Fig. 11.13 Boiling points and melting points for H₂O, H₂S, H₂Se, and H₂Te plotted as a function of molecular weight. The unexpectedly high values for H₂O are caused by strong hydrogen bonds.

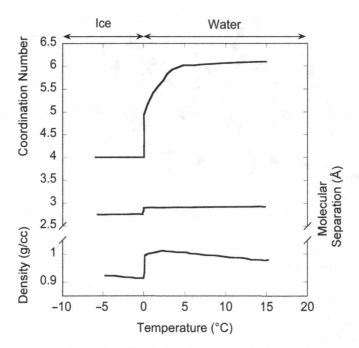

Fig. 11.14 The melting of ice. Schematic diagram showing (a) the change in coordination number, (b) the change in molecular separation, and (c) the corresponding change in density as ice is heated from –5 °C to 15 °C. Adapted, with permission, from D. Tabor, *Gases, Liquids, and Solids*, Penguin Books, Baltimore MD (1969). Fig. 77, p. 205.

sulfide, selenide, and telluride increase smoothly with molecular weight, but the freezing and boiling points of water are far higher than the extrapolation, indicating a change in intermolecular bonding. The weak van der Waals forces are reinforced by stronger hydrogen bonds in water and ice. These bonds must be broken to achieve changes of state.

Additional evidence for hydrogen bonds comes from thermal measurements. The latent heat of melting for ice is much higher than other common substances, which is why it is a good refrigerant; the high specific heat of water is the reason why the day–night temperature changes in oceans is about 1 °C, compared with 15–20 °C over the continents. That's one reason why we enjoy summer holidays by the seaside. The high entropy of ice and water are responsible for the larger specific heat values (see Chapter 24).

The melting of ice involves changes in bond length and near neighbor coordination. In ice, the directional hydrogen bonding leads to tetrahedral coordination with four neighboring water molecules. As the temperature is raised, thermal expansion causes a small increase in bond length, but the coordination remains unchanged up to the melting point of 0 °C. This causes a small decrease in density, as in other solids (Fig. 11.14).

A radical change takes place at the melting point (0 °C). The O–O distance jumps from 2.76 Å in ice to 2.9 Å in water. At the same time, the coordination number increases from four to a value between 4.5 and 5.5 in liquid water (Fig. 11.13). This leads to a 10% jump in density on melting, reaching a maximum at 4 °C. Note that the more efficient packing with about five neighbors around each water molecule is responsible for the increased density in liquid water. This is abnormal behavior. Most materials lose density in transforming from the solid to the liquid state (see Chapter 20).

Because water is an excellent solvent for ionically bonded solids, natural waters are generally dilute solutions with variable amounts of suspended particles and colloidal matter. *Rainwater* is saturated with atmospheric N_2, O_2, and CO_2 plus a nucleus of solid matter on which the rain drops condense. It is slightly acidic, causing it to corrode metal pipes, as well as fragile stones such as marble.

Rivers, lakes, and other forms of *surface waters* are variable in nature, but are often high in suspended mineral particles and organic matter. *Ground waters* are filtered free from dissolved solids but have a high proportion of dissolved minerals. Ca^{2+} and Mg^{2+} are especially troublesome because they react with soap to form an insoluble scum. "Hard" waters such as these are "softened" with ion-exchange media such as zeolites to replace Ca^{2+} with Na^+.

Polluted water contains contaminants, making it unfit for human consumption. Water-borne diseases such as cholera, dysentery, and typhoid are associated with the human wastes in sewage. The simplest and best test is a bacteriological examination for *B. Coli*, a sure sign of pollution by sewage. To ensure sterility, drinking water supplies are filtered and chlorinated to remove large organisms and kill the remainder.

Salt water: Most salts are soluble in seawater. The reason water is so effective as a solvent is its high dielectric constant, more than 80, and the ease with which water molecules cluster around ions, forming hydrated ions. The large dipole moment of the water molecule is responsible for this behavior. In accordance with Coulomb's Law, the attractive forces between anions and cations are inversely proportional to the dielectric constant. Therefore, the attractive force between Na^+ and Cl^- is greatly reduced in water, allowing salt crystals to easily dissociate. One liter of seawater contains about 35 g of dissolved salts: 55% Cl^-, 31% Na^+, 8% SO_4^{2-}, 4% of Mg^{2+}, 1% Ca^{2+} and 1% K^+.

11.7   Organometallic Molecules

There are also families of molecules which have both some combination of "inorganic" and "organic" character. Some of these can be used as precursors to manufacture glasses and ceramics via a "sol–gel" process. A good example of this is tetraethylorthosilicate (TEOS).

With the help of a catalyst, TEOS is converted to silicic acid and ethyl alcohol

$$Si-(OCH_2CH_3)_4 + 4H_2O \rightarrow Si(OH)_4 + 4(OHCH_2CH_3)$$

and silicic acid is condensed to amorphous silica gel by dehydration

$$Si(OH)_4 \rightarrow SiO_2 + 2H_2O.$$

Other organometallics are useful as *coupling agents*. Interfacial bonding is extremely important in polymer composites. High strength requires an efficient transfer of external stress from the polymer matrix to the filler particles or fiber reinforcement. Surface-active coupling agents with two different functional groups form chemical bonds to both the polymeric resin and the filler. Many different alkylsilanes and organotitanates have been developed for this purpose. Silane coupling agents bond oxides to polymers. For bonding epoxies, styrenes, and nylon to glass, a typical molecule is shown as follows:

The coupling agents are converted to the silanol form by reacting with water

$$Cl(CH_2)_3Si(OCH_3)_3 + 3H_2O \rightarrow Cl(CH_2)_3Si(OH)_3 + 3(CH_3OH).$$

Silanol then reacts with the surface of glass in the presence of water. Coupling agents bond to ceramics at one end through an ionic bond (i.e. the O) and to polymers at the other end through a covalent bond.

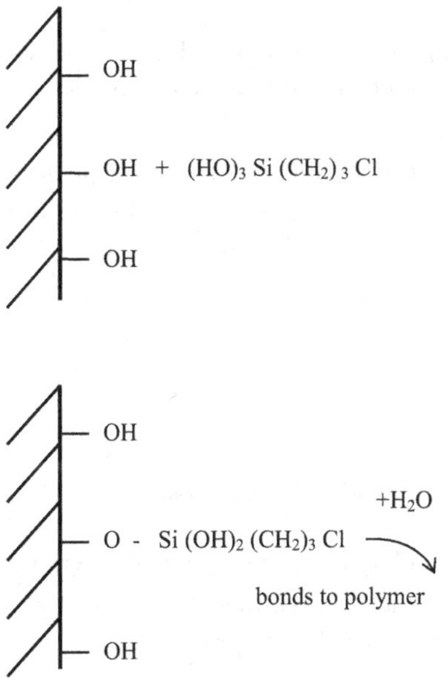

OH

OH + $(HO)_3 Si (CH_2)_3 Cl$

OH

OH

$O - Si (OH)_2 (CH_2)_3 Cl$ $\xrightarrow{+H_2O}$

bonds to polymer

OH

11.8 Problems

(1) Describe the crystal structure of ice I.
   (a) Draw a schematic showing the molecular arrangements, labeling the O–H distances both within a molecule and between molecules.
   (b) Describe the coordination of all of the atoms involved.
   (c) Why is ice considered to be a high-entropy solid?
(2) Describe how the bonding in ice affects the density and melting point.
(3) Outline the process of glass formation from tetraethylorthosilicate $Si–(OCH_2CH_3)_4$.
(4) Methane III is an orthorhombic crystal with space group *Cmca*. The lattice parameters are $a = 11.7079$ Å, $b = 8.1893$ Å, and $c = 8.1842$ Å. The atom positions are given below, in terms of fractions of the atomic coordinates.

Atom	x	y	z
C	0.749 80	0.500 00	1.000 00
H	0.801 00	0.579 50	0.933 60
H	0.698 60	0.566 30	1.079 60
C	0.000 00	0.729 60	0.230 30
H	0.000 00	0.605 90	0.199 20
H	0.000 00	0.741 60	0.357 40
H	0.072 80	0.785 50	0.182 30

   (a) Plot a unit cell of this material; make sure to use covalent radii for the atoms. Also, ensure that all of the molecules are complete.

(b) Calculate the theoretical density.

(c) Determine the C–H distance within a molecule and the C–C distance between molecules. Compare the latter distance with the interatomic distance between C atoms in diamond.

(d) Estimate the hardness of this crystal on Moh's scale, and justify your choice.

(5) Crystalline benzene (C_6H_6) has lattice parameters of $a = 7.46$ Å, $b = 9.666$ Å, and $c = 7.034$ Å. Its space group is *Pbca*. The atomic positions are given by:

C (0.9431,   0.1387,   0.9946)
C (0.8665,   0.0460,   0.1264)
C (0.9226,   0.9075,   0.1295)
H (0.9024,   0.2447,   0.9823)
H (0.7591,   0.0790,   0.2218)
H (0.8629,   0.8369,   0.2312) and all symmetry-related positions.

(a) Draw a benzene molecule (small apparent distortions in the molecule are due to uncertainties in the lattice parameters). Show the C–C and C–H bonds. I would recommend repairing molecular fragments so that you show only complete molecules.

(b) Show the arrangement of several benzene molecules (use covalent radii for atoms).

(c) Redraw this using van der Waals radii and show as space filling, so that you see the atoms at full sizes.

(d) Explain, based on your understanding of bonding and structure–property relations why the melting temperature of benzene is lower than that of polyethylene.

(6) Hydrogen bonding

(a) Describe the mechanism behind hydrogen bond formation. Give an estimate of typical binding strength in kJ/mole, and the oxygen–oxygen separation needed to begin forming hydrogen bonds with oxygen.

(b) Describe Keesom forces.

(c) Calculate the Keesom energy for ice at $0\,°C$. In SI units the Keesom energy is given by $U_L = -\frac{2}{3}\frac{u^4}{(4\pi\varepsilon_0)^2 k_B T r^6}$

where $\varepsilon_0 = 8.85 \times 10^{-12}$ $C^2/(Nm^2)$ and $k_B = 1.38 \times 10^{-23}$ Nm/K. Assume the dipole moment of water is $u = 6.2 \times 10^{-30}$ Cm.

How does this energy compare to that of a hydrogen bond? Discuss your results.

12 Polymers

12.0 Introduction

Polymers consist of long chains of covalently bound atoms; they are macromolecules with aspect ratios that generally exceed 100. The differences between hard plastics, elastomers, and fibers depend to a large extent on the strength of the intermolecular bonds holding the chains together. Some polymers are cross-linked by covalent bonds, others by hydrogen bonds, and some by weak van der Waals forces.

Polymers are made from monomers, which react to form the basic repeat units of the polymer chain, e.g. $-(CH_2-CH_2)_n-$ with the number of repeat units denoted by the subscript n. The molecular weight of the polymer is given by $n \times m$, where m is the molecular weight of the repeat unit.

The interatomic bonds in polymers can be divided into primary forces (~50 kcal/mole) and secondary forces (<10 kcal/mole). Primary bonds, typically covalent carbon–carbon bonds, are similar to those in diamond (1.54 Å long) but interconnected in one dimension rather than three.

12.1 Basic Polymer Terminology

Polymers are long-chain molecules generally with a molecular weight of 1000 amu or more, composed of small repeat units. As an example of this, consider the structure of a single chain of polyethylene $-(CH_2-CH_2)_n-$ shown in Fig. 12.1. The backbone of the structure is a chain of C atoms singly bonded to two other C atoms. The coordination requirements of the carbon are then satisfied by completing two additional bonds to sidegroups. For polyethylene, the sidegroups are H atoms. The chains can be either arranged in extended arrangements – the extended zig-zag conformation, where the chain appears to be straight viewed from the top – or it can be configured in a coiled fashion. The coiling occurs because there is little energy penalty for rotating the chain around the axis of the C–C single bonds. When the polymer backbone twists by rotation around a C–C bond, the *polymer conformation* is said to have changed. Within the chain, the C–C bonds (~1.53 Å) are very close to the bond length in diamond (1.54 Å). The C–H bond is ~1.1 Å in length. For polyethylene with a molecular weight of ~50 000 g/mol, if each chain were stretched out, it would have a length of 4500 Å and a width of ~3Å for an aspect ratio of ~1500.

(a)

(b)

(c)

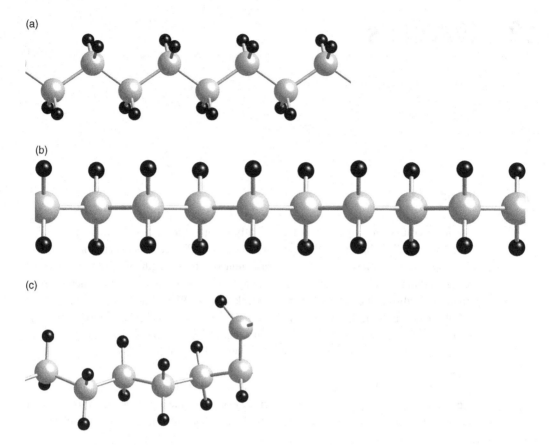

Fig. 12.1 Structure of a single polyethylene chain: (a) extended chain viewed from side, (b) extended chain viewed from top, and (c) bent conformations of a polyethylene chain. The larger atoms are C, the smaller atoms are H.

There is strong covalent bonding within the backbone chain of a polyethylene molecule, but the bonding between the chains is van der Waals in character. As a result, many of the properties will be different when measured in different directions; this is referred to as anisotropy. For example, the mechanical stiffness is large when it is difficult to deform the bonds with an applied stress. Polymers are much stiffer parallel to the chain direction than perpendicular to it in an oriented sample.

Depending on the synthesis of the molecules and their processing, the chains in a polymer may be linear, branched, or interconnected to form a three dimensional network (see Fig. 12.2). Cross-linking forms covalent bonds between chains, making it impossible for chains to slide past each other and therefore the sample cannot flow. Cross-linking to a high degree increases the stiffness and decreases the flexibility of the polymer.

The fact that there is a low energy cost to twist or coil the polymer chain randomly means that polymers are often partially or fully amorphous, where amorphous means that there is no long-range order to the atomic arrangements. Even in polymers with some degree of crystallinity, the crystallites are separated by amorphous regions and no polymer is 100% crystalline – even in single crystals, due to the chain folds. This is

Fig. 12.2 Schematics of (a) linear (straight), (b) branched, and (c) cross-linked polymer chains.

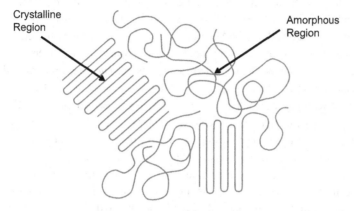

Crystalline Region

Amorphous Region

Fig. 12.3 Schematic showing amorphous and crystalline regions in a polymer.

shown schematically in Fig. 12.3. The degree to which polymers crystallize depends on the rate of cooling and on the nature of the polymer chain. Simple polymers like polyethylene are as much as 95% crystalline when cooled slowly or when processed with strong orientation to align the polymer chains. In the crystalline regions of a polymer, the crystallites are composed of chain folded lamella and are often on the order of a few hundred ångströms in thickness. Therefore, one molecular chain may pass through many crystalline and amorphous regions. This feature of polymers gives many semicrystalline polymers high strength because of these tie chains.

All polymers experience a *glass transition temperature*, T_g, rather than crystallizing as the temperature is decreased. T_g is the temperature above which the chains can be moved with respect to each other (i.e. the chains have translational diffusion); below T_g, the chains are essentially frozen into position. This is referred to as a glass transition temperature, because the frozen-in structure has no long-range translational order, as would be characteristic of a crystal. Instead, short-range order only is retained, as in many liquids. A readily visualized picture of the chains of an amorphous polymer would be as strands in a bowl of cooked angel hair pasta (with really thin, really long noodles).

More information on glasses and the glass transition temperature are given in Chapter 18.

Relationships between the characteristic temperatures T_m, T_g, and T_{cmax} are of considerable practical interest. T_m is the melting point of the polymer crystal, T_g the glass transition temperature, and T_{cmax} the crystallization temperature for which the crystal growth rate is largest. It is difficult to predict from structural considerations alone, but T_g generally lies between 50 and 70% of the melting point T_m, using the absolute temperature scale. The T_g/T_m ratio is usually closer to 0.5 for polymers with large sidegroups, such as polyisoprene and polystyrene which are more easily entangled. T_{cmax} is usually half-way between T_g and T_m where the resistance arising from viscosity of the polymer melt and the thermodynamic driving force for crystallization are balanced.

Thermosets, thermoplastics, and rubbers differ in both structure and physical properties. *Thermoplastics* flow when heated above a characteristic softening temperature where they are processed. This temperature is often 50 °C above T_g for an amorphous polymer and 20–30 °C above T_m if the polymer is semicrystalline. Upon cooling, the material becomes hard and tough. A thermoplastic can be repeatedly heated and softened because it consists of separated chain molecules with vanishingly few cross-links except those induced by degradation during thermal processing. *Thermoset* polymers differ from thermoplastics because of their ability to cross-link when cured at high temperature, when functionalized with highly reactive groups, or when exposed to radiation. Cross-linking allows the polymer to maintain its mechanical integrity at temperatures above the T_g of the backbone or melting point of the crystals. Epoxy resins are thermoset polymers which cross-link through rapid chemical reactions at room temperature.

In polymers that have more than one type of sidegroup, the arrangement of the sidegroups must also be considered. As an example, consider polypropylene, which is formed when methyl groups are substituted for one hydrogen on each repeat unit. Figure 12.4 shows two stereoisomers of polypropylene.

Different possibilities are illustrated in Fig. 12.5. If the stereochemistry of the repeat units is random, the polymer is called *atactic*. In *syndiotactic* polymers the stereochemistry alternates, and in *isotactic* polymers all of the units have the same stereochemistry. Atactic polymers crystallize little if at all because of their irregular structure. Syndiotactic polymers tend to have lower crystallinity than their isotatic form due to steric interference in syndiotactic molecules. As an example, *polyvinylchloride (PVC)* $-(CH_2-CHCl)_n-$ is widely used in various structural applications as conduits, siding, and pipes. Commercial PVC is atactic. Defects in the repeat unit can also occur which change the spacing of sidegroup R. These are referred to as head-to-head or tail-to-tail defects, depending on the spacing.

Crystallinity is important for mechanical strength and temperature stability. The tacticity of a polymer is determined during the polymerization, and cannot be changed without breaking bonds. As an alternative, isotactic polymers are also referred to as composed of *meso diads (two monomer units)*. For *racemic* polymers, the stereochemistry is alternating. Syndiotactic polymers have all racemic diads. Therefore, atactic polymers would have random diads.

Designed catalysts are often required to synthesize polymers in a stereoregular fashion. Without this regularity, crystallinity is not possible.

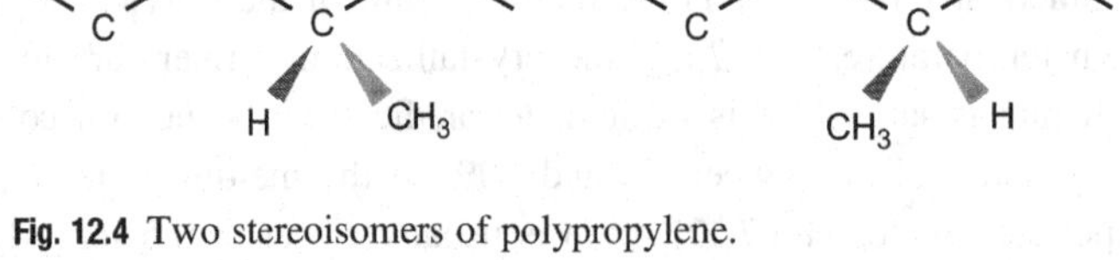

Fig. 12.4 Two stereoisomers of polypropylene.

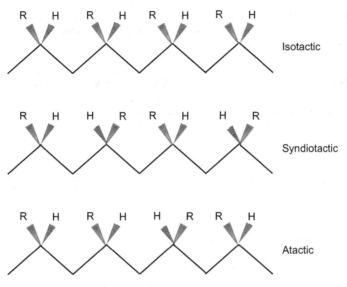

Fig. 12.5 Different polymer tacticities.

12.2   Systematics of Semicrystalline Polymers with Singly Bonded C–C Backbones

Many polymers are largely amorphous, but a number of commercially important polymers, including polyethylene, polypropylene, the nylons, and polyethylene terepthalate (PET) crystallize well. In the vast majority of semicrystalline polymers, in the polymer crystal the chains are arranged with adjacent chains parallel to each other.

Consider polymers where the chain is held together by C–C single bonds. The characteristic geometry of C participating in four single covalent bonds is governed by the sp^3 hybrid orbital, which put the four bonds to C in tetrahedral coordination. A good example of this is *polyethylene* (PE), a highly crystallizable polymer used in pipes, sheets, and fibers. It has a simple linear repeat unit $(-CH_2-CH_2-)_n$ with a glass transition temperature T_g below room temperature and a moderately high crystalline melting point of 125 °C. As shown in Fig. 12.6, in crystals of polyethylene, the C–C chain is a stretched to its full extent. This regularity in the arrangement of the backbone is essential to forming a crystal, and allows the chains to be packed together efficiently. All of the chains are aligned parallel to each other, creating an orthorhombic unit cell, as shown in Fig. 12.7. High-density polyethylene (HDPE) has long linear chains and a high degree of crystallinity. Low-density polyethylene (LDPE) is highly branched. This lowers the degree of crystallinity as well as the density because the chains cannot pack together effectively. High-density polyethylene is used in plastic bottles, food packaging, corrosion-resistant packaging, pipes, and electrical conduits. Low-density polyethylene is used in moisture barriers, plastic films, packaging, and in electrical insulation.

In some other crystalline polymers, the –C–C– backbone spirals around the chain axis. A good example of this is *polytetrafluoroethylene*, $(-CF_2-CF_2-)_n$, also known as *Teflon* (see Fig. 12.8). The driving force for the coiled conformation is often associated

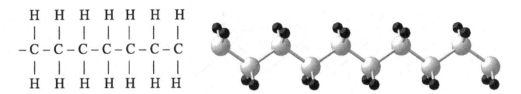

Fig. 12.6 Two schematics of a polyethylene chain showing the C–C backbone.

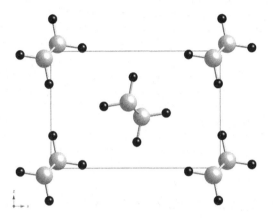

Fig. 12.7 A unit cell of polyethylene looking end-on down the chains. The chains are arranged in a herringbone fashion to increase the packing density.

with the need to pack relatively bulky sidegroups around the chain. As illustrated, when the sidegroups are shown using their full van der Waals radii, the hydrogen atoms are small enough to allow the chain to stay in a planar zig-zag arrangement. In contrast, in poly(tetrafluoroethylene), the fluorine sidegroups are large enough that a planar chain cannot be achieved. Teflon is a tough flexible material with the lowest coefficient of friction of any known polymer.

Polypropylene (PP) is a polyolefin with a slightly more complex molecular structure.

$$(-CH_2 - \underset{\underset{H}{|}}{\overset{\overset{CH_3}{|}}{C}} - CH_2 - \underset{\underset{H}{|}}{\overset{\overset{CH_3}{|}}{C}} -)_n$$

It has a T_g below room temperature ($-18\ ^{\circ}$C) and a moderately high T_m ($176\ ^{\circ}$C). Atactic polypropylene cannot be crystallized, since the stereochemistry of the methyl sidegroups along the chain is random. It is sometimes used as a rubbery modifier to asphalts and composites. In contrast, isotactic polypropylene can be 60–80% crystalline by volume. Using van der Waals radii, the methyl groups are about 4.0 Å in size, larger than the distance between carbons in the extended conformation. As a result, the material cannot form a planar zig-zag with the chains in the extended conformation. Instead, to accommodate the size of the methyl sidegroups, chains twist on their own

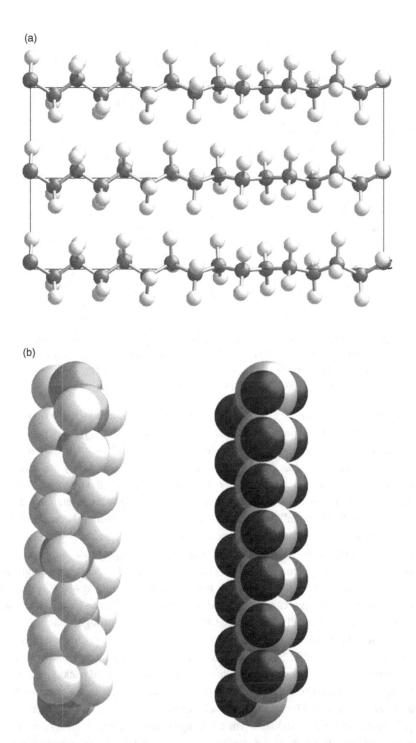

Fig. 12.8 (a) Crystal structures of Teflon, showing the spiral –C–C– backbone; C is the filled circle, F is the lighter circle. (b) A comparison of Teflon (left) with polyethylene (right) with the atoms shown at their full van der Waals radii. While the small hydrogen atom in polyethylene can be fitted along a straight backbone, the larger fluorine atom in Teflon requires the backbone to twist.

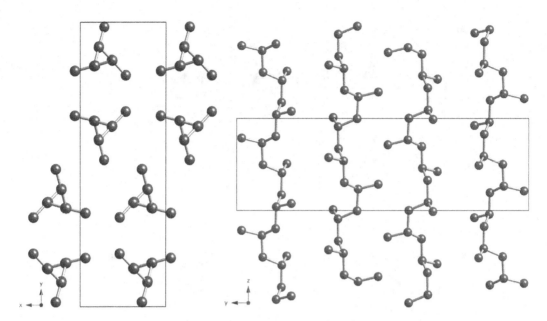

Fig. 12.9 Monoclinic crystal structure of isotactic polypropylene, with only the carbon atoms shown for clarity. In the view down the chains (left) it is clear that right- and left-handed spirals pack together, and that high packing densities are achieved by pointing the protruding methyl groups from one chain towards open spaces on adjacent chains.

axis, in this case forming a 3_1 screw axis in the crystalline regions, as shown in Fig. 12.9. Both right- and left-handed spirals appear in the crystal, with right handed spirals interacting primarily with left-handed spirals, and vice versa. *Wallach's Rule* says that racemic crystals (e.g. those with a mixture of chiralities) are usually denser than the corresponding chiral one. In real crystals, defects in the orientation of the spiral chains are often observed, so that "up" and "down" chains are interchanged.

The size of the crystalline regions of isotactic polypropylene is on the order of the wavelength of light. Thus, light is scattered by the differences in refractive index at crystallite boundaries, producing a translucent white appearance familiar from many milk jugs. Polypropylene is used in dishwasher-safe food containers, high-durability carpeting and ropes, as the flip tops of bottles, in some piping systems, as the dielectric in high-energy-density capacitors, and in non-woven fabrics and warm base-layer clothing. Commercial PP with aligned fibers or films are stronger than steel.

Consider next polystyrene. Atactic polystyrene is amorphous, with a glass transition temperature of ~90–100 °C. It is of considerable commercial importance, being utilized in styrofoam, clear plastic wedding champagne cups, laminated packaging, and in many block copolymer thermoplastic elastomers. The amorphous material is relatively rapidly attacked by solvents. To improve the solvent resistance as well as the thermal stability, polystyrene can be crystallized when it is synthetized in either the syndiotactic or the isotactic forms. In practice, the syndiotactic form crystallizes more rapidly, and so is of more commercial importance. Syndiotactic polystyrene has numerous polymorphs. Both α and β (shown in Fig. 12.10) have extended zig-zag chains along the backbone; the large aromatic rings on every repeat unit project radially from the axis of the

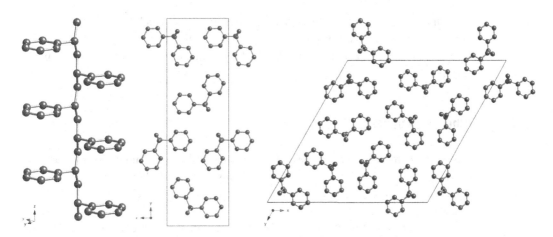

Fig. 12.10 Syndiotactic polystyrene. (Left) View of the polymer chains from the side; the chains are syndiotactic, such that the stereochemistry is alternating. (Middle) Crystal structure of the orthorhombic β syndiotactic polystyrene. (Right) Trigonal unit cell of α syndiotactic polystyrene. For all three structures, only the C are shown, for clarity.

(a)

(b)

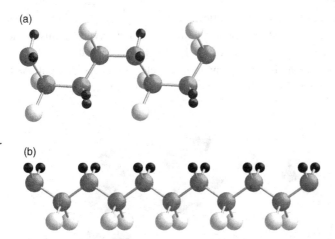

Fig. 12.11 The (a) α and (b) β forms of $-(CH_2-CF_2)_n-$. Carbon is the large filled circle, hydrogen the small dark circle, and fluorine is white.

backbone. The γ form has helical chains. Acronitride–butadiene–styrene (ABS) copolymers are used in products that require more toughness than polystyrene and are the principal material used in Lego bricks.

Phase transformations between different backbone conformations are also observed, as in poly(vinylidene fluoride), PVDF (see Fig. 12.11). The alpha form, α-PVDF, has a rumpled chain with a "two-up, two-down" geometry (more formally called the trans-trans-trans conformation). The beta form has an extended all-gauche chain conformation, and is favored by stretching. Such transitions can have marked effects on the properties; in the case of polyvinylidene fluoride, the piezoelectric coefficients are zero in the α phase and finite in the β phase. This will be discussed in detail in Chapter 28.

12.3   Other Backbones

There are numerous other types of backbones utilized in industrially important polymers, some of which have aromatic groups or multiple bonds between carbons along the backbones, others of which include other atoms such as O or N into the backbone.

Millions of tons of plastics are used annually in building and construction. Typical of these structural plastics are *phenolic resins* and epoxy resins. Cured phenolic resins are used in the production of brake linings, plywood sheets, and abrasive wheels. They have outstanding heat resistance and good dimensional stability. Cross-linked resole resin is produced by dehydrating an ether condensation product made from phenol (C_6H_5OH) and formaldehyde (H_2CO); a section of its molecular structure is shown in Fig. 12.12.

Aromatic thermoplastic polyesters such as *polyethylene terepthalate* (PET) are used for the blow molding of plastic bottles, for biaxially oriented films (Mylar) and as Dacron fiber. These are heteropolymers with carbon and oxygen atoms in the polymer chain, as shown in Fig. 12.13. Because PET tends to crystallize rather slowly, the crystallites are significantly smaller than the wavelength of visible light. As a result of this, PET bottles (e.g. disposable water or soda bottles) are transparent.

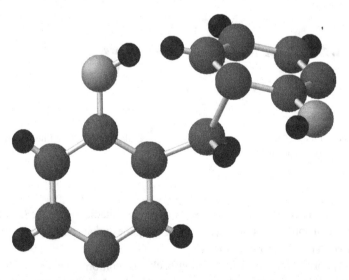

Fig. 12.12 Section of the structure of a phenolic resin. C atoms are show as the large darker spheres, O as the lighter large spheres, and the H as the smallest atoms. C shown with missing neighbors would connect to other phenol groups through a CH_2 intermediary.

Fig. 12.13 Structure of polyethylene terephalate.

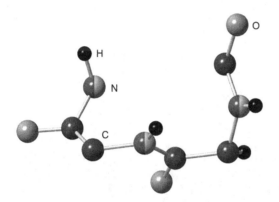

Fig. 12.14 Polypeptide backbone, C–C–N–C–C–N.

Peptide chains make up the backbone of collagens, the natural polymer important in skin and leather. The backbone of a single chain is shown in Fig. 12.14; the repeat group is C–C–N. In collagens, three of these peptide chains spiral around a common axis. Cross-linking between the chains can be accomplished by exposure to ultraviolet light, including sunlight.

As an example of a crystalline nylon, consider nylon-6: $-(NH(CH_2)_5CO)_n-$. The carbons in the backbone appear in both 3- and 4-coordination, using sp^3 and sp^2 hybrid orbitals, respectively. As shown in Fig. 12.15, the N in the backbone has trigonal planar coordination and the oxygens have an asymmetric linear coordination, with one strong double bond to a carbon atom, and a hydrogen bond to an adjacent NH group. H, of course, is also linear.

Not all polymers have carbon in the backbones. Commercial siloxane polymers (sometimes called silicones) have an O–Si–O–Si–O linkage similar to silicate minerals and ceramics: 1.6 Å bond lengths, and O–Si–O angles of 109.5°. Polydimethylsiloxane has two methyl (CH_3) groups complete the coordination for each of the Si atoms (see Fig 12.16).

Silicone polymers are used as elastomers and resins. Silicone elastomers are lightly cross-linked and are often reinforced with silica filler. They have excellent low-temperature flexibility and high-temperature stability. Important applications include wire and cable insulation as well as medical and surgical materials. Silicon rubbers, varnishes, and paints are made from silicone resins which are more heavily cross-linked

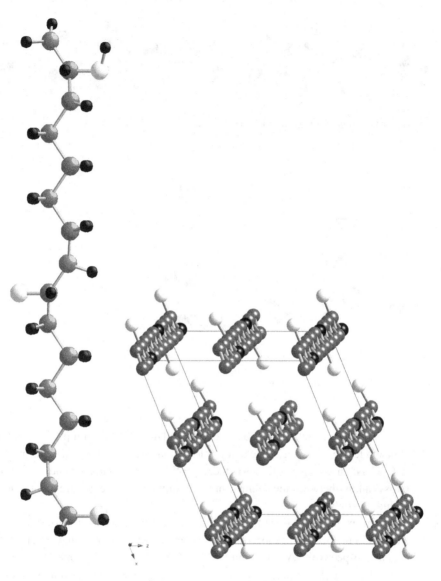

Fig. 12.15 Nylon-6: (left) single chain and (right) alignment of multiple chains. The hydrogens are not shown in the unit cell for simplicity.

than the elastomers. At elevated temperature (typically above 250 °C), the siloxane chains form rings.

12.4   Rubber

Crystalline solids and glasses generally obey Hooke's law, but rubber and other elastomers do not. The elastic stiffness of rubber is smaller by factors of 1000 to

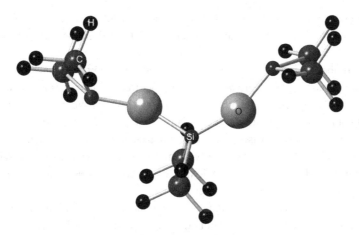

Fig. 12.16 Section of the structure of polydimethylsiloxane. The silicone backbone is similar to that observed in single chain silicates.

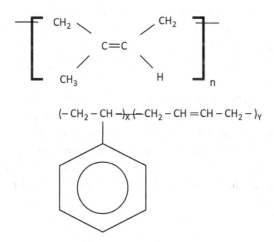

Fig. 12.17 Chemical repeat units of (top) polyisoprene, and (bottom) styrene–butadiene rubber.

10 000 than most other solids. Deformations of 500% are possible, compared to 1% or less in many other solids. The two main types of rubber are natural or synthetic polyisoprene and styrene–butadiene rubber (SBR). Schematics of the structure are depicted in Fig. 12.17.

Rubber consists of long, flexible molecular chains. The molecules have a backbone of non-collinear single-valence bonds that undergo rotation as a result of thermal agitation. The chains are coiled in random fashion, and under tension are pulled into partial alignment. If the molecular chains are completely separate as in uncrosslinked natural rubber, the material would be a viscous liquid. To make a rubbery solid, the chains must be cross-linked at a relatively few number of points to form a loose three-dimensional network of crumpled chain-like molecules. Such a soft rubbery solid can undergo large deformations and recover completely when mechanical stresses are removed.

Natural rubber occurs as latex, a milky secretion of many different tropical plants. The most important is *Hevea brasilienis* (the rubber tree) which gives a raw rubber with 1–2% protein, 2–4% resins, and a latex containing 30–40% rubber by weight. Millions of tons of natural rubber are used in heavy-duty tires, vibration dampers, and in springs and bearings. Vulcanization is carried out by masticating a small percentage of sulfur into the rubber at 140 °C. This improves the thermal and mechanical durability by establishing cross-links between the hydrocarbon chains. Resistance to abrasion is further enhanced with carbon black fillers.

Synthetic polyisoprene has a similar chemical structure to natural rubber but does not contain the proteins and fatty acids found in natural rubber. It is preferred in rubber bands, baby bottles, motor mounts, and shock absorbers. Noted for its long fatigue life and resistance to abrasion, polyisoprene has very good elasticity but rather poor resistance to motor oil and gasoline.

Styrene–butadiene rubber (SBR) is the most common and cheapest form of synthetic rubber. SBR is a petroleum by-product. It has better abrasion resistance than polyisoprene and is used in automobile tire treads where good wear and good grip are essential. Other applications include shoe soles, brake pads, cable insulation, and pharmaceutical products. The principal disadvantages are poor fatigue resistance and relatively low strength.

The most elastic type of rubber is *butadiene rubber* (BR). BR has low hysteresis, good flexibility, and high abrasion resistance. It is used in blends with natural rubber and SBR. Polybutadiene has several different forms structurally isomeric forms (two of which are shown in Fig. 12.18) with very different properties. The amorphous *cis*-form is very rubbery with a melting point of 1 °C. The polymer chains pack together much better in the *trans*-form, which has a melting of 146 °C, but the material has lost the useful rubbery characteristics.

12.5   Block Copolymers

Block copolymers are a class of macromolecules produced by joining two or more chemically distinct polymeric segments. Segregation of these blocks on a molecular scale of 50 to 1000 Å can produce complex nanostructures with highly desirable properties.

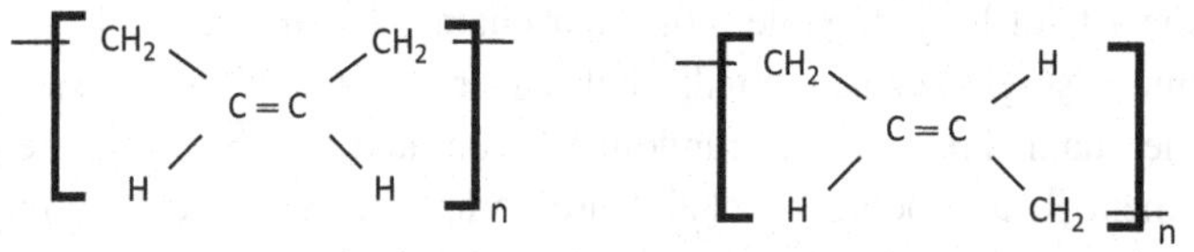

Fig. 12.18 Two forms of butadiene rubber: (left) amorphous *cis*-form, and (right) *trans*-form.

Everyday examples include upholstery foam, asphalt additives, and adhesive tape. Bedding upholstery made from polyurethane foams are composed of segmented copolymers known as thermoplastic elastomers that combine high-temperature resilience and low-temperature flexibility. In a similar way, block copolymers are blended with asphalt to avoid cracking at low temperatures, and rutting at high temperatures.

There are three main types of polymers: (1) those with flexible and crystallizable chains (PE, PP); (2) cross-linked amorphous networks of flexible chains such as cured rubbers and phenol–formaldehyde; and (3) stiff, rigid chains such as the polyimide family, used for high-temperature insulation and heat shields. Block copolymers utilizing two or all three of these types have proven effective in enhancing the physical and chemical properties of polymeric bodies. Crystalline domains embedded in a viscous matrix, as in Dacron, are used in the production of films and fibers. Neoprene, a moderately cross-linked polymer with some crystallinity, is an oil-resistant rubber with good chemical properties. Heat-resistant materials are made from partially cross-linked rigid chains. These block copolymers are under development for jet engines and plasma technology.

Combining all three ideas – crystalline domains reinforced by cross-linked rigid chains – provides materials of high strength with good thermal properties. Buildings, automobiles, and aircraft require properties such as these.

12.6   Chemical Properties

Oil- and Water-Soluble Polymers

"Like dissolves like" is the general rule, and this is true for polymers as well as for smaller organic molecules. Two water-soluble polymers are shown in Fig. 12.19. Solubility is enhanced by electronegative groups such as –OH, –O–, –NH, and –NH_2 that form hydrogen bonds with water molecules.

Water solubility and swelling can be minimized by replacing the polar groups with non-polar groups, or by cross linking the polymer chains.

Most water-soluble polymers such as polyvinyl alcohol are amphiphilic, or contain both hydrophobic and hydrophilic groups. The molecules assume a conformation to allow the electronegative hydrophilic groups better access to the water molecules. At organic–aqueous interfaces the non-polar portions of the chain face toward the organic phase and polar groups face away – like small-molecule surfactants.

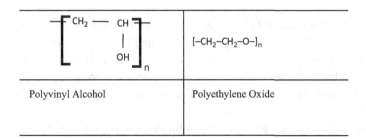

Fig. 12.19 Water-soluble polymers possess electronegative groups such as hydroxyl or oxygen.

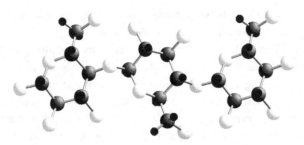

Figure 12.20 Backbone of cellulose Iα. The C is the large dark atom, oxygen is the larger light sphere, and H is the smallest atom shown.

Polymers are soluble in water when a sufficient number of polar groups are present. Common wood (cellulose) contains many hydroxyl groups, causing it to swell in water. The structure of the cellulose chain is shown in Fig. 12.20. The backbone of cellulose involves C_5O rings bonded together through additional O atoms. It does not dissolve because of high crystallinity of the stiff backbone. Grafting on additional polar groups will increase the solubility of cellulose in water, as in carboxymethyl cellulose.

Linear amorphous polymers with non-polar side groups tend to be soluble in oil. Polyisobutylene is a good example.

$$(-CH_2 - \underset{\underset{CH_3}{|}}{\overset{\overset{CH_3}{|}}{C}} -)_n$$

Permeability

The "like permeates like" rule also applies to the transmission of gases through polymer films. Cellulose and other polymers with polar sidegroups are permeable to water vapor, but polyisobutylene is not. On the other hand, gaseous hydrocarbons such as propane will pass through polyethylene but polyvinylalcohol films are impervious.

The permeability coefficient for inert gases such as N_2 depends on crystallinity. It is higher for low-density polyethylene (LDPE) than the more crystalline form HDPE.

Foamed Polymers

Elastomers, thermosetting polymers, and thermoplastics can all be expanded into foams with more than 95% porosity. Lightweight packing materials, flotation devices, and ceiling tiles are made from porous polystyrene. Foamed polyurethanes are manufactured as flexible upholstery and rigid reinforcement for composite airplane wings. Foamed polymers provide excellent insulation because of their very low thermal conductivities.

A number of techniques have developed for the foaming process.

(1)   Gases evolved during the crosslinking process can be trapped during hardening.

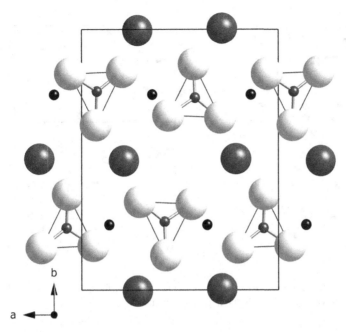

Fig. 12.21 When heated, sodium bicarbonate easily decomposes into sodium carbonate water vapor and carbon dioxide. It is an excellent blowing agent used in making porous materials and in preparing foods. The structure consists of isolated CO_3 carbonate groups and deformed NaO_6 octahedra arranged in pleated sheets parallel to (101) cleavage planes. The weak hydrogen bonds between carbonate groups are easily ruptured, liberating carbon dioxide gas. Oxygen is the large lighter circle, Na is the large gray circle, C is the small dark grey circle shown with three neighbors, and H is the small filled dark circle.

(2) CO_2 and other gases can be mechanically mixed into the molten polymer, forming bubbles.

(3)   Blowing agents such as sodium bicarbonate release gas into the mix during heating (Fig. 12.21):

$$2NaHCO_3 \rightarrow Na_2CO_3 + H_2O \uparrow + CO_2 \uparrow.$$

The bubbles of carbon dioxide are useful in foaming thermoplastics.

Flame-Retardant Polymers

Many polymeric hydrocarbons are readily combustible and can serve as a fuel source. Others like Teflon (PTFE) burn in oxygen-rich atmospheres. The oxidation resistance is improved by using polymers containing carbonate (CO_3) or phosphate (PO_4) groups that are already fully oxidized. Further improvement is achieved by incorporating flame retardant fillers such as ATH (alumina trihydrate). ATH releases steam when heated (Fig. 12.22).

Aluminum trihydrate forms a layer structure resembling that of $Ca(OH)_2$ except that one-third of the cation sites are empty. Al is in octahedral coordination with six oxygens. The

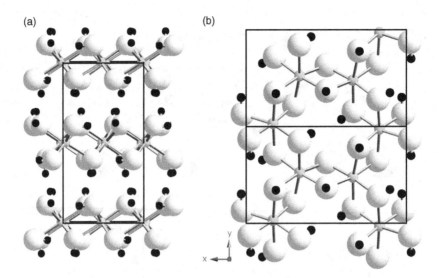

Fig. 12.22 Structure of gibbsite Al(OH)₃: (a) looking parallel to the layers, (b) perpendicular to one layer. Al is shown in gray and the OH⁻ group in white.

octahedral share edges within the $Al(OH)_3$ layers. Hydrogen bonds are formed within and between the layers. When heated, the hydroxide decomposes to the oxide and water:

$$2Al(OH)_3 \rightarrow Al_2O_3 + 3H_2O.$$

12.7   Problems

(1) Rationalize why the mechanical properties of amorphous and cross-linked polystyrene differ, as shown in the graph.

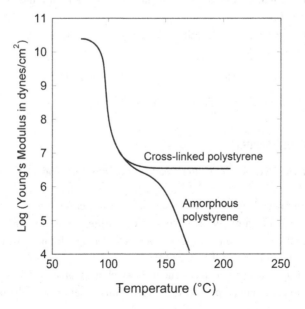

(2) Explain the difference between a thermoplastic and a thermosetting polymer. Explain the mechanism that causes this phenomenon.

(3) Draw (schematically) a syndiotactic chain and an isotactic chain.

(4) What is T_g, and what does it signify?

(5) How are properties such as viscosity, melting points, and hardness affected by the degree of cross-linking?

(6) Draw the unit cell of polyethylene.

(7) Shown below is a part of the crystal structure of monoclinic isotactic polypropylene $(-CH_2-CHCH_3-)_n$. For simplicity, only the C atoms are shown.

  (a) What is meant by isotactic?

  (b) Explain why the chains twist (in contrast to the extended zig-zag chain conformation of crystalline polyethylene).

  (c) Label the handedness of the chains shown.

  (d) Along which direction would you expect this crystallite to be mechanically stiffer? Detail your reasoning.

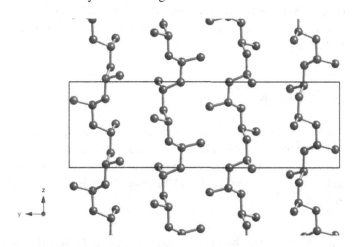

(8) Polyethylene is also orthorhombic, with space group *Pnma*, $a = 7.417$ Å, $b = 2.534$ Å, and $c = 4.939$ Å and

C	(0.0400,	0.2500,	0.0620)
H	(0.1850,	0.2500,	0.0290)
H	(0.0110,	0.2500,	0.2780).

  (a) Plot a unit cell, showing the C–C and C–H bonds

  (b) What is the C coordination?

  (c) Determine the C–C and C–H bond lengths.

  (d) What is the C–C–C bond angle?

  (e) Plot somewhat more than a unit cell looking down the *b* axis, so that you can see the arrangement of the chains.

  (f) Compare the chain arrangement to that in PTFE.

13　Pauling's Rules, Bond Valence and Structure-Field Maps

13.0　Introduction

It is useful to be able to predict the structures that will be adopted by an ensemble of atoms, as this often provides insight into the properties that should be manifested. Several approaches to this problem have been adopted over the years. This chapter covers several of the more successful approaches: Pauling's rules for ionically bonded solids; the bond valence approach, which can be applied to ionically and covalently bonded crystals; and structure-field maps.[1]

13.1　Pauling's Rules

Pauling's rules are a simple, yet remarkably robust approach to predicting the crystal structures of ionically bonded solids based on the formal charges on the ions and on the ion sizes. The underlying assumptions are that the ions can be treated as hard spheres, and that electroneutrality should be maintained in a small volume – ideally around every anion.

Pauling's rule #1: The cation coordination is determined by the radius ratio: $\frac{r_{cation}}{r_{anion}}$

To justify this, consider the ionic configurations shown in Fig. 13.1. Since the internal energy of the crystal is the bonding energy per volume, it is intuitively appealing that the stability of the crystal should rise as the number of cation–anion bonds is increased, *providing cation–anion contact is maintained.* Cations with large radius ratios can fit many anions in their coordination polyhedra, while smaller cations can fit fewer. The distance between cation and anion is equal to the sum of their ionic radii. The $\frac{r_{cation}}{r_{anion}}$ radius ratio limits where cation coordination numbers change are simply a matter of geometry. A given polyhedron becomes unstable when the anions are touching, and the cation and anion are no longer in contact. Table 13.1 gives the limiting radius ratios and the predicted coordination geometries.

---

[1]　Portions of this chapter are adapted (with permission) from R. E. Newnham, "Phase diagrams and crystal chemistry," in *Phase Diagrams: Materials Science and Technology, Volume V,* Academic Press, New York (1978).

Table 13.1 Limiting radius ratios for Pauling's rule #1

Radius ratio	Coordination polyhedron	Cation coordination number
$\frac{r_{cation}}{r_{anion}} \geq 1$	Close-packed	12
$0.732 \leq \frac{r_{cation}}{r_{anion}} < 1$	Cubic	8
$0.414 \leq \frac{r_{cation}}{r_{anion}} < 0.732$	Octahedral	6
$0.225 \leq \frac{r_{cation}}{r_{anion}} < 0.414$	Tetrahedral	4
$0.155 \leq \frac{r_{cation}}{r_{anion}} < 0.225$	Triangular	3
$\frac{r_{cation}}{r_{anion}} < 0.155$	Linear	1 or 2

(a)   (b)   (c)

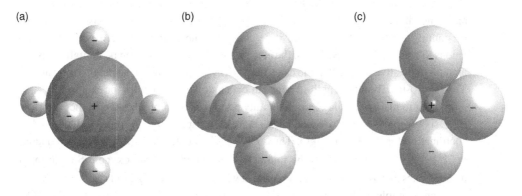

Fig. 13.1 Diagrams illustrating the physical basis of Pauling's first rule. The total attractive energy between a cation and its anion neighbors increases with the number of cation–anion bonds, and is inversely proportional to the distance between cation and anion. Configuration (a) is unstable because there are too few anion neighbors; (b) shows a stable situation in which as many anions as possible have been bonded to the cation, while maintaining cation–anion contact. Case (c) is unstable because there are too many anions for the size of the cation, which would make the cation–anion distance unnecessarily large. In (c), one anion neighbor has been removed so that the cation is visible.

To illustrate the rule, consider magnesium oxide: the radius of Mg^{2+} is 0.86 Å and O^{2-} is 1.26 Å, using the Shannon–Prewitt crystal radii. The radius ratio is 0.68, for which the predicted cation coordination is octahedral. There is one key caveat to add in considering Pauling's first rule. As was described in Chapter 9, various tables of ionic radii exist. Most widely in use are Shannon–Prewitt's ionic radii. Note that these assign more of the cation–anion bond length to the anion radii, with the result that the cation radii are smaller. One consequence is that use of these radii will tend to underestimate the cation coordination number via Pauling's first rule. This is particularly true for small, highly charged cations. Somewhat better results with the predicted coordinations of small cations are achieved with the Shannon–Prewitt crystal radii. Conversely, the crystal radii tend to overpredict the coordination of the *larger* cations. A compromise position is to use an average of the two sets of radii. Not too surprisingly, Pauling's table of ionic radii tends to work fairly well in predicting cation coordination with Pauling's rules.

Cation coordination numbers are not always determined unambiguously by the first rule. When $\frac{r_{cation}}{r_{anion}}$ is near one of the limiting radius ratios, the cation is found in different coordination numbers with that anion in different crystal structures. For example, Al^{3+} occurs in both octahedral and tetrahedral sites in oxides, B^{3+} is observed in either triangular or tetrahedral coordination. Some of the other factors that can govern preferred coordination numbers are the propensity for covalency (many d^{10} ions prefer more covalent bonding, and as a result are often found in tetrahedral coordination), distortions of coordination polyhedra, and crystal field theory (see Chapter 14).

Pauling's rule #2: The anion coordination is governed by local electroneutrality

The electrostatic bond strength, EBS, that can be provided from a cation to any one of its anion neighbors is given by: $EBS = \frac{cation\ charge}{cation\ coordination}$. This is essentially a means of accounting for the fact that the positive charge of the cation is shared with all of the anions to which it is bonded. Thus, the electrostatic bond strength can be thought of as the amount of positive charge that the cation can "deliver" per bond to the neighboring anion. The electrostatic bond strength of Mg^{2+} in octahedral coordination, for example, is $\frac{2}{6}$.

In order to maintain local electroneutrality around each anion, the sum of the incoming EBS should balance the anion charge. As an example, rule #1 predicted octahedral coordination for the Mg in MgO. Thus, $EBS = \frac{2}{6}$. To determine the anion coordination, $n_{Mg}\left(\frac{2}{6}\right) - 2 = 0$, where n_{Mg} is the number of Mg bonded to the oxygen. Simple math demonstrates that $n_{Mg} = 6$, which is consistent with the observed rocksalt structure of MgO.

As an example of Pauling's second rule, consider the case of forsterite, Mg_2SiO_4. One begins the calculation by estimating the cation coordination from radius ratios.

$$\frac{r_{Si}}{r_O} = \frac{0.4\ \text{Å}}{1.3\ \text{Å}} = 0.31 \Rightarrow \text{4-coordination (tetrahedral) for the Si};\ EBS_{Si} = \frac{4}{4}.$$

$$\frac{r_{Mg}}{r_O} = \frac{0.8\ \text{Å}}{1.3\ \text{Å}} = 0.62 \Rightarrow \text{6-coordination (octahedral) for the Mg};\ EBS_{Mg} = \frac{2}{6}.$$

Pauling's second rule gives $n_{Si}\left(\frac{4}{4}\right) + n_{Mg}\left(\frac{2}{6}\right) - 2 = 0$, where n_{Si} and n_{Mg} are the numbers of Si and Mg ions bonded to a given oxygen. The (–2) in the equation is the charge of the O ion. Both n_{Si} and n_{Mg} must be integers (it doesn't make sense to bond to a fraction of an atom), and both should be greater than zero (so that they are bonded into the structure). This equation can be satisfied if $n_{Si} = 1$, and $n_{Mg} = 3$. This solution agrees with the observed coordinations shown in Fig. 13.2.

Thus, Pauling's first and second rules predict the cation and anion coordinations, respectively.

Pauling's rule #3: Ionic crystals favor corner, rather than edge- or face-sharing of polyhedra. When edges or faces are shared, the shared edge/face tends to shorten

Pauling's third rule states that shared faces or shared edges decrease the stability of ionically bonded crystals. Corner-, edge-, and face-sharing of polyhedra were described in Chapter 9. Pauling's third rule can readily be visualized as a means of keeping the

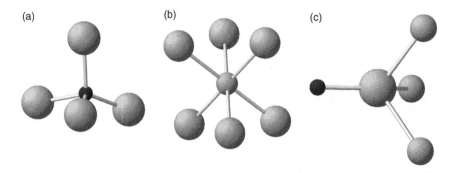

Fig. 13.2 Coordinations of (a) Si, (b) Mg, and (c) O in forsterite.

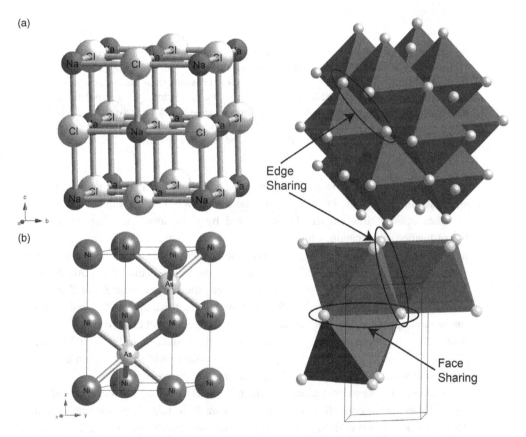

Fig. 13.3 Comparison of the crystal structures of (a) NaCl and NiO with edge-sharing of cation polyhedra and (b) niccolite (NiAs) with edge- and face-sharing of cation polyhedra.

cations as far apart as possible, in order to reduce electrostatic repulsion. As a comparison, consider the structures of NiO and NiAs; both are shown in Fig. 13.3. NiO has the rocksalt structure, in which the NiO_6 octahedra edge-share. In contrast, NiAs entails both edge- and face-sharing of Ni polyhedra. The Ni–Ni distance is 2.674 Å in NiAs,

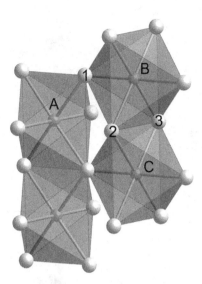

Fig. 13.4 A section of the rutile TiO_2 crystal structure, showing corner- and edge-sharing of the Ti-centered octahedra. When polyhedra edge-share, the cations are closer together than when they corner-share. Thus, the Ti–Ti distance is 3.569 Å between atoms marked A and B, and 2.958 Å between atoms marked B and C. Pauling's third rule states that when polyhedra edge-share, the shared edge tends to contract. Thus, the O–O distance between atoms marked 2 and 3 is 2.534 Å, while the distance between atoms 1 and 2 is 2.778Å. Contraction of the shared edge allows the two Ti atoms B and C to move further apart.

and 2.948 Å in NiO. The face-sharing is energetically unfavorable in ionic compounds and is stabilized by orbital overlap across the shared face. Pauling's third rule suggests that the rocksalt structure should be favored by more ionic AX compounds with 6-coordination of the cation, while more metallic or covalent bonding favors NiAs. Thus, many transition metal $A^{2+}X^{2-}$ oxides have the rocksalt structure, while the less ionic sulfides, selenides, arsenides, and tellurides of the same cations have the NiAs structure.

Another piece of Pauling's third rule states that when edges or faces are shared, the shared edge/face tends to contract. Rutile is an excellent example of this point. Rutile can be envisaged as a series of diamond chains running parallel to the c axis (Fig. 13.4). Along this axis, the polyhedra edge-share. One chain connects to the next by corner-sharing. Electrostatic repulsion is reduced by increasing short cation–cation distances, which in this case is accomplished by shortening the shared edge.

There are exceptions to Pauling's third rule. Rocksalt itself has edge-shared octahedra. Likewise, in corundum, the Al^{3+} octahedra have both face- and edge-sharing. This is observed even though the Sc_2S_3 structure, which consists of edge-shared octahedra, would, in principle, be possible.

Pauling's rule #4: Highly charged cation polyhedra, in particular, tend towards corner-sharing, in structures with more than one cation

The fourth rule can also be justified on the basis that the structure will tend to keep small, highly charged cations as far apart as possible. That is, rule #3 is most likely to be true when the cations are small and highly charged. According to the fourth rule, cations of high valence and small coordination tend not to share more than one anion.

Pauling's rule #5: Crystal structures are as simple as possible

The fifth (or parsimony) rule states that the number of different polyhedra is small. In practice, this means that the anions in a crystal will generally satisfy rule #2 in the same way, unless there is some other reason to believe this will not be the case. That is, the structure stability is increased if electroneutrality is fulfilled around each anion, rather than one anion being overbonded, and its neighbor being underbonded. One consequence of this is that ternary compounds will be less common than binary compounds. For example, in the CaO–FeO–SiO_2 system, there are six binary compounds and one ternary compound.

Here are some examples of the application of Pauling's rules.

NaCl, rocksalt

For $NaCl$, $\frac{r_{Na}}{r_{Cl}} = \frac{1.2\ \text{Å}}{1.7\ \text{Å}} = 0.71 \Rightarrow 6$ coordination. The electrostatic bond strength for the Na is then $\frac{1}{6}$. Pauling's second rule yields $n_{Na}\left(\frac{1}{6}\right) - 1 = 0$, which is true if $n_{Na} = 6$. Thus, octahedral coordination of both the cation and anion are predicted, in agreement with the known rocksalt crystal structure.

Al_2O_3, corundum

For Al_2O_3, $\frac{r_{Al}}{r_O} = \frac{0.53\ \text{Å}}{1.3\ \text{Å}}$ to $\frac{0.675\ \text{Å}}{1.3\ \text{Å}} = 0.41$ to $0.52 \Rightarrow$ either 4- or 6-coordination. The electrostatic bond strength is then either $\frac{3}{4}$ or $\frac{3}{6}$. Application of Pauling's second rule gives $n_{Al}\left(\frac{3}{4} \text{ or } \frac{3}{6}\right) - 2 = 0$. In this case, the only solution for which n_{Al} is an integer is $n_{Al} = 4$, which occurs only if the Al is 6-coordinated. Thus, Pauling's rules correctly predict both the cation and anion coordination for corundum.

$MgSiO_3$, enstatite

An example of a structure where not all of the anions satisfy Pauling's second rule in the same way is enstatite, $MgSiO_3$. As was shown above, radius ratios suggest that that Mg should be octahedrally coordinated to O, while the Si is 4 coordinated. The second rule then gives $n_{Si}\left(\frac{4}{4}\right) + n_{Mg}\left(\frac{2}{6}\right) - 2 = 0$. One means of satisfying this equation is for $n_{Si} = 1$, and $n_{Mg} = 3$, as was the case for forsterite. However, for enstatite, this solution turns out to be inconsistent with the chemical formula. Consider the following bond-counting argument: $\left(\frac{6\ \text{Mg–O bonds}}{\text{Mg}}\right)\left(\frac{\text{Mg}}{3\ \text{O}}\right) = \left(\frac{2\ \text{Mg–O bonds}}{\text{O}}\right)$. The first term on the left-hand side comes from the predicted Mg coordination, while the second term comes from the chemical formula (there are 3 O for every Mg in the chemical formula). However, the product conflicts with the solution $n_{Mg} = 3$. Thus, there must be two ways in this structure of achieving local electroneutrality. Some of the O have only two Si neighbors ($n_{Si} = 2$, $n_{Mg} = 0$), while others have one Si and three Mg neighbors ($n_{Si} = 1$, $n_{Mg} = 3$). It is a good practice, after calculating the cation and anion coordinations using Pauling's first and second rules, to make sure that the result is consistent with the chemical formula.

$Al_2SiO_4F_2$, topaz

Topaz is a hard gemstone with the composition $Al_2SiO_4F_2$. Since the radii of F^- and O^{2-} are very similar, the radius ratios are the same for cation–O and cation–F bonds. Pauling's first rule predicts that Si^{4+} is 4-coordinated, while Al^{3+} could be either 4- or 6-coordinated.

Rule #5 states that the structure should be as simple as possible (in this case meaning as few different polyhedra as possible). Then every oxygen ion will have the same coordination to Si and Al, as will each F.

Rule #2 for O gives $n_{Si}\left(\frac{4}{4}\right) + n_{Al}\left(\frac{3}{4} \text{ or } \frac{3}{6}\right) - 2 = 0$. The analogous relation for F is $n'_{Si}\left(\frac{4}{4}\right) + n'_{Al}\left(\frac{3}{4} \text{ or } \frac{3}{6}\right) - 1 = 0$. Integer values for the number of Al bonded to each O or F are possible only if the Al is octahedrally coordinated. Given this, n_{Si} has three possible values: 0, 1, 2. If $n_{Si} > 2$, electroneutrality cannot be achieved around each oxygen.

It can be shown that only $n_{Si} = 1$ is possible. If $n_{Si} = 0$, then every Si is surrounded by four fluorines. Double-checking against the chemical formula gives $\left(\frac{4 \text{ Si--F bonds}}{\text{Si}}\right)\left(\frac{\text{Si}}{2\text{F}}\right) = \left(\frac{2 \text{ Si--F bonds}}{\text{F}}\right)$. Electroneutrality would then predict: $2\left(\frac{4}{4}\right) + n'_{Al}\left(\frac{3}{6}\right) - 1 = 0$, which is impossible. Thus, $n_{Si} \neq 0$.

If $n_{Si} = 2$, then $n_{Al} = 0$. In this case, every aluminum is completely surrounded by fluorine. From the chemical formula, $\left(\frac{6 \text{ Al--F bonds}}{\text{Al}}\right)\left(\frac{\text{Al}}{\text{F}}\right) = \left(\frac{6 \text{ Al--F bonds}}{\text{F}}\right)$. Inserting this into Pauling's second rule for F yields $n'_{Si}\left(\frac{4}{4}\right) + 6\left(\frac{3}{6}\right) - 1 = 0$, which cannot be solved for positive integers of n'_{Si}.

The only remaining possibility is $n_{Si} = 1$. Pauling's second rule for the O then gives $1\left(\frac{4}{4}\right) + n_{Al}\left(\frac{3}{6}\right) - 2 = 0$, which can be solved if $n_{Al} = 2$. Since the Si:O ratio is 1:4, $\left(\frac{1 \text{ Si--O bonds}}{\text{O}}\right)\left(\frac{4\text{O}}{\text{Si}}\right) = \left(\frac{4 \text{ Si--O bonds}}{\text{Si}}\right)$. Thus, all four anions around Si are oxygen, making $n'_{Si} = 0$. Then, Pauling's second rule for the F becomes: $0\left(\frac{4}{4}\right) + n'_{Al}\left(\frac{3}{6}\right) - 1 = 0$, and $n'_{Al} = 2$.

Combining these results, we find that each Si is coordinated to four O, each Al to four O and two F, every O to one Si and two Al, and F to two Al. Thus all coordinations are correctly predicted.

The following points should be noted in utilizing Pauling's rules for structure prediction.

(1)   Pauling's rules apply to ionically bonded materials only. This is self-evident if you consider elemental Si. In this case, the radius ratio would be $\frac{r_{Si}}{r_{Si}} = 1$, which would imply 12-coordination. Experimentally, of course, Si has the diamond crystal structure, in which each Si atom is tetrahedrally coordinated. This is because Si has covalent, rather than ionic bonding.

(2)   Pauling's rules essentially treat ionic crystal structures as packing of charged billiard balls of different sizes. Deviations from Pauling's rules are often indicative of covalency in the structure.

(3)   While the general rule that larger ions tend to prefer larger coordination numbers is true, coordination polyhedra are not inviolate, especially for the larger cations. That is, the structure is often governed by the smaller, higher-charged cation polyhedra, and the larger cations fit into available interstices. Thus, Ca^{2+} appears in tetrahedral coordination with O^{2-} in $CaSiO_4$, in octahedral coordination in $CaCO_3$, in cubic coordination in $(Zr_{1-x}Ca_x)O_{2-x}$, and 12-coordination in $CaTiO_3$. This is in some ways a manifestation of the fact that ions aren't hard spheres. The deformability of the electron clouds of larger cations tends to accommodate a wider array of coordinations.

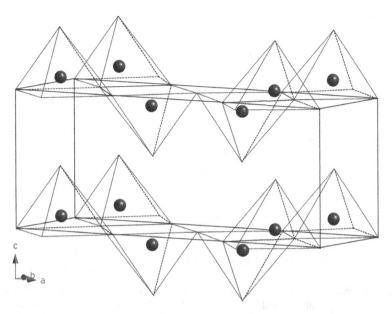

Fig. 13.5 Crystal structure of V_2O_5, showing the square pyramidal 5-coordination for the V ions. The oxygens are located at each of the apices of the V coordination polyhedra.

(4) Pauling's rules work less well when the coordination polyhedra are irregular. For example, Mo^{6+}, V^{5+}, and sometimes W^{6+} are too small for octahedral coordination, and too big for tetrahedral coordination. This can lead to distorted coordination polyhedra. Figure 13.5 shows one of the crystal structures of V_2O_5, where it is apparent that the effective coordination number of the V^{5+} is five. In other cases, the coordination polyhedra themselves are undistorted, but the cation displaces from the central position, so that some of its neighbors are closer than others.

13.2   Bond Valence

The beauty of Pauling's rules is their simplicity and their general applicability. Based on no more *a priori* knowledge than a set of radii and the ionic charges, it is often possible to predict the atom coordinations. Furthermore, since nature tends to reuse the same crystal structures, once the composition and the atom coordinations are known, the crystal structure itself can often be determined.

Pauling's rules provide a reasonable, but not perfect fit to the experimental structure data for ionically bonded materials. However, it would be useful to have a more general formalism that better described less ionic materials, as well as materials with less symmetrical coordination polyhedra.

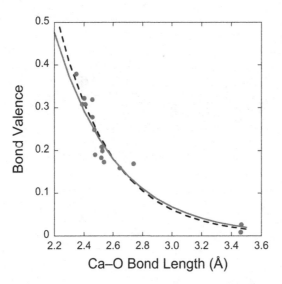

Fig. 13.6 Calculated values for the bond valence (symbols) of the Ca^{2+}–O^{2-} bond plotted with two different fits to the data. The dashed line corresponds to $r_{ij}^0 = 1.967$ Å and $B = 0.37$. The solid line is for $r_{ij}^0 = 1.896$ Å and $B = 0.41$. Redrawn with permission from I. David Brown, *The Chemical Bond in Inorganic Chemistry: The Bond Valence Model*, Oxford University Press, New York (2002), American Chemical Society. Fig. 3.1, p. 27.

One approach to this is to assign empirically a *bond valence,* v_{ij}, that is a function of the separation distance between the atoms i and j. This serves in much the same way as the electrostatic bond strength in Pauling's rules, but is more general in that it recognizes the fact that bonding becomes stronger (and hence the amount of electric flux contributed by the atom to the bond) as atoms approach more closely. Two different functional forms for the distance dependence of the valence are in use: $v_{ij} = \left(\frac{r_{ij}^0}{d_{ij}}\right)^n$ or $v_{ij} = \exp\left(\frac{r_{ij}^0 - d_{ij}}{B}\right)$, where r_{ij}^0 is a normalization value for the bond length of a particular pair of bonded atoms. When the observed separation distance between the atoms, $d_{ij} = r_{ij}^0$, then $v_{ij} = 1$. Note that n and B are constants for a given bond that describe how the bond valence scales with distance. Figure 13.6 illustrates the relationship between bond valence and bond length for the Ca–O bond; data were taken from Brown's book on bond valence.[2] The line fit through the data is an approximation which tends to somewhat underestimate the strength of the shortest bonds and overestimate the importance of the longest bonds.

Table 13.2 illustrates the correlation between Pauling's rules and the bond valence approach. It is worth noting here that the "valence" can be either the nominal valence of an ion, or alternatively, the number of electrons participating in a covalent bond. Thus the valence for a single bond is one, while the valence of a double bond should be two, etc.

[2] I. David Brown, *The Chemical Bond in Inorganic Chemistry: The Bond Valence Model*, Oxford University Press, New York (2002).

Table 13.2 Comparison between Pauling and bond valence approaches

Pauling's rules	Bond valence approach
Electrostatic bond strength $= \frac{\text{cation charge}}{\text{cation coordination}}$ Formal anion charge	Bond valence $= v_{ij}$ Valence of the ion or the number of electrons participating in the bonding (to account for covalent multiple bonds) $= V_i$
$\sum EBS = \|anion\ charge\|$	$V_i = \sum_j v_{ij}$, where the sum is over all of the neighbors of atom i

Table 13.3 Example bond valence parameters

Bond type	r_{ij}^0 (Å)	B (Å)
$Ag^{1+}-O^{2-}$	1.842	0.37
$Ag^{1+}-S^{2-}$	2.119	0.37
$Al^{3+}-O^{2-}$	1.620	0.37
$Al^{3+}-S^{2-}$	2.13	0.37
$Al^{3+}-F^-$	1.545	0.37
$Al^{3+}-Cl^-$	2.032	0.37
$Al^{3+}-Br^-$	2.20	0.37
$Al^{3+}-I^-$	2.41	0.37
$Al^{3+}-N^{3-}$	1.79	0.37
$Al^{3+}-P^{3-}$	2.24	0.37
$B^{3+}-O^{2-}$	1.371	0.37
$Ba^{2+}-O^{2-}$	2.285	0.37
$Ca^{2+}-O^{2-}$	1.890	0.41
$Cr^{2+}-O^{2-}$	1.73	0.37
$Cr^{3+}-O^{2-}$	1.724	0.37
$Cr^{4+}-O^{2-}$	1.81	0.37
$Cr^{5+}-O^{2-}$	1.76	0.37
$Cr^{6+}-O^{2-}$	1.794	0.37
$Fe^{2+}-O^{2-}$	1.734	0.37
$Fe^{3+}-O^{2-}$	1.759	0.37
$K^{1+}-O^{2-}$	2.132	0.37
$Na^{1+}-Cl^-$	2.15	0.37
$Pb^{2+}-O^{2-}$	1.963	0.49
$Si^{4+}-O^{2-}$	1.624	0.37
$Ti^{4+}-O^{2-}$	1.815	0.37

Table 13.3 gives bond valence parameters for a number of different bonds. It is clear that r_{ij}^0 scales relatively well with the expected bond length. Thus, r_{ij}^0 increases progressively for the F^-, Cl^-, Br^-, I^- series. For many oxides, B is approximately 37 pm.

As an example of the use of bond valence sums, consider the V_2O_5 structure shown in Fig. 13.5. The V_2O_5 structure has one distinct V position, and three O positions, O(I), O(II), and O(III). Table 13.4 shows the observed bond lengths, and the bond valence,

Table 13.4 Bond lengths and bond valences for V_2O_5

V–O bond length (Å)	v_{ij}
1.583	1.793
1.781	1.050
1.882	0.799
1.882	0.799
2.026	0.541
2.801	0.067

calculated using $r_{ij}^0 = 1.799$ Å and $b = 0.37$. It is clear that there is one O neighbor that is particularly close; this is the one at the apex of the pyramid. Four O neighbors form the base of the pyramid, at bond lengths ranging from 1.78 to 2.03 Å. The last oxygen listed in the table is the distance from a V in one layer to the closest oxygen in the layer below. If the bond valences for all of these bonds are summed, $\sum 1.793 + 1.050 + 0.799 + 0.799 + 0.541 + 0.067 = 5.049$, very close to the nominal V valence of 5. O(I) is bonded to one V at a distance of 1.583 Å, and another at 2.801 Å. Summing the contributions from these two produces a net valence of 1.859, slightly below the nominal valence of 2– for the O. O(II) has 2 V neighbors at distances of 1.882 Å, and a third at 2.026 Å. $\sum 2(0.799) + 0.541 = 2.14$, again, close to the oxygen valence. Finally, O(III) has two neighbors at a distance of 1.781 Å, producing a bond valence of 2.10.

The bond valence method requires more *a priori* knowledge than do Pauling's rules, but is also more powerful. Bond valence sums can often be used to assess the correctness of a proposed crystal structure; if the sums differ significantly from the valence of the atom, the structure is likely to be in error. They can also be used to determine atom positions that are difficult to resolve by X-ray diffraction. For example, H is so light that it is a poor scatterer of X-rays. If, following determination of a structure, there is an anion that doesn't obey the sum rule well, H is probably bonded to it. Likewise, bond valence sums distinguish between Al^{3+} and Si^{4+} positions in aluminosilicates. This method also handles distorted coordination polyhedra well, and eliminates some of the artificial changes in ionic radii with coordination number.

The bond valence method has also been used to estimate the bond strengths of complex ions such as SiO_4^{4-}. Values for these are shown in Fig. 13.7. In general, it is found that the most stable compounds are formed between anions and cations of comparable bonding strengths. Thus, Mg^{2+} and SiO_4^{4-} both form bonds of 0.33 valence units. These readily combine to make forsterite, Mg_2SiO_4, which is believed to form a large portion of the Earth's mantle. In contrast, the compound K_3PO_4 entails bonds between PO_4^{3-} complex ions with a bond strength of 0.25 valence units and K^+, with a bond strength of 0.13 valence units. This material has comparatively poor stability because of its poor valence match. Therefore, it is deliquescent, the material will dissolve itself in water absorbed from the atmosphere. This is a consequence of the fact that both ions would sooner bond to H_2O. Thus, the bond valence approach allows

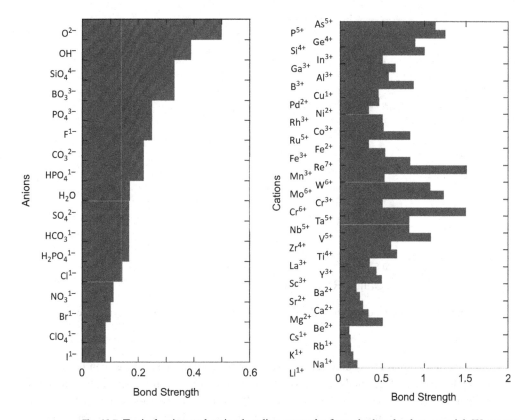

Fig. 13.7 Typical anion and cation bonding strengths from the bond valence model. Water can serve as either a cation or an anion. Data from I. David Brown, *The Chemical Bond in Inorganic Chemistry: The Bond Valence Model*, Oxford University Press, Oxford (2002).

one to develop a set of rules which are comparable to Ramberg's rules to predict which compounds should form from a mixture of ions.

13.3   Structure-Field Maps

Another useful way of correlating radii with structure type is the *structure-field map*. Several different types of structure-field maps have been developed. Since atomic coordination is a strong function of the relative sizes of atoms, radii are often used on the axes. Construction of structure-field maps is most straightforward when all of the bond types on the diagram are comparable, so that choice of which radii to utilize is unambiguous. As an example, consider the AX compounds shown in Fig. 13.8. For the most part, the different structure types group into different regions of the map. The CsCl structure, which entails 8-coordination for both cations and anions are found for the largest cations. NaCl and NiAs have 6-coordination for all atoms, and are found at smaller cation radii. The zincblende and wurtize structures, both of which possess tetrahedral coordination for the atoms, are found for the smaller cations.

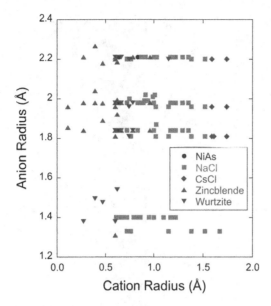

Fig. 13.8 Structure field map for AX compounds plotted using Shannon–Prewitt ionic radii.

A clearer distinction between the NiAs- and NaCl-family structures can be achieved by recognizing that bond types also influence structure. In a survey of ~200 compounds with the rocksalt structure and 46 with the NiAs structure, it is found that the former is favored by a significant difference in electronegativity between the constituent atoms. The preponderance of the compounds which adopt the rocksalt structure have electronegativity differences exceeding 0.8 on Pauling's scale. Exceptions occur primarily in compounds with lone-pair electrons, such as the tin and lead chalcogenides. The NiAs structure is favored by much smaller electronegativity differences, usually <0.7 on Pauling's scale. The exceptions to this guideline are found primarily in the third-row transition metal chalcogenides.

A map for fluorides and oxides of composition A_2BX_4 is shown in Fig. 13.9. The phenacite structure, in which both cations have 4-coordination is adopted by the smallest ions. In the $BaAl_2O_4$ structure, the aluminum atoms sit on tetrahedral sites. The Ba atoms are far too large for this, and are expelled into the largest cavities in the structure. Spinel offers sits for atoms with 4- and 6-coordination.

Structure-field maps are useful in predicting unknown structures and phase transformations. Compounds near one of the boundaries will often undergo transformation to the other phase on excursions of temperature or pressure. Numerous examples of this are given by Muller and Roy.[3] Structure-field maps also help to emphasize the observation that nature has its own sense of parsimony; relatively few crystal structures are common for any particular stoichiometry.

[3] O. Muller and R. Roy, *The Major Ternary Structural Families*, Springer-Verlag, New York (1974).

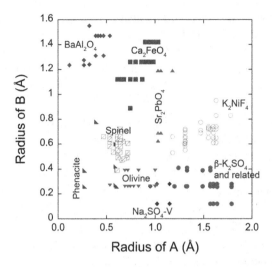

Fig. 13.9 Structure-field map for A_2BX_4 oxides and fluorides. Reasonable separation into distinct phase fields is observed. Redrawn with permission from O. Muller and R. Roy, *The Major Ternary Structural Families*, Springer-Verlag, NY 1974. American Chemical Society, fig. 6, p. 76.

13.4   Problems

(1)   Consider Be_2SiO_4.

    (a)   Apply Pauling's rules to predict the coordination of all atoms in the structure.

    (b)   Using the structure-field map shown in Fig. 13.9 what should the structure be?

(2)   Predict the coordination of each atom in $LiNbO_3$ using Pauling's rules. Radii are given in Appendix A.

(3)   Predict the coordination of each atom in $SrTiO_3$ using Pauling's rules. Radii are given in Appendix A.

(4)   In Diaspore (AlOOH) there are two O sites. The Al is 6-coordinated, with O(1)–Al distances of 1.85, 1.85, and 1.86 Å. The O(2)–Al distances are 1.97, 1.97, and 1.98 Å. Recall that the measured bond length, d_{ij} can be expressed as:

$$d_{ij} = r_O - b\{\ln(v_{ij})\}.$$

For the Al–O bond, $r_O = 1.644$ Å, and b is 0.38. Calculate the bond valence sums and specify to which atom you believe the H is bonded.

(5)   In PbO, $r_{Pb} = 0.98$ Å and $r_O = 1.38$ Å.

    (a)   On the basis of your knowledge of crystal chemistry, predict the coordination of the atoms.

    (b)   Experimentally, the crystal structure of PbO at ambient temperatures is <u>not</u> the one predicted by Pauling's rules. Explain the observed crystal structure and the reason for the discrepancy.

    (c)   What is the consequence of the structure difference on melting point?

(6) Red PbO is a tetragonal (space group 129: *P4/nmm*, with $a = 3.947$ Å, $c = 4.988$ Å) and Pb at (0, 0.5, 0.2385) and O at (0, 0, 0).

    (a) Draw a unit cell, showing the Pb–O bonds.

    (b) Illustrate the coordinations of the Pb and O ions. How can you justify the Pb^{2+} coordination given that Pb^{2+} is a large ion, and in other structures adopts coordinations as high as 12 with O^{2-}?

    (c) The bond valence for the Pb^{2+}–O^{2-} bond can be given by $d_{ij} = r_O - b\{\ln(v_{ij})\}$, where $r_O = 1.963$ and $b = 0.49$. Calculate how much bond valence is contributed by each of the Pb^{2+} near neighbors. How much is contributed by the next near neighbors?

(7) Corundum is the mineral name for α-Al_2O_3. It is rhombohedral with space group *R3c* (#161); $a = 4.763$ Å, $c = 13.00$ Å (using hexagonal axes). Al atoms sit at (0, 0, –0.35) and (0, 0, 0.35). The O are at (0.31, 0, 0.25).

    (a) Plot a unit cell, showing the bonds between the Al and O atoms.

    (b) What are the coordinations of the Al and O atoms?

    (c) Replot the unit cell showing the coordination polyhedra around the Al and discuss the presence of corner, edge, and face sharing.

    (d) For the Al^{3+}–O^{2-} bond, $r_O = 1.620$Å and b is 0.37. Calculate the bond valence sums for Al^{3+} and O^{2-}.

(8) The NiAs structure has space group 186: *P6₃mc* with $a = 3.4392$ Å, $c = 5.3484$ Å, with Ni at (0, 0, 0) and As at (2/3, 1/3, 1/4) and all symmetry related positions.

    (a) Draw a unit cell.

    (b) What are the nearest neighbor coordinations of the Ni and As atoms? What is the closest Ni–Ni distance? This is so close to the distance between two metallically bonded Ni that the coordination of Ni often includes the two nearest Ni neighbors. As a result, the NiAs structure often has a metallic component to its bonding. This contrasts strongly with materials with the NaCl structure (a competitor), which are typically electrical insulators.

(9) In $AgNbO_3$, the Ag ion has a radius of 1.29 Å, the Nb ion has a radius of 0.78 Å, and the O ion has a radius of 1.21 Å. On the basis of this information, predict the coordination of each ion in the structure.

(10) In α-$NaFeO_2$, the Na ion has a radius of 1.0 Å, the Fe ion has a radius of 0.645 Å, and the O ion has a radius of 1.4 Å.

    (a) On the basis of this information, predict the coordination of each ion in the structure.

    (b) Based on the coordination numbers, what structure might this be a derivative of?

(11) BN is polymorphic; a cubic form comparable to sphalerite exists; in addition a hexagonal form is characterized by (space group 186: *P6₃mc* with $a = 2.504$Å, $c = 6.6612$Å) and B at (1/3,2/3,1/4) and N at (1/3, 2/3, 3/4) and all symmetry related positions.

    (a) Draw several unit cells of this structure showing the bonds that connect the B and N ions.

    (b) What is the coordination number for the B and N atoms?

    (c) Contrast this structure to that of graphite.

(d) Brown reports that the bond valence for the B–N bond is given by: $v_{ij} = \exp\left(\frac{1.47\text{Å}-d_{ij}}{0.37}\right)$. Use the experimental bond distances to calculate the bond valence for B–N bonds within and between layers.

(12) **(a)** Describe what is meant by bond valence and the relation between bond valence and Pauling's rules for structure prediction in ionically bonded compounds.

(b) Illustrate how the bond valence for a particular bond changes as a function of separation distance.

(c) For Ti^{4+}–O^{2-}, $r_O = 1.78$ Å and $b = 0.43$. One Ti atom has four oxygen neighbors at 1.9419 Å and two at 1.9755 Å. The oxygen has two close neighbors and one more distant neighbor. Calculate the bond valence sums for Ti and O.

(13) In Cu_2O, the Cu–O distances are 1.849 Å. For the Cu^+–O^{2-} bond, $v_{ij} = \exp\left(\frac{1.61\text{Å}-d_{ij}}{0.37}\right)$. On the basis of this information, do the following.

(a) Determine the coordination of both atoms.

(b) The Shannon–Prewitt radius for Cu^+ is 0.46 Å. Presume that the material can be correctly treated using Pauling's rules. Estimate the coordinations for both ions based on this information.

(c) This answer obtained in part (a) is consistent with the observed crystal structure. Justify the observed coordinations in terms of the characteristics of the ions involved.

14   Crystal Field Theory

14.0   Introduction

While the most significant contribution to atomic bonding arises from overlap of the outermost orbitals of the approaching atoms, in some cases incompletely filled inner d and f orbitals also interact with neighbors. Depending on the atoms surrounding a transition metal atom, there are a variety of different possibilities for the d orbitals.

(1)   The transition metal can give electrons away (e.g. to behave as a d^0 ion), at which point it acts largely as a spherical ion much like an alkali or alkaline-earth cation.

(2)   Electrons in the d orbitals can remain confined on the transition metal core, where they interact comparatively weakly with the surroundings. This interaction, however, can be strong enough that the degeneracy in the energy levels of the d electrons is lifted, so that some orbitals have higher energies than others due to the *crystal field* experienced.

(3)   The d orbitals interact with the adjacent atoms to form extended states, including bands.

This chapter focuses primarily on the second possibility.

14.1   Orbital Shape

As is described in numerous chemistry textbooks, orbitals have a particular shape based on the probability of finding electrons in various locations around the atom. These shapes can be calculated explicitly from solutions of the quantum mechanical equations for the hydrogen atom, and are presumed to apply to the other elements on the periodic table. Figure 14.1 illustrates the shapes of the s, p, and d orbitals.

It is important, to recall, as was described in Chapter 9, that when atoms ionize, they tend to lose electrons from incompletely filled shells on the outside of the atom. Practically, this means that for the 3d transition metal series, the 4s electrons are typically removed from the atom before the 3d shell. Table 9.3 shows the different number of d electrons retained in various ions. The key point to recognize in understanding crystal field theory is that the remaining d or f electrons extend a reasonable distance from the nucleus of the atom, and as a result, interact electrostatically with the neighboring atoms in solids. This can lift the degeneracy of the energy levels for these electrons, as is described in the next section.

14.2   Spherical and Non-Spherical Fields

Recall that electrons in the outermost shells of the atom, especially the s and p electrons are the ones most likely to participate in bonding. Electrons in the d and f shells are often buried far enough inside the atom/ion radius that they do not participate in the bonding. Nonetheless, one would expect the average energy of the electrons in the inner orbitals to rise if the atom is surrounded by a sphere of negative charge due to the repulsion associated with like charges being brought into proximity. The electrons in the d or f orbitals are negative; so is the sphere of negative charge. In reality, of course, the neighbors (also called *ligands*) do not produce a sphere of charge, but are instead arranged in some coordination polyhedron. For many transition metals, tetrahedral and octahedral coordination geometries are common.

Consider the geometry shown in Fig. 14.2, in which Ti^{3+} is in octahedral coordination with O^{2-}. Ti^{3+} has a single d electron left following ionization. If the electron

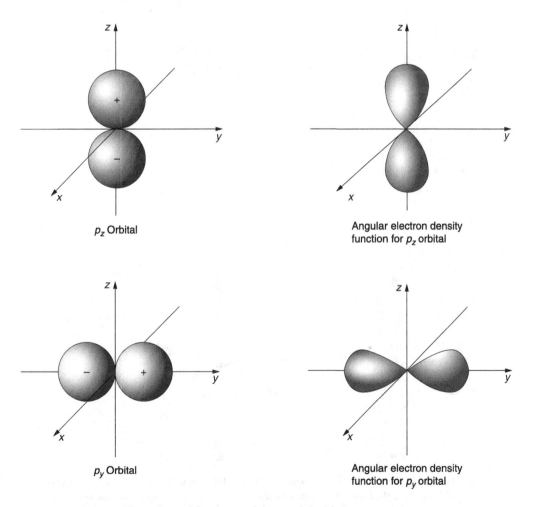

Fig. 14.1 Illustrations of the shapes of the p and d orbitals.

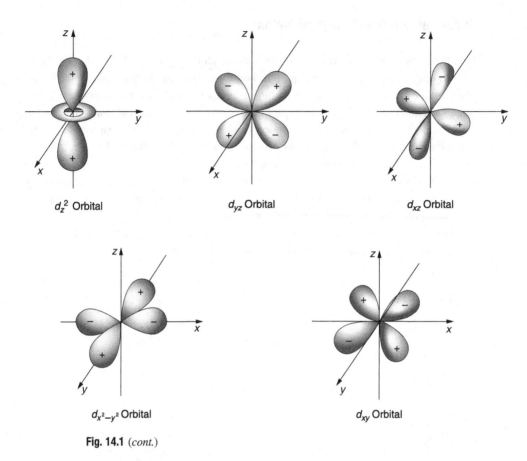

d_{z^2} Orbital d_{yz} Orbital d_{xz} Orbital

$d_{x^2-y^2}$ Orbital d_{xy} Orbital

Fig. 14.1 (*cont.*)

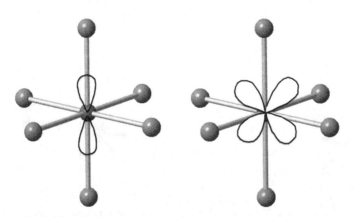

Fig. 14.2 Orientation of the (left) d_{z^2} and (right) d_{xz} orbitals with respect to Ti^{3+} in octahedral coordination. An electron in the former points directly at the neighbor, while the latter points between neighbors.

Table 14.1 Typical Dq values for some transition metal ions in octahedral coordination with H_2O

Atom	Dq (cm^{-1})	Dq (eV)	Atom	Dq (cm^{-1})	Dq (eV)
Ti^{3+}	2030	0.252	V^{2+}	1180	0.146
V^{3+}	1800	0.223	Cr^{2+}	1400	0.174
Cr^{3+}	1760	0.218	Mn^{2+}	750	0.093
Mn^{3+}	2100	0.260	Fe^{2+}	1000	0.124
Fe^{3+}	1400	0.174	Co^{2+}	1000	0.124
Co^{3+}	1910	0.237	Ni^{2+}	860	0.107
			Cu^{2+}	1260	0.156

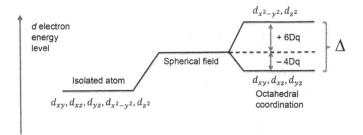

Fig. 14.3 Splitting of d electron energy levels in an octahedral crystal field.

occupies the d_{xy}, d_{xz}, or d_{yz} orbitals, then the energy of the electron is comparatively lower, since it does not point directly at one of the negatively charged neighbors. In contrast, if the electron is in the d_{z^2} or $d_{x^2-y^2}$ orbital, then its energy is increased, as the lobe of charge from the orbital points directly towards its anion neighbor. The relative energy level for the d electrons is shown in Fig. 14.3. The split in energy for an octahedral field is given by $10Dq = \Delta$, where $D = \frac{35Ze^2}{a_o^5}$, a_o is the separation distance between the two atoms, Ze is the nuclear charge of the ligand, and q comes from the integral of the radial wavefunction for the electron.

Table 14.1 gives reference values for Dq for several transition metal ions in octahedral coordination with H_2O. Recall that the total energy split is $10Dq$, putting the energy splits into the visible energy range for many of the cases shown.

It is sensible that the magnitude of the crystal field splitting should depend on a number of factors, including the following:

- The strength and type of bonding to the neighboring atom: The stronger the attractive forces, the shorter the interatomic separation distance, the more the interaction with the inner electron shells, and the larger the crystal field splitting. Indeed, as shown via the equation above, the splitting increases with the fifth power of the separation distance. Furthermore, the type of bonding matters. Recall that covalent bonding is characterized by electron sharing between atoms. This moves the bonding electrons closer to the buried d or f shells. Thus covalency increases the crystal field splitting.

  As an example, the magnitude of the energy level split is 25% larger when a transition metal is bonded to the more covalent NH_3, than when it is bonded to

Table 14.2 Scaling factors for crystal field splitting values for selected ligands relative to oxides

CN^-	1.7
NH_3	1.25
H_2O, O^{2-}	1.00
F^-	0.9
Cl^-	0.8
Br^-	0.76
S^{2-}	0.7

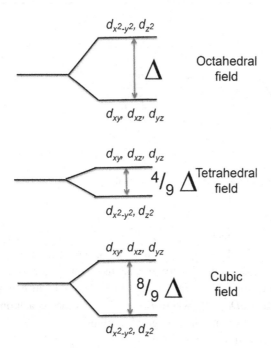

Fig. 14.4 Relative crystal field splitting for octahedral, tetrahedral, and cubic crystal fields.

either H_2O or O^{2-}. Scaling factors for other ligands are given in Table 14.2. Conversely, the more ionic bond with the F^- ion decreases the crystal field splitting by a scaling factor of 0.9.

- The number and arrangement of nearest neighbors: The arrangement of the neighbors will govern which orbitals interact most strongly with the neighboring atoms. For tetrahedral or cubic coordinations, for example, the energy level splits are smaller, since the orbitals do not point as directly at neighboring atoms. The relative magnitude of the crystal field splitting is shown in Fig. 14.4.

If the magnitude of the crystal field splitting is comparatively small, then the electrons populate the d orbitals in a manner consistent with Hund's rules to maximize spin (see Fig. 14.5). For the most part, O^{2-} ligands preserve the high spin state of the transition metal cation. Stronger ligands, such as CN^- and NH_3 increase the splitting, and can

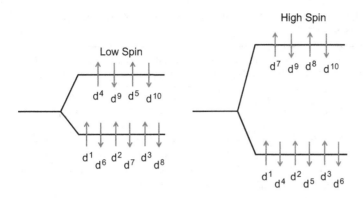

Fig. 14.5 Illustrations of the low and high field spin states.

favor low spin complexes in which the electron spins pair in the low energy level before the upper level is filled.

A few notes about the low and high spin states follow.

(1) The d^5 electron configuration is comparatively stable unpaired. As a result, the high spin configurations tend to be favored unless the crystal field splitting is quite strong.

(2) Transition metals in the 4d and 5d series are more likely to form low spin configurations than 3d transition metals. Recall from Chapter 9 that the 4d and 5d orbitals extend to a larger fraction of the overall atom size (e. g. farther from the nucleus) than the 3d series. Thus, they are more likely to interact strongly with the neighboring atoms, producing larger crystal field splits. As a general rule $\Delta(5d) > \Delta(4d) > \Delta(3d)$. This has numerous consequences in terms of color and for magnetic properties.

(3) Low spin states are less likely when the atoms are in tetrahedral coordination, relative to octahedral coordination, since the crystal field splitting is smaller.

14.3 Crystal Field Stabilization Energy and Site Occupancy

For a transition metal ion in octahedral coordination, each electron in the lower energy orbital stabilizes the energetics of the crystal by $4Dq$, whereas each electron in the higher crystal field level destabilizes the energy by $6Dq$. For octahedral coordination the crystal field stabilization energy, E_{cfse}, is

$$E_{cfse} = 11.962 \frac{J}{\text{mole}} [4 \times e^- \text{ in lower level} - 6 \times e^- \text{ in upper level}] \times Dq(oct)$$

for Dq in cm^{-1}.

An example of the importance of this is shown in Fig. 14.6, where the variation in lattice energies is shown for several 3d transition metal halides (note that the negative of the lattice energy is plotted). There is a general trend that the magnitude of the lattice energy is increased going across the transition metal row, due to the decrease in radius.

Table 14.3 Relative stabilization energies of ions in octahedral and tetrahedral coordination for some 3d transition metal species

Number of d electrons	Atom	Stabilization octahedral (kJ)	Stabilization tetrahedral (kJ)	Octahedral site preference energy (kJ/mole)
1	Ti^{3+}	96.6	64.4	32.2
2	V^{3+}	173.6	120	53.6
3	V^{2+}	168.2	36.4	131.8
	Cr^{3+}	251.0	55.6	195.4
4	Cr^{2+}	100.4	29.3	71.1
	Mn^{3+}	150.2	44.3	105.9
5	Mn^{2+}	0	0	0
	Fe^{3+}	0	0	0
6	Fe^{2+}	47.7	31.4	16.3
	Co^{3+}	188	109	79
7	Co^{2+}	71.5	62.7	8.8
8	Ni^{2+}	122.6	78.9	95.4
9	Cu^{2+}	92.9	27.6	65.3
10	Zn^{2+}	0	0	0

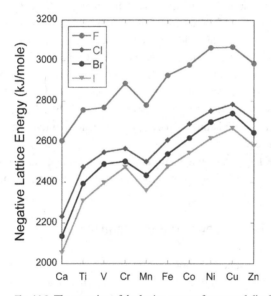

Fig. 14.6 The negative of the lattice energy for several divalent transition metal halides. Adapted, with permission, from P. George and D. S. McClure, "The Effect of Inner Orbital Splitting on the Thermodynamic Properties of Transition Metal Compounds and Coordination Complexes", in *Progress in Inorganic Chemistry vol. 1*, Ed. F. A. Cotton, Interscience Publishers, NY, 1959, Figure 2, p. 394.

Moreover, local anomalies are observed due to the crystal field stabilization, producing the "bumps" in the curve. Similar trends are observed for the divalent oxides.[1]

---

[1]   Data taken from P. George and D. S. McClure, "The effect of inner orbital splitting on the thermodynamic properties of transition metal compounds and coordination complexes," in *Progress in Inorganic Chemistry, vol. 1*, ed. F. A. Cotton, Interscience Publishers, NewYork (1959).

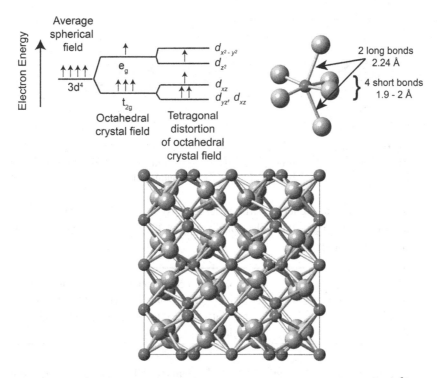

Fig. 14.7 Electron energy levels for bixbyite, Mn_2O_3. The energy levels of the Mn^{3+} ion can be lowered if the MnO_6 octahedron is distorted to move two of the oxygen atoms further away, producing the structure shown at the bottom.

A comparable formalism can be written down for atoms in tetrahedral coordination. Table 14.3 compares the stabilization of ions in tetrahedral and octahedral coordination. It can be seen by comparing the stabilization energies for the tetrahedral and octahedral sites that Ni^{2+} and Cr^{3+} have strong preferences for octahedral coordination, while d^0, d^5, and d^{10} ions have no site preference based on E_{cfse}. These values can be used to assess the likely occupancy of various transition metal atoms in compounds where more than one site is available, such as the spinels. It should be noted, however, that other factors can also influence the likely site occupancy. For example, d^{10} ions, for which there is no E_{cfse}, have a tendency towards tetrahedral coordination, due to the tendency towards covalency. This will be discussed again in detail in Chapter 29.

Jahn–Teller Distortions

Sometimes additional distortions develop due to *Jahn–Teller* effects. Consider a d^4 ion, such as Mn^{3+} or Cr^{2+} in octahedral coordination. When these ions are in a low spin configuration, the electron distribution is as shown in Fig. 14.7; there is one electron in the higher energy level. Such materials can be further stabilized if the one electron in the orbital pointed directly at the anion neighbor can be moved further away, as is the case for bixbyite Mn_2O_3. That is, the energy can be lowered if the coordination polyhedron is distorted, so that instead of having six equidistant neighbors, two are moved further

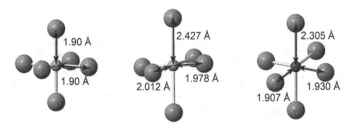

Fig. 14.8 Geometry of the octahedral coordinations for (left) Cr in CrF_3 with no Jahn–Teller distortion and a symmetrical coordination geometry, (middle) Cr in CrF_2, and (right) Cu in CuF_2 in which the Jahn–Teller distortions push two of the neighbors significantly further away from the cation.

away. Such Jahn–Teller distortions lead to asymmetric coordination geometries, and additional splitting of the d electron energy levels. The ions for which this is important include the following.

d^4	Cr^{2+}, Mn^{3+}	In octahedral coordination: octahedra distort along the z axis,
d^9	Cu^{2+}	usually by elongating the bonds along this axis
d^3, d^4		In tetrahedral coordination: tetrahedral groups are elongated
d^8, d^9		or flattened

Thus, as shown in Fig. 14.8, $Cu^{2+}F_2$, and $Cr^{2+}F_2$ have a distorted rutile crystal structure and asymmetric coordination polyhedra. In contrast, $Cr^{3+}F_3$ has regular octahedral coordination since the d^3 ion doesn't lower in energy by additional distortion of the coordination polyhedra.

14.4   Color and Phosphors

Because the splitting of the d or f energy levels often corresponds to energy differences in the visible range, transition of electrons from one level to another produces color. Indicating drierite, for example, is used to tell when the drying media, $CaSO_4$, has started to exhaust its capacity. The key reaction is

$$CaSO_4 \overset{H_2O}{\rightleftharpoons} CaSO_4 \cdot 2H_2O.$$

The $CaSO_4$ is doped with the transition metal Co. Note that Co in four coordination (as it is in $CaSO_4$: Co) is pink, while Co is six coordination (as it is in $CaSO_4 \cdot 2H_2O$) is blue. This gives a visual indication of the state of the reaction.

The oxidation state of the cation will also affect the colors. For example, in silicate glasses, Fe^{2+} leads to a blue-green color (easily observed by looking at many window glasses edge-on), while Fe^{3+} produces a yellow tint. Indeed, iron is one of the most common colorants in minerals as well, since it is the only transition metal ion among the eight most common elements in the Earth's crust.

In the same vein, consider the importance of the interatomic separation distance on the crystal field splitting by comparing the colors of Cr_2O_3 and ruby (Al_2O_3 doped with

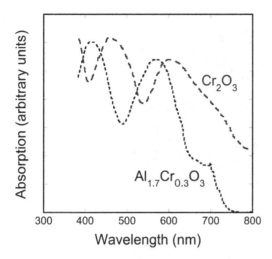

Fig. 14.9 Relative absorption coefficients for Cr_2O_3 and heavily doped ruby. The Cr–O separation distances are close to 1.9 Å in ruby and close to 2 Å in Cr_2O_3. Redrawn from V. O. Schmitz-Du Mont and D. Reinen, *Farbe und Konstitution bei anorganischen Festoffen. III Die Lichtabsorption des dreiwertigen chroms*, Z. fur Electrochemie Bd. 63 Nr. 8 978-987 (1959) Abb .5. Copyright Wiley-VCH Verlag GmbH & Co. KGaA. Adapted with permission.

Cr^{3+} on the Al^{3+} site). In both cases, the color is produced by absorption associated with promoting an electron from the lower-lying d energy level in the Cr^{3+} ion to the higher level. In Cr_2O_3, the Cr^{3+} ion absorbs in the blue and red, producing a green color. It can be seen in Fig. 14.9 that as the atom separation distance decreases, the crystal field splitting rises, and the energy of the absorption is shifted to higher energies (shorter wavelengths). The stronger crystal field in ruby (due to the shorter Al^{3+}–O^{2-} bond) produces absorption in the violet and green. The transmission of red and some blue produces the purplish pink color of ruby. This link between near neighbor distance and optical properties means that ruby crystals can be used as *in situ* probes of the pressure achieved in high-pressure cells.

In other Cr^{3+}-containing minerals, the balance between the two transmission windows is poised so delicately that the color perceived depends on the spectrum of the illumination. Alexandrite, for example, is a pale green in daylight (the output of the Sun peaks in the green) and deep red in candlelight, where more of the incident radiation is at longer wavelengths.

In *phosphors* (also called luminescent materials), electrons excited to higher energy levels give away this energy as packets of light. This light is in addition to any blackbody radiation produced due to the material being at a finite temperature. Phosphors are often wide band gap compounds that have been doped with rare-earth or transition metal compounds to provide light emission in the visible range. This means that they typically can accommodate cations that take 4-, 6-, or 8-coordination. Typical doping concentrations are on the order of 1 mole%. This allows light emitted in one portion of the material to escape without being reabsorbed by another ion, where the excited electron might return to a ground state without radiating the light. The decrease in the amount of emitted light as the dopant concentration is increased is sometimes known as concentration quenching.

Phosphors are used widely in applications ranging from coatings for light sources, X-ray fluorescence, to lasers. Crystal field theory often determines the wavelengths of

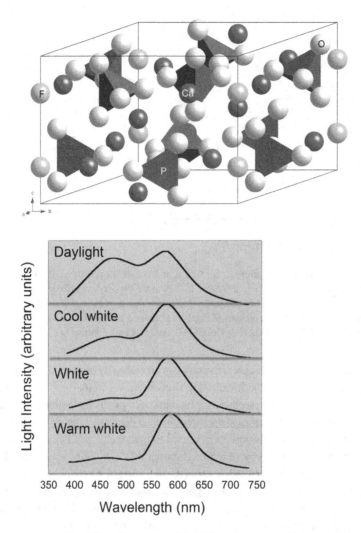

Fig. 14.10 (Top) Crystal structure of $Ca_5(PO_4)_3F$. (Bottom) Light output from a series of phosphors with different color outputs. Adapted, with permission, from data from Sylvania. K. H. Butler, *Fluorescent Lamp Phosphors: Technology and Theory*, The Pennsylvania State University Press, University Park, PA 1980 Fig. 8.9, p. 108.

light that are emitted by the phosphor. Optical properties are described in more detail in Chapter 27; that chapter also discusses several X-ray phosphors.

When an electron is excited to a higher energy state, two families of selection rules govern the likelihood of a transition to a lower energy state via emission of a photon.

(1) The spin selection rule states that transitions between levels with different spin states have a low probability of occurring.

(2) The parity selection rule makes electronic transitions between levels with the same parity less likely to occur. Some crystal fields reduce the importance of this selection rule.

The more strongly forbidden a transition is, the longer an electron will stay in the excited state before releasing the energy radiatively.

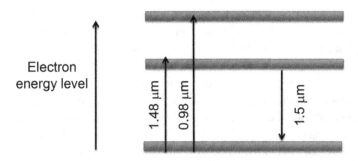

Fig. 14.11 Portion of the energy level diagram for an Er-doped silica fiber, showing absorption lines at 0.98 and 1.48 μm that can be used for pumping the laser, and an emission line at 1.5 μm.

As an example of the use of a phosphor, consider the operation of fluorescent lamps. Most fluorescent lamps have glass tubes that are filled with a small amount of mercury gas. Electrical excitation raises the mercury to higher energy levels, leading to emission of ultraviolet (UV) light that is not directly useful for lighting. Thus, the glass tube is coated on the inner surface with a phosphor coating. The phosphor is designed to absorb the UV radiation, and re-emit it in the visible. Similar approaches are used to make "white LEDs". A widely used coating is a $Ca_5(PO_4)_3F$ doped with Sb^{3+} and Mn^{2+}. The Sb^{3+} absorbs the ultraviolet radiation. Two pathways then follow: first the Sb^{3+} can re-radiate the light downshifted to the blue portion of the electromagnetic spectrum. Alternatively, it can transfer some energy to the Mn^{2+}, which radiates in the yellow/orange. The latter process is called charge transfer, since the host absorbs the exciting light, and transfers the energy to the atom responsible for the radiation. The more covalent the bonding in the host material, the lower the energy of the charge transfer absorption. The relevant crystal structure of $Ca_5(PO_4)_3F$, along with typical emission spectra is shown in Fig. 14.10. For such phosphors, the output color shifts towards the red wavelengths as the concentration of Mn rises. If the F is partially replaced with Cl, both emission bands shift to longer wavelengths.

The ability to tailor the color of a phosphor based on the crystal chemistry of the host structure is very important, and drives major research efforts.

Fiber optic lasers are also enabled by the energy differences associated with crystal field splitting. SiO_2 fibers are doped with around 500 parts per million of Er^{3+}, yielding the energy diagram shown in Fig. 14.11. It is notable that, because the 4f shell is buried inside the 5s and 5p shells, the energy levels of the emitted radiation vary comparatively little from host-to-host. Electrons from Er^{3+} can be excited to higher energy levels by semiconductor lasers. Stimulated emission occurs at 1.5 μm, close to the 1.55 μm band used as one of the two key transmission wavelengths for optical communication.

14.5   Problems

(1)   Numerous materials are isomorphs of rocksalt (meaning that they have the same crystal structure, but a different chemical composition). Consider the isomorph NiO.

Describe, based on your understanding of crystal field theory, which d orbitals are expected to be low in energy, and which should be higher energy levels? Rationalize your choice based on the coordination.

(2)   Ruby is Cr-doped α-Al_2O_3, in which the Al and Cr are in 6-coordination with oxygen; the oxygen is tetrahedrally coordinated.

   (a)   Draw a schematic diagram showing the d electron energy levels relative to the spherical field, as well as the occupancy of the orbitals. Assume a high spin state.

   (b)   Why does the introduction of Cr result in color in ruby? α-Al_2O_3 itself is a transparent, wide band gap insulator.

   (c)   It is experimentally observed that the optical transitions in ruby are dependent on the applied hydrostatic pressure. Why should this be?

(3)   Draw a sketch of how you think the lattice energies would vary along the series ScF_3, TiF_3, VF_3, CrF_3, MnF_3, FeF_3, CoF_3, NiF_3, and GaF_3. Explain your reasoning.

(4)   (a)   Describe the factors that control the magnitude of the crystal field.

   (b)   Consider Cr_2O_3, in which the Cr is in octahedral coordination. Draw a schematic diagram showing the d electron energy levels relative to the spherical field, as well as the occupancy of the orbitals.

   (c)   Give one property that is affected by the crystal field in this compound, and explain why.

(5)   Explain how crystal field theory influences color in oxides and site occupancy in spinels.

(6)   How should the magnitude of the crystal field splitting change as a function of the length of the bonds and why?

15   Solid Solutions and Phase Diagrams

15.0   Introduction

Phase diagrams are the maps showing the most stable form of matter for a given set of conditions (pressure, temperature, chemical composition, etc.).[1] They can be determined experimentally or, when adequate thermodynamic data are available, by computation. In relating phase diagrams to crystal chemistry, we seek an atomistic understanding of the geometry of the diagram and of the thermodynamic parameters on which the diagram is based. Among the questions to be considered here are the following:

- Can the number of intermediate phases in a composition diagram be predicted?
- Which structure types will occur?
- What determines solid solution limits?
- Where is the deepest eutectic?

Such questions can be approached at several levels of sophistication. Here, we adopt a crystallographic point of view, attempting to relate thermochemical observations to atomic structure. The aim is to develop physical insight and to recognize trends, rather than to explain every observation. This chapter will concentrate on guidelines provided by atomic polarizability and radii. It is important to recognize that real materials can be quite complex, and prediction of phase diagrams is complicated by the fact that in some cases the cohesive energies of competing phases are very similar.

Excellent compilations of existing diagrams are available from NIST-ACerS and ASM.[2] Only a small fraction of the total number of phase diagrams have been worked out. There are approximately 90 elements on the periodic table. The number of phase equilibrium diagrams is given by $\frac{n!}{m!(n-m)!}$, where n is the total number of chemical elements, and m is the number of elements involved in a given system. Thus, for binary combinations, there are $\frac{90!}{2!(88)!} = 4005$ systems. For ternaries $\frac{90!}{3!(87)!} = 117\,480$ and for quaternaries $\frac{90!}{4!(86)!}$. A daunting task.

---

[1] Sections of this chapter were originally published in Newnham, R.E., "Phase diagrams and crystal chemistry," Chapter 1 in *Phase Diagrams for Materials Science and Technology*, ed. A. M. Alper, Academic Press, New York, pp. 1–73 (1978). Reproduced here by permission.

[2] ACERS-NIST Phase Equilibria Diagrams, American Ceramic Society and National Institute for Standards and Technology, Columbus, OH, 2013; ASM Alloy Phase Diagram Database, ASM International, Materials Park, OH.

Thus, the simplicity of approach to be taken here is a strength as well as a weakness. A useful theory is not only accurate but easy to use and of general applicability as well. Arguments based on the simple concepts discussed here can be quickly applied to a large number of situations, even when only incomplete information is available.

15.1   Miscibility and Compound Formation

Miscibility refers to the ability for one material to "dissolve" into another, without causing a change in crystal structure. In the solid state, this is often called solid solution.[3] Substitutional solid solution is favored when the constituents of the end members are similar in size, charge, electronegativity, and end-member crystal structure, as will be described in detail later in this chapter. Intermediate compounds are favored the more the end members differ in these four factors.

The importance of ionic size can be illustrated with the oxide binary phase diagrams shown in Fig. 15.1. Solid solutions form when the ions are similar in size; hence the oxides of Ni^{2+} (0.7 Å) and Mg^{2+} (0.72 Å) form a complete substitutional solid solution. The Ca^{2+} ion (1.00 Å) ion is larger than Ni^{2+}, and MgO–CaO are only partially soluble. A deep eutectic and only very limited solubility occur in the SrO–MgO binary. Solid solubility is negligible in the remaining three diagrams, as compound formation develops. The BaO–NiO and SrO–BeO binaries show two and four intermediate phases, respectively. Thus, the tendency towards compound formation increases with size mismatch, as the extent of substitutional solid solution decreases. As will be discussed in more detail below, substitutional solid solution occurs when one atom replaces another in a crystal structure. Interstitial and vacancy solid solutions behave differently.

The influence of valence on oxide phase diagrams is less obvious, but in general, the number of intermediate phases appears to increase with the difference in valence, unless the material is able to compensate for the charge imbalance in some way. Al^{3+} (0.53 Å), Mg^{2+} (0.72 Å), and Ti^{4+} (0.605Å) are generally found in six coordination with oxygen. Spinel ($MgAl_2O_4$) is the only intermediate phase between MgO and Al_2O_3, where the valence difference is one. In the MgO–TiO_2 binary, there are three compounds, showing an increased tendency towards compound formation with valence difference. Large differences lead to a large number of intermediate phases and deep eutectics. As a result, the Li_2O–MoO_3 system used as a flux in growing single crystals is an important example, with numerous intermediate phases, despite the fact that Li^+ (0.74 Å) and Mo^{6+} (0.60 Å) are similar in size.

Solid solutions between ions with different valences are less common because of the importance of electrical neutrality. Only a very few stable structures (including the rocksalt, fluorite, and perovskite structures) tolerate defect concentrations of more than a few percent. These will be discussed more extensively in the section on vacancy solid solutions.

---

[3] Note that "solid solution" is often used with a slightly relaxed definition, in which the main structural features are the same, but displacive phase transitions may lead to differences in space group across the phase diagram.

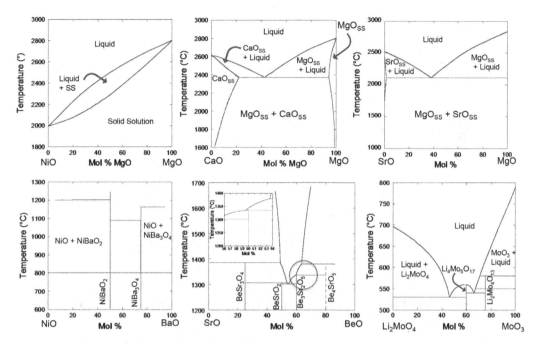

Fig. 15.1 Six binary phase diagrams illustrating the importance of ionic size and charge. Complete solid solution occurs in the MgO–NiO system, where the cations are similar in size. This gives way to extensive compound formation when one cation is small and the other large. The inset in the SrO–BeO phase diagram shows that the phase rule is not violated near $Be_3Sr_2O_5$. Diagrams are redrawn, with permission, from those in the ACERS-NIST Phase Equilibria Diagrams, American Ceramic Society and National Institute for Standards and Technology, Columbus, OH.

Trends in Binary Phase Diagrams of the Elements

Empirical rules have been devised to explain the various features of phase diagrams. A study of binary systems between two solid elements yields some interesting correlations regarding the positioning of the solubility limits, the existence of inter-mediate compounds, and eutectic compositions. Figure 15.2 shows the Fe–Ge phase diagram.[4] Fe accepts about 12% Ge in solid solution, but there is virtually no solid solution on the Ge side. Several intermediate compounds exist, mainly on the Fe side, whereas the lowest eutectic temperature is closest to Ge. This is a feature common to many binary phase diagrams: the lowest eutectic and the largest solid solution are on opposite sides.

A survey of hundreds of binary diagrams reveals the following trends.

(1)   The lowest eutectic is usually nearest the end member with the lowest melting point. These include the weakly bonded metals such as the alkali and alkaline-earth elements, and the semimetals with covalent bonds and lower coordination numbers. Group B elements in columns IIB–VIB such as tellurium, cadmium,

[4] H. Okamoto, "Fe–Ge (iron–germanium)," *J. Phase Equil. Diff.*, **29** (3), 292 (2008).

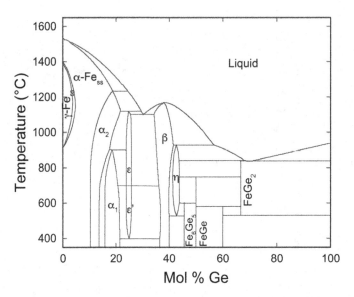

Fig. 15.2 Phase diagram for Fe–Ge. Adapted, with permission, from H. Okamoto, "Fe–Ge (iron–germanium)," *J. Phase Equil. Diff.*, **29** (3), 292 (2008).

and bismuth are representative. These elements, with their unusual bonding patterns have great difficulty accepting normal metals into solid solution. As a result, segregation into a two-phase region takes place instead.

(2)   More extensive solid solutions are observed at the other end of the binary phase diagram. Typical metals such as the transition metal elements generally have more symmetric crystal structures with high coordination numbers, stronger bonds, and higher melting points. The densely packed structures of these metals readily accept most other elements into solid solution. The lowest eutectic point is seldom nearest this end member.

Using data from hundreds of binary phase diagrams, Okajima and Sakao[5] have devised a listing of solid elements that describes their tendency to form solid solutions or eutectics. Elements near the top of the list form extensive solid solution, while those at the bottom are nearest eutectics. As shown in Table 15.1, titanium is at the top of this list, and tellurium is at the bottom. The solid solubility limit in higher-ranked elements is generally larger than lower-ranked elements. Examples are shown in Fig. 15.3, where four binary diagrams follow this ordering sequence.

Note that elements like Ti that are near cross-overs in the periodic system are ranked especially high. The existence of mixed 4s, 4p, and 3d orbitals with similar energy makes many different bonding geometries possible. Many other elements are readily accepted into solid solution. Transition metal elements in the 5d and 4d series with strong bonds and high melting points are also near this end of the scale, far removed

---

[5] K. Okajima and H. Sakao, "Empirical rules on solid solubility limits and activities in binary alloys," *Trans. Japan Inst. Metals,* **16** (9), 557–568 (1975).

Table 15.1 Arrangement of the elements related to their ability to form solid solutions with other elements in binary phase diagrams. Ti accepts most elements into solid solution, Te accepts the fewest

Ti	Ni	V	Sb	Zn
Ir	Co	Re	As	Bi
Ta	Pt	B	Ge	Ga
Os	Th	Cr	Si	Hg
Zr	Nb	Fe	Be	K
Pd	Ru	Mn	Ca	Na
U	W	Cu	Ba	Se
La	Mo	Al	Li	Te
Ce	Rh	Mg	In	
Pu	Hf	Tl	Sn	
Au	Ag	Pb	Cd	

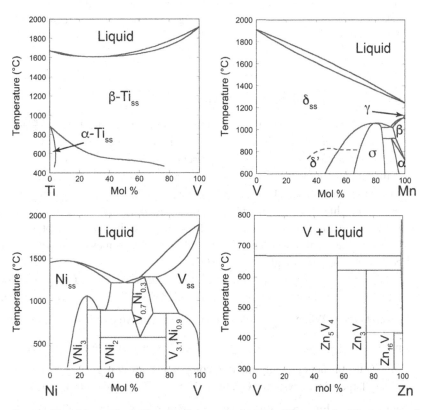

Fig. 15.3 Four binary phase diagrams for metals showing differences in the propensity for solid solution formation. All of the elements shown are from the 3d transition metal series. Figures adapted, with permission, from ASM Alloy Phase Diagram Database, ASM International, Materials Park, OH.

from low eutectic points. Iron and other common 3d elements are nearer to the middle of the scale.

The ranking in Table 15.1 is correct for more than 90% of binary phase diagrams amongst elements, and is only slightly out-of-line for the remaining few.

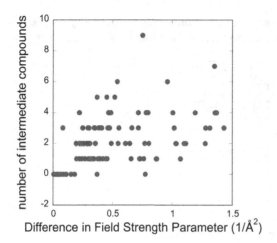

Fig. 15.4 Correlation between the number of compounds formed and the difference in the cation field strength parameter for oxide systems. Large differences in field strength lead to extensive compound formation. In the $CaO-Al_2O_3$ system, for instance, there are four intermediate compounds.

Dietzel's Correlation for Phase Diagrams in Oxides

In the theory of ionic crystals, Coulombic fields are of the form $\frac{charge}{distance^2}$, a quantity sometimes referred to as the field strength parameter. In applying this to inorganic salts, the field strength parameter can be represented as $\frac{Z}{d^2}$, where Z is the cation valence and d is the interatomic distance, the sum of the cation and anion radii. The basic idea is that each cation attempts to shield itself from other cations, thereby reducing the Coulombic repulsion. Shielding is accomplished by surrounding the cation with anions, and the field strength parameter is a measure of this effect. Using this concept, correlations can be established with the extent of immiscibility in ionic melts, and with the number of compounds in binary and ternary systems.

Dietzel[6] showed that the number of intermediate compounds is proportional to the difference in field strength of the two cations. These ideas have been extended by several others. When $\Delta\left(\frac{Z}{d^2}\right)$ is less than 10%, extensive of complete solid solution takes place. As $\Delta\left(\frac{Z}{d^2}\right)$ increases, a progression occurs: first simple eutectics are favored, then subsolidus or incongruently melting compounds develop. Finally, when the difference is very large, binary systems with many intermediate compounds occur. Example of this are shown in Fig. 15.4. Binary systems with $\Delta\left(\frac{Z}{d^2}\right) > 40\%$ exhibit very limited solid solution. For the oxide systems analyzed, liquid immiscibility was most common when $50\% < \Delta\left(\frac{Z}{d^2}\right) < 100\%$. Similar principles appear to govern ternary systems. Among silicate ternaries, the field strength parameter between the non-Si^{4+} cations determines

[6] A. Dietzel, *Z. Electrochem.*, **48**, 9 (1942).

the number of compounds. No compounds form when $\Delta\left(\frac{z}{d^2}\right)$ is below 0.05–0.07, while up to three or four compounds appear when $\Delta\left(\frac{z}{d^2}\right)$ lies between 0.7 and 0.8. Such predictions are less reliable for ions with large polarizabilities.

Model Systems

Model systems are systems in which the radius ratios and crystal structures are similar, but where the valences differ. For example Zn_2SiO_4 is a model structure of Li_2BeF_4, with doubled valences. Other model pairs include BeF_2–SiO_2, LiF–MgO, MgF_2–TiO_2, CaF_2–ThO_2, $KMgF_3$–$SrTiO_3$, $RbBF_4$–$BaSO_4$, and CdI_2–$ZrSe_2$. The lower valence compounds generally have lower hardnesses, lower melting point, and lower refractive indices, together with increased chemical reactivity. As might be expected, phase diagrams involving model structures are often similar. Melting points increase with larger valence.

15.2   Types of Solid Solutions

Three different families of solid solutions are known: substitutional, interstitial, and vacancy solid solutions. The most common type of solid solution is the substitutional solid solution in which one atom substitutes for another in a crystal structure. Some of the crystallographic restrictions limiting the substitution are considered in this section. The restrictions are somewhat different for interstitial solid solutions and other defect solid solutions which differ from simple substitution.

Solid solutions can have a profound effect on materials properties, and are widely used to tune the hardness, the band gap, emission wavelengths, and many other responses. The impact of this is discussed throughout the book.

Substitutional Solid Solution

Atoms sometimes substitute for one another in crystals, forming a solid solution: a homogeneous crystal of variable composition. For example, as discussed above, NiO and MgO form a complete substitutional solid solution across the phase diagram. Magnesium and nickel are distributed nearly at random over the octahedral interstices in the rocksalt structure, as shown in Fig. 15.5. In the same way, forsterite (Mg_2SiO_4) and fayalite (Fe_2SiO_4) form a complete solid-solution series. Both end members and all intermediate compositions possess the olivine structure. Oxygen ions make up a close-packed array, with Si^{4+} occupying tetrahedral interstices, and Mg^{2+} and Fe^{2+} in octahedral interstices.

As described above, substitutional solid solution is favored when the end members are similar to each other. That said, some structures are more stable than others, and tolerate extensive atomic substitution. Many examples of mixed crystals occur in the three common metal structures, as well as in the spinel, perovskite, and rocksalt families. On the other hand, quartz and diamond crystals are much less tolerant to

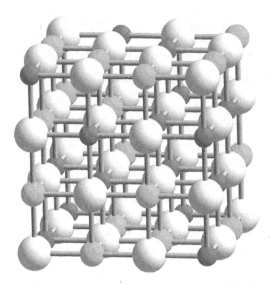

Fig. 15.5 Substitutional solid solution in the rocksalt crystal structure. The anions are shown as the large, light colored spheres; two different types of atoms are distributed randomly on the cation sites.

substitution. As discussed in Chapter 7, solid solutions between semiconductors are widely used to tailor band gaps. There is generally little solid solution in molecular crystals because different molecules have different sizes and shapes.

The extent of substitutional solid solution is also a function of temperature in many systems. Often, the solubility limits increase with temperature because of the entropy of mixing. The large entropy arising from atomic disorder tends to stabilize mixed crystals at high temperatures.

As a rule of thumb, while extensive cation substitutional solid solution is common, it is less common for anions. Of the anion substitutional solid solutions, interchange between OH^- and F^- is the most common. Cl^-–Br^- solid solutions are also common. However, when Cl^- or Br^- replace F^- the large size difference tends to drive ordered superstructures, rather than solid solution. Usually, solid solution between oxides and sulfides is limited. This may be because the more extended p orbitals for S, Se, and Te favors more covalent or metallic character to the bonding, compared to the more ionic bonding present in oxides. The result tends to be new compounds, rather than solid solutions.

Unit cell dimensions vary smoothly with composition in solid solution series. For a cubic crystal, the lattice parameter can be represented by: $(a_{SS})^n = (a_1)^n c_1 + (a_2)^n c_2$ where a_{SS}, a_1, and a_2 are the lattice parameters of the solid solution and the two end members 1 and 2. Mole fractions c_1 and c_2 are the respective concentrations and n is an arbitrary power describing the variation. Vegard suggested that, for many substances, $n = 1$. This can be rationalized in the following way: the lattice parameters should scale approximately as the cube root of the volume. For additive volumes: $n = 3$ is known as Retger's law. Accurate experimental values are needed to determine n, because solid solutions seldom form if a_1 and a_2 differ by more than 15%. In general, Vegard's law is observed in many oxide materials, though deviations are observed in some solid solution systems with immiscibility domes.

Vegard's law is less well followed in many metal alloys. Thus, while the lattice parameters change smoothly with composition, in many cases, the interatomic distances are approximately 10% shorter than the sum of the metallic radii. Cases where the interatomic distances slightly exceed the sum of the metallic radii are also known. There appear to be at least two main contributors to the contraction in bond lengths relative to a straight-line interpolation in many metallic solid solutions. First, there is a finite amount of charge transfer from one atom to another due to the finite electronegativity differences. That is, the bond is not 100% metallic. The ionic component of the bonding favors the shorter bond. Secondly, because the two end members have different compressibilities, in an alloy, one metal expands and the other is compressed until the electron densities equalize. This also tends to reduce the bond length.

In substitutional solid solutions, guest atoms often do not have exactly the same crystallographic coordinates as host atoms. Instead, sometimes the atom displaces slightly in its coordination polyhedra. For example, ruby, $Al_{2-x}Cr_xO_3$ is a solid state laser material. In dilute ruby, Cr occupies the Al site, but is displaced by 0.1 Å along c. Trivalent Cr is larger than Al^{3+}, and the displacement leads to more reasonable interatomic distances. This type of off-center substitution is likely to occur when the site has variable parameters (such as the z position coordinate of Al in $\alpha\text{-}Al_2O_3$. Size and bonding differences between host and solute atom are important in governing this type of behavior.

Miscibility Limits for Substitutional Solid Solution

Hume-Rothery showed that solid solubility is very restricted when atomic radii differ by more than 15%. This limit is true in the most tolerant structures; in other structures, 5–10% might be a better bound. In many metals, if the difference in radii exceeds 15%, less than 1% substitutional solid solution is typical, as shown in Fig. 15.6.

While size is clearly a key factor in governing the extent of solid solution in metals, other factors such as valency and electronegativity are important as well. Darken–Gurry plots are made by plotting metallic radius against electronegativity for various elements. Predictions regarding miscibility were made by drawing an ellipse about the solvent element with diameters ±0.4 in units of electronegativity difference, and ±15% size difference for solute and solvent. Elements falling within the ellipse form extensive solid solution, while those outside do not. The Darken–Gurry plot for iron shown in Fig. 15.7 is typical.[7] Of the 20 elements within the ellipse, 19 are more than 5% miscible in iron, while 28 out of 36 outside the ellipse are less than 5% miscible. When this idea is extended to other metals (using 5% as the dividing line between extensive and restricted solid solution), >3/4 of the predictions are correct. There is some indication that size is more important than electronegativity, since more than 90% of the elements falling outside the 15% size limits show poor solubility. On the other hand, only ~50% of these within the size limits show extensive miscibility.

---

[7] Data collected by J. T. Waber, K. A. Gschneidner, M. Y. Prince, and A. C. Larson, *Trans. Metal. Soc. AIME*, **227** (3), 717 (1963).

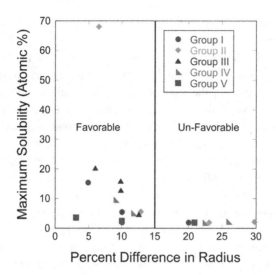

Fig. 15.6 Solubility of different elements into Mg showing the importance of relative size. Those atoms that are 15% difference in size show less than 1% solubility. Adapted, with permission, from J. C. Phillips, "Chemical bonds in solids," in *Treatise on Solid State Chemistry, Vol. 1: The Chemical Structure of Solids*, ed. N. B. Hannay, Plenum Press, New York (1973). Fig. 11, p. 28.

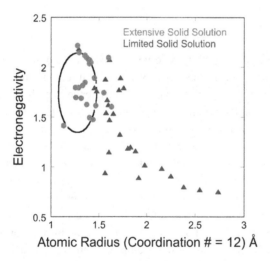

Fig. 15.7 A Darken–Gurry plot for metallic elements alloyed with iron. Solid circles denote elements with a solubility of >5% in either of the polymorphic modifications of iron. Triangles represent elements which do not form extensive solid solutions with iron. Iron itself is located at the center of the ellipse, showing that solubility is greatest between atoms of similar size and electronegativity. Redrawn from R. E. Newnham, "Phase diagrams and crystal chemistry," in *Phase Diagrams for Materials Science and Technology*, ed. A. M. Alper, Academic Press, New York (1978). Figure 5, p. 11. With permission by Elsevier.

The importance of atom size in governing solid solution holds for nonmetals as well as metals. Here, the difference in bond length may be a better indicator than the difference in radius alone. Consider the binary phase diagrams shown in Fig. 15.1. Complete solid solution occurs in the NiO–MgO solid solution, where bond lengths differ by only 1%, but not in the NiO–CaO system, where the Ca–O bonds are 15% longer than the Ni–O bonds. Miscibility is negligible in the remaining systems, where the size differences are even larger.

Navrotsky[8] pointed out that unit cells with larger volumes may be better at tolerating substitutional solid solution. Thus, in CaO–MgO (rocksalt structure) there is limited solid solution. In $CaCO_3$–$MgCO_3$ (calcite structure) there is extensive solid solution, and in $Ca_3Al_2Si_3O_{12}$–$Mg_3Al_2Si_3O_{12}$ (garnet structure) there is complete solid solution. In each case, the substitution involves Mg^{2+} and Ca^{2+}, so the change in bond lengths is comparable. This suggests that structures with larger volumes, like the garnets, can more readily absorb the strain associated with ion substitution elsewhere in the lattice.

Coupled solid solutions are also possible, in which two elements are replaced simultaneously. A good example of this occurs in the feldspar family, in anorthite ($CaAl_2Si_2O_8$)–albite ($NaAlSi_3O_8$) solid solutions. In feldspars, both the Al and the Si are in 4-coordination, while the alkali and alkaline-earth elements fill larger cavities in the structure. For each Ca^{2+} that is replaced by a Na^+, one Al^{3+} must be replaced with a Si^{4+} for charge balance purposes. The chemical formula of the resulting solid solution can be written as $Ca_{1-x}Na_xAl_{2-x}Si_{2+x}O_8$.

Clustering and Short-Range Ordering in Substitutional Solid Solutions

In alloys, guest atoms interact with each other as well as with matrix atoms. For dilute mixtures, the substituting atoms interact primarily with host atoms; as the concentration of the guest rises, guest–guest interactions become more important. They tend to repel one another in monovalent matrices and attract each other in multivalent matrices. This results in short-range order in monovalent atoms and clustering in multivalent atoms. Short-range order promotes compound formation. In contrast, clustering promotes phase separation.

Consider an alloy of composition RX with a close-packed structure. In a random solid solution, each atom position has equal probability of being occupied by R or X and each atom is surrounded by six R and six X atoms, on the average. If the atoms differ sufficiently in scattering power, the numbers and species of near neighbors can be experimentally determined by X-ray diffuse scattering measurements. Studies of a number of intermetallic systems have shown that departures from randomization can be substantial. In the Cu–Au, Ag–Au, and Au–Ni binaries, short-range order exists in which unlike atoms have a higher probability of being neighbors than like atoms. Another type of deviation occurs in the Al–Ag and Al–Zn systems, one in which like atoms tend to be neighbors and unlike atoms begin to segregate. This is called clustering. Solid solutions can therefore be thought of as a range of configurations,

[8] A. Navrotsky, *Physics and Chemistry of Earth Materials*, Cambridge University Press (1994).

tending toward clustering and phase segregation on one side, and extending toward short-range order and eventually long-range order (compound formation) on the other. All real solutions exhibit either clustering or short-range ordering to some degree, though many are close to being random, especially at high temperatures.

Since the bonding forces are strongest for near neighbors, the internal energy can be crudely considered as resulting from energies associated with neighboring pairs.[9] Let W_{RX}, W_{RR}, and W_{XX} be the energies for the neighboring pairs RX, RR, and XX. For a perfect solid solution of composition RX, there will be twice as many RX pairs as RR or XX pairs. The internal energy is then proportional to $2W_{RX} + W_{RR} + W_{XX}$. For an RX system with segregated R and X phases, the total internal energy is proportional to $2W_{RR} + 2W_{XX}$ and for one with long-range order it is $4W_{RX}$. To including short-range order and clustering, these results can be generalized to an energy of $U = 4W_{RX} + 2(W_{RR} + W_{XX})(1 - S)$, where S is an ordering parameter ranging from 0 (complete segregation) to 1 (long-range order). $S = 1/2$ is an ideal solid solution in which R and X are distributed at random. Clustering and short-range order correspond to $S < 1/2$ and $S > 1/2$, respectively. If $2W_{RX} < W_{RR} + W_{XX}$, the energy is minimized for $S > 1/2$, a situation favoring order because of strong attractive forces between R and X atoms. Clustering occurs if $2W_{RX} > W_{RR} + W_{XX}$. This discussion presupposes that the internal energy can be written as a sum of pair energies (a supposition that rarely holds for oxides), that the number of nearest neighbors is the same in all places, and that $T = P = 0$, so that the Gibbs free energy is equal to the internal energy.

Interstitial Solid Solutions

As the name implies, in interstitial solid solutions, atoms are placed into interstices in the host material, without removing any atoms. Steel, an alloy of Fe and C, is the most commercially important example of an interstitial solid solution, as was discussed extensively in Chapter 10. In steel, the carbon atoms go into interstices, where they form strong bonds to neighboring Fe atoms. When a dislocation hits the carbon atom, it has to break the bond in order to continue moving. This greatly reduces dislocation motion in steel, making steel much harder than Fe.

The limits of interstitial solid solution are often governed by the amount of structural distortion that arises in putting comparatively small ions in the interstitial sites. While most materials will tolerate some distortion of this type, when the distortions become too large, formation of additional phases is often preferred. Indeed, many carbides and boride compounds can be regarded as ordered phases in which the C or B is interstitial in a metal lattice.

In ionically bonded materials, interstitial solid solutions tends to form either in host materials that already have a propensity for accommodating interstitial defects, or in materials with large cavities in the structure. For example, when YF_3 dissolves into CaF_2, the Y adopts the Ca site. Two of the fluorines are also accommodated on the host lattice, but the third occupies the large interstitial cavity in the fluorite crystal structure.

---

[9] J. C. Slater, *Introduction to Chemical Physics*, McGraw Hill, New York (1939).

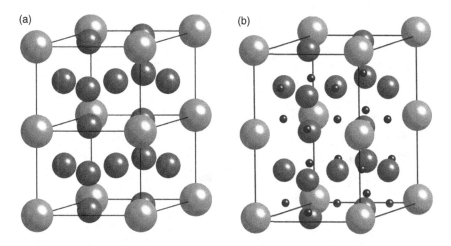

Fig. 15.8 Crystal structures of (a) LaNi$_5$ and (b) LaNi$_5$H$_6$, where the larger sphere is La, the intermediate size is Ni, and the small spheres are H.

Gas storage in crystals is another interesting use of interstitial sites. Gaseous hydrogen has many applications, but it is not easy to store in a safe and economical fashion. It can be held as a compressed gas, or as a liquid at temperatures <20 K, but these methods are expensive and dangerous. Alternatively, the hydrogen can be stored as a hydride.

Intermetallic compounds such as LaNi$_5$ and its alloys are capable of incorporating large amounts of hydrogen, converting to the hydride LaNi$_5$H$_6$, as shown in Fig. 15.8. Here, two phases, rather than an interstitial solid solution with variable composition, are important. The hydride and LaNi$_5$ are in equilibrium with each other, at a given temperature and pressure, and the hydrogen content can be varied within wide limits at an equilibrium pressure that is nearly constant. When LaNi$_5$ is placed in contact with H$_2$ at a pressure slightly greater than the equilibrium pressure, and the temperature is lowered to compensate for the heat liberated during the reaction, H$_2$ is absorbed by LaNi$_5$ until it is entirely converted to hydride. If gaseous H$_2$ is then allowed to escape from the vessel, the pressure decreases rapidly to the equilibrium pressure, and remains there while the hydrogen drains away from the hydride. Only when the hydride disappears does the H$_2$ gas pressure drop below the equilibrium value. At an external pressure of only 4 atm, the density of hydrogen in LaNi$_5$H$_6$ is equivalent to 1000 atm. Moreover, selective absorption by the intermetallic compounds rejects other gases, resulting in purification of the hydrogen gas.

One of the interesting features about trapped molecules in interstitial sites is that in some ways they behave like a gas, and in other ways like a solid. Crystals such as cordierite (Mg$_2$Al$_4$Si$_5$O$_{18}$) contain about one cavity in a volume of 100 Å^3. When all the cavities are occupied by gas molecules (often H$_2$O or CO$_2$ in mineral specimens) the density of molecules is equivalent to 20 atm pressure, and yet the molecules are never in contact with one another. The degree of interaction between molecule and cage ranges from tight bonding through hindered rotation to free rotation. The infrared spectrum of water in cordierite shows all the sharp overtone and combination bands of water vapor

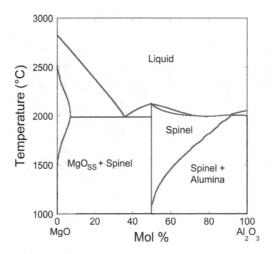

Fig. 15.9 $MgO-Al_2O_3$ phase diagram. The nominal composition of spinel is $MgAl_2O_4$; a wide solid solution regime is observed with Al_2O_3. Adapted, with permission, from ACERS-NIST Phase Equilibria Diagrams, American Ceramic Society and National Institute for Standards and Technology, Columbus, OH, 2013.

with one important difference – the spectra depend on the polarization vector, showing that the molecules are oriented in the cages.[10] Trapped gases constitute an unusual state of matter, a dense, non-interacting gas with preferred orientation. Trapped molecules are also interesting geologically. In tight cages like those of cordierite, the molecules have little chance of escaping. Like insects trapped in amber, they were present when the mineral was formed, and are therefore representative of the fluids and gases of the past.

Vacancy Solid Solutions

Vacancy, or omission, solid solutions develop when the substitution results in missing atoms in the crystal structure. For example, there is extensive (though incomplete) solid solution in the $MgAl_2O_4-Al_2O_3$ (see Fig. 15.9) and $CaO-ZrO_2$ binaries. The spinel–alumina solid solution is stable because cation vacancies are tolerated. One of the metastable polymorphs of alumina, γ-Al_2O_3, has a structure resembling spinel, but with cation vacancies. Thus the solid solution extending from $MgAl_2O_4$ toward Al_2O_3 can be written as $Mg_{1-x}Al_{2+(2x/3)} \square_{x/3}O_4$, emphasizing the cation vacancies. Substitution of a few percent CaO in zirconia leads to retention of the cubic zirconia structure to room temperature. Here, charge balance is achieved by removing an oxygen for each Ca^{2+} that substitutes for a Zr^{4+}, i.e. $Zr_{1-x}Ca_xO_{2-x}\square_x$.

Massive non-stoichiometry can also develop in systems such as NiTe and $NiTe_2$, and other transition metal chalcogenide systems, because of the compatibility of the end-member structures. $NiTe_2$ has the CdI_2 structure, while NiTe is isostructural with NiAs, as shown in Fig. 15.10. $NiTe_2$ has trigonal symmetry, while NiTe has a hexagonal unit

---

[10] E. F. Farrell and R. E. Newnham, "Electronic and vibrational absorption spectra in cordierite," *Am. Mineral.*, **52** (3–4), 380 (1967).

(a)   (b)

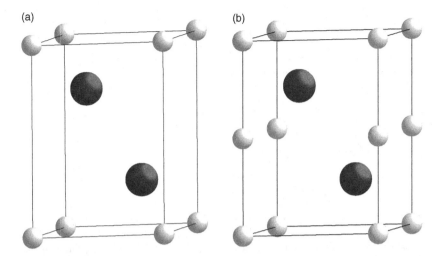

Fig. 15.10 Crystal structures of (a) $NiTe_2$ and (b) NiTe, where Ni is the smaller sphere, and Te is the larger.

cell with similar lattice parameters. An intermediate composition midway between NiTe and $NiTe_2$ could be described as $NiTe_2$ with 25% anion vacancies, or NiTe with 50% anion interstitials.

A homogeneous array of defects in found in high temperature titanium monoxide, with the rocksalt structure. The stability range extends from TiO to $TiO_{1.3}$. X-ray diffraction and density measurements reveal that even in "stoichiometric" TiO, more than 15% of the atomic sites are vacant. Fewer than half of the Ti are coordinated to six oxygens. This type of behavior is in stark contrast to some of the other rocksalt-type oxides. The defect concentration in CoO is only 3×10^{-3} at 1400 °C, while that of MgO is below the limits of detectability, less than 10^{-10} at 1700 °C. Ti^{2+} behaves differently from Co^{2+} and Mg^{2+} because of the overlapping d orbitals. The d electrons are delocalized in conduction bands, giving added stability to the crystal, and providing a source or sink for the electrons involved in the non-stoichiometric behavior. The d orbitals of CoO are more contracted because of increased nuclear charge, and do not overlaps with neighboring metal ions. The absence of non-stoichiometry in MgO stems from the inaccessibility of higher oxidation states and the high energy required to force Mg^{2+} into interstitial sites.

Although the point defect description is valid at low concentrations, defect interactions and clustering often become evident at concentrations of 10^{-4} or higher. Defect conglomerates can lead to coherent intergrowths and non-stoichiometric phases. For example, wustite $Fe_{1-x}O$, has a defect rocksalt structure, but with clusters, rather than isolated cation vacancies. Tetrahedral sites begin to fill as octahedral sites empty, forming Fe^{3+} clusters as oxidation proceeds. Figure 15.11 shows a defect cluster of this type. The oxygen sublattice is continuous throughout the host structure and the defect cluster. The clusters tend to produce long-range order, generating superlattice structures.

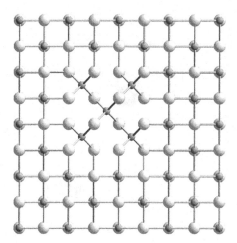

Fig. 15.11 Defect cluster in FeO showing some tetrahedrally coordinated iron atoms.

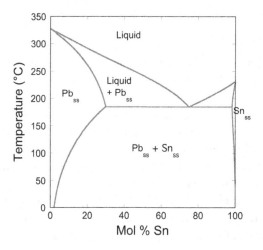

Fig. 15.12 Pb–Sn alloy system used as a low-melting solder. Figure adapted, with permission, from ASM Alloy Phase Diagram Database, ASM International, Materials Park, OH.

15.3   Solder

The phase diagram of the lead–tin (Pb–Sn) system is shown in Fig. 15.12. It is a simple eutectic system with low melting temperatures, which is extremely useful in solders. Likewise, the low melting temperature made pewter (75% Sn, 25% Pb) a useful metal alloy for household utensils during the Colonial Period. The end members in the Pb–Sn binary system have different crystal structures, as shown in Fig. 15.13. Lead has a face-centered cubic structure, space group $Fm3m$, with a = 4.9505 Å and 12 equal bond lengths of 3.501 Å. Tin has a less symmetric tetragonal structure with six neighbors, four at 3.016 Å and two at 3.175 Å. Pb is larger than Sn. In the Pb–Sn phase diagram, the Pb structure will accept 20% Sn at the eutectic temperature (183 °C) but only 2% Pb

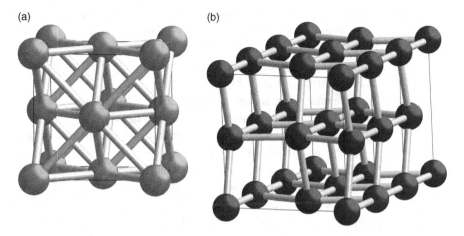

Fig. 15.13 (a) FCC structure of Pb, and (b) structure of white tin showing three unit cells.

can be accommodated in tin. This behavior is common. It is easier to replace a large atom with a small one, than the reverse.

With the advent of legislation requiring reduction in the use of lead, Pb–Sn solders have been largely replaced by Sn–Ag–Cu, usually with more than 96% Sn. The unusual crystal structure of tin helps produce deep eutectics, which is essential in this application.

In much the same way, deep eutectics are helpful in production of many glasses, as well as in sintering aids for powder processing. Deep eutectics are characteristic of systems in which the end member crystal structures differ significantly.

15.4   Problems

(1) Explain why the Al_2O_3–Cr_2O_3 phase diagram has the shape it does. Figure adapted, by permission, from one in E. M. Levin, C. R. Robbins, and H. F. McMurdie, *Phase Diagrams for Ceramists*, American Ceramic Society, Columbus, OH (1964).

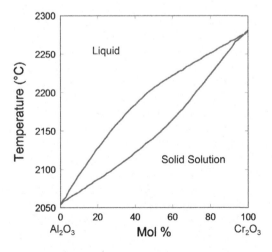

(2)   Pyrrhotite, $Fe_{1-x}S$ is a vacancy solid solution (sometimes called an omission solid solution) in which up to 20% of the Fe atoms can be missing. Write a chemical formula showing how charge is balanced in such a material.

(3)   Explain why it might be easier to form an oxygen interstitial in cubic ZrO_2 (fluorite structure) than in Al_2O_3.

(4)   Draw the phase diagram between MgO and CaO, labeling all of the regions.

    (a)   Why does the diagram have the shape it does?

    (b)   Draw a picture of the microstructure of a solid material with the average composition $Mg_{0.33}Ca_{0.67}O$.

(5)   Would you expect there to be complete substitutional solid solution between Mg and Cu? Explain your reasoning. The Mg–Mg bond length = 3.2 Å, while the Cu–Cu bond length is 2.56 Å. The electronegativities are: $\chi_{Mg} = 1.31$, $\chi_{Cu} = 1.90$.

(6)   Figure 15.7 shows a Darken–Gurry plot for metallic elements alloyed with iron. Solid circles denote elements with a solubility of >5% in either of the polymorphic modifications of iron. Triangles represent elements that do not form extensive solid solutions with iron.

    (a)   Locate where iron would appear in the Darken–Gurry plot.

    (b)   Explain why the extent of substitutional solid solution in metals depends on atom size and electronegativity.

16 Defects

Crystalline materials generally possess a number of different defects that can be characterized in terms of their dimensionality. Point defects refer to mistakes that happen at a single point in the lattice, often a missing atom (a vacancy) or an extra atom inserted into an interstice in the crystal structure (see Fig. 16.1). Line defects involve a row of "mistakes" in the crystal structures; edge and screw dislocations are important examples. Planar defects include flaws such as grain boundaries and stacking faults. Volume defects refer to inclusions of second phases or porosity. An understanding of defects is important, as defects have a profound influence on a multitude of material properties. It is important to note that the latter three types of defects are not required to have finite concentrations in strain-free materials. Thus, single crystals can be produced. However, a finite concentration of point defects is favored in all crystals at ambient temperatures, as a means of introducing entropy into solids.

16.1 Point Defects and Kröger–Vink Notation

In elemental solids, point defects are relatively simple. For a material at thermodynamic equilibrium, the concentration of defects rises exponentially with temperature, where the activation energy for defect formation is the enthalpy. As shown in Fig. 16.2, it is more energetically costly to form vacancies in elemental solids that melt at high temperatures. This is associated with the observation that the energy required to create a vacancy should be related to the energy required to break a bond. At the melting temperature, many bonds are broken, so it is not surprising that an increase in melting temperature is strongly correlated to an increase in the energy required to form the defects. Similar correlations are observed in multicomponent materials as well.

The formation of vacancies has been confirmed by a variety of techniques. For example, the size of a crystal increases more rapidly near the melting temperature than can be accounted for based on thermal expansion of the bonds alone. This extra size can be attributed to an increase in the vacancy concentration at elevated temperatures. Likewise, there are signatures for defect formation in the specific heat.

In multicomponent materials, the situation is somewhat more complicated, as the defects can appear on either site of the lattice, producing changes in local composition and charge. A compact means of describing the defect is *Kröger–Vink notation*; see Fig. 16.3. The main part of the three-part symbol describes the species in question. Elemental symbols are used to describe specific atoms, a "V" is used to denote a

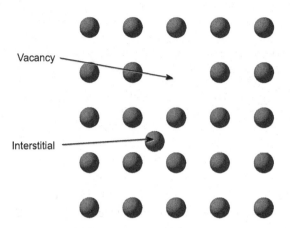

Fig. 16.1 Array of atoms in a crystalline solid showing vacancy and interstitial point defects.

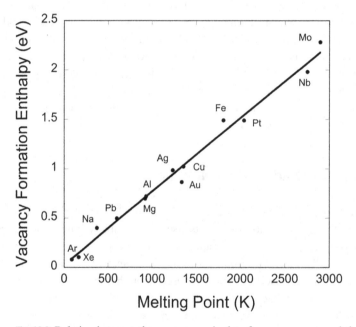

Fig. 16.2 Relation between the energy required to form a vacancy and the melting temperature in some elements. A similar relation is also found in compounds. Adapted, with permission, from D. M. Smyth, *The Defect Chemistry of Metal Oxides*, reprinted by Xi'an Jiaotong University Press, 2000.

vacancy. The subscript in the symbol refers to the position of the defect in the material. Again, elemental symbols are used to denote sites associated with that atom in the perfect crystal. The superscript refers to the effective charge *with respect to the lattice*. To distinguish the effective charge from the true charge of ions, positive charges are denoted by $^\bullet$, while negative charges are shown with $'$. An uncharged defect is marked with a superscript x. The superscripts are repeated to show higher charges. For example, in an ionic material such as MgO, the species $V_O^{\bullet\bullet}$ denotes a doubly charged oxygen

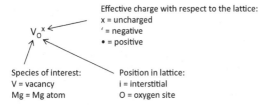

Fig. 16.3 Kröger–Vink notation for point defects.

vacancy. The V shows that the species is a vacancy; the subscript O shows that the defect is on an oxygen site, while the two superscripts refer to the fact that the material was "expecting" an O^{2-} ion on the oxygen site. Since there is nothing on that site, the defect is positively charged with respect to what the lattice was expecting. Likewise, $Ag_i^{\bullet}$ denotes a singly charged silver ion on an interstitial site, while Mg'_{Al} is a Mg^{2+} ion on an Al^{3+} site.

It is important to note that because it is possible to trap electrons and holes on defects, defects will not necessarily be fully charged. For example, if one electron was trapped on the oxygen vacancy, the defect would be $V_O^{\bullet}$; if there were two electrons trapped, the defect would be written as V_O^{x}. Predicting the charged state of a defect is non-trivial; experimental measurements are often used to determine this.

16.2   Stoichiometric and Non-Stoichiometric Point Defects

A multicomponent material is *stoichiometric* if for $A_xB_yC_z$, the ratio of A:B:C is exactly x:y:z. Several types of stoichiometric defects exist; two of these are illustrated in Fig. 16.4. A *Schottky* defect develops when anions and cations are missing in the ratio of the chemical formula. One can imagine such a defect developing if atoms that were in the interior of the crystal moved to new sites on the surface, leaving vacancies behind. Thus, for MgO:

$$Mg_{Mg}^{x} + O_O^{x} \rightleftharpoons V_{Mg}^{''} + V_O^{\bullet\bullet} + Mg_{Mg,\,surface}^{x} + O_{O,\,surface}^{x}.$$

This same reaction can also be written as $nil \rightleftharpoons V_{Mg}^{''} + V_O^{\bullet\bullet}$. For Al_2O_3, a fully ionized Schottky defect is $nil \rightleftharpoons 2V_{Al}^{'''} + 3V_O^{\bullet\bullet}$. Such defects will make minor changes in the density of the material.

A *Frenkel* defect develops when an atom that was originally on a site in the lattice is displaced towards an interstice, so that the defects form in pairs of a vacancy plus an interstitial. Thus, for a Frenkel defect on the Ag site in AgI:

$$Ag_{Ag}^{x} + I_I^{x} \rightleftharpoons V_{Ag}^{'} + Ag_i^{\bullet} + I_I^{x}.$$

The likelihood of formation of Frenkel defects increases when there are interstices large enough to fit the displaced atom, and when the displacement does not bring atoms with like charges into contact. Thus, in general, structures such as the fluorite structure, which have large open cavities, are more susceptible to Frenkel defects than those based on a close-packed structure.

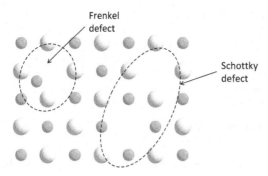

Fig. 16.4 Illustration of Schottky and Frenkel defects in a rocksalt structure.

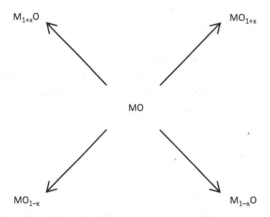

Fig. 16.5 Non-stoichiometric metal oxide (MO) compounds.

In general, stoichiometric defects may change diffusion coefficients and ionic conduction (see Chapter 25), but they do not necessarily introduce free e^- or $h^\bullet$. As a result, there is little change in the electronic conductivity with formation of stoichiometric defects.

A non-stoichiometric defect is one in which the ratio of atoms is changed relative to the chemical formula. Such defects lead to a dependence of the defect concentration on the atmosphere used during processing, as well as to the generation of electronic carriers. The latter leads to changes in the observed electrical conductivity, which is one of the means frequently used to characterize non-stoichiometric defects.

As shown in Fig. 16.5, there are numerous ways to make a single material non-stoichiometric. Note that $M_{1+x}O$ is not the same as MO_{1-x}, as the former may be non-stoichiometric as a result of metal interstitials, while the latter may have oxygen vacancies. Thus, the densities would be expected to differ. Which defect is most likely to form depends on a number of factors, including the sizes of the interstices available in the material.

In writing either stoichiometric or non-stoichiometric defect formation reactions for a crystalline solid, the following rules must be obeyed:

(1) Conservation of mass: Do not create or destroy atoms.

(2) Conservation of charge: The bulk of an ideal crystal is electrically neutral. Therefore, defects develop in electrically neutral combinations. More simply: Do not create or destroy electrons.

(3) Conservation of structure: Defect generation or elimination does not change the stoichiometric ratio of sites in the structure. Therefore, for compound $A_xB_yC_z$, the ratio of A:B:C sites should be x:y:z. This does not mean that the number of sites in the crystal will be unchanged by the reaction, only that on either side of the reaction that the number of sites should appear in the ratio of the chemical formula.

(4) Conservation of electronic states: The total number of electronic states in a system is a function of the electronic states of the component atoms. Thus, this quantity should also be preserved.

This is best seen with a series of examples. Consider the simple oxide MgO. Formation of an oxygen vacancy would be given by the following reaction:

$$Mg_{Mg}^\times + O_O^\times \rightleftharpoons Mg_{Mg}^\times + V_O^{\bullet\bullet} + \frac{1}{2} O_{2(g)} + 2e^-. \tag{16.1}$$

In writing such a reaction, it is often useful to start by putting the defect you are trying to form on the product side of the equation. Then look at the subscript; if it is a site (i.e. an Mg or an O in this example), then you will need to site balance on this side of the equation. For MgO, on either side of the equation, the ratio of Mg: O sites must appear in the ratio of the chemical formula (1:1). Thus, the $Mg_{Mg}^\times$ appears on the right-hand side.

This Mg is a real atom, so it must have come from somewhere. There are two possibilities; it could have either come from the gas phase or the material. A quick check would suggest that the vapor pressure of Mg is very close to zero under most experimental conditions, which rules out the gas phase. Thus, the Mg atom must have come from an Mg site.

This means that there is now an Mg site on the left-hand side of the equation. Now, site balance the left-hand side. Since there is an Mg site, there must also be an oxygen site, which should be filled with an oxygen atom if you start from a perfect material. The oxygen atom on the left-hand side must now be placed on the right-hand side of the equation. As the sites already appear in a 1:1 ratio, it is preferable to send the oxygen off into the gas phase. This is a reasonable thing to do, because oxygen has a finite vapor pressure under many conditions.

At this point, the reaction is mass and site balanced. The final step is to charge balance. The two electrons balance the effective positive charges on the oxygen vacancy. One can thus envisage formation of an oxygen vacancy in this material as occurring when an oxygen atom is removed from the material, and given off to the gas phase, liberating two electrons in the process. The defect chemistry reaction functions much as other chemical reactions; Le Chatelier's principle applies. Thus, as the oxygen partial pressure (p_{O_2}) of the processing ambient is decreased, the reaction would be expected to proceed towards the right-hand side, in order to liberate oxygen into the gas phase. Increasing the oxygen partial pressure should decrease the concentration of oxygen vacancies.

One can also write an equilibrium constant for the defect formation. In the dilute limit, when the activities of the species can be replaced by their concentrations, the equilibrium constant would be given by $K_{V_O^{\bullet\bullet}} = \left[V_O^{\bullet\bullet}\right]n^2 p_{O_2}^{1/2}$, where the square brackets denote concentrations and n is the concentration of conduction electrons. When $V_O^{\bullet\bullet}$ is the principal defect present, the electroneutrality condition gives $2[V_O^{\bullet\bullet}] = n$. Substituting this into the equilibrium constant yields $K_{V_O^{\bullet\bullet}} = \frac{n^3}{2} p_{O_2}^{1/2}$ and hence $n = (2K_{V_O^{\bullet\bullet}})^{1/3} p_{O_2}^{1/6}$.

Other defect chemistry reactions can be obtained in much the same way, so that the reaction is charge, mass, and site balanced. For MgO the following are other possibilities:

$$\frac{1}{2}O_{2(g)} \rightleftharpoons V_{Mg}^{''} + O_O^\times + 2h^\bullet \tag{16.2}$$

$$Mg_{Mg}^\times + O_O^\times \rightleftharpoons Mg_i^{\bullet\bullet} + \frac{1}{2}O_{2(g)} + 2e^- \tag{16.3}$$

$$\frac{1}{2}O_{2(g)} \rightleftharpoons O_i^{''} + 2h^\bullet. \tag{16.4}$$

It is interesting to consider the asymmetry of equations (16.1) and (16.2). Why was the formation of a metal vacancy not written as: $Mg_{Mg}^\times + O_O^\times \rightleftharpoons O_O^\times + V_{Mg}^{''} + Mg_{(g)} + 2h^\bullet$? The answer is that it is quite unlikely under normal processing for there to be significant levels of Mg volatility. In many instances, the only volatile species is associated with the atom on the anion site, though exceptions certainly exist at elevated temperatures in materials containing Pb, Bi, etc., where the cation volatility is high.

Before leaving the point of balancing a defect chemistry reaction, it is worth considering how reactions (16.1)–(16.4) differ as the stoichiometry changes. As an example, consider TiO_2 as the host composition. In this case, site balance mandates that on either side of the reaction Ti and O sites should appear in the ratio 1:2. The results for formation of anion vacancies, cation vacancies, cation interstitials, and anion interstitials, respectively, are:

$$Ti_{Ti}^\times + 2O_O^\times \rightleftharpoons Ti_{Ti}^\times + O_O^\times + V_O^{\bullet\bullet} + \frac{1}{2}O_{2(g)} + 2e^- \tag{16.5}$$

$$O_{2(g)} \rightleftharpoons V_{Ti}^{''''} + 2O_O^\times + 4h^\bullet \tag{16.6}$$

$$Ti_{Ti}^\times + 2O_O^\times \rightleftharpoons Ti_i^{\bullet\bullet\bullet\bullet} + O_{2(g)} + 4e^- \tag{16.7}$$

$$\frac{1}{2}O_{2(g)} \rightleftharpoons O_i^{''} + 2h^\bullet. \tag{16.8}$$

It is important to keep track of site balance, because the partial pressure dependences of the defect concentrations can change for different stoichiometries. The relative concentrations of the different types of defects will depend on the equilibrium constants for their formation in a particular host material, as well as on the processing ambient.

One final note to make is that the existence of the charged ionic defects can lead to unexpected behavior on doping multicomponent materials. Aliovalent doping can be compensated either electronically (by the formation of free electrons or holes) or ionically, by changing the concentration of point defects. For example, when $PbZr_{1-x}Ti_xO_3$ is donor doped with La, the large La atom occupies the Pb site, forming a positively charged defect. Owing to the comparatively high volatility of Pb at typically processing temperatures, a typical compensation would be $[La_{Pb}^{\bullet}] = 2[V_{Pb}'']$. Similarly, many acceptor dopants are compensated by formation of oxygen vacancies, e.g. $2[V_O^{\bullet\bullet}] = [Fe_{Ti}']$. In contrast, for some doping concentrations, La doping of $BaTiO_3$ produces electronic compensation: $[La_{Pb}^{\bullet}] = n$. The electron produced greatly increases the electrical conductivity of the $BaTiO_3$.

16.3 Defect Clustering

Although the point defect description is valid at low concentrations, defect interactions and clustering often become evident at concentrations of 10^{-4} or higher. Defect conglomerates can lead to coherent intergrowths and non-stoichiometric phases. For example, wüstite, $Fe_{1-x}O$, has a defect rocksalt structure, but with clusters, rather than isolated cation vacancies. Tetrahedral sites begin to fill as octahedral sites empty, forming Fe^{3+} clusters as oxidation proceeds. Figure 16.6 shows one possible defect cluster. The oxygen sublattice is continuous throughout the host structure and the defect cluster. Koch clusters intergrow with the rocksalt structure, but are not electrostatically neutral and must be compensated by additional Fe^{3+} ions in the immediate vicinity. The clusters tend to produce long-range order, generating superlattice structures.

Defect clusters occur in other nonstoichiometric compounds as well. In VO_{2+x}, oxygens are displaced from their normal sites to give an interstitial complex. A 2:1:2 cluster is typical, with two displaced oxygen atoms and one additional oxygen occupying two kinds of low-symmetry interstitial positions, associated with two oxygen

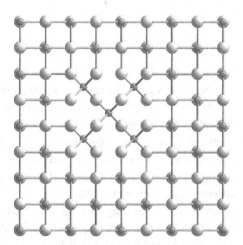

Fig. 16.6 Schematic showing one of the possible defect clusters in $Fe_{1-x}O$. The small spheres are iron atoms, the larger spheres are oxygen. Looking down the c axis of the rocksalt structure.

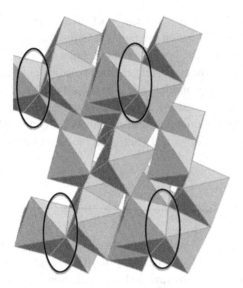

Fig. 16.7 Crystal structure of Ti_5O_9, one of the Magneli phases. Several face-shared octahedra are marked with ovals. Ti are at the center of the octahedra.

vacancies. Other types of defect clusters occur in CaF_2–YF_3 and MH_2–MH_3 mixed crystals. The vanadium carbides form defect rocksalt structures similar to those in wüstite; V_6C_5 and V_3C_7 contain clusters of vacant sites ordered in spirals along the body diagonal directions.

Excellent examples of coherent intergrowths occur in the Magneli phases with composition Ti_nO_{2n-1}, as shown in Fig. 16.7. These are shear structures with rutile-like regions joined by lamellae of edge-sharing octahedra. For large n, the shear planes are widely spaced, so that the driving force for ordering is small, and the compounds order only sluggishly. Random fluctuations in shear plane spacing occur under these circumstances, giving rise to non-stoichiometry.

16.4   Line Defects

The two best known types of line defects are edge dislocations and screw dislocations (see Fig. 16.8). These constitute end members types; dislocations with intermediate characters also exist. The edge dislocation can be imagined as one or more extra half planes of atoms inserted into the structure. A screw dislocation, on the other hand involves displacing one portion of the structure with respect to the other, creating a step on the surface. A simple way of visualizing this is to imagine cutting half way through a crystal, slipping one piece with respect to the other, and re-attaching.

There are several important points to note regarding dislocations. First, one can define the Burgers vector, $\vec{b}$, as the vector that would be required to close the circuit in a path around the dislocation. This is illustrated for an edge dislocation in Fig. 16.9. For an edge dislocation, $\vec{b}$ is perpendicular to the dislocation line; for a screw dislocation, $\vec{b}$ is parallel to the dislocation line.

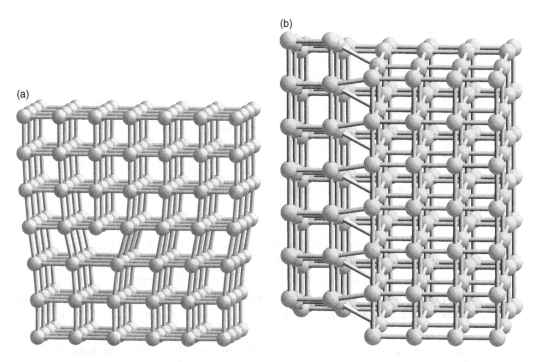

Fig. 16.8 Schematics of (a) edge and (b) screw dislocations in a simple cubic metal.

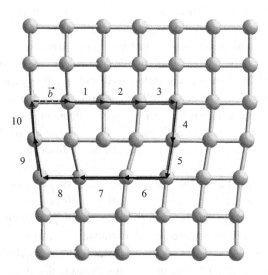

Fig. 16.9 Schematic showing a circuit around an edge dislocation. An equal number of steps are taken right (in this case steps 1–3) and left (steps 6–8), as well as down (steps 4 and 5) and up (steps 9 and 10). The Burger's vector, $\vec{b}$, is the vector required to close the circuit. Two conventions for definition of $\vec{b}$ are employed; it can be drawn either from position 10 to 1 or 1 to 10.

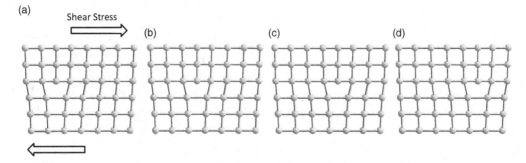

Fig. 16.10 Motion of an edge dislocation in response to a shear stress. Parts (a)–(d) show increasing times.

A second key point is that there is a considerable level of distortion built into the crystal near the dislocation; bond lengths and bond angles are changed. It is energetically costly to put in a dislocation in a perfect crystal. The larger the magnitude of the Burgers vector, the more distortion results, and the higher the strain energy density associated with the dislocation. The strain energy per unit length, E, is given by $E \propto \alpha G b^2$, where G is the shear modulus, b is the magnitude of the Burgers vector, and α typically ranges from 0.5 to 1. Thus, the larger the length of the Burgers vector, the more energetically costly the dislocation will be. This tends to favor dislocations with small Burgers vectors.

The process of slip in metal depends heavily on the motion of dislocations. The logic behind this is straightforward. If slip proceeded by sliding one plane of atoms past the other in a one-step process, this would require all of the bonds holding the two layers together to be broken simultaneously. This would necessitate a lot of energy. Instead, it is easier for slip to occur by dislocation motion, as illustrated in Fig. 16.10. As can be seen there, for a dislocation to move over one atom spacing, only a row of bonds, rather than a plane of bonds, needs to be broken. This significantly reduces the energy barrier for slip to occur. In many ways, the process is analogous to sliding a heavy carpet across a floor. If one takes the edge of the carpet and pulls, in order for the carpet to move as a unit, the entire frictional force between the carpet and the floor must be overcome. However, the carpet can be moved much more readily if one pushes a wrinkle.

There are several factors that contribute to the population of dislocations, including mechanical and thermal stresses, precipitation of vacancies during cooling, growth of second-phase particles, and interfaces. The role of mechanical stresses can be investigated by bending a copper wire. If the initial concentration of dislocations is low, then dislocation motion can proceed unimpeded, and the wire can be plastically deformed easily. If the wire is bent repeatedly, however, it will eventually become more and more stiff. This is a consequence of the fact that additional dislocations are nucleated during the deformations. Eventually, the concentration of dislocation climbs to the point that they impede each other's motion. That is, the accumulation of defects results in the need to apply a larger stress to produce plastic deformation. This process is referred to as *work hardening*. If the same piece of wire is heated to an elevated temperature, so that some atomic motion is possible, many of the dislocations can be annealed out (recall that dislocations are not required by thermodynamics in unstressed materials). The result is that the wire will recover to its initial behavior.

Precipitation hardening entails strengthening a material by formation and dispersion of hard second-phase particles. These precipitates also act to impede dislocation motion.

An excellent example of the role of interfaces on dislocations occurs during the growth of epitaxial films. Consider two materials with the same crystal structure, but somewhat different lattice parameters. If one of these materials is used as the substrate, and a film of the second material is grown on the top, then at very small film thicknesses, it may be possible for the film to be coherently strained to the substrate, so that it has the same in-plane lattice parameters. However, as the film thickness grows, eventually the stored elastic energy is large enough that dislocations will nucleate to provide stress relief.

Screw dislocations are important in the growth of crystals. When an atom attaches to a surface during growth, it can decrease its energy the most if it makes many bonds, so that it's coordination becomes more like that of an atom in the bulk of the material. Thus, it is more favorable to attach the atom at a step or a kink on the surface, rather than on a terrace on the surface. This is shown in Fig. 16.11. As growth continues, this will lead to sidewise motion of the steps on the surface. Eventually, the step can grow itself out of existence. Further growth will then require nucleation of a new island on the flat surface.

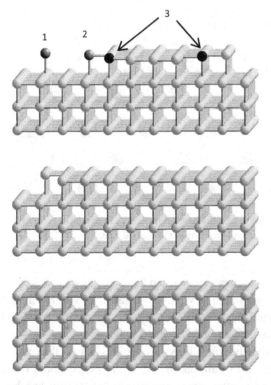

Fig. 16.11 Surface of a material showing attachment of adatoms on a (1) terrace, (2) step, or (3) at a kink. Progressive growth of atoms on a surface on a perfect material leads to elimination of the steps, as shown in the subsequent images.

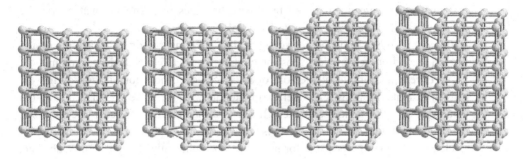

Fig. 16.12 Growth of a crystal from a screw dislocation. Time is increasing left to right in the figure. The step at the screw dislocation serves as a nucleation site for continued growth of the material.

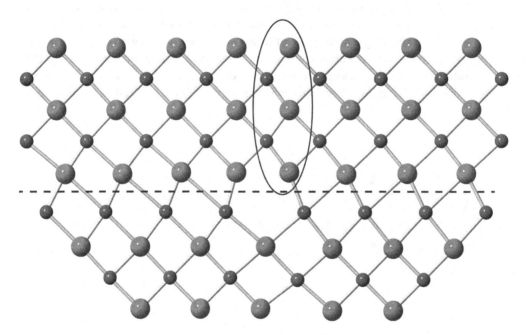

Fig. 16.13 Schematic representation of a dislocation in a rocksalt crystal. The extra half planes of atoms are circled. These two planes terminate at the dashed line.

In contrast, a screw dislocation provides a step that is retained on subsequent growth, as shown in Fig. 16.12. As a result, it serves as a favorable site for growth of the crystal. Indeed, the existence of spiral staircase-like features on surfaces of crystals is one of the factors that led to an understanding of screw dislocations.

As a rule of thumb, ionically bonded crystals are less plastic than metals because dislocations are more difficult to form and move. Dislocations should be electrostatically neutral. As a result, it is often necessary to insert more than one half plane of atoms. An example of this is shown in Fig. 16.13 for the case of a crystal with the rocksalt structure.

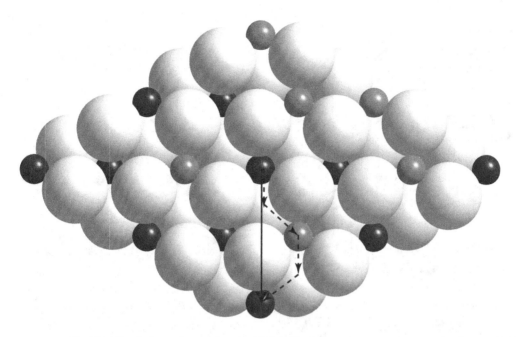

Fig. 16.14 Basal plane of corundum (α-Al$_2$O$_3$). The large oxygen atoms are nearly close-packed. Two layers of Al are shown in different tones of gray to denote that they are at different heights. The full Burger's vector is shown as the solid arrow. Slip occurs through a series of partial dislocation motions shown as dashed arrows.

The motion of dislocations can be hindered by impurities which tend to collect near dislocations. The difference in size between impurity atoms and host atoms reduces the strain energy associated with the dislocation. Slip and plastic deformation is then made more difficult because of the attractive forces between impurities and dislocations. CaCl$_2$ concentrations of 10^{-4} in NaCl are sufficient to double the yield stress.

An intimate relationship between deformation characteristics and crystal structure exists in alumina. Plastic deformation becomes important above 1000 °C, when slip occurs parallel to the close-packed (0001) plane. The minimum translation distance in the basal plane to give registry of the crystal structure is about 5 Å, a very large Burgers vector. In moving the dislocation, oxygen ions are required to move past other oxygens in the adjacent close-packed layer. This is accomplished by "partial" dislocation motion, avoiding direct superposition of the anions by zig-zag movements. Such complex dislocation structure requires synchronized movements of Al^{3+} and O^{2-} ions, which is possible only at high temperatures and low strain rates (see Fig. 16.14). Other slip systems are activated at still higher temperatures.

16.5   Planar Defects

Planar defects include mistakes along a plane, such as stacking faults, some twin structure, and grain boundaries. Grain boundaries mark the area where two different crystallites come into contact with each other. Many important engineering materials are

(a) (b)

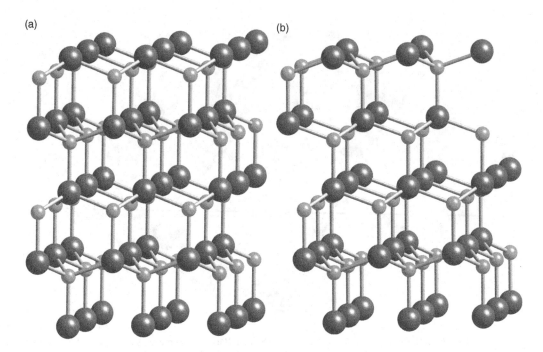

Fig. 16.15 Side view of the crystal structure of (a) wurtzite and (b) wurtzite with a stacking fault.

polycrystalline, so that they are composed of consolidated masses of small crystals, which are called *grains*. In general, there is no crystallographic relationship between the grains on either side of the grain boundary, and so the grain boundary marks a plane in which there is some disorder in the atomic positions.

Better defined relationships are obtained between the two portions of a crystal structure separated by a stacking fault. A stacking fault is observed when the normal layering sequence of a crystal is interrupted by an error. Good examples of systems prone to stacking faults are covalently bonded semiconductors in which the zincblende and wurtzite structures are similar in energy. As was shown in Chapter 9, both of the structures contain puckered hexagonal rings of atoms; in wurtzite the six rings stack on top of each other, in zincblende, they are offset. Figure 16.15 shows an example of wurtzite with and without a stacking fault; the stacking fault in this case consists of one layer of zincblende-like stacking in the structure. Growth twins of this type are difficult to remove, as strong bonds need to be broken in order to reconstruct the crystal. Stacking faults are common in materials with layered crystal structures, particularly when different polymorphs of the material are known.

Another example of a two-dimensional defect with well-defined crystallographic arrangements is the *twins* that can occur in some materials. An example of this is shown in Fig. 16.16 for the case of a twin in tetragonal $BaTiO_3$.

Both stacking faults and twins illustrate the point that the most favorable two-dimensional defects are those that lead to the smallest changes in the atom coordinations at the defect. It is clear from Figs. 16.15 and 16.16 that these types of planar defects do not introduce dangling bonds (in contrast to grain boundaries, where larger structural

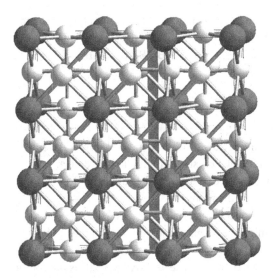

Fig. 16.16 A twin plane (marked in gray) in tetragonal $BaTiO_3$. Ba atoms are the large filled circles, O is white, and the Ti atoms are the small filled circles near the center of the unit cell. The Ti atoms are shifted up to the left of the domain wall, and down to the right of the wall. The structural distortion is very small.

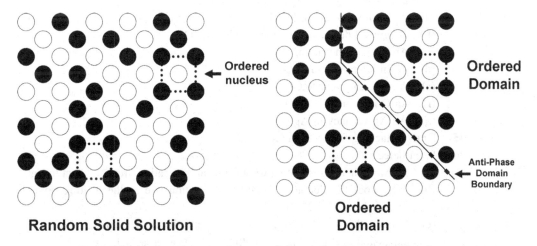

Fig. 16.17 Illustration of the development of an antiphase boundary arising from ordering.

distortions and anomalous coordinations are observed). Since there are only certain ways in which the pieces of the structure around the defect can be matched in this way, such defects tend to occur along well defined crystallographic planes.

Antiphase boundaries mark a plane between two domains of material in which the material on either side of the defect has the same crystal structure and orientation, but where one piece of the crystal appears to have been displaced with respect to the other. This is illustrated in Fig. 16.17, where the boundary separates ordered domains for which the origin has been shifted by half of a repeat unit in the horizontal and vertical directions on one side of the boundary. Antiphase boundaries are also observed in metals as a mistake in the stacking sequence of the material. For example, in a HCP metal, an antiphase boundary interrupts the stacking sequence, and would appear in a transmission electron microscope as ABABAB | BABABAB.

16.6   Problems

(1) Draw a schematic of an edge dislocation in a metal. Show the Burgers vector.

(2) Draw a (100) plane of a NaCl crystal with a Schottky defect.

(3) Consider UO_2. The radius of U is 1.00 Å, the radius of O is 1.35 Å.

    (a) Predict the atom coordinations.

    (b) Give an isomorph of this material.

    (c) Explain why it might be easier to form an oxygen interstitial in UO_2 (fluorite structure) than in Al_2O_3.

    (d) Suppose you wanted to increase the oxygen diffusion coefficient of this material. Write two defect chemistry reactions that would enable this, and explain your reasoning. One of the two reactions should involve doping with another ion.

(4) Write defect chemistry equations for the following. Assume full ionization of the defects.

    (a) A Schottky defect in NaCl.

    (b) A Frenkel defect on the oxygen site in UO_2.

    (c) An oxygen vacancy in MgO.

    (d) A Zn interstitial in ZnO.

    (e) A Na vacancy in NaCl.

    (f) An O interstitial in BaO.

    (g) An Ag vacancy in Ag_2O.

(5) Draw a stacking fault in an FCC metal.

(6) Point defects.

    (a) Draw a schematic showing what is meant by a cation Frenkel defect in NaCl.

    (b) Why does the Schottky defect concentration rise as the melting point in approached?

    (c) Explain the link between defect concentration and the measured thermal expansion coefficient as the melting point is approached.

    (d) Write the defect formation reaction for the Schottky defect in CaF_2. Assume full ionization of defects.

(7) Consider CeO_2. The Ce–O bond length is 2.343 Å. For the Ce^{4+}–O^{2-} bond, $v_{ij} \approx \exp\left(\frac{2.078 - r_{ij}}{0.37}\right)$.

    (a) Predict the coordinations of the Ce^{4+} and O^{2-}.

    (b) Write the defect formation reaction for fully ionized cerium vacancies in this compound.

    (c) Describe one material property that would be influenced by the existence of the non-stoichiometric defect in part (b), and explain the mechanism by which this occurs.

(8) Write formation reactions for the following defects, assuming full ionization of defects. Where needed, use your knowledge of crystal chemistry to estimate likely compensation mechanisms.

(a) An oxygen vacancy in TiO_2.

(b) An anion interstitial in CaF_2.

(c) A metal vacancy in CaO.

(d) A Schottky defect in $LiNbO_3$.

(e) Y_2O_3 doping in ZrO_2.

(9) Shown below is a graph relating the enthalpy of formation for Schottky defects in a variety of alkali halides. (Data are average values taken from W. Bollmann, "Formation enthalpy of Schottky defects and anti-Frenkel defects in CaF_2-type crystals," *phys. stat. sol. (a)*, **61**, 395 (1980).)

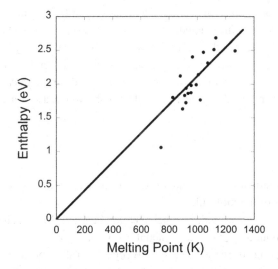

(a) Explain why there is a correlation with melting point.

(b) Write the defect formation reaction for the Schottky defect in $PbCl_2$. Assume full ionization of defects.

(c) Suppose NaCl is heated in air. Write the defect reaction for the formation of a likely non-stoichiometric defect under these conditions.

(d) Write the formation reaction for a likely defect that would develop if the salt were heated instead under a high chlorine gas partial pressure. Assume full ionization of defects.

17 Gases and Liquids

When heated, most solids melt into a liquid, and then at still higher temperatures, boil to form a gas. Some of the structure–property relations in gases and liquids are discussed in this chapter.

In a solid, the atoms are fixed in position with relatively little vibrational motion (a few percent of the bond length). In a liquid, the molecules flow by one another and take the shape of the container. In a gas, the molecules move freely with relatively few interactions.

For many engineers – mechanical, electrical, materials, and civil – solids are much more important than gases and liquids, but this is not true for chemical, aeronautical, biological, and environmental engineers who deal with fluid phases on a routine basis.

17.1 Gases

The principal characteristics of a gas are very small molecular interactions and random translational motion of the molecules.

The 3–30–300–3000 rule provides a quick picture of a gas at 1 atmosphere and 300 K: the molecules are about 3 Å in diameter, about 30 Å apart, traveling at speeds of about 300 m/s, and undergoing collisions every 3000 Å (Fig. 17.1). Two other "magic numbers" of interest to engineers are "–0.03 and –3". They correspond to the drop in atmospheric pressure and temperature for every 1000 ft of altitude. If at sea level the pressure is p, it decreases by about 3% to $(29/30)\,p$ at 1000 feet, and the temperature drops by about 3 °C. These decreases continue throughout the troposphere to an altitude of about 8 miles.

The mean free path between collisions can be estimated as follows. Let d be the effective diameter of the molecule, l be the distance traveled per unit time, and n molecules per cm^3. We make the simplifying assumption that all molecules are at rest except for the reference molecule.

In traveling a distance l, the molecule will have swept out a volume $\pi d^2 l$, and will collide with $\pi d^2 l \cdot n$ other molecules. The mean free path, λ, is the distance travelled divided by the number of collisions:

$$\lambda = \frac{l}{\pi d^2 l\, n} = \frac{1}{\pi d^2\, n} \approx 3000\,\text{Å},$$

where n is equal to Avogadro's number (6.023×10^{23} molecules/mole) divided by the volume occupied by one mole of gas (22.4 liters/mole), or about 3000 molecules in a million cubic ångströms. This corresponds to a typical intermolecular distance of 30 Å.

Table 17.1 Interatomic distances and bond angles for important gases. Most of the data were measured by electron diffraction and molecular spectroscopy

		Bond	Distances (Å)	Bond angles
Ammonia	NH_3	N–H	1.01	H–N–H: 107°
Carbon dioxide	CO_2	C–O	1.16	O–C–O: 180°
Carbon monoxide	CO	C–O	1.12	
Hydrogen	H_2	H–H	0.74	
Hydroxyl	OH	O–H	0.97	
Methane	CH_4	C–H	1.09	H–C–H: 109.5°
Nitrogen	N_2	N–N	1.10	
Nitrous oxide	N_2O	N–N	1.13	N–N–O: 180°
		N–O	1.19	
Oxygen	O_2	O–O	1.21	
Ozone	O_3	O–O	1.28	O–O–O: 117°
Water	H_2O	O–H	0.96	H–O–H: 104.5°

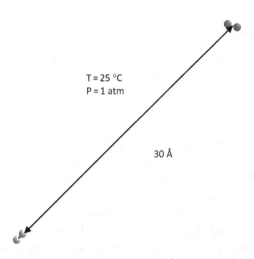

T = 25 °C
P = 1 atm

30 Å

Fig. 17.1 An instantaneous view of two molecules in air. The intermolecular distances are about ten times larger than the size of a molecule. Under standard conditions there are 6.023×10^{23} molecules in a volume of 22.4 liters, or about 30 molecules in a million cubic ångströms. A typical intermolecular distance is about 30 Å.

The root mean square speed of a typical gas molecule can be estimated from its thermal energy:

$$\sqrt{v^2} = \frac{3RT}{M}$$

where R is the universal gas constant ($8.314 \frac{J}{K\,mole}$), T is the absolute temperature, and M the molecular weight. For O_2 it is 425 km/s, for H_2 it is 1690 km/s, and for Hg vapor an order of magnitude less.

A listing of important gases in the atmosphere is given in Table 17.1. Most of the bond lengths are very short, near 1 Å. Some, like O_2 and N_2, are shorter than those in many other organic molecules because of their double- and triple-bond character.

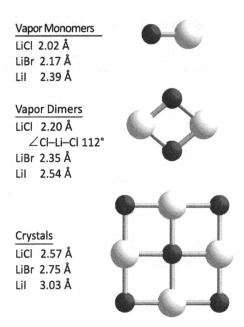

Vapor Monomers
LiCl 2.02 Å
LiBr 2.17 Å
LiI   2.39 Å

Vapor Dimers
LiCl 2.20 Å
    ∠Cl–Li–Cl 112°
LiBr 2.35 Å
LiI   2.54 Å

Crystals
LiCl 2.57 Å
LiBr 2.75 Å
LiI   3.03 Å

Fig. 17.2 Bond lengths and bond angles in the vapor species associated with lithium halide crystals. LiCl, LiBr, and LiI have the rocksalt structure. Bond lengths in crystals are generally longer than those in the vapor state.

A similar shortening takes place in the vapor species found over inorganic compounds. Interatomic distances for lithium halide monomers, dimers, and crystals are given in Fig. 17.2. In a LiCl crystal, for example, the bonds are 10% longer in the crystal than in dimers, and 20% longer than the monomer molecules. The increase can be rationalized on the basis of structure. In the crystalline state (rocksalt structure) each Li ion is bonded to six chlorines in octahedral coordination. The negative chlorine ions tend to repulse one another, lengthening the Li–Cl bond. There is less Li–Li and Cl–Cl repulsion in LiCl dimers, and none in the monomers. Similar trends are observed among alkaline earth oxides. Vapor bondlengths are 20–30% shorter than those of oxide crystals.

17.2   Vapor Pressure and Latent Heat

If a liquid or solid is sealed in a container and the air is evacuated above it, some of the condensed phase will evaporate, and some of these molecules will later condense back on to the surface. Equilibrium is reached when the rates of evaporation and condensation are equal. The pressure exerted by the vapor under these conditions is known as the *vapor pressure*. It increases with increasing temperature, up to the boiling point.

The *boiling point* is the temperature at which the vapor pressure equals the external pressure, usually one atmosphere. If the external pressure is raised, the boiling point is raised. Normally boiling takes place from the liquid state to the vapor state but in some materials the evaporation process occurs directly from the solid to the gas phase. This is called *sublimation*. Solid carbon dioxide (dry ice) sublimes at −79 °C.

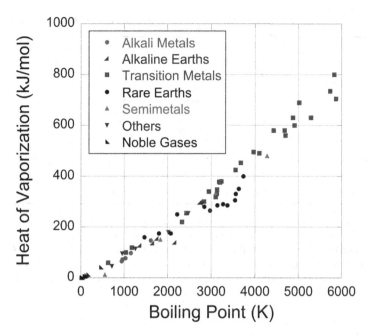

Fig. 17.3 Latent heats of boiling for a number of elements. There is a strong correlation between latent heat and boiling point.

The molecular interpretation of vapor pressure is straightforward. Since the molecules in a liquid have a wide range of thermal energies, only those above a certain value are capable of escaping. Those of lower energy are incapable of overcoming the attractive forces to neighboring molecules. It follows that the vapor pressure p is controlled by the latent heat of evaporation, L.

The latent heat, in turn, is directly proportional to the boiling point. Fig. 17.3 shows the latent heat for many chemical elements plotted as a function of T_{boil}. Among metals, the 4d and 5d transition metal elements have the highest latent heats and the highest boiling points. The d electrons are heavily involved in the molten metals, just as they are in the solid phases. Latent heats for the alkali metals are an order of magnitude smaller, and those of the gaseous elements are about a hundred times smaller.

17.3 Triple Points and Critical Points

A *P–T diagram* (Fig. 17.4) is obtained when the vapor pressure (P) is measured over a wide range of temperatures (T). Solid and vapor coexist along the *sublimation* curve; liquid and vapor on the *vaporization* curve; and liquid and solid phases on the *fusion* curve. The three curves intersect at a *triple point*, and the vaporization curve terminates at the *critical point*.

All three phases – solid, liquid, and vapor – coexist at the *triple point* (Fig. 17.4). For water the triple point is located at a pressure (P_t) of 611.7 Pa and a temperature (T_t) of 273.16 K. Triple points for oxygen, nitrogen, and other gases are listed in Table 17.2.

Table 17.2 Triple points and critical points for five common gases of interest to engineers

	N_2	O_2	H_2	Ar	CH_4
M (g/mol)	28.01	32.00	2.02	39.95	16.04
T_t (K)	63.15	54.36	13.8	83.81	90.69
P_t (kPa)	12.46	0.15	7.04	68.95	11.70
ρ_t (g/mL)	0.87	1.31	0.08	1.42	0.45
T_c (K)	126.20	154.58	32.98	150.66	190.56
P_c (MPa)	3.39	5.04	1.29	4.86	4.59
ρ_c (g/mL)	0.31	0.44	0.03	0.53	0.16
T_{boil} (K, 1 atm)	77.35	90.19	20.28	87.29	111.67
T_m (K, 1 atm)	63.14	54.35	13.94	83.78	90.67

Note: M is the molecular weight. T_t, P_t, and ρ_t are the temperature, pressure and density at the triple point. T_c, P_c, and ρ_c are the corresponding data for the critical point. T_{boil} and T_m are boiling points and melting points measured at one atmosphere. Note that $T_t \approx T_m$ and $T_c \approx T_m + T_{boil}$ in accordance with Prudholme's rules. The boiling point (T_{boil}) is approximately two thirds of the critical temperature (T_c), and the densities of ρ_t and ρ_c are proportional to the molecular weights M.

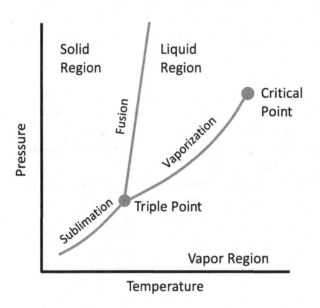

Fig. 17.4 Pressure–temperature (P–T) diagram showing the stability regions of solid, liquid, and vapor phases, and the location of the triple point and the critical point.

The *critical point* is a limiting point at which the density of the liquid equals the density of the vapor phase, in other words, where the two fluid phases become indistinguishable from one another. For water, the critical point is located at a temperature of 647 K and a pressure of 22 MPa (218 atmospheres). Under these conditions the density of water is 0.40 g/cm^3, or about 40% of the normal value. Experimental results for other fluids are given in Table 17.2.

17.4   Liquids

Liquids are the halfway house of matter: part way between the periodic certainty of a solid and the wild uncertainty of a gas. This naturally raises the question: "Is a liquid more like a solid or a gas?"

For most materials, under common temperatures and pressures, liquids are more like solids. The *density* of a typical liquid is only a few percent less than that of the solid, but orders of magnitude larger than that of a gas. The *compressibility* of liquids is greater than that of solid but much, much less than that of a gas. The *heat of evaporation* (the energy required to convert liquid into gas) is far larger than the *heat of fusion* (the energy required to convert a solid into a liquid). In NaCl, for example, the heat of evaporation is 39 000 cal/mole compared to 6690 cal/mole for the heat of fusion.

Moreover, X-ray and neutron diffraction patterns indicate that the short range order of liquids and solids of the same composition are very similar. That is, the atomic coordination and bond lengths are nearly the same for the crystal and the melt, whereas the diffraction patterns of gases are very different.

All these experiments indicate that liquids are generally similar to solids. The situation is very different at high temperatures and high pressures where the properties of liquids and gases converge at the critical point (Fig. 17.4). Here the two fluids develop an incoherent structure with about 30% of the normal density.

A molecular view of a melt was first put forward by Bernal. When solids melt they generally expand by about 10% in volume. Remembering that in a typical metal each atom has eight or 12 near neighbors, we may think of the liquid state as a solid with one missing neighbor (Fig. 17.5). The melting process consists mainly of producing a few holes in the structure which otherwise resembles the solid structure. This, however, is only an instantaneous view of the structure since the atoms are continually shifting positions.

The missing-neighbor model explains some of the properties of liquids. Rapid atomic motion is one. The slightly elongated interatomic distances, atomic vacancies, and larger thermal energies make atomic motion easier and lower the viscosity. Molecules are continuously shifting in position and occasionally one is able to squeeze by its neighbors.

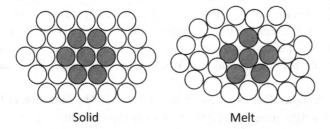

Solid   Melt

Fig. 17.5 Melting involves introducing a few small changes in the crystal structure: longer bondlengths, lower coordination numbers, local disorder, and larger vibrations. Adapted, with permission, from D. Tabor, *Gases, Liquids, and Solids*, Penguin Books, Baltimore MD (1969). Fig. 72, p. 200.

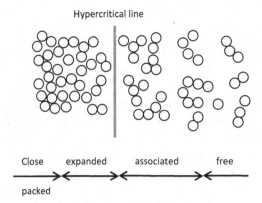

Hypercritical line

Close   expanded   associated   free

packed

Fig 17.6 As the temperatures and pressures approach the critical point, the structure of the liquid expands, increasing the distances between molecules and lowering the coordination number. The liquid and gaseous phases become indistinguishable. Adapted, with permission, from D. Tabor, *Gases, Liquids, and Solids*, Penguin Books, Baltimore MD (1969). Fig. 74, p. 202.

The process becomes easier and easier at higher temperatures. The viscosity of the liquid drops and the coordination number decreases to an average of three or four (Fig. 17.6) at the critical point. The coherent liquid structure is altered to that of an incoherent gas.

17.5   Types of Liquids

Six types of liquids are illustrated in Fig. 17.7. In a *simple monatomic liquid* such as argon, the atoms are held together by weak van der Waals forces. It has a low melting point (84 K), a low boiling point (87 K), low density, and low electrical conductivity. Liquid argon is stable over a very narrow temperature of 3K. Other inert gases behave in a similar manner (Fig. 17.8).

Carbon tetrachloride is representative of *simple molecular liquids*. The CCl_4 molecules have tetrahedral geometry (point group $\overline{4}3m$) with covalent C–Cl bonds 1.77 Å in length. Like simple monatomic liquids, the CCl_4 molecules in the liquid state are held together by weak van der Waals forces. The melting point is –23 °C and the boiling point is 77 °C, somewhat higher than argon. The melting points and boiling points increase with molecular weight (Fig. 17.9) just as they do with monatomic liquids, but the stability range of the liquid is much larger. Molecular liquids are more stable than monatomic liquids because the van der Waals bonds are enhanced by quadrupole contributions.

Water is an example of an *associated molecular liquid*. The formation of hydrogen bonds between neighboring molecules leads to higher melting and boiling points than would be expected. Figure 11.13 compared the data for H_2O with those of H_2S, H_2Se, and H_2Te. The exceptionally high values for water (0 °C and 100 °C) are indicative of strong hydrogen bonds. The behavior of ice and water were discussed in Chapter 11.

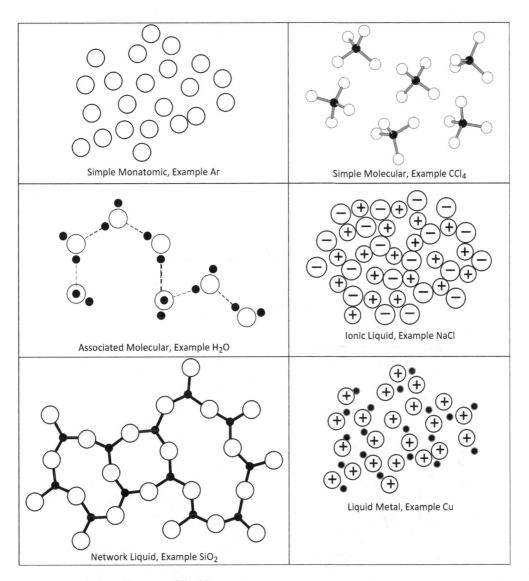

Fig. 17.7 Six types of liquids.

Molten salts such as NaCl form *ionic liquids* that are electrically neutral on a local scale but possess long-range Coulomb forces. Mobile cations and anions lead to high ionic conductivity. Sodium chloride melts at 801 °C and boils at 1465 °C. Alkali-halide melting points are plotted as a function of bond length in Fig. 4.3. There are clear trends within each group: fluorides are more stable than chlorides, bromides and iodides. Lithium salts have lower than expected melting points because they fail to make good contact with their six anion neighbors in the rocksalt structure (due to the small size of Li^+). In general, the alkali halides melt at much lower temperatures than oxides with the rocksalt structure.

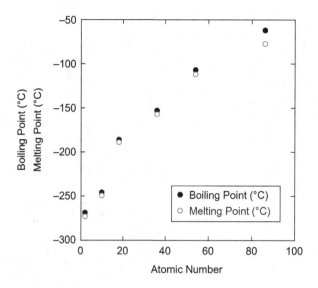

Fig. 17.8 Boiling and melting points of inert gases.

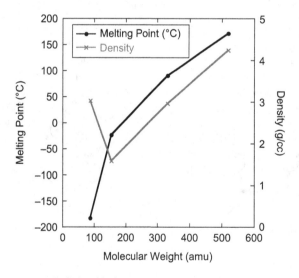

Fig. 17.9 Relationship between melting point and molecular weight in a series of carbon tetrahalide gases.

Sometimes a bonding change occurs in ionic crystals with polyatomic ions. Some groups such as NH_4^+ or SO_4^{2-} persist in the melt, but others dissociate. The pyrophosphate $P_2O_7^{4-}$ anion breaks up into PO_4^{3-} and PO_3^- ions, while the AlF_6^{3-} group in cryolite (Na_3AlF_6) coverts to AlF_4^- tetrahedra and two F^- ions on melting.

Liquid metals like molten copper (melting point 1085 °C, boiling point 2562 °C) possess mobile electrons that conduct both electricity and heat extremely well. As with solid metals, bonding takes place between the negatively charged electrons and the ion cores. Since most metals have close-packed crystal structures, liquid metals usually

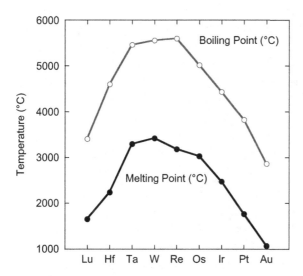

Fig. 17.10 Melting and boiling points of 5d transition metals.

have lower density than the solid form. Typically they differ by 2–8%. A few metals such as tin, antimony, silicon, and bismuth have more open structures with smaller changes in volume on melting.

The highest melting and boiling points are found among the 5d and 4d transition metals where the d electrons contribute to bonding. Elements like tungsten and platinum are used as lamp bulb filaments and as crucibles (Fig. 17.10). Note the wide stability ranges for both the liquid and solid phases among 5d transition metal elements. Alkali metals have only one electron to participate in bonding and therefore the melting points are much lower, as was shown in Fig. 4.4.

Polymeric liquids contain extensive network bonding leading to the formation of molecular and ionic glasses. Among molecular compounds, the tendency for glass formation is especially noticeable for large molecules, such as glucose, which become entangled as the melt cools, and are therefore unable to crystallize. Branching and cross-linking further reduce the mobility of the molecules producing glass-like systems.

Similar phenomena occur in ionic glasses such as SiO_2, B_2O_3, P_2O_5, and GeO_2, and in closely related chalcogenide glasses (S, Se, Te, As_2S_3). In molten silica, adjacent SiO_4 tetrahedra are linked by shared oxygens to form the network. Polymeric liquids have broad, diffuse melting ranges and are very viscous. Fluxes such as Na_2O lower the melting point of SiO_2 and reduce the viscosity by breaking up the network.

17.6   Latent Heats and Volume Changes

Under normal conditions, as pointed out earlier, a liquid is more like a solid than a gas. This means that the heat of evaporation is much greater than the heat of fusion. For all the elements, the heat of fusion is about ten times smaller. Compare Fig. 17.11 with Fig. 17.3 for numerical values. The heat of fusion for monatomic elements is

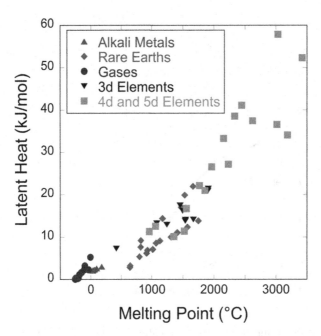

Fig. 17.11 The latent heat of fusion is an order of magnitude smaller than the latent heat of evaporation. Strongly bonded metals like tungsten have high melting points and high latent heats.

proportional to the melting point. 4d and 5d elements have large latent heats, far larger than other metals.

The *volume changes on melting* are generally rather modest. Since many solids have close-packed crystal structures, the density usually decreases on melting. In metals the densities differ typically by 2–8% (Fig. 17.12). A few metals such as tin, antimony, and bismuth have more open structures and only small changes in volume on melting. Most of the exceptions are semimetals with significant covalent contributions to the bonding.

More substantial volume changes take place in the alkali halide family. The percent increase in volume at the melting point is plotted against radius ratio in Fig. 17.13. Salts with small radius ratios show large changes in volume (up to 30%) indicative of lower coordination numbers in the melt.[1]

Most organic crystals with weak van der Waals bonding between molecules also expand on melting, typically by 8–15%.

Aluminum trichloride exhibits a very large increase (83%) in volume and a very large entropy of fusion (36.5 e.u.) on melting. The fusion process in $AlCl_3$ is accompanied by a major change in bonding and structure. Solid $AlCl_3$ has a monoclinic crystal structure based on hexagonally close-packed layers of large Cl^- ions. (Fig. 17.14a). The smaller Al^{3+} cations occupy octahedral interstitial sites in the close-packed array. Each Al is bonded to six Cl, and each Cl to two Al.

[1] Data in the figure are from A. K. Galwey, "A view and a review of the melting of alkali metal halide crystals: Part 1. A melt model based on density and energy changes," *Journal of Thermal Analysis and Calorimetry*, **82** (2005), 23–40, and from the Shannon–Prewitt ionic radii.

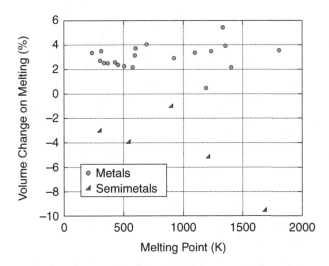

Fig. 17.12 Volume expansion on melting for a number of different metals. Most metals expand by a few percent on melting. Normal metals with high coordination numbers increase more than semimetals such as Si, Ge, Ga, Bi, and P with partially covalent bonds and lower coordination numbers.

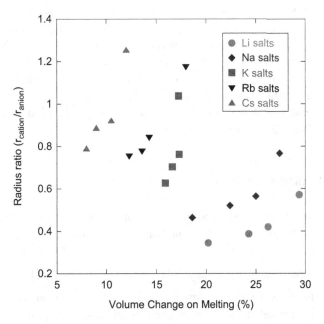

Fig. 17.13 Alkali-halide crystals show significant expansions during melting. The changes are especially large in lithium and sodium salts.

There is an unusually large change in structure at the melting point (192 °C). The molten state consists of Al_2Cl_6 dimers (Fig. 17.14b) with aluminum in tetrahedral coordination. It is a molecular liquid held together by van der Waals forces that is far less dense (1.30 g/cm^3) than the crystalline state (2.48 g/cm^3).

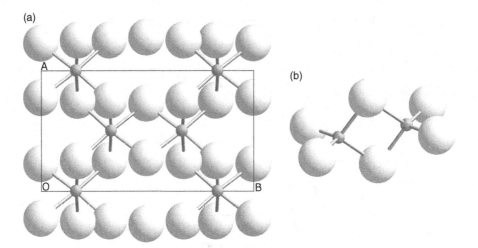

Fig. 17.14 An unusually large increase in volume takes place when $AlCl_3$ melts. The close-packed crystalline state (a) is far denser than the liquid state (b) as the intermolecular bonding changes from strong ionic forces to weak van der Waals bonds. Al = filled circles, Cl = open circles.

Not all materials expand on melting. Covalently bonded crystals like Si often have open structures with low coordination numbers (tetrahedral) and low density (2.33 g/cm^3). Both the coordination number and the density increase when Si melts. The density of molten silicon is 2.57 g/cm^3 at the melting point (1414 °C). Bonding changes also occur in semiconductors where the conductivity generally increases on melting, producing a metallic molten state. In germanium and silicon the 4-coordinated diamond structure collapses to give a coordination number of about 8 in the melt.

A similar decrease in volume takes place when ice melts. The open 4-coordinated structure of ice is less dense than water (see Chapter 11).

17.7   Viscosity and Surface Tension

Viscosity retards the flow of liquids and gases. The force required to overcome viscosity is proportional to the velocity gradient normal to the flow direction (Fig. 17.15). If the velocity gradient is dv/dz, the shear stress (force/unit area) f is $f = \eta \, dv/dz$, where η is defined as the coefficient of viscosity. It is measured in units of $\left[\dfrac{\frac{N}{m^2}}{\frac{m}{s \cdot m}} \right] = \dfrac{N \cdot s}{m^2} = $ poise.

Liquid viscosity arises from the intermolecular forces that create an internal friction in flowing fluids. Thermal fluctuations reduce this friction, causing the viscosity to decrease exponentially as the temperature is raised (Fig. 17.16).

In contrast, viscosity increases when pressure increases. Under pressure, the molecules are pushed closer together, making it more difficult to create a pressure gradient. This increase in viscosity is a great advantage in lubricating heavily loaded gears and other surfaces in danger of seizure. The viscosity of a lubricant may increase a million times under high stress. This is helpful as, otherwise, the liquid could be squeezed out, leading to the possibility of gears freezing up.

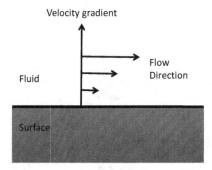

Fig. 17.15 Fluid flow is often slower near a surface because of viscosity.

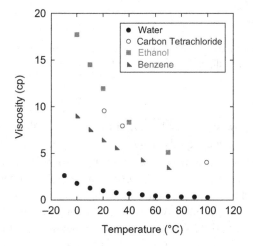

Fig. 17.16 The viscosity of water, carbon tetrachloride, ethanol, and benzene decreases markedly with temperature.

Viscosity in the molten state is a measure of the resistance to flow. In polymeric materials there is a threshold above which the chains entangle, greatly increasing the viscosity. Melt viscosity is linearly proportional to molecular weight below the threshold, and then rises rapidly as entanglement takes place. For polyisobutylene the threshold molecular weight is about 600, corresponding to chain lengths of about 10 molecular units.

$$\left(\begin{array}{c} \text{H} \quad\ \text{CH}_3 \\ |\qquad | \\ -\ \text{C}\ -\ \text{C}\ - \\ |\qquad | \\ \text{H} \quad\ \text{CH}_3 \end{array}\right)_n$$

Polyisobutylene is an amorphous elastomer with a low T_g of –70 °C. It is used in adhesives, as a flexibilizing additive for rigid polymers, and as an oil additive.

Most industrial polymers have average chain lengths well above the critical threshold where the physical properties stabilize.

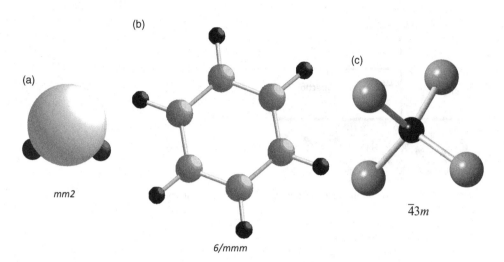

(b)

(a)

mm2

(c)

$\overline{4}3m$

6/mmm

Fig. 17.17 Low symmetry molecules like (a) H_2O show large changes in viscosity with temperature. Water molecules possess dipole moments while (b) benzene and (c) carbon tetrachloride do not.

Organic molecules having low symmetry behave differently than those with high symmetry. The structure of organic liquids made up of low-symmetry molecules undergo major changes with temperature as the molecules absorb thermal energy and begin to move more rapidly. This has a major effect on physical properties such as viscosity. According to Waller's rule, the ratio of the viscosities at the freezing and boiling points is low for molecules of high symmetry, and high for molecules of low symmetry. The viscosities of water, benzene, and carbon tetrachloride (Fig. 17.16) illustrate Waller's rule. The symmetries of H_2O, C_6H_6, and CCl_4 are shown in Fig. 17.17.

Surface Tension

Small raindrops are nearly spherical in shape, and since a sphere is the geometric form with the smallest surface area per unit volume, it is apparent that the surface of a liquid has higher energy than the bulk. Liquids attempt to lower energy by reducing surface area.

The excess energy of a free surface has a simple molecular interpretation. Molecules inside the liquid are bonded to neighboring molecules in all directions, but those on the surface are not. Except for a few vapor molecules, there are no chemical bonds on the outer surface. As a result, the surface molecules are pulled in toward the inside of the fluid. The surface area can only be increased by pulling molecules to the surface; this asymmetry in the molecular forces is responsible for the surface energy.

Roughly half the near-neighbor bonds are missing for molecules located on the outer surface of the liquid. This incomplete bonding corresponds to about half the latent heat of vaporization, giving rise to the surrounding vapor phase.

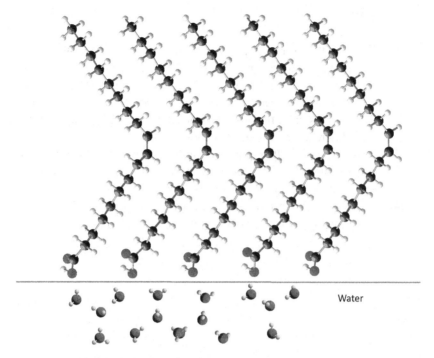

Fig. 17.18 Molecules of surfactant form a monolayer on surface of water.

Surfactants take advantage of these dangling bonds on the surface of a liquid. For example, organic molecules with polar end groups such as oleic acid ($C_{17}H_{33}COOH$, Fig. 17.18) which are normally insoluble in water will spread out over the surface of pure water. If the amount of surfactant added to the water surface is less than a certain amount there is very little reduction in the surface tension of the water. However, if the amount added exceeds a certain critical value, the surface tension falls rapidly to a lower value.

Commercially, surfactants are compounds that reduce surface tension when added to water solutions. The surfactants of soap and synthetic detergents perform the cleaning action through the reduction of surface tension. The surfactants used in the cleaning solution are generally long chains. One end of the chain is *hydrophilic* (water-loving) and the other end is *hydrophobic* (water-hating). The hydrophobic portion of the surfactant molecule is usually a *hydrocarbon* containing eight to 18 carbon atoms in a straight or slightly branched chain. The hydrophilic functional group is most often an alkylbenzene sulfonate made from petroleum or an alkyl sulfate from animal fat. A surface-active anion widely used in detergents is illustrated below.

```
H    H H H H H  H H H H H H     O
 \   | | | | | | | | | | | |     |
H – C - C–C– C – C– C– C– C– C– C– C– C– O – S – O⁻
 /   | | | | | | | | | | | |     |
H    H H H H H  H H H H H H     O
```

17.8   Problems

(1) Calculate the average separation distance between molecules in air at ambient pressures and temperatures. Use the ideal gas law as an approximation.

(2) **(a)** Compare the structures of SiO_2 (tridymite) and H_2O (ice). Sketch the structures as accurately as you can, showing the coordination of the atoms.

   (b) Discuss the differences in bonding between SiO_2 and H_2O. How does this difference affect their melting points?

(3) Rank order the melting points of Ar, ZrO_2, Cu, and H_2O. Justify your reasoning based on the bonding involved. If more than one type of bonding is present, be specific about which you believe is broken on melting.

(4) Rationalize why there might be a dependence of the melting point on molecular size in hydrocarbons, as shown in the graph.

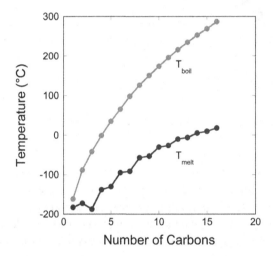

18 Glasses

Glasses can be defined as liquids that become rigid without crystallizing as they are cooled. These liquids are very viscous above the melting point (T_m) and become rigid at the glass transition temperature (T_g). Cooling slowly between T_m and T_g can either cause devitrification or lead to a number of different glassy states depending on dwell times and temperatures.

Viscosity, or the resistance to fluid flow, is governed by two factors: one is the number and type of chemical bonds which must be broken to allow the molecular groups to migrate; and the other is the available space into which these clusters may move. The stronger the bonds to be broken, the higher the melting point, and the larger the mobile clusters, the higher the viscosity. As rigid glass is heated, thermal expansion provides the free volume required for flow. The resulting decrease in viscosity with increasing temperature is due to the increase in free volume and the decreasing size of the migrating clusters.

Silicate-based glasses are ubiquitous in windows, packaging, tableware, and in displays, among other applications. Amorphous semiconductors, phase change memories, and infrared optics drive use of chalcogenide glasses. Many polymers also have a non-crystalline component; these are discussed in detail in Chapter 12, and will not be described further here.

18.1   Non-Crystalline Solids

Crystals are, by definition, solids with atomic arrangements that are periodic in three dimensions. Well-crystallized solids have clear peaks in X-ray, electron, or neutron diffraction patterns. Other solids have no long-range order in the atomic positions, and are referred to as *amorphous* or *glassy*. For amorphous materials, diffraction measurements do not show clear peaks, but tend to have a few broad and ill-defined humps.

There is a continuum of degrees of order between the extremes of crystalline and amorphous, with subtleties being discovered regularly as new synthesis and characterization techniques emerge. Some materials possess order in only one or two dimensions. Graphite sheets, for example, can readily be displaced with respect to each other, so that the material is fully ordered in the plane, but the out-of plane translational symmetry is lost. Clays show similar behavior. Some liquid crystals show periodicity only along a single axis. That is, rod-like molecules arrange in sheets where the periodicity between layers is well-defined, but the molecule position within the layers is random.

While there is no long-range order in amorphous solids, the short range order (including the bond lengths, bond angles, and coordination numbers) is often very similar to that of crystals of the same compositions. However, these structural units are arranged with inherent randomness. Amorphous solids are thus always high-entropy ones.

Thermodynamically, below the melting temperature, materials should crystallize into the solid phase via a nucleation and growth process. This necessarily requires that atoms be able to move through the melt to reach their lowest energy configuration. However, in cases where it is difficult to rearrange the positions of atoms or atom clusters in the melt, or if the melt is cooled so quickly that atoms do not have time to reach an equilibrium site, an amorphous state can be frozen in. As a result, rapid cooling rates tend to favor production of amorphous materials. "Splat cooling" for example, allows quenching rates of 10^6 deg/s. In this technique, a melt is typically poured in a thin stream between cooled counter-rotating rollers. Metallic glasses are often prepared using this technique. Even faster cooling rates are achieved when atoms are condensed from a vapor source onto a substrate. Thus, deposition of thin films from a vapor phase often leads to amorphous layers that will crystallize on heating.

Rapid chemical changes can also lead to amorphous solids. Many sol–gel processes, for example, begin with the cations of interest in a solvent. Rapid removal of the solvent and any organic groups will often produce an amorphous layer. Additional energy has to be provided for the atoms to have high enough diffusion coefficients to reach the crystalline state.

Crystallization can also be destroyed *post facto*. Many materials, especially those with comparatively open crystal structures, can be amorphized by mechanical stresses, including those from shock waves, or from grinding, by exposure to some high-energy radiation, or by bombardment by high-energy particles. The mineral zircon, $ZrSiO_4$, for example, often contains a small amount of uranium as an impurity. Decay products from uranium amorphize these minerals over time; the degree of amorphous character can be used to date the rocks. Even modest levels of radiation-induced damage can be problematic in silicon semiconductors, where displacement of atoms from their crystalline sites produces dangling bonds and trap states that change transistor and device performance.

18.2   Glasses

Most of the materials described in common parlance as "glasses" are produced by cooling at a modest rate from the processing temperature. Characteristic of such materials are strong bonds and difficulty in rearranging the constituent units into the crystalline form. The larger the group of ions that must be moved into the correct position, the more strong bonds that must be broken, and the more complex the crystal structure, the more likely that an amorphous solid will result. In practice, this implies

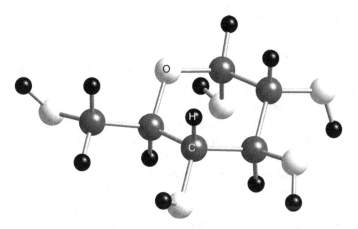

Fig. 18.1 Glucose molecule. Carbon is dark gray, oxygen is lighter gray, and hydrogen is black.

that melts with very high viscosities are more likely to produce glasses by conventional processing. The critical cooling rate for a given composition thus depends on the structure of the melt.

Consider, for example, the relative difficulty in moving a spherical ion, or a molecule with a complex shape through the melt. If the structure of the crystal is simple, then a spherical atom can readily find an attachment point to grow the crystal. In contrast, glass formation is more likely for large, irregularly shaped molecules such as glucose (Fig. 18.1), toluene, or glycerol, which become entangled as the melt cools, and are therefore unable to crystallize. Low-symmetry molecules require more time to arrange themselves in crystalline order, since they have to be aligned precisely to the crystalline matrix. An ordered state is even more difficult to reach if the "molecules" themselves are not ordered. Many long-chain polymers are examples of this, due to the presence of branches, cross-links, or insufficient stereo-regularity.

Common window or speciality glasses contain some network-formers such as SiO_2, B_2O_3, P_2O_5, and GeO_2, as well as modifiers such as Na^+, Ca^{2+}, or K^+. Random networks form most easily when the cations are triangularly or tetrahedrally coordinated and the cation polyhedra are corner-shared. This, and other ideas common to glasses that readily form by cooling from a melt are summarized in Zachariasen's rules. Only the first four rules need to be satisfied to make oxide glasses from the melt.

Zachariasen's rules for oxide glasses are as follows:

- Each oxygen is linked to not more than two cations.
- The coordination number of oxygen ions about the network cation is small, typically four or less.
- Oxygen polyhedra share corners, not edges or faces.
- At least three corners of each polyhedron should be shared.
- Modifier ions, if present, are large and have low charge.

The first rule ensures sufficient flexibility around the bridging oxygens to allow the wide variety of geometries observed in amorphous materials. In silica (SiO_2), flexibility is also promoted by variations in the Si–O–Si angle, as will be discussed further in Chapter 19.

The second rule implies that the electrostatic bond strengths of network-forming cations will be high. This serves both to create a strong network with bonds that are not easily broken, and to satisfy local electroneutrality around each oxygen with only one or two bonds to network formers, in keeping with rule #1.

The third rule recognizes the fact that by sharing corners, rather than edges or faces, the highly charged network-forming cations are kept far apart. This lowers the repulsive Coulomb forces, while leaving sufficient flexibility in the strongly bonded network. Molecular groups consisting of edge- and face- shared units tend to have rigid geometries.

The fourth rule promotes cross-linking to give a three-dimensional network. Discrete SiO_4^{4-} units are easily re-arranged in the melt phase, but as a larger Si–O–Si network develops, the melt viscosity rises. When two corners on each silicate tetrahedra are shared with other network formers, both short chains and rings are present; this is characterized by a Si:O ratio of 1:3. Materials with this Si:O ratio are mediocre glass-formers. When three of the corners are shared, a continuous three-dimensional network can form. This occurs at Si:O ratios of 2:5, a ratio characteristic of a good glass-former.

The fifth rule reflects the fact that large cations with low charges (like K^+, Ca^{2+}, and Na^+) will tend to break up the network and may induce crystallization. Such ions are collectively classed as modifier ions. They tend to decrease the melting point of the glass and lower the melt viscosity, as is discussed further below. They thus greatly simplify manufacture of oxide glasses by reducing the melt viscosity (improving melt homogeneity) and lowering melt temperatures, though at the cost of increased thermal expansion coefficients and degraded chemical stability in acidic environments. Smaller alkali cations are more likely to produce phase separation in the melt.

18.3   Structure of Oxide Glasses

The local structures of silica glass and molten silica are very similar to the crystalline forms. Silicon atoms are bonded to four oxygens, and oxygen to two silicons, forming a three-dimensional network of silica tetrahedra linked together by all four corners. As a result, molten silica is a very viscous liquid near the melting point and consequently silica glass is very difficult to shape.

The structure of silica glass in two dimensions is illustrated in Fig. 18.2. As can be seen, there are many 5-rings in the silicate network, in contrast to the 6-rings of the cristobalite and tridymite forms, and the 4-rings of the feldspar ladder. The predominance of 5-rings can be traced to the fact that the tetrahedral O–Si–O angle of 109.5° is very close to the pentagonal angle of 108°. Thus, pentagonal rings form in the molten phase and the strong Si–O bonds retain this configuration in the glassy form of silica.

To loosen up this structure and lower the viscosity, modifier ions such as Na^+ and Ca^{2+} are incorporated into the structure. This is done by adding soda (Na_2O) and lime (CaO) to the molten silica. This has the effect of lowering the melting point and lowering the viscosity (see Fig. 18.3).

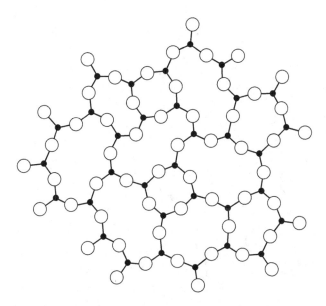

Fig. 18.2 In silica glass each silicon (small filled circles) is bonded to four oxygens (open circles) and each oxygen to two silicons. The corner-linked SiO_4 tetrahedra form a three-dimensional network represented here in two dimensions.

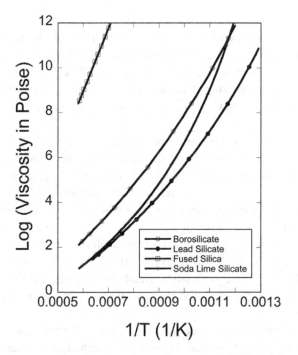

Fig. 18.3 The viscosity of several commercial silicate glasses plotted as a function of temperature. Data from D. C. Boyd and D. A. Thompson, "Glass," *Kirk-Othmer: Encyclopedia of Chemical Technology Volume 11,* 3rd edition, pp. 807–880, John Wiley & Sons Inc. (1980). Figure 8, page 825.

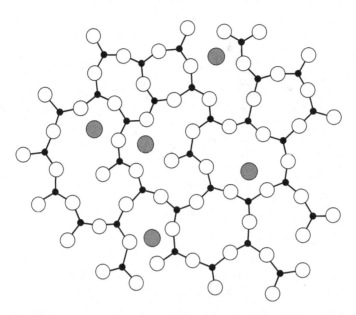

Fig. 18.4 Two-dimensional sketch illustrating the structure of soda–lime silicate glass. The small filled circles are silicon, the larger filled circles are modifier ions, and the open circles are oxygen.

The Na^+ and Ca^{2+} break up the silicon–oxygen network (Fig. 18.4). There are now two types of oxygen ions, bridging and non-bridging. The bridging oxygens are bonded to two silicons, the non-bridging to only one. When fewer than three corners of the silica tetrahedra are shared, the network breaks up and the melt becomes very fluid. When cooled, the low-viscosity melt crystallizes rather than forming a glass.

The effectiveness of an alkali flux in decreasing the viscosity and lowering the melting point increases with bond length and decreases with bond strength. Cs_2O is a better flux than Li_2O. The bonds between cesium and oxygen are easily ruptured and the resulting oxygen breaks up the –Si–O–Si– link into non-bridging –Si–O O–Si–. This weakens the network and lowers the melting point. Lithium has lower fluxing power because it tends to hang on to its oxygens.

Soda–lime silica glasses have higher durability than alkali silicate glasses because of the stronger alkaline-earth–oxygen bonds. Soda (Na_2O) and lime (CaO) are inexpensive and easy to melt when combined with silica. Nearly 90% of the glasses produced commercially are soda–lime glasses. Other commercial glasses of engineering interest are listed in Table 18.1.

Zachariasen's rules apply best to oxide glasses formed by normal furnace cooling from the melt. Many other materials that do not follow Zachariasen's rules for glass formation, including metals, polymers, and ceramics, can nonetheless be produced in amorphous form either through chemical techniques, vapor deposition, or other rapid quenching methods.

Table 18.1 Chemical compositions (in mole%) of several commercial glasses[1,2,3]

Type	SiO$_2$	Al$_2$O$_3$	CaO	MgO	Na$_2$O	K$_2$O	Other	Property
Window glass	72	0.1	9.2	5.6	13	-	-	Low cost
Container glass	74	0.8	11	0.3	13	0.2	-	Low melting temperature
Silica glass	99.5+	-	-	-	-	-	-	Optical fibers Low thermal expansion coefficient
96% silica	97.1	0.2	-	-	<0.2	<0.2	2.5B$_2$O$_3$	Easier processing than pure silica glass
Borosilicate	83.3	1.3	-	-	4.1	0	11.2 B$_2$O$_3$	Good chemical and heat resistance
Leaded glass	62.9	2.6	-	10.3	-	-	8.5 PbO 13.6 B$_2$O$_3$	High refractive index
Aluminosilicate	57	20.5	5.5	12	1	-	4 B$_2$O$_3$	Low expansion coefficient
Flat panel display glass	65–70	9–12	5–10	1–2	<<1	<<1	5–10 B$_2$O$_3$ 0–5 BaO	Low alkali content

Glass-forming elements satisfying Zachariasen's rules include boron, silicon, germanium, and phosphorus. The electrostatic bond strength (= cation valence/cation coordination) for these network formers is near unity. This ensures that Pauling's electrostatic valence rule is satisfied for an oxygen linked to two such network formers.

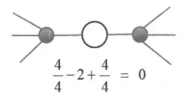

$$\frac{4}{4} - 2 + \frac{4}{4} = 0$$

In addition to network formers with high bond strength, and modifiers (Na$^+$, K$^+$, Li$^+$, Ca^{2+}, Mg^{2+}, etc.) with low bond strength, there are intermediates with bond strengths near 0.5. Important intermediates include Ti^{4+}, Pb^{2+}, Zr^{4+}, Zn^{2+}, and Al^{3+}. Intermediates enter the network when sufficient glass-forming elements are present. They can impart special properties to the glass such as the high refractive index of titanium- and lead-containing glasses.

Boron is the most useful of the alternate glass-formers. It is usually used in combination with silicon, sodium, and aluminum to form borosilicate glasses with about 75 weight % SiO$_2$ and 10–15% B$_2$O$_3$.

---

[1] Private communication, Carlo Pantano.

[2] D. R. Uhlmann and N. J. Kreidl, *Optical Properties of Glass*, The American Ceramic Society, Westerville, OH (1991).

[3] http://glassproperties.com/glasses/

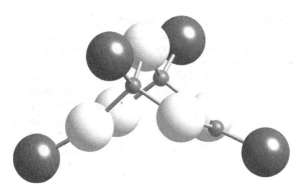

Fig. 18.5 Clinographic projection of the idealized structure of the $[B_4O_5(OH)_4]^{2-}$ ion in borax $(Na_2[B_4O_5(OH)_4] \cdot 8H_2O)$: Small filled circle – B, white – O, gray – OH$^-$.

Borax $(Na_2B_4O_5(OH)_4 \cdot 8H_2O)$ is a typical mineral source for borate glasses. Borate minerals come from the famous Death Valley region of California. The $[B_4O_5(OH)_4]^{2-}$ complex in borax illustrates the tetrahedral and triangular coordinations of boron (Fig. 18.5). The structure is held together by hydrogen bonds extending from one borate group to other borate groups and to the water molecules surrounding the Na$^+$ ion. Hydrogen bonds are easily ruptured, imparting a low decomposition temperature to the mineral.

The crystal structure of B_2O_3 consists of triangularly coordinated boron linked together to form larger rings, which are, in turn, loosely bonded to identical complexes, forming a ribbon-like structure (Fig. 18.6). The melting point of B_2O_3 is only 450 °C which makes it an excellent flux.

In borosilicate glasses, B^{3+} takes either triangular or tetrahedral coordination depending on the presence of modifier ions. In an unmodified borosilicate, boron is in triangular coordination, satisfying the bonding requirements of the oxygen ions. When modifier ions are added, however, the boron atoms adopt tetrahedral coordination. Approximately one boron is converted for every alkali ion added. In this case, the oxygen ions are bonded to one silicon, one boron, and two sodium atoms, again satisfying Pauling's electrostatic valence rule (see Fig. 18.7). This complexity in structure of modified borosilicate glasses broadens the working range by helping to control the viscosity.

18.4   Structure–Property Relations

Inorganic glasses can be thought of as oxide polymers with cross-linked network structures and different packing densities. The properties of glasses show clear trends with cross-link density, and with packing efficiency and bond strength.

Several of the structural units that make up oxide glasses are pictured in Fig. 18.8. Arsenic sesquioxide (As_2O_3) contains shallow trigonal pyramids with As–O bond lengths of 1.80 Å and O–As–O angles of 100°. A lone pair of electrons is appended to the arsenic ion at the apex of the pyramid. Each oxygen ion is bonded to two As^{3+} ions linking the pyramidal groups together in As_2O_3 glass.

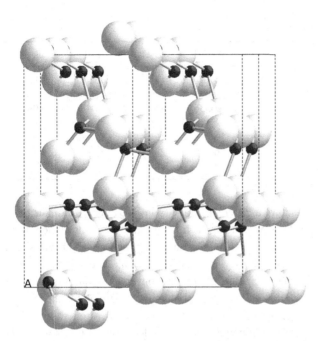

Fig. 18.6 The structure of crystalline B_2O_3. Triangular BO_3 groups are connected together in ribbon-like 10-fold rings. The structure of B_2O_3 glass is similar. B are the small filled circles, O are the larger circles.

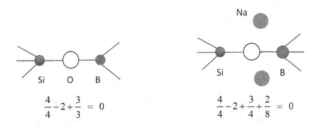

Fig. 18.7 Pauling's electrostatic valence rule in modified and unmodified borosilicates.

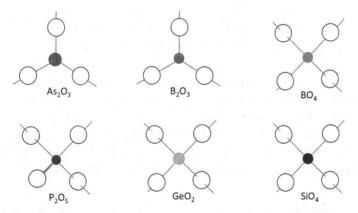

Fig. 18.8 Network-forming ions in oxide glasses: the basic building blocks in commercial glasses.

Table 18.2 Properties of similar oxide glasses[4]

	As_2O_3	B_2O_3	P_2O_5	GeO_2	SiO_2
Bond energy (kcal/mol)	70	110	88	104	106
Density (g/cm^3)	3.70	1.81	2.24	3.63	2.20
Packing density (oxygens/Å^3)	0.0336	0.0470	0.0475	0.0416	0.0439
Network connectivity	3	3	3	4	4
Cross-link density	1	1	1	2	2
T_g (°C)	160	247	270	800	1200
Activation energy for viscous flow (kcal/mol)	23	40	41.5	75	180

The molecular units have a different geometry in B_2O_3 glass. Here, the triangular BO_3 groups are flat rather than pyramidal. Each planar group has three short B–O bonds of 1.37 Å, and an O–B–O angle of 120°. The borate triangles in B_2O_3 glass are connected to one another through the three oxygen ions. When alkali or alkaline-earth ions are added to B_2O_3, some of the triangular BO_3 groups are converted to tetrahedral BO_4 groups.

In phosphate glass (P_2O_5), the structural units are deformed tetrahedra. Each P^{5+} ion is bonded to four oxygens, but one of the bonds is a short, strong double bond, 1.39 Å long. The other three oxygen neighbors are single bonds (1.62 Å long). These three oxygens form links to neighboring phosphate groups in amorphous P_2O_5. The O–P–O angle between singly-bonded oxygens is 99°, rather than the ideal 109.5° angle for regular tetrahedra.

The near-neighbor bonds in GeO_2 and SiO_2 glasses are more symmetric. Here each cation is bonded to four oxygens arranged in a regular tetrahedron with bond angles of 109.5°. Each of the four oxygens is bonded to two cations. The bond lengths are 1.61 Å for SiO_2 and 1.75 Å for GeO_2.

Oxide glasses contain two kinds of oxygens, those that link together two of the network formers (B, Si, ...) and those that do not. The latter may be bonded to a network modifier such as Na^+ or be part of a hydroxyl group. If there are just two bridging oxygens per network ion, the network resembles a linear polymer. Each additional bridging oxygen acts as cross-link between these chains. The *cross-link density* per network former is therefore equal to the number of bridging oxygens minus two.

The oxygen packing density can be estimated from the measured density and the molecular weight. Packing densities in Table 18.2 are expressed as the number of oxygens per cubic angstrom. The packing densities of oxide glasses are very similar despite the large differences in specific gravities. Structure–property relationships between these parameters and the physical properties will be discussed next.

On heating, oxide glasses transform from a rigid, brittle solid to a viscoelastic solid. The highly cross-linked network structure gradually loosens as temperature rises and a number of the –O– cross-links fail. The *glass transition temperature* (T_g) is largely controlled by the cross-link density. Comparing the five oxide glasses in Table 18.2, the transformation temperatures in GeO_2 and SiO_2 are much higher than those in As_2O_3,

[4] N.H. Ray "Composition–property relationships in inorganic oxide glasses," *J. Non-Crystalline Solids*, **15** 423–434 (1974).

Table 18.3 Bond strengths and coordination numbers for modifier ions in oxide glasses

Modifier cation	Average coordination number	Bond strength to oxygen (kcal/mol)
Li	4	36
Na	6	20
K	9	13
Rb	10	12
Cs	12	10
Mg	6	37
Ca	8	32
Sr	8	32
Ba	8	33

From N. H. Ray, "Composition–property relationships in inorganic oxide glasses," *J. Non-Crystalline Solids*, **15**, 423–434 (1974).

B_2O_3, and P_2O_5, and so are the cross-link densities. Silica and germania glasses have twice the number of cross-links.

Tightness of packing also has an influence on the transformation temperature. The number of oxygens per unit volume is a measure of the packing density. P_2O_5 and B_2O_3 have higher transformation temperatures and closer packing than As_2O_3. SiO_2 exceeds GeO_2 for the same reason.

The addition of alkali or alkaline-earth oxides to the glass also has a profound effect on the transformation temperature. Table 18.3 lists average coordination numbers and bond energies for common modifier ions. In most oxide glasses, modifiers generally lower the cross-link density and the transformation temperature. Boron oxide glasses are the exception. In this case, modifiers convert some of the triangular BO_3 groups to BO_4 tetrahedra, doubling the number of cross-links from one to two. This raises the glass transition temperature.

In oxide glasses which do not contain boron, T_g decreases when modifier cations are added to the glass. In alkali silicate glasses, the smaller cations like Li^+ and Na^+ cause much larger decreases than the larger K^+ ion. Potassium has a higher coordination number, which helps to prevent collapse of the silicate network.

Melt viscosity is also controlled by the cross-link density. Flow rates in viscous melts involve the temporary rupture of chemical bonds and the activation energy for this process depends on bond energies and the number of cross-links. Activation energies for oxide glasses are listed in Table 18.2. Cross-link densities and activation energies for As_2O_3 and B_2O_3 are similar and much lower than the tetrahedrally bonded GeO_2 and SiO_2 glasses. The tightness of packing also affects the viscosity. The activation energies for B_2O_3 and P_2O_5 are higher than As_2O_3, and SiO_2 is larger than GeO_2 for this reason.

Silicate glasses modified with alkali or alkaline-earth cations are less viscous than pure SiO_2. The activation energies for divalent cations are larger than the monovalent alkali

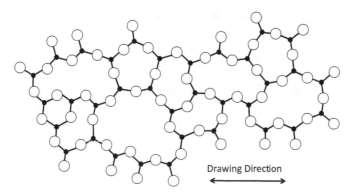

Drawing Direction

Fig. 18.9 Schematic sketch of the structure of fiber glass.

ions. B_2O_3 glasses behave differently because of the change in bonding. Sodium additions lead to an increase in cross-link density as boron changes from triangular to tetrahedral coordination. This causes an increase in the activation energy for viscous flow.

Thermal expansion behaves in a similar manner. Heavily cross-linked glasses like SiO_2 have very low thermal expansion coefficients. The addition of alkali ions lowers the number of cross-links and raises the thermal expansion coefficient.

Cation mobility is important in the processing of oxide glasses. Na^+ diffuses more easily than the larger K^+ ions, and the more highly charged Ca^{2+} ions. The diffusion coefficient of Na^+ in common soda-lime window glass is about 100 times larger than Ca^{2+}. This also has important consequences in terms of ionic conductivity, as will be described in Chapter 25. The mobility of the network formers is generally lower than the mobility of the modifiers, since the bonding holding the atoms in place is so strong for small, highly charged cations.

The exceptional *tensile strength* of glass fibers brings up the question of the difference between the structure of fibers and that of bulk glass. The fiber spinning process involves extruding molten glass under stress through a nozzle. Measurements carried out as a function of stress and temperature indicate that the structure is more open at high temperature and more anisotropic under larger drawing forces (Fig. 18.9). The alignment of –Si–O–Si–O– chains along the fiber axis makes a strong contribution to the tensile strength, and the openness of the structure in the radial direction enhances the flexibility of the fiber.

18.5   Glass Transition Temperatures

At one time, glass was defined as a supercooled liquid, but it is important to realize that there is a glass transition temperature (T_g), which differentiates the glassy state from the liquid state. Properties such as thermal expansion, refractive index, and rigidity change rapidly near T_g. At temperatures above T_g, the glass is viscous and plastic, and it changes slowly to the liquid state with comparatively small changes in the atomic structure. The liquid states of glass-forming oxides are very viscous compared to other inorganic materials (Table 18.4).

Table 18.4 Melting points and melting viscosities for glass-forming melts and normal liquids

	Melting point (°C)	Viscosity (poise)
Normal liquids		
H_2O	0	0.02
Zn	420	0.03
LiCl	613	0.02
Fe	1535	0.07
Glass-forming melts		
As_2O_3	309	10^6
B_2O_3	450	10^5
GeO_2	1115	10^7
SiO_2	1710	10^7

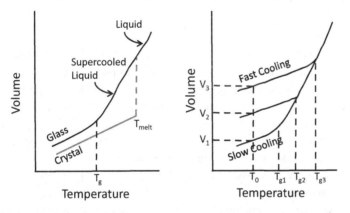

Fig. 18.10 Volume–temperature relationships for glasses, liquids, supercooled liquids and crystals.

Glass transition temperatures can be measured by cooling the molten glass from high temperature while recording the volume contraction coefficient (Fig. 18.10). A sharp change in slope is noted between the liquid state and the rigid glassy state. The value of T_g depends on cooling rate, as also indicated in Fig. 18.10. Rapidly cooled melts have higher glass transition temperatures than slowly cooled glasses of the same composition. Oxide and molecular glasses are characterized by a small range of T_g.

18.6   Immiscible Glasses and Glass-Ceramics

One-component oxide glass melts consisting entirely of a network former (SiO_2, B_2O_3, . . .) invariably solidify into homogeneous glass. Glass melts corresponding to a stable chemical compound also solidify to a homogeneous glass, but glass melts of

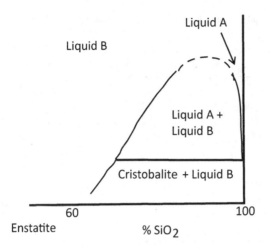

Fig. 18.11 Phase diagram for the magnesia–silica system showing phase separation in the molten state and the resulting metastable glasses. The vertical axis is temperature.

composition between two stable compounds tend toward phase separation. Transition metal ions and other additives are often partitioned in phase-separated glasses.

Liquids and glasses tend to phase separate when there are bonding problems. Oil and water do not mix because the bonding in these two liquids is quite different. Oil is a hydrocarbon with covalent bonds along the backbone and van der Waals bonds between molecules. Water is much more ionic with hydrogen bonds between molecules. *Like bonds to like* is the rule underlying phase separation.

In glass-forming systems based on SiO_2, GeO_2, or B_2O_3, phase separation occurs for modifier ions such as Mg^{2+}, Fe^{2+}, Y^{3+}, and Zr^{4+}. As an example, consider the MgO–SiO_2 system where glass formation takes place for silica-rich compositions (Fig. 18.11). Phase separation takes place in glasses in the range between 60 and 90 mole% SiO_2. The cause is local charge imbalance.

The three diagrams in Fig. 18.12 illustrate the bonding situation in homogeneous glasses corresponding to compositions SiO_2, $MgSiO_3$, and to an intermediary composition between the two. Electrostatic bond strengths ($= \frac{cation\ charge}{coordination}$) for Si–O and Mg–O are 4/4 and 2/6, respectively. Stable glass melts are obtained for SiO_2 and $MgSiO_3$ where local charge balance is obtained for every oxygen ion, but not for compositions in between, where some oxygens have too much charge and others too little. Phase separation takes place under these circumstances. Similar immiscibility phenomena occur in other oxide glasses containing modified ions with electrostatic bond strengths between 0.25 and 0.8.

Pyrex and Vycor are commercial borosilicate glasses that make use of phase separation. There are three regions of immiscibility in the Na_2O–B_2O_3–SiO_2 system. Vycor glass owes its unique properties to a spontaneous segregation of a nearly pure silica glass from a phase rich in boron and sodium. The latter phase is three-dimensionally connected and can be removed by chemical attack, leaving a highly porous silica glass which is easily densified at lower temperatures.

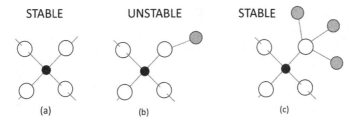

STABLE   UNSTABLE   STABLE

(a)   (b)   (c)

Fig. 18.12 Molecular units representing local bonding in (a) SiO_2 glass and (c) $MgSiO_3$ glass, both of which are homogeneous. Pauling's second rule is satisfied for all oxygens in both melts. An unstable molecular unit for the intermediary composition is shown in (b) where one of the oxygens lacks sufficient neighboring cations. Si is shown as the small filled circle, O as the open circle, and Mg as the larger filled circle.

Pyrex is another borosilicate glass of the composition 80% SiO_2, 13% B_2O_3, 4% alkali, and 2% Al_2O_3. Because boron can convert from 3- to 4-coordinated on addition of the alkali, there are few non-bridging oxygens in the glass. Consequently, Pyrex resists attack and can be used in applications requiring chemical inertness. The primary reason for adding boron to Vycor or Pyrex is to lower the melting and working temperatures of the glass. Pyrex has the lowest liquidus temperature (774 °C) of any high-silica-content glass, explaining its ability to withstand devitrification.

Glazes are protective glass coatings formed on ceramics that typically have low melting temperatures. Salt glazes on ceramic sewer pipes are obtained by simply throwing common salt into the furnace kiln at about 1300 °C. The salt volatilizes and reacts with the surface layers of the fired ceramic to give an aluminosilicate glass with excellent chemical resistance. Better quality glazes are obtained using fluxes such as B_2O_3, Na_2O, K_2O, CaO, PbO, and ZnO. These usually require a second firing at lower temperatures.

Titanium dioxide (rutile) is often used as a nucleating agent leading to the devitrification (crystallization) of oxide glasses. This is a different type of immiscibility used in making glass-ceramics. Commercial glass-ceramics are polycrystalline ceramics made by the controlled devitrification of glass. When properly nucleated, the glass-ceramics consist of fine-grained crystallites and residual glass with very little porosity, and high strength.

From a commercial viewpoint, the most important glass-ceramics are based on the Li_2O–Al_2O_3–SiO_2 system. These low thermal expansion glass-ceramics have a wide variety of applications including opaque and transparent cookware, mirror blanks, hot plates and heat exchangers. Glasses are prepared in the 1200–1300 °C temperature range, shaped near 1000 °C, and crystallized (cerammed) at lower temperatures using TiO_2 as a nucleating agent. The number of rutile seeds controls the grain size and transparency of the glass-ceramics. The ceramic grains generally have crystal structures which are stuffed derivatives of the β-quartz or β-spodumene structures. Al and Si occupy the tetrahedral sites with Li in interstitial positions (Fig 18.13).

18.7   Metallic Glasses

A number of glass-forming amorphous alloys have been discovered in the past few decades. Measurements have shown that some glassy alloys are twice as strong as steel,

(a)   (b)

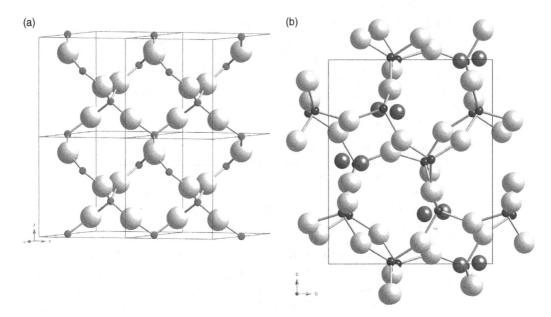

Fig. 18.13 Drawings of (a) β-quartz looking perpendicular to the c axis and (b) β-spodumene (keatite) structures. The tetrahedral ions (Al and Si) are shown as the small filled circles, O as the large light circle, and Li (in spodumene) is shown as the larger darker circle.

have greater corrosion and wear resistance, and are tougher than ceramics. They also retain many of the advantages of glasses, including the potential for melt forming.

When conventional alloys cool from the melt they rapidly solidify into a polycrystalline state with grains of different sizes and shapes. Grain boundaries are the boundaries between crystallites of different crystallographic orientation. Often, they act as weak regions where fractures form and corrosion begins. In contrast, the liquid phase of glassforming alloys can be undercooled far below the melting point, freezing in a vitreous solid without crystallization. This eliminates the grain boundaries.

The pure metallic elements have low viscosities in the liquid phase ($\sim 10^{-3}$ poise) and hence have a very strong tendency to crystallize on cooling. More complex compositions tend to favor glassy metals. Metallic glasses based on Pb, Au, Zr, Cu, Mg, Fe, Co, Ti, Ca, Pt, and several rare earths have been reported. Two of the most successful metallic glass compositions are based on pentary alloys in the Zr–Ti–Cu–Ni–Be and Zr–Ti–Cu–Ni–Al systems. Known commercially as Vitreloy, aerospace materials up to 10 cm thick have been cast in silica containers. Vitreloy 1 ($Zr_{41.2}Ti_{13.8}Cu_{12.5}Ni_{10.0}Be_{22.5}$) and Vitreloy 4 ($Zr_{46.75}Ti_{8.25}Cu_{7.5}Ni_{10}Be_{27.5}$) contain five elements of different sizes, which leads to a very dense molten phase of high viscosity, as shown in Fig. 18.14. Above T_g in the supercooled region, these Zr-based glasses remain stable and soften into a malleable viscous state at 400 °C, more like a thermosetting polymer than a metal. Thermoplastic shaping and forming can be carried out as easily and cheaply as in polymers. Compared with steel and other structural materials, Vitreloys have high elastic limits and high mechanical strengths. Bulk metallic glasses require slow crystallization

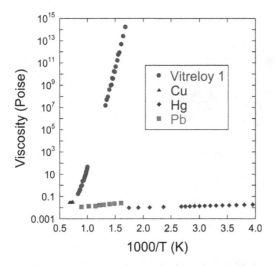

Fig. 18.14 Comparison of the temperature dependence of the viscosity between a metallic glass (Vitreloy 1) and three elemental liquid metals.

kinetics leading to a stabilized supercooled liquid state and high glass-forming ability. The following empirical rules promote such behavior.

(1) Multicomponent alloys of three or more elements increase the complexity and size of the crystalline unit cell. This reduces the energetic advantage of an ordered structure and promotes glass formation.

(2) As in silicate glasses, metallic glasses close to deep eutectics form a liquid that is stable at low temperatures. This tends to make it easy to quench the melt, bypassing crystallization. That is, the low eutectic temperature reduces the mobility of the atoms in the liquid state, so that it becomes more difficult to undergo the structural re-arrangement required for crystallization.

(3) Large differences in size of the constituent elements leads to higher packing density and smaller free volume in the liquid state. As a result, there are strong interactions between the atoms, with the smaller element diffusing rapidly via interstitial sites, and producing a very viscous liquid. The metallic elements in Vitreloy range in radius from 1.1 to 1.6Å.

A good working model, which describes many of the structural features of metallic glasses, is the *efficient cluster packing model*. The clusters in question are the first coordination polyhedron for the solute atoms. The solute atoms tend to avoid each other, bonding in preference to the solvent atoms. These clusters are then packed as efficiently as possible. Fig. 18.15 illustrates the idea.

Metallic glasses exhibit a number of useful properties that depend explicitly on the lack of long range ordering. For example, because they are not crystalline, metallic glasses do not have slip systems and dislocations, as are characteristic of crystalline metals. As is discussed in Chapter 30, slip offers the primary mechanism for plastic deformation in metals, as one plane "slides" over another via motion of dislocations. The lack of slip

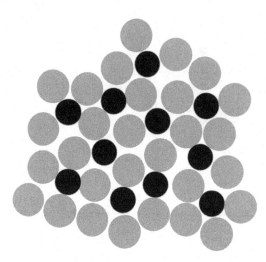

Fig. 18.15 Schematic showing in two dimensions the efficient cluster packing model for a binary metallic glass. The darker atoms are the solute atoms, the lighter are the solvent. It is apparent that the solute atoms tend to be shielded from each other by the solvent atoms.

systems in the amorphous metals enables strain levels of more than 2% without plastic deformation, enabling alloys with very high yield strengths. The large region of elastic response is useful in springs and in some sporting equipment such as golf clubs.

18.8   Problems

(1) BPO_4 is tetragonal with space group $I\bar{4}$ (space group 82) with lattice parameters $a = 4.332$ Å, and $c = 6.64$ Å. The B is at $(0, 0.5, 0.25)$, P is at $(0, 0, 0)$, and O is at $(0.14, 0.26, 0.13)$.

    (a) Draw a unit cell of this material.

    (b) What are the coordination numbers of each of the ions involved?

    (c) Would such a material be a potential glass former, based on Zachariasen's rules?

(2) Would you expect the composition $KAlSi_3O_8$ to be a good glass former, based on Zachariasen's rules? Explain your reasoning.

(3) Explain the structural features common to metallic glasses.

(4) What structural features produce a high viscosity melt? How does melt viscosity influence glass formation?

19   Silica and Silicates

As silicon and oxygen are the two most abundant elements in the Earth's crust, it is important to understand the crystal chemistry of silicate minerals. This chapter will describe the crystal structures of several important materials based on SiO_2 itself, and will lay out a classification scheme for other silicates. These materials are both industrially important, and good examples of more widely observed physical phenomena. For example, SiO_2 undergoes both displacive and reconstructive phase transitions.

19.1   Classification of Silicates

In the vast majority of silicates, the silicon atoms are surrounded by four oxygen atoms in a tetrahedral arrangement, as shown in Fig. 19.1. With an electrostatic bond strength of 1, oxygen ions may be coordinated with only two silicon atoms in silica. This low coordination number makes close packed silica structures impossible for SiO_2, and in general silicates have more open structures than those discussed previously. Because Si^{4+} ions are highly charged they tend to stay as far apart as possible, leading to corner linking of the tetrahedra. Silicate minerals are frequently classified structurally on the basis of the way in which the silica tetrahedra link together. Table 19.1 shows the typical classification scheme. In isolated silicates (also called orthosilicates) SiO_4^{4-}, tetrahedra are independent of one another; in paired silicates (also called pyrosilicates) $Si_2O_7^{6-}$ ions are composed of two tetrahedra with one corner shared; in metasilicates, SiO_3^{2-}, two corners are shared to form a variety of ring or chain structures; in layer structures, $(Si_2O_5)_n^{2n-}$, layers are made up of tetrahedra with three shared corners; in the various forms of silica, SiO_2, four corners are shared.

In a wide range of silicates, the Si–O bond distance has been found to vary between 1.57 and 1.72 Å near room temperature. The O–Si–O bond angle ranges from 98° to 122°, centered near the tetrahedral bond angle of 109.5°. On the other hand, the Si–O–Si bond angle varies much more widely, and typically adopts values between 120 and 180°; the most common bond angle is ~139°.[1]

Note that sometimes, Al^{3+} substitutes for Si^{4+} in tetrahedral sites, but never more than 50%. Over a wide variety of materials, the Si–O bond length is ~1.62 Å, typically making this the shortest, strongest bond in the silicate crystal structure. As a result,

---

[1] F. Liebau, *Structural Chemistry of Silicates: Structure, Bonding and Classification*, Springer-Verlag, New York (1985).

Common

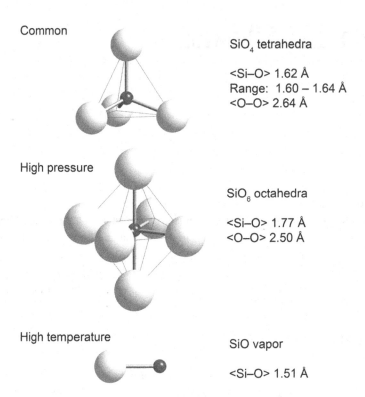

SiO$_4$ tetrahedra

<Si–O> 1.62 Å
Range: 1.60 – 1.64 Å
<O–O> 2.64 Å

High pressure

SiO$_6$ octahedra

<Si–O> 1.77 Å
<O–O> 2.50 Å

High temperature

SiO vapor

<Si–O> 1.51 Å

Fig. 19.1 Under normal pressure and temperatures, silica exists in several polymorphic forms made up of SiO$_4$ tetrahedra. At very high pressures (near 100 kbar) silicates often transform to denser structures made up of SiO$_6$ octahedra. Short-lived Si–O vapor species are observed at high temperatures.

silicates tend to cleave between the silicate groups, leaving the strong Si–O bonds intact. Thus, amphiboles cleave into fibers; mica cleaves into sheets.

19.2   Polymorphs of Silica

Silica (SiO$_2$) is one of the most useful of all engineering materials. Silica refractories, silica glass, silica sand, silica bricks, silica insulators, silica prisms, silica resonators and clocks, silica gel, silica fibers, silica dielectrics in microelectronics, and silica coatings are all in common use.

Silica occurs as an amorphous form when cooled quickly enough from the melt (see Chapter 18), and as a large number of crystalline polymorphs: quartz, tridymite, cristobalite, coesite, keatite, and stishovite. The polymorphic forms all have the same chemical fomula (SiO$_2$), but different crystal structures.

Phase diagrams are maps showing the most stable forms of matter for a given set of conditions (e.g. pressure, temperature, and chemical composition). The pressure-temperature phase diagram for silica contains a number of stable phases, including molten silica, cristobalite, tridymite, and α- and β-quartz. In addition, there are high pressure forms

Table 19.1 Classification schemes for silicates

Type	Si:O ratio	% sharing	Schematic	Example
Isolated tetrahedra	1:4	0%		Mg_2SiO_4, forsterite
Tetrahedral pairs	2:7	25%		$Sc_2Si_2O_7$, thorveitite
Ring silicates	1:3	50%		$Be_3Al_2Si_6O_{18}$, beryl
Single-chain silicates (pyroxenes and pyroxenoids)	1:3	50%		$MgSiO_3$, enstatite
Double-chain silicates (amphiboles)	Variable, often 4:11	62.5% in amphiboles		$Ca_2Mg_5(Si_4O_{11})_2(OH)_2$, tremolite
Layer silicates (micas and clays)	2:5	75%		$Al_2Si_2O_5(OH)_4$, kaolinite (china clay) $KAl_2(AlSi_3O_{10})(OH)_2$, muscovite mica
Network silicates	1:2	100%	See Section 19.2	SiO_2, quartz $KAlSi_3O_8$, orthoclase

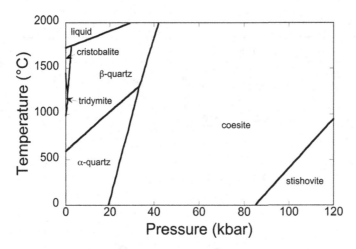

Fig. 19.2 Phase diagram for SiO_2. Redrawn from *Phase Diagrams for Ceramists 1969 Supplement*, The American Ceramic Society, Columbus, OH (1969). Fig. 2271. Printed with permission of The American Ceramic Society, www.ceramics.org.

such as coesite and stishovite which are seldom encountered in engineering applications. All of the polymorphs of silica except stishovite are built up of SiO_4 tetrahedral units (Fig. 19.2). Stishovite is a high-pressure form of silica that has the rutile (TiO_2) structure. It has a much higher density than quartz, or other low-pressure forms of silica. In stishovite, each silicon atom is bonded to six oxygens, arranged in an octahedron.

Cristobalite, tridymite, and quartz are three important polymorphs of silica found in silica refractories. Their structures consist of corner-sharing tetrahedra with Si–O bond lengths of 1.6 Å and bond angles close to 109.5° for O–Si–O and 150–180° for Si–O–Si.

A key feature of the cristobalite structure is the puckered hexagonal ring. Imagine the tetrahedra arranged so that they alternate pointing up and pointing down. In the cristobalite structure (Fig. 19.3), these rings are staggered in position. In contrast, in the tridymite structure, the rings are superposed in consecutive layers (Fig. 19.4).

The transformation from cristobalite to tridymite is reconstructive in nature because it is necessary to break Si–O bonds and reassemble the network. Hence the transition is slow and sluggish. Transformations from cristobalite or tridymite to quartz are also reconstructive. This permits easy quenching of the cristobalite and tridymite phases to room temperature during ceramic processing.

All three forms, however, undergo displacive phase transformations on cooling, in which the high-temperature β forms deform in structure to α forms. At the transition temperatures, silicon and oxygen atoms displace slightly (about 0.1 Å) in position without rupturing any chemical bonds. These transitions are rapid in nature and cannot be avoided. The α–β transformations cause problems during processing because of the large thermal stresses and strains accompanying the phase changes.

The quartz structure (Fig. 19.5) is more complex than those of cristobalite or tridymite. If consists of –Si–O–Si–O–Si–O–Si– helices which spiral around the optic axis. The spiral can be either right-handed or left-handed. Mineral specimens of quartz are 50% of each, and are frequently twinned, with intergrowth of the two handednesses.

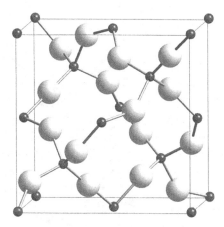

Fig. 19.3 Structure of high (β) cristobalite. The cubic unit cell, space group $P2_13$ contains eight corner-linked silica tetrahedra.

(a)

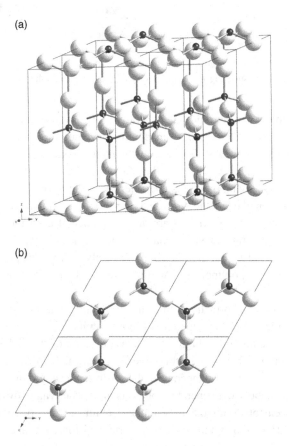

(b)

Fig. 19.4 Idealized high (β) tridymite structure contains corner-linked silica tetrahedra (a) looking perpendicular to the c axis, and (b) down the c axis. The hexagonal unit cell belongs to space group $P6_3/mmc$, with $a = 5.03$ Å, $c = 8.22$ Å.

(a)

(b)

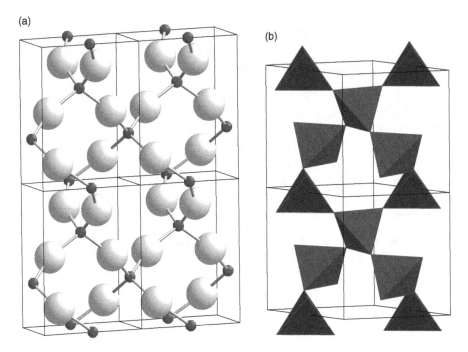

Fig. 19.5 Crystal structure of high temperature (β) quartz: (a) four unit cells with O shown in white and Si in gray, and (b) structure showing the connectivity of the silicate tetrahedra.

Synthetic crystals employed in electronics are generally grown from seed crystals in the right-handed form. Piezoelectric resonators made from synthetic quartz crystals are widely used as timing devices and frequency standards. Note the orientation of silicon–oxygen bonds in the quartz structure. They favor hexagonal prism faces as the major morphological faces, giving quartz its characteristic of hexagonal crystals.

The differences between the α- and β-quartz structures are shown in Fig. 19.6. Considering the six oxygen atoms arrayed around the corner of each unit cell, it is clear that in α-quartz, the arrangement is characteristic of a three-fold screw axis, while in β-quartz, a six-fold screw axis is observed. Only a slight displacement is required to convert from one polymorph to another. Hence, the phase transformation is displacive.

For reconstructive phase transformations, there are no required relationships between the symmetries of the high- and low-temperature polymorphs. Nonetheless, in many cases, the high-temperature phase has higher symmetry, because of the higher entropy associated with the vibrational modes in this case. In contrast, for displacive phase transitions, the symmetry of the low-temperature phase is often a subgroup of the high-temperature form. That is, the low-temperature phase is often a derivative by distortion, in which atom displacements from the parent phase eliminate one or more of the symmetry elements. The α–β transitions in quartz, cristobalite, and tridymite (Table 19.2) are examples.

High-temperature forms generally have lower density and higher symmetry, while pressure stabilizes the denser polymorphs.

Table 19.2 Symmetry and density of the silica polymorphs

Polymorph	Stability range	Density (g/cm³)	Space group	Si site symmetry and equivalent directions
High cristobalite	1470–1723 °C	2.22	$Fd3m$	$\overline{4}3m$ (24)
High tridymite	867–1470 °C	2.17	$P6_3/mmc$	3m (6)
High quartz	573–867 °C	2.51	$P6_22$	222 (4)
Low quartz	< 573 °C	2.65	$P3_12$	2 (2)
Low cristobalite	metastable <200 °C	2.33	$P4_122$	2 (2)
Coesite	30 kbar, low T	2.89	$B2/b$	1 (1)

(a)   (b)

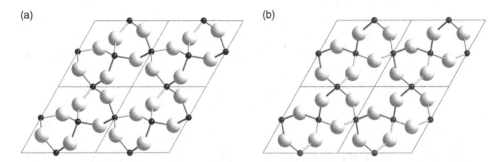

Fig. 19.6 Comparison between α- and β-quartz. In these figures, the structures are viewed looking down the c axis, and Si is shown in dark gray and O in lighter gray. The screw axes are located at the corners of the unit cell. It is clear that the two structures are closely related, and can transform from one to the other with a change in temperature, but without requiring any bond breaking. The stable form of quartz under atmospheric conditions is α-quartz, which is trigonal, space group $P3_12$, with $a = 4.91$ Å, $c = 5.40$ Å. Among minerals, α-quartz is the dominant form of SiO_2. Note that β-quartz is stable above 573 °C. It is hexagonal, space group $P6_22$.

The kinetics of the phase transition depend on the energy barrier between the two polymorphs. Because reconstructive phase transitions require breaking bonds, they are energetically expensive, and often require considerable time at temperature for the reaction to complete. Solvents can be used to help break the strong bonds more readily, and so increase the kinetics of the reaction. For example, in order to transform quartz to tridymite, Si–O bonds must be broken. Because these bonds are quite strong, the process is slow. Brick manufacturers speed up the transformation by adding a small amount of lime to produce a liquid phase.

Unusual atom coordinations and chemistries are found in high-temperature gases, these are sometimes used as vapor-phase sources for material processing. For example, amorphous silicon monoxide can be condensed from chilled SiO gas (Fig. 19.1). Thin layers of this are used as protective coatings against scratching and oxidation in some optical components like mirrors and lenses. The higher volatility of SiO relative to SiO_2 is often used in the semiconductor industry to clean Si surfaces from oxygen contamination at elevated temperatures.

19.3   Stranded Phases and Stuffed Derivatives

Phase transitions with slow reaction kinetics typically require nucleation and growth of the new phase from the parent phase.[2] This process is slow. As a result, if, in the materials' processing, insufficient time at temperature is provided to complete the reaction, metastable phases can persist. These are sometimes referred to as stranded phases. Stranded phases can be produced by rapid changes of temperature, pressure, or composition. For example, tridymite is stable only above 867 °C, but is easily quenched to room temperatures, where it remains for geologic times without converting to the thermodynamically stable α-quartz polymorph. Strong Si–O bonds must be broken for this phase transformation to proceed, creating a large energy barrier. Because of the slow kinetics of the phase transformation, tridymite is retained metastably by *thermal stranding* associated with a rapid temperature change. *Barically stranded* phases are phases where a rapid change in pressure does not allow the thermodynamically stable phase to be produced. As a result of baric stranding, coesite and stishovite, two of the high-pressure polymorphs of SiO_2 can be found as minerals under ambient pressure conditions. Rapid dehydration can produce *compositionally stranded* phases.

The stability of polymorphs can be altered by changes in the material composition. As an example, consider the high temperature forms of SiO_2, including cristobalite, tridymite, and β-quartz. In each of these materials, there are large open spaces due to the low-density network silicate structures. At high temperatures, where atoms have large vibration amplitudes, the structures remain propped open. However, at lower temperatures, the structures tend to collapse around large internal cavities to reduce their size. This type of displacive phase transformation can be prevented if the cavities are filled by large alkali or alkaline-earth atoms. To preserve electroneutrality, this "stuffing" is accompanied by substitution of some Al^{3+} for Si^{4+} in the network.

The stuffed derivatives of the silica polymorphs have the general composition $MAlSiO_4$, in which half the silicons are replaced by aluminums, and the alkali ions M^+ occupy interstices. The larger the size of the interstice in the host structure, the larger the alkali ion that can be accommodated. Thus, tridymite, which has the largest interstice, can host the large potassium ion in the stuffed derivative kalsilite ($KAlSiO_4$), as shown in Fig. 19.7. Nepheline ($KNa_3Al_4Si_4O_{16}$) has a crystal structure very similar to kalsilite with an ordered arrangement of sodium and potassium ions. It is also the most common of the stuffed derivatives of silica and an important fluxing agent used in the densification of ceramics. The weak Na–O and K–O bonds in nepheline lower the melting point and the melt viscosity relative to SiO_2.

Smaller alkali ions prefer other SiO_2 structures. Carnegieite (α-$NaAlSiO_4$) is a cristobalite derivative (Fig. 19.8). The intermediate cavity size in the cristobalite can accommodate the Na^+ ion. The structure of $LiAlSiO_4$ (Fig. 19.9) is based on the quartz structure. Synthetic $LiAlSiO_4$ – often referred to as β-eucryptite – has a very small thermal expansion coefficient. It is used to make ceramics with excellent thermal shock

---

[2] Portions of this section are adapted (with permission) from R. E. Newnham, "Phase diagrams and crystal chemistry," in *Phase Diagrams: Materials Science and Technology, Volume V*, Academic Press, New York (1978).

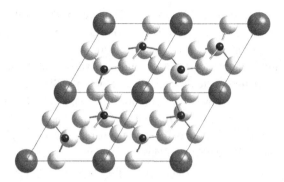

Fig. 19.7 Four unit cells of kalsilite, viewed looking down the c axis. Kalsilite (KAlSiO$_4$) is a stuffed derivative of the tridymite structure, with the potassium ions in the large channels. Oxygen is white, silicon and Al are the small filled circles, and potassium is the large filled circle.

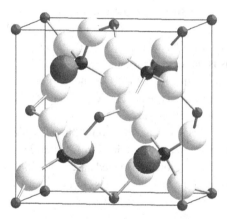

Fig. 19.8 High-temperature NaAlSiO$_4$ has the carnegieite structure with sodium ions in interstitial sites of the cristobalite structure. Si is the smallest filled circle, Al is the intermediate filled circle, Na is the large filled circle, and O is white.

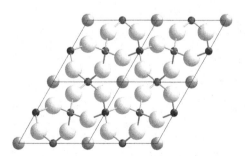

Fig. 19.9 The crystal structure of β-eucryptite (LiAlSiO$_4$), a stuffed derivative of β-quartz, with Li$^+$ ions occupying interstitial sites. Li is the largest filled circle, oxygen is white, Si is the smallest filled circle, and Al is the intermediate-size filled circle.

resistance. Note that the cavity size and coordination number of the alkali ions in the stuffed derivatives increase in the order: eucryptite, carnegeite, nepheline, and kalsilite, just as the ionic radii increase in going from Li^+ to Na^+ to K^+.

While the examples given above are for compounds in which half of the Si^{4+} has been replaced by Al^{3+} (so that there is one alkali atom per Si) high-temperature polymorphs can be stabilized by lower concentrations of alkali ions. Even a modest amount of alkali can act to prop open (or buttress) the high-temperature phase. This will lower the phase transition temperature to the low-temperature polymorph, and can prevent the transition to the higher density, collapsed phase on cooling. Mineral samples of tridymite and cristobalite tend to have more impurities than quartz specimens. This may help explain why the metastable phases are observed under ambient conditions.

19.4   Isolated and Pair Silicates

At the opposite end of the connectivity spectrum from network silicates are orthosilicates, also called isolated silicates. In such materials, none of the oxygens bridge between silicate tetrahedra. As a result, the discrete SiO_4^{4-} groups are bonded together through the other atoms in the structure. Characteristic of this geometry are Si:O ratios in the chemical formula of 1:4, as is the case for forsterite, Mg_2SiO_4.

The unit cell of forsterite is shown in Fig. 19.10. Olivine is based on a distorted version of hexagonal close-packing of the oxygen sublattice, with half of the octahedral sites filled by Mg atoms and the Si atoms in some of the tetrahedral sites. Each oxygen thus has one Si and three Mg neighbors. The calcium isomorph, γ-Ca_2SiO_4, is formed during the hardening of Portland cements, as is described in Chapter 21.

There are many other crystal structures showing isolated silicate tetrahedra, including the garnets, zircon, kyanite, topaz, and the silicate spinels. In many cases, the oxygen

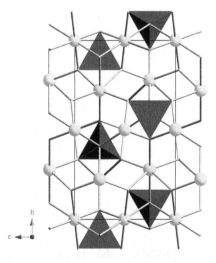

Fig. 19.10 Crystal structure of forsterite (also called olivine) showing the isolated SiO_4^{4-} tetrahedra and the Mg–O bonds which hold the structure together. The Mg ions are shown as light-colored spheres. O are at the ends of the bonds from the Mg.

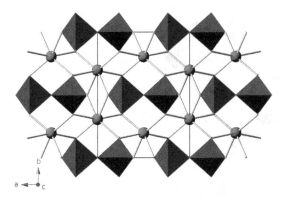

Fig. 19.11 Crystal structure of $Sc_2Si_2O_7$, thortveitite. The Sc ions are shown as solid spheres, the silicate tetrahedra are all paired, oxygen ions reside at each corner of the silicate tetrahedra.

sublattices of these materials are relatively densely packed; close-packed oxygen sublattices are less common as the connectivity of the silicate tetrahedra rise. As a result, many of the network silicates have comparatively low densities.

In pair silicates, two silicate tetrahedra share one oxygen between them, to form bowtie-like structures, as shown in Fig. 19.11 for thortveitite, $Sc_2Si_2O_7$. Here, one out of four oxygens per tetrahedra is shared with an adjacent tetrahedra, yielding 25% sharing of the oxygen, and a characteristic Si:O ratio of 2:7. Thortveitite is important as an ore of Sc. Paired silicates of this type are also observed in some cements.

19.5   Ring Silicates

In ring silicates (also called cyclosilicates), the SiO_4 tetrahedra share two corners with other tetrahedra within the ring. The structures are often labeled by the number of tetrahedra within the ring. For example, 3-, 4-, 6-, 8-, 9-, and 12-membered rings have all been reported in crystalline silicates; 5- and 7-membered rings have been reported in glasses. Typical ring structures are shown in Fig. 19.12. In each of the rings shown, the characteristic Si:O ratio is 1:3, as can be demonstrated by a simple exercise in counting. One ring is bonded to the next by the other ions in the crystal structure.

Beryl, $Be_3Al_2Si_6O_{18}$ provides an excellent example of a ring silicate. As shown in Fig. 19.13, the Be and Si ions in beryl are tetrahedrally coordinated with oxygen, while the Al is octahedrally coordinated with oxygen. Two different types of oxygen are found in the structure: some of the oxygens have only two Si neighbors, while other oxygens are triangularly coordinated to one Al, one Si, and one Be each. The structure is shown looking down the c axis. The 6-rings stack on top of each other, rotated by 30° about the c axis with respect to each other. Beryl is important as an ore of Be, and as a gemstone (Cr-doped beryl is emerald). As will be discussed in Chapter 24, the beryl isomorph cordierite ($Mg_2Al_4Si_5O_{18}$) in which the Al adopts tetrahedral coordination, and the Mg is octahedrally coordinated, is important for its low thermal expansion coefficient.

Branched versions of ring silicates also exist, in which the silicate ring has other tetrahedra projecting from the ring.

Fig. 19.12 Some of the known silicate ring structures. The 3-ring is from benitoite, the 4-ring from taramellite, the 6-ring from beryl, and the 12-ring from traskite.

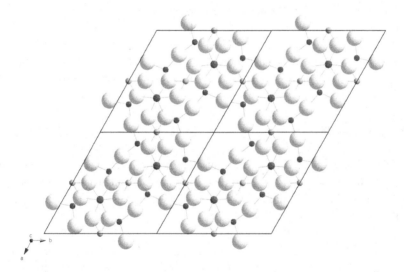

Fig. 19.13 Four units cells of beryl looking down the c axis. The Si atoms are show as the large white spheres; the Si are the small tetrahedrally coordinated cations participating in the characteristic 6-rings. The Be atoms are the small, lighter-colored atoms in tetrahedral coordination. The Al atoms are 6-coordinated. Only atoms with z coordinates from $0.1c$ to $0.9c$ are shown for clarity.

19.6   Chain Silicates

Chains can be considered as ring structures with infinite members, so it is not surprising that the characteristic Si:O ratio of single chain silicates is identical to that of ring silicates, i.e. 1:3. Such single chains can be straight, as in the pyroxenes, or helical, as in

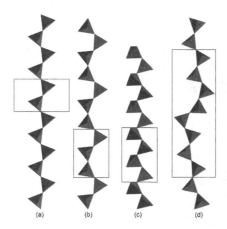

Fig. 19.14 Chain configurations for some single-chain silicates. The boxed areas show the repeating unit for the chain. (a) Diopside, (b) sorensenite, (c) krauskopfite, and (d) pyroxferroite.

Fig. 19. 15 Crystal structure of enstatite, $MgSiO_3$, showing the single silicate chains and the octahedral Mg holding them together.

the pyroxenoids. Figure 19.14 shows a few of the chain configurations for single-chain silicates. An example of a pyroxene is shown for the case of enstatite in Fig. 19.15. It can be seen that all of the chains run parallel to each other in the crystal structure, and that there are no Si–O–Si bonds *between* chains. Instead, the chains are held together by much weaker Mg–O bonds. As a result, the strength of the mineral is parallel to the chains. Pyroxenes and pyroxenoids contain very rigid and energetically stable arrays of edge-shared octahedra to which the tetrahedral chains adjust themselves. The number of different cations of the right size and valence to form such arrays of edge-shared octahedra is very large and includes several ions (Na^+, Mg^{2+}, Mn^{2+}, Fe^{2+}, Fe^{3+}, Al^{3+}) of high abundance. The twisting of the single chain silicates in rhodonite ($MnSiO_3$) to accommodate the MnO_6 octahedra is shown in Fig. 19.16.

More complex chain structures are also known, in which double chains or tubular chains develop; a few of these are shown in Fig. 19.17. One of the more common

(a)

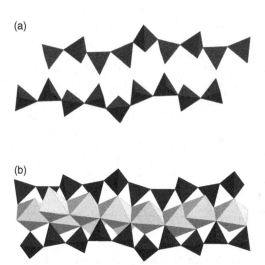

(b)

Fig. 19.16 Chain structure of the pyroxenoid rhodonite (MnSiO$_3$): (a) shows the twisted tetrahedral single chains, while (b) illustrates how these chains wrap around the MnO octahedra.

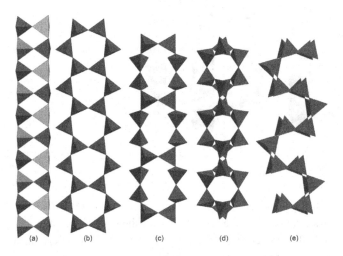

Fig. 19.17 Examples of more complex chain silicates. (a) Sillimanite at 1000 °C – in this structure there are also alumina in the tetrahedra sites, (b) tremolite – an amphibole, (c) xonotlite, (d) narsarsukite, and (e) tuhualite. Adapted, by permission, after F. Liebau, *Structural Chemistry of Silicates: Structure, Bonding, and Classification*, Springer-Verlag, New York (1985).

families of double chains are the amphiboles, in which silicate 6-rings are linked along a single axis. For this geometry, half of the silicate groups have two oxygen shared with other silicate groups; the other half have three silicate tetrahedral neighbors. In that case, the amount of oxygen sharing is $62\% = \frac{75\% + 50\%}{2}$. Characteristic of this arrangement is an Si:O ratio of 4:11. Other double-chain silicates have different amounts of O sharing, and different characteristic Si:O ratios. As is the case with ring and single-chain silicates, branched versions of these more complex chains are also known.

Some of the minerals that are used as asbestos belong to the amphibole family. Here, the fibrous morphology of the crystallites mimics the aligned chains in the crystal structure. Asbestos is a good thermal insulator, so that it can be used in building insulation, or applications where heat resistance is needed. Unfortunately, with the amphiboles, the crystallites are often quite fine, with the result that they can hang suspended in air, where they can be breathed in, and ultimately become embedded in lung tissue.

19.7   Layer Silicates

Clay minerals are hydrated layer silicates, which may contain other components such as Al, Ca, Mg, Fe, Na, K, etc. The basic structure of all the clay minerals is a composite sheet composed of a layer of linked SiO_4 tetrahedra and a layer of octahedrally coordinated cations, in which each Si atom is surrounded by four O atoms; the octahedral cation has six O^{2-} or OH^- neighbors. Figure 19.18 illustrates the tetrahedral and octahedral layers. Clay is used in making fine china, whitewares, and refractories. In addition to these "conventional" uses, clay is also used to coat paper and as a filler in plastics, rubber, paint, and fertilizers. Intercalated clays have additional atoms or molecules residing between the layers; the layers themselves are left largely undisturbed by this process. As this occurs, the layers are pushed apart, so that the interlayer spacing changes, as is detected in diffraction experiments. Intercalated or pillared clays can be used to prepare catalysts, as well as composite structures where the size of one of the phases can be maintained at nanometer dimensions.

In nature, clays are sedimentary minerals formed principally as a weathering product of feldspars and other igneous and metamorphic minerals. This can occur when granite, gneiss, and other rocks are covered by marshes and bogs containing acidic water. Atmospheric oxygen is excluded under these conditions, and K^+, Na^+, and Ca^{2+} are slowly leached from the feldspar to form partially hydrated aluminosilicates such as kaolinite (china clay) $Al_2Si_2O_5(OH)_4$. The dissolved alkalies and silica are washed away and clay remains as china clay or kaolin. The mineral name "kaolin" is derived from the Chinese word for "high ridge." Other finely divided minerals containing iron or silica together with organic matter, are often found in sedimentary deposits of clay. These impurities have a large effect on the color, plasticity, and firing characteristics of a clay body. Typically, clay crystals are tiny hexagonal platelets 1–10 μm across and less than 0.1 μm thick. The shape of the particles is related to the layered crystal structure of kaolinite.

Classification of Layer Silicates

The layer silicates are classified by the stacking of the octahedral and tetrahedral layers. When the octahedral layer only is present, the minerals brucite and gibbsite result. Two

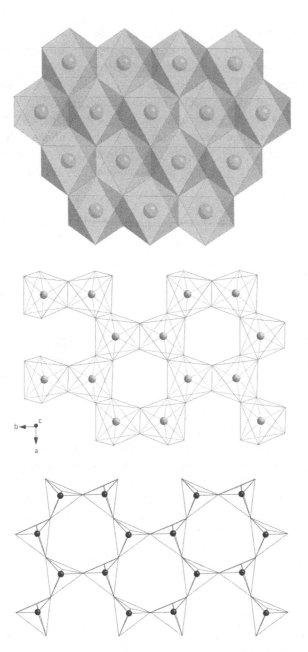

Fig. 19.18 (Top) Octahedral layer in which all octahedral sites are filled, e.g. a trioctahedral layer; (middle) octahedral layer in which 2/3 of octahedral sites are filled; and (bottom) tetrahedral layer.

principal types of octahedral layers exist, those based on the structure of brucite, $Mg(OH)_2$, in which all of the octahedral sites are filled, and those based on the structure of gibbsite $Al(OH)_3$ in which 2/3 of the octahedral sites are filled. These are referred to as trioctahedral and dioctahedral layers, respectively. In both cases, the OH^- groups terminate the top and bottom of the layer structures, as shown in Fig. 19.19 for brucite.

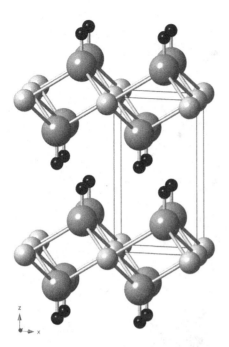

Fig. 19.19 Crystal structure of brucite from side. The smallest atoms are H, the largest are O, and the intermediate size atoms are Mg.

As a result, the chemical formula is sometimes re-written as $(HO)_3Mg_3(OH)_3$ or $(HO)_3(Al_2\square)(OH)_3$, where the box is an empty octahedral site. The layers are held together by hydrogen bonding.

Chrysotile (= serpentine), $Mg_3Si_2O_5(OH)_4$, the chief constituent of asbestos, has a brucite-like layer superposed on the silica tetrahedral layer (see Fig. 19.20). The structural formula is thus $O_3Si_2O_2(OH)Mg_3(OH)_3$. Chrysotile forms slender fibers because of the size mismatch between the octahedral and tetrahedral layers; the 2 Å Mg–O bond length is slightly too large to match the silicate layer. The overall morphology thus resembles rolled-up cigar wrappings. The dioctahedral analog, kaolinite $Al_2Si_2O_5(OH)_4$ (Fig. 19.21), is sometimes written as $Al_2\square Si_2O_5(OH)_4$ in order to emphasize the fact that 1/3 of the octahedral sites are vacant. Because the shorter Al–O bond length (~1.9 Å), these layers are a better size fit to the tetrahedral layers, and so tend to stay flat. In both serpentine and kaolinite, the layers are bonded by a hydrogen bond.

Talc $(Mg_3Si_4O_{10}(OH)_2)$ and pyrophyllite $(Al_2Si_4O_{10}(OH)_2)$ have triple-layer structures with two silica tetrahedral layers sandwiching the octahedral layer, as shown in Fig. 19.22. Because the O–H bond distance within a layer is shorter than the Si–O bond distance, the hydroxyl groups are buried in the layer, leaving O on both surfaces. As a result, the bonding between layers is van der Waals, yielding a Moh's hardness of 1 for talc.

Mica family minerals such as muscovite $(KAl_2(AlSi_3O_{10})(OH)_2)$ and phlogopite $(KMg_3(AlSi_3O_{10})(OH)_2)$ differ only slightly in structure from talc and pyrophyllite; they have the same octahedral and tetrahedral layers arranged in a triple sandwich, but

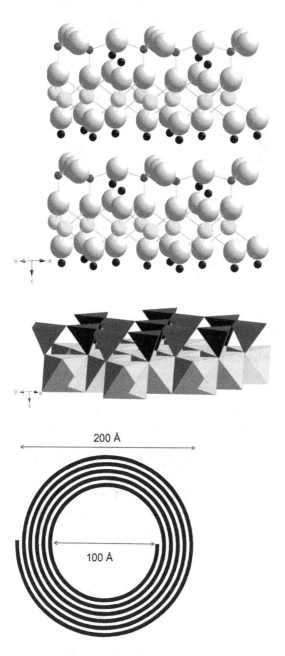

Fig. 19.20 Crystal structure of serpentine $Mg_3Si_2O_5(OH)_4$. (Top) Ball and stick view of two layers. The oxygen are the large open circles, the Si the small tetrahedrally coordinated filled circle, H, the small linearly coordinated filed circle, and Mg, the lighter-colored circle of intermediate size. (Middle) Polyhedral view of one layer. (Bottom) Curling of serpentine layers to form needles.

with K^+ ions between layers. An example is shown in Fig. 19.23). Charge compensation is accomplished by replacing 25% of the Si^{4+} ions with Al^{3+}. The addition of the ionic bonding increases the strength of the interlayer bonding relative to talc. Micas are readily recognized by the flake-like morphology and easy cleavage.

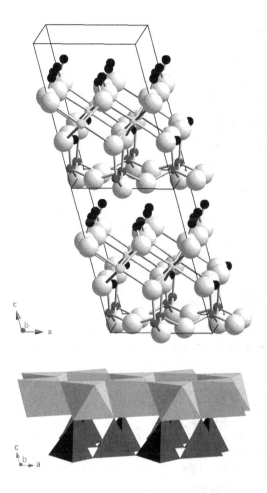

Fig. 19.21 Crystal structure of kaolinite $Al_2Si_2O_5(OH)_4$. (Top) Ball and stick view of two layers. The oxygen are the large open circles, the Si the small tetrahedrally coordinated atoms, H, the small linearly coordinated dark atoms, and Al, the lighter-colored octahedrally coordinated atoms. (Bottom) Polyhedral view of one layer, with the H atoms hidden.

Weak interlayer bonding is the key feature of layer silicates. Not only are they weak mechanically (talc is the softest of all silicates), but they decompose easily under either chemical attack or with temperature change.

As with other silicates, the tetrahedral layer in layer silicates sometimes distorts to accommodate octahedral cations of different sizes. This can be accomplished by cooperative rotation of the tetrahedra around an axis normal to the layers, so that the six-fold symmetry is reduced to three-fold symmetry; by corrugation of the tetrahedral layer; or even by periodically interrupting the stacking sequence. These possibilities are described by Liebau.[3]

---

[3] F. Liebau, *Structural Chemistry of Silicates: Structure, Bonding and Classification*, Springer-Verlag, New York (1985).

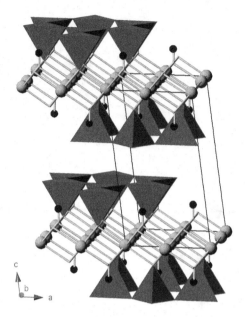

Fig. 19.22 Crystal structure of talc, showing the tetrahedral–octahedral–tetrahedral stacking sequence. O atoms are hidden for clarity.

Fig. 19.23 Crystal structure of muscovite mica showing the tetrahedral–octahedral–tetrahedral stacking, along with the K ions between the layers. O atoms are hidden for clarity.

Plasticity – the ability of a material to change shape under mechanical stress and to retain this shape change when the stress is removed – is a key property of clay. When feldspar is transformed into clay, it changes from a brittle, non-plastic mineral into a highly deformable, plastic mineral. Primary clays, or raw clays, are seldom pure kaolin; they generally contain small grains of quartz, iron minerals, and undecomposed feldspar. But, when the clay is carried away by water, the tiny clay particles are kept in suspension and the heavier impurity grains are left behind. Pure, white, secondary kaolins are formed in this way.

Water plays a key role in the plastic behavior of clay. All clays contain water, some is incorporated as hydroxyl groups in the crystal structure of kaolinite, and some is dispersed between the layers as water molecules. The plasticity is controlled by four factors.

(1)   The layer-like structure of kaolinite with weak hydrogen bonds between layers.
(2)   The small, often submicron, sizes of the clay particles.
(3)   The presence of free water molecules that lubricate molecular slippage between the silicate layers.
(4)   The careful control of acidity to maintain the optimum electrostatic forces between particles.

The ability of clays to become mechanically strong during drying is intimately related to the crystal structure. When mixed with water, the interlayer hydrogen bonding between kaolinite layers relaxes, allowing water molecules to slip between the layers. Weak, new hydrogen bonds are formed between the water molecule and the hydroxyls of the alumina octahedral layer and the oxygens of the silica tetrahedral layer (Fig. 19.24). The water molecules lubricate the surfaces and the kaolin layers slide past

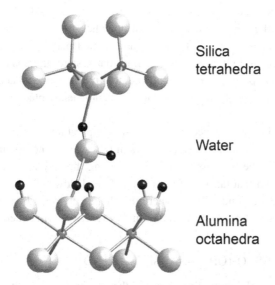

Silica
tetrahedra

Water

Alumina
octahedra

Fig. 19.24 Water molecules lubricate the sliding action of kaolin layers. Hydrogen bonds extending from the hydroxyls of the alumina layer link to the water molecules, which, in turn, form hydrogen bonds to each other and to the oxygens of the silica tetrahedra in the next kaolinite layer.

one another with ease. When the shear stress is removed, the hydrogen bonds reform, locking in the shape change. The sheet-like crystallites can move in the (001) plane without losing the cohesion maintained by the electrical forces present in the watery solution. Because of this, the plastic clay–water paste can be shaped, and will retain the form given it.

As the water molecules evaporate during drying or firing, the lubricant between the clay layers disappears, and hydrogen bonds between the octahedral and tetrahedral layers lock the kaolinite layers into position. Of course, not all the tiny crystallites in clay are parallel to one another. In a given volume, the number of clay particles in direct contact with each other depends on the size of the crystallites. The best plasticity is obtained with very finely divided clays such as ball clay and bentonites. Here the particle sizes are in the nanometer range, between 0.01 and 0.1 μm in size.

The structure of the clay–water system is very sensitive to pH.

If a stiff plastic clay paste is prepared by stirring clay and water, the pH is generally slightly acidic. Adding a few drops of an alkaline solution such as sodium carbonate will convert the plastic paste into a liquid slip of cream-like consistency. When more sodium carbonate is added, the liquid slip eventually turns back into sticky paste. If acetic acid is added to lower the pH, the stiff paste is again converted to a slip, and if still more acid is added, the slip will return to the original consistency of a stiff paste. This is very important in the slip-casting method used to make pottery, earthenware, and china.

What causes this behavior? The control of particle flocculation and viscosity involves the electrostatic interactions between the OH^- and H^+ ions in the water and the surface charges on the tiny clay crystallites. In acidic conditions (pH ~ 6), the crystals flocculate in a diffuse manner referred to as a "house-of-cards" arrangement (Fig 19.25). Optically, the clay–water system appears isotropic under these conditions. A much denser type of flocculation occurs under basic conditions. Under the action of van der Waals forces, the platy crystallites pull together to form a face-to-face flocculated state. The c axes of adjacent crystals are aligned parallel to one another to give an anisotropic system. In between the two flocculated states there is a pH range where the forces between particles are neutralized to give a deflocculated state in which the clay crystals are highly dispersed. Gel formation takes place under these conditions.

When clay–water bodies are heated above room temperature, several chemical reactions take place. Elimination of the water occurs in two stages. The first transformation is the evaporation of the water molecules surrounding the clay particles. This is basically a drying operation that takes place at temperatures below 100 °C.

The second stage occurs around 500 °C, where the chemically bound water evaporates; kaolinite loses water and decomposes into a highly reactive form called metakaolin, $Al_2Si_2O_7$:

$$Al_2Si_2O_5(OH)_4 \rightarrow Al_2Si_2O_7 + 2H_2O\uparrow.$$

Metakaolin is a very poorly crystallized relic of the kaolin structure. It retains the tetrahedrally bonded Si_2O_5 silica layer and a remnant of the alumina layer. The $AlO_2(OH)_4$ octahedra are converted to a deformed tetrahedral layer.

Face to Face Flocculation

Dispersed

Edge to Face Flocculation

Fig 19.25 Clay flocculates in an edge-to-face pattern in acidic surroundings and face-to-face in basic solutions. A deflocculated state occurs at intermediate pH values.

Further heating above 975 °C causes metakaolin to dissociate into mullite and silica:

$$3Al_2Si_2O_7 \rightarrow Al_6Si_2O_{13} + 4SiO_2.$$

Mullite forms as needle-like crystals in a silica matrix. The needle morphology is due to the chains of AlO_6 octahedra which form parallel to the c axis (Fig. 19.26). These chains are cross-linked by SiO_4 and AlO_4 tetrahedra. As a mineral mullite is very rare, but it is a common component in ceramic whitewares and refractories. Interlocking mullite fibers lend strength to the ceramic bodies. Similar edge-sharing octahedral chains are formed in rutile, as was shown in Chapter 9. Rutilated quartz crystals and ruby contain tiny rutile needles that impart a star-like optical phenomena.

Clay Products

Important clay products include bricks and tiles, refractories, stoneware, earthenware, porcelain, and china, sanitary ware, high-voltage insulators, and low-loss capacitors.

Common *bricks* and *tiles* consist of practically 100% clay with only small additions of other raw materials. Standard building bricks (US size: 19.3 × 9.3 × 5.7 cm) are wire-cut in the plastic state with a moisture content of 15–20% moisture. Bricks are fired at temperatures sufficiently high to achieve compressive strengths of 1500 to 7000 psi, according to building requirements. Complete densification is not desirable, because porosity improves the thermal insulation properties. *Roofing tiles* are made by extruding a stiff plastic clay through a die and wire-cutting the tile segments.

The word *refractory* refers to a heat-resistant material with a high melting point. Clay-based refractories, sometimes called fireclays, are composed of clay and silica, and

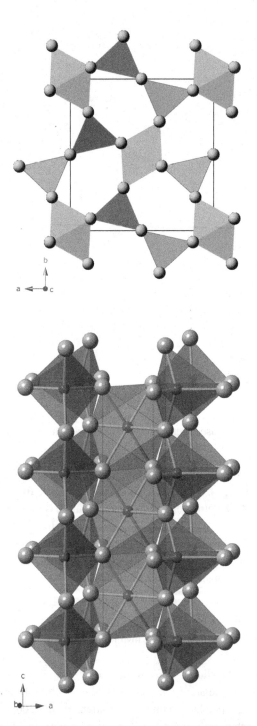

Fig. 19.26 Two views of a somewhat simplified view of the crystal structure of mullite showing edge-sharing chains of AlO_6 octahedra, along with the tetrahedra that bind the chains together. (Top) Looking down on edge-sharing chains. (Bottom) Side view of edge-sharing chains.

are fired at temperatures in excess of 1600 °C. Fireclay refractories are used for boiler furnaces, blast furnaces, lime kilns, cement kilns, and for various metallurgical furnaces for the melting and heat treatment of iron, steel, and non-ferrous metals.

Stoneware is a very hard, dense, and almost impervious ceramic with a glassy texture when broken. It is similar to porcelain, but is made from plastic clays rather than china clay. Sewer pipes, paving tiles, chemical apparatus, vases, and other decorative objects are made from stoneware. Plastic stoneware clays contain enough fluxes to vitrify at temperatures of 1250 to 1300 °C. An ideal stoneware clay is a highly plastic clay containing about 20% finely divided feldspar or mica.

Earthenware is mainly used in the manufacture of dishes and other tableware, wall tiles, and sanitary ware. It is made by mixing plastic clays, china clays, flint (SiO_2), and feldspar. The mixture gives a hard but porous ceramic when fired between 1150 and 1250 °C. This is called the bisque firing. A second firing, the glost firing, is carried out at lower temperature (1050 to 1150 °C) to provide a glassy, impervious coating. Earthenware glazes consist of a glassy frit made from borax ($Na_2B_4O_5(OH)_4 \cdot 8H_2O$) and other low-melting materials.

Porcelain is a white, translucent ceramic used as dinnerware, electrical insulators, and laboratory equipment. A typical porcelain for household china is made from china clay (40%), ball clay (10%), feldspar (30%), sand (18%) and dolomite (2%). All porcelain ceramics are fired to the point where all the pores are filled with a glassy bond. The feldspar melts above 1200 °C, and begins to dissolve the other constituents to form the glassy bond. Additional firings are used to provide attractive decoration and protective coatings.

Sanitary ware is a term used to describe bathroom fixtures such as lavatories, sinks, and bathtubs. Most sanitary ware is made of vitreous china covered with a hard glaze. The compositions are similar to porcelain with clay, feldspar, and flint sand being the major raw materials. Production of large pieces is carried out by the slip-casting method.

Electrical porcelain must be impervious to moisture, have good mechanical properties, high electrical resistivity, and excellent breakdown strength. The composition of high-voltage insulators is similar to hard porcelain, but the color and translucency is not so important. For transmission line insulators, both the volume resistivity and the surface resistivity are important. Porcelain has excellent bulk resistivity and a high breakdown strength at room temperature and low frequencies. However, the properties degrade with increasing temperature and frequency, as the alkali ions become electrically active. Magnesium-based ceramics made from finely divided talc and clay have much lower dielectric loss at high frequencies. Surface resistivities are improved by glazes or silicone coatings. Polymer composites are emerging as a competitive material for high-voltage insulators of this type.

Acid Decomposition of Silicates

Some of the observed variations in the reactivities and modes of dissolution of silicates can be related to structure. As described previously, most silicates can be classified in accordance with the way in which the tetrahedra are linked together.

The action of sulfuric or hydrochloric acid on silicate minerals generally produces one of the following results.[4]

(I)   The silicate structure suffers complete breakdown with the dissolution of both the metal cations and the silica tetrahedra. A silica gel may form subsequently because of the tendency of the tetrahedra to polymerize in aqueous solutions.

(II)   In other silicates, acid attack results in only partial decomposition in which the cations dissolve but siliceous residue remains behind.

(III) Lastly, there are those silicates which show little or no reaction at all with the acid.

Stoichiometric dissolutions occurs in case I, where the metal cations and silica tetrahedra dissolve simultaneously. Many orthosilicates (independent tetrahedra) and pyrosilicates (paired tetrahedra sharing a common oxygen) dissolve stoichiometrically. Examples include forsterite Mg_2SiO_4, fayalite Fe_2SiO_4, willemite Zn_2SiO_4, hemimorphite $Zn_4Si_2O_7(OH)_2 \cdot H_2O$, and akermanite $Ca_2FeSi_2O_7$. Since the silicate groups in these minerals are small, they decompose readily. The only exceptions are minerals like zircon ($ZrSiO_4$) and phenakite (Be_2SiO_4) where the ions linking the silicate groups are either highly charged (Zr^{4+}) or very small (Be^{2+}).

Chain, layer, and network silicates are generally much less soluble because the infinite silicate groups are not easily broken down into smaller units. Among the class III insoluble minerals are the chain silicates enstatite $MgSiO_3$, diopside $CaMg(SiO_3)_2$, and spodumene $LiAl(SiO_3)_2$. Other insoluble minerals are the layer silicates kaolinite $Al_2Si_2O_5(OH)_4$, muscovite $KAl_2(AlSi_3O_{10})(OH)_2$, and talc $Mg_3(Si_4O_{10})(OH)_2$, and network silicates like quartz SiO_2, and albite $NaAlSi_3O_8$. Replacement of silicon by aluminum and ferric ions can, in some instances, weaken the silicate linkage to allow dissolution. If the framework Al:Si ratio exceeds 2:3, the silicate structure is usually decomposed under acid attack.[5]

Materials that dissolve leaving a silica residue (class II) include many modified silica glasses, chrysocolla (amorphous $CuSiO_3 \cdot 2H_2O$) and chamosite $Fe_3Si_2O_5(OH)_4$. Here the alkali, alkaline-earth, or transition metal elements are removed easily since they are less well bound into the network. The terms "incongruent dissolution" or "leaching" have been used to describe this type of dissolution.

19.8   Silicate Minerals under Pressure

Pressure transformations of silicates are of great geophysical interest. Petroleum and mining engineers are well aware of the changes in the Earth's outer mantle with depth (Fig. 19.27). Silicates are the most abundant minerals in the outer crust, but major changes in composition and structure occur in the lower mantle below 50 km. Transformations involving a change from tetrahedral coordination to octahedral coordination ($^{IV}Si^{4+}$ to $^{VI}Si^{4+}$) take place at high pressures, leading to increases of density of 10% or more, accompanied by major changes in the speed of seismic waves.

[4] B. Terry, "The acid decomposition of silicate minerals," *Hydrometallurgy* **10**(2), 135–171 (1983).

[5] K. J. Murata, "Internal structure of silicate minerals that gelatinize with acid," *Amer. Min.*, **19**, 545–562 (1943).

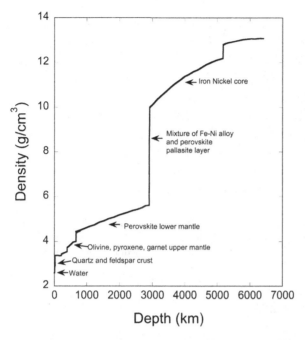

Fig. 19.27 Numerous discontinuities in densities, some of which accompany pressure-induced transitions within the earth's crust. Adapted from data in A. M. Dziewonski and D. L. Anderson, "Preliminary reference Earth model," *Physics of the Earth and Planetary Interiors*, **25**, 297–356 (1981). Drawn from Data in Table II.

Several important phase transformations in silicate minerals take place in the Earth's lower mantle. These transitions can be divided into two groups: those that involve a change in cation coordination, and those that do not. Under pressure, Mg_2SiO_4 changes from the olivine structure to the spinel structure, but the coordinations of the Mg^{2+} and Si^{4+} cations remain unchanged. At still higher pressures, Mg_2SiO_4 undergoes a second transformation to a two-phase state made up of a perovskite ($MgSiO_3$) and periclase (MgO). The coordination of both Mg^{2+} and Si^{4+} increases in the perovskite structure.

Chain silicates such as enstatite ($MgSiO_3$) and wollastonite ($CaSiO_3$) undergo similar changes with pressure. Enstatite adopts the ilmenite structure at 250 kbar with a 50% increase in density, and wollastonite converts to the perovskite structure at 160 kbar. Both transformations involve a change in coordination for silicon from tetrahedral to octahedral.

Deeper in the Earth's crust, high-pressure polymorphs are favored, as the pressure acts to reduce free volume in the crystal structure. As an example of considerable geological importance, consider the effect of pressure on potash feldspar ($KAlSi_3O_8$). At low pressures, the silicate tetrahedra are arranged in a corner-sharing three-dimensional network structure, as shown in Chapter 2. The K^+ ions occupy the large cavities. However, above 120 kbar potash feldspar ($KAlSi_3O_8$) converts to the hollandite structure. As shown in Fig. 19.28, this structure has the Al and Si in octahedral, rather than tetrahedral, coordination geometries. As a result, the density increases from 2.55 to 3.84 g/cm^3, as the amount of free volume is significantly reduced.

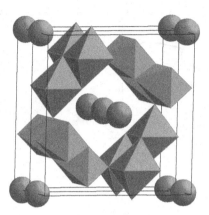

Fig. 19.28 Potash feldspars transform to the hollandite structure with silicon changing from tetrahedral to octahedral coordination. Hollandite is a stuffed derivative of the rutile structure. Potassium is shown as isolated circles; the silicate octahedra are also shown.

19.9   Weathering of Silicates

Cement, plaster, bricks, and several other important building materials are made from processed sedimentary rocks. Sedimentary minerals such as calcite, kaolinite, montmorillonite, and gypsum are widely used in the building industry.

During the sedimentation process, the original constituents of the Earth's crust, igneous rocks, are attacked by air and water. Of the common minerals in igneous rocks, only quartz is highly resistant to the weathering process. Most other minerals are transformed into sedimentary minerals that are more stable under the conditions found at the Earth's surface.

The primary process is the chemical breakdown of silicate minerals. The ease with which this happens depends on the structure of the silicate groups. Olivine and other silicates with separate $(SiO_4)^{4-}$ groups are very susceptible to attack. Chain silicates, pyroxenes, and amphiboles, are more easily broken down than layer and network silicates.

During the weathering of feldspars (see Chapter 2 for the crystal structure), the alkali (Na^+, K^+) and alkaline-earth (Ca^{2+}) ions go readily into solution, and the oxygens in the aluminosilicate framework are partially replaced by hydroxyl ions. Al is coordinated to six OH^- ions, while Si retains its tetrahedral configuration. When first set free, these groups are in solution, but then form colloidal aggregates, and later small clay crystals. A similar gelation process takes place during the setting of cements (see Chapter 21).

The behavior of various chemical elements during this hydration process depends on ionic size and valence. Three general groups of ions can be distinguished: (1) alkali and other ions of large size and low charge readily dissolve in water, and are carried away in a leaching process; (2) the intermediate ion sizes generally precipitate as oxides, hydroxides, and colloidal clay particles; (3) small, highly charged ions form complex ions in water.

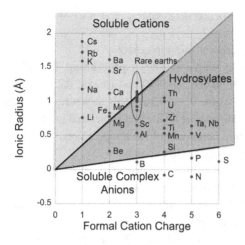

Fig. 19.29 Shannon–Prewitt ionic radii and formal cation charge for rock-forming minerals. During the sedimentary processes involving interactions with water, large ions with low charge are generally soluble, while those of small size and high charge form complex ions in water. Intermediates are usually precipitated as hydrates. Note that some cations have negative Shannon–Prewitt ionic radii since they partially embed themselves in the adjacent oxygen electron shells. Adapted, by permission, from K. Bjorlykke, *Petroleum Geoscience: From Sedimentary Environments to Rock Physics*, Springer, New York (2010).

Since electrostatic potential is proportional to charge/distance, the cation valence divided by the cation radius is sometimes referred to as the *ionic potential*. As shown in Fig. 19.29, ions with an ionic potential less than three are soluble in water, and those with an ionic potential greater than 10 form complex anions. Ions between 3 and 10, including Al^{3+} and Si^{4+}, form fine-grained precipitates like clay and chert.

Many of the sedimentary minerals are used in making hydraulic cements. Silica-rich sediments are classified as *resistates* because quartz is the most resistant to the weathering process. Kaolinite and montmorillonite are two of the *hydrolysate* layer silicates that are key raw materials for Portland cement. The third important raw mineral source for cement is limestone (massive calcium carbonate). Limestone is a sedimentary mineral formed from *precipitation* of oceanic waters saturated with $CaCO_3$.

19.10   Problems

(1)   β-Quartz is hexagonal with space group $6_2 22$ (#180), with $a = 5.01$ Å, $c = 5.47$ Å. Si atoms sit at (0.5, 0, 0); the O ions are at (0.2, 0.8, 0.8333). Several other SiO_2 polymorphs are known, including derivatives of the diamond and wurtzite structures.

    (a)   Plot a unit cell showing the Si–O bonds. Make sure you have looked at the structure in both 001 and 100 projections.

    (b)   What are the coordinations of the Si and O?

(c)   Contrast the polyhedra corner-, edge-, and face-sharing with that for corundum.

(d)   Describe the α–β phase transition in quartz.

(2)   Cristobalite is a cubic polymorph of SiO_2 (space group 198: $P2_13$, with $a = 7.20$ Å) and Si at $(-0.008, -0.008, -0.008)$ and $(0.255, 0.255, 0.255)$, and O at $(0.125, 0.125, 0.125)$ and $(0.44, 0.34, 0.16)$. This structure is slightly less symmetrical than the usual idealized cristobalite that is pictured in many textbooks, but is closer to the real structure observed at high temperatures.

(a)   Draw a unit cell, showing the Si–O bonds.

(b)   Describe how this structure relates to that of diamond. (Hint: try making the O atoms invisible, and plotting more than one unit cell.)

(c)   Calculate the first three terms in the Madelung constant for cristobalite.

(3)   Olivine, $(Mg_{1-x}Fe_x)_2SiO_4$ is orthorhombic with $a = 10.26$ Å, $b = 6.00$ Å, and $c = 4.77$ Å. Its space group is *Pnma*. The atomic positions are Mg(1) at $(0, 0, 0)$; Mg(2) at $(0.2775, 0.25, -0.01)$; Si at $(0.0945, 0.25, 0.426)$; O(1) at $(0.092, 0.25, 0.767)$; O(2) at $(0.449, 0.25, 0.219)$; and O(3) at $(0.163, 0.0365, 0.277)$

(a)   Draw a unit cell, showing the connectivity of the silica tetrahedra.

(b)   Determine the Mg coordination.

(c)   Redraw the structure showing only the oxygen atoms. In what (distorted) stacking sequence are the oxygens arranged?

(d)   Describe the type of silicate connectivity that this crystalline version of this material has, illustrating with figures. Based on your understanding of Zachariasen's rules, would this material be a good glass-former? Be sure to describe your reasoning.

(4)   Compare the structures of SiO_2 (tridymite) and H_2O (ice).

(a)   Sketch the structures as accurately as you can, showing the coordination of the atoms.

(b)   Discuss the differences in bonding between SiO_2 and H_2O. How does this difference affect their melting points?

(5)   Describe the systematics of silicate tetrahedra connectivity in silicate minerals. Illustrate the various types of connectivity, and for each case, give the characteristic Si:O ratio.

(6)   Consider brucite, a trigonal crystal with space group $\bar{3}ml$, $a = 314.20$ pm, and $c = 476.60$ pm. The atoms are at the following locations:

Mg $(0,0,0)$
O $(2/3, 1/3, 0.2216)$
H $(2/3, 1/3, 0.4303)$.

(a)   What is the separation distance between the O ion at $(1/3, 2/3, 0.7784)$ and the one at $(2/3, 1/3, 0.2216)$?

(b)   Draw a side view of this structure (you do *not* need to show a unit cell, just the general arrangement of the atoms in a layer). On a second drawing, show how the structure is used as the basis for kaolinite.

(c)   Describe the interlayer bonding in kaolinite and talc.

(8) The results of a recent structure refinement for kaolinite $Al_2Si_2O_5(OH)_4$ are shown in the figure below.

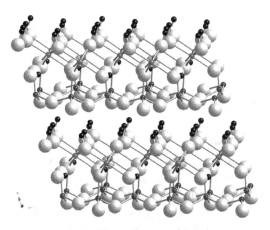

(a) Describe the connectivity of the silicate polyhedra in this material.

(b) Based on your knowledge of ionic radii, coordination number, etc., label an example of each type of atom in the structure.

(c) Describe the expected anisotropy in the mechanical stiffness for this material.

(d) One of the oxygens in the structure is 3-coordinated, with Al neighbors at 1.868 Å and 1.884 Å, and a H neighbor. Given the following bond valence relations, predict the primary O–H bond distance for this atom.

For Al–O bonds: $v_{ij} = \exp\left(\frac{1.644\text{Å}-d_{ij}}{0.38}\right)$.

For primary H–O bonds: $v_{ij} = \exp\left(\frac{0.907\text{Å}-d_{ij}}{0.28}\right)$.

20 Phase Transformations

The molecular mechanisms of phase transformations are described in this chapter.[1] A phase transformation is a transition between polymorphs at a constant composition. The polymorphs can differ in terms of bond type, primary coordination, packing density, or order/disorder character. Examples of reconstructive and displacive transitions are presented, along with a comparison with the more robust changes of state associated with melting and boiling. The atomistic and electronic origins of entropy are explored by enumerating the many causes of disorder in crystals, liquids, and liquid crystals. Pressure-induced phase transformations are associated with differences in densities of the two phases. A few examples of the uses of phase transformations finish the chapter.

20.1 Phase Transformations in Solids

Phase transformations can be classified in numerous ways. For example, in thermodynamics, a first-order phase transition is one in which there is a discontinuous change in entropy or volume at the transition, while a second-order transition entails a discontinuity in the specific heat, thermal expansion, or isothermal compressibility. In kinetics, phase transitions are classified in terms of reaction rate as sluggish or rapid. From a crystal chemistry perspective, phase transitions are classified as reconstructive or displacive based on how the structure is changed.

Reconstructive transformations are those that require breaking of primary chemical bonds, and reassembly of the atoms into a new crystal structure (Fig. 20.1). Because the internal energy of solid is the bonding energy per volume, breaking of bonds requires both energy and time to accomplish. Formation of a new phase within the old one requires nucleation and growth, and hence necessitates diffusion for transport. As a result, reconstructive phase transformations are often slow, and it is possible to retain a phase outside of its thermodynamic stability limits. This can be accomplished by quenching a material – i.e. by rapidly changing either the temperature (or the pressure) so that kinetic limitations trap the metastable phase. Reconstructive phase transitions are always first order in character, and are typically associated with large latent heats due to

---

[1] Portions of this chapter are adapted (with permission) from R. E. Newnham, "Phase diagrams and crystal chemistry," in *Phase Diagrams: Materials Science and Technology, Volume V*, Academic Press, New York (1978).

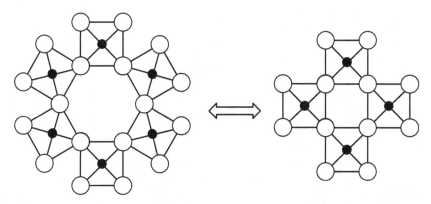

Fig. 20.1 Primary chemical bonds are broken in reconstructive phase transformations. Large atomic displacements and large latent heats are observed and the transformations are often sluggish, and easily quenched to temperatures and pressures outside their normal stability range. Adapted, by permission, after M. J. Buerger, "Crystallographic aspects of phase transitions," in *Phase Transformations in Solids*, eds. Smoluchowski, Meyer, and Weil, Wiley, New York (1951).

the change in the internal energy. The two polymorphs are independent structures; as a result, there is no requirement that there be any symmetry relationship between the two phases.

In contrast, displacive phase transformations entail a distortion of the crystal structure without breaking any primary chemical bonds, as illustrated in Fig. 20.2. Instead, some of the atoms displace slightly in the unit cell. Because no diffusion is required, displacive phase transitions are both rapid and difficult to prevent. As a result, it is unusual to be able to quench in a higher temperature phase to lower temperatures. In many cases, there is a group–subgroup relationship for the symmetry of the two phases. The resulting transitions tend to have a small latent heat, and can be either first or second order in character.

One mineral family that shows both displacive and resconstructive phase transformations is the feldspars $KAlSi_3O_8$ (orthoclase), $NaAlSi_3O_8$ (albite), and $CaAl_2Si_2O_8$ (anorthite). The reconstructive phase transition in the system entails the arrangement of Al and Si amongst the tetrahedral sites. Disorder is favored at elevated temperatures, so above ~1260 K in albite, the Al and Si are disordered on the tetrahedral sites. Thus, for orthoclase, any given tetrahedral site has a 25% probability of being occupied by an Al, and a 75% probability of being occupied by a Si. On cooling, the Al and Si order per the *Al avoidance rule*. An Al in tetrahedral coordination has a Pauling electrostatic bond strength of 3/4. An oxygen atom in the framework has bonds to only two high bond strength tetrahedral cations. An oxygen with two Al neighbors does not have enough incoming positive charge to achieve local electroneutrality. An oxygen with at least one silicon neighbor comes closer to obeying Pauling's rules, and hence is more stable. This tends to drive ordering at lower temperatures such that the aluminums avoid being near neighbors. Such a transition requires strong Al–O and Si–O bonds to be broken and reformed, and so is a reconstructive phase transformation with a latent heat of 12.2 kJ/mole. The displacive phase transformation is associated with a cage slumping,

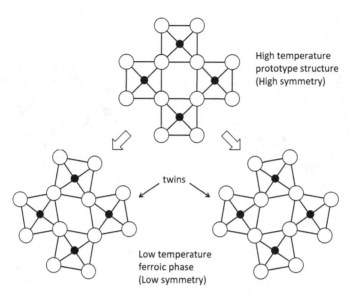

High temperature
prototype structure
(High symmetry)

twins

Low temperature
ferroic phase
(Low symmetry)

Fig. 20.2 In *displacive phase transformations* the primary chemical bonds remain unbroken; atomic displacements and latent heats are generally small. Twin structures, including ferromagnetics, ferroelectrics, and ferroelastics, are common. Adapted, by permission, after M. J. Buerger, "Crystallographic aspects of phase transitions," in *Phase Transformations in Solids*, eds. Smoluchowski, Meyer, and Weil, Wiley, New York (1951).

and does not require any bond breaking. Both Na^{1+} and Ca^{2+} have ionic radii of ~1 Å. At elevated temperatures, the thermal oscillations of these ions are large enough to "fill" the cavities. At lower temperatures, the cages shear around the smaller cations as the oscillation amplitude drops. Because no bonds are broken, the displacive phase transformation has a latent heat of only 1.7 kJ/mole. For larger ions such as K^{1+} or Ba^{2+}, this displacive transition does not take place. These two transitions are illustrated in Fig. 20.3.

Other reconstructive transformations entail a change in the coordination number of the atoms, and hence the internal energy. As a result, coordination changes are usually associated with a large latent heat. For example, in CsCl (8-coordinated atoms), there is phase change to the rocksalt crystal structure (6-coordinated atoms) at 460 °C; the latent heat is 7.5 kJ/mole. The associated crystal structures are described in Chapter 9.

Transformations proceed by a variety of structural mechanisms. The *martensitic* transformation in iron from FCC to BCC does not require long-range diffusion, and so is rapid. The transformation can be visualized by presuming that the FCC structure expands in the *a–b* plane of the FCC lattice, and contracts parallel to *c* (Fig. 20.4). Martensitic transformations are also important in several shape memory alloys. In contrast, zirconium, titanium, thallium, lithium, and beryllium all convert from HCP to a BCC structure with increasing temperature; this is sometimes visualized as a combination of deformation of the unit cell coupled with shear. Transformations from FCC to HCP structures occur by shearing the close-packed (111) plane of FCC. This can change the stacking sequence from ABCABC to ABABAB.

Another mechanism that is important in some displacive phase transformations is *cooperative rotation of polyhedra*. This was illustrated in Chapter 19 for the α–β phase

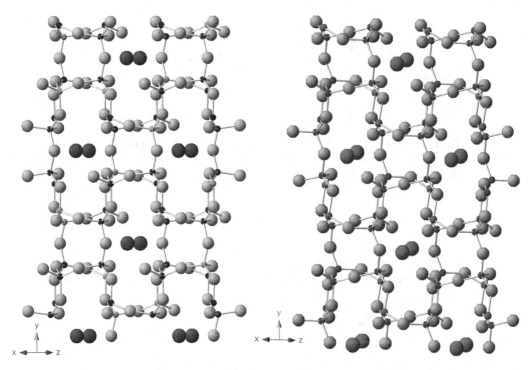

Fig. 20.3 High- and low-temperature crystal structures of albite. (Left) High-temperature form. The Na ions (large spheres, shown without any bonding) are better centered in the large cavities in the aluminosilicates framework. The Al and Si (shown as light and dark tetrahedrally coordinated atoms) randomly occupy the tetrahedral sites. The Al and Si order via a reconstructive phase transformation at lower temperatures. (Right) In the low-temperature form, the Na ions have displaced in their cavities, accompanied by a shear of the aluminosilicates framework.

transformation in quartz. A second example is the tilt transitions in perovskites. The perovskite structure has the general chemical formula ABX_3. In an ideal cubic perovskite, the A atom is 12 coordinated to X, the B atom is octahedrally coordinated to X, and the X atom has four A neighbors and two B neighbors. As shown in Fig. 20.5, in the ideal oxide perovskite where $X = O$, the *tolerance factor, t*: $t = \frac{r_A + r_O}{\sqrt{2}\,(r_B + r_O)} = 1$. Tolerance factors above one are characteristic of tetragonal distortions of the unit cell. For tolerance factors slightly below one, in many cases a rhombohedral distortion is observed. However, for smaller A site ions, the tolerance factor drops further still, and the materials have a tendency towards a family of tilt transitions in which the octahedra rotate with respect to each other. $SrTiO_3$, for example, undergoes such a tilt transition at low temperatures, as illustrated in Fig. 20.6.

Order–disorder phase transformations are very common in metallic systems. The classic example is copper–gold alloys with compositions near Cu_3Au. At elevated temperatures, these Cu–Au alloys form solid solutions with the FCC structure. Copper and gold atoms occupy the lattice sites at random in a disordered manner. If the Cu_3Au alloy is annealed at a critical temperature near 390 °C, the Cu atoms segregate into face-

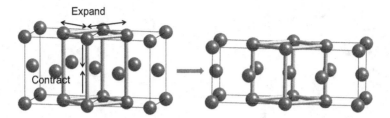

Fig. 20.4 (Left) Undistorted FCC cell. (Right) Expansion in the a–b plane, coupled with contraction parallel to c produces a BCC structure. The thicker lines show comparable portions of the structure. The thinner lines show the original FCC unit cell.

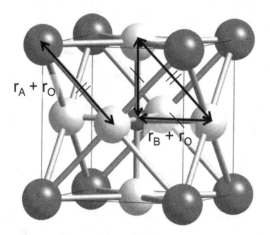

Fig. 20.5 Ideal cubic ABO_3 perovskite unit cell with A atoms on the corners, O at the face centers, and B at the unit cell center.

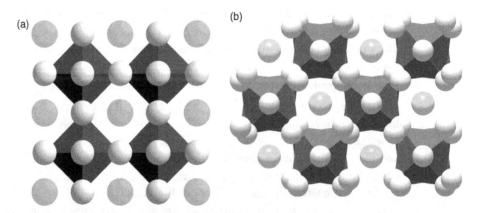

Fig. 20.6 At high temperatures $SrTiO_3$ has (a) the cubic perovskite structure. (b) The low-temperature tetragonal structure of $SrTiO_3$ has a doubled unit cell made up of corner-sharing TiO_3 octahedra. Alternating clockwise and counterclockwise rotations of the octahedra take place below the transition temperature of 110 K.

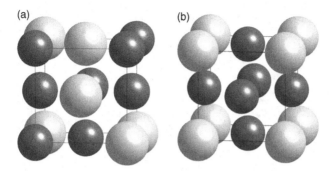

Fig. 20.7 The order–disorder transformation in Cu_3Au. (a) High-temperature disordered phase, and (b) low-temperature ordered phase. Large atoms are Au; smaller atoms are Cu.

center positions while Au migrates to the corner sites (Fig. 20.7). This is a cubic–cubic reconstructive phase transformation in which many near-neighbor chemical bonds are ruptured. Similar transitions have been observed in more than 60 alloy systems.

Order–disorder phenomena tend to have slower kinetics in ionically bonded solids than in metals; in metals ordering can occur if near neighbors exchange places. When the bonding is ionic, however, it is unlikely that a cation will exchange places with an anion. Thus the atom jump distance to the next-near neighbor site is longer, lowering the probability of a successful jump. The Al–Si ordering in feldspars are a good example.

Some phase transformations are associated with Jahn–Teller distortions of the coordination polyhedra (see Chapter 14 for an introduction). Consider the mineral hausmannite: Mn_3O_4. At high temperatures above the phase transformation, hausmannite is cubic with the spinel structure (Chapter 29). On cooling through the transition at 1170 °C, it develops a large tetragonal distortion. Mn_3O_4 contains one divalent and two trivalent manganese ions. The outer electron configurations are Mn^{2+}: $3d^5$ and Mn^{3+}: $3d^4$. Trivalent manganese is a Jahn–Teller ion occupying an octahedral site within the spinel structure. Its crystal field energy is lowered by forming four short bonds and two longer ones. This distorts the octahedra, and drops the crystallographic symmetry from cubic to tetragonal (Fig. 20.8).

20.2   Entropy, Disorder, and Structure

The stability of a polymorph can be described in terms of the free energy of the solid; the phase with the lowest free energy is thermodynamically stable under those conditions. Phase transformations occur at temperature, pressure, or field conditions at which two polymorphs have the same free energy.

The free energy of an unstressed material without an applied electric or magnetic field can be described as $G = H - TS = U + PV - TS$, where G is the Gibbs free energy (the accessible energy which can be transferred during a chemical reaction) H is enthalpy, T is temperature, S is entropy, U is internal energy, P is pressure, and V is volume.

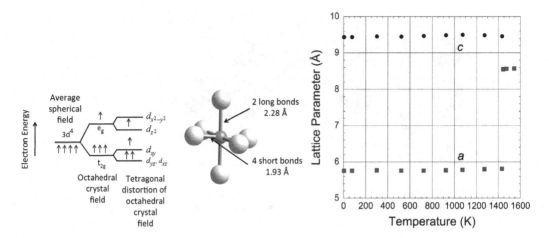

Fig. 20.8 (Left) The phase transformation in Mn_3O_4 is caused by a Jahn–Teller distortion in the outer electron shell of Mn^{3+}. Distortion of the Mn^{3+}-centered octahedron by elongating along one axis lowers the free energy of the system at low temperatures. (Right) An undistorted octahedron is recovered above the phase transition temperatures, as shown by the lattice parameters as a function of T. (Data *from Landolt–Bornstein Tables, Volume 42.*)

The roles of electrical and magnetic fields will be described in more detail for ferroelectric phase transitions in Chapter 28 and ferromagnetic transitions in Chapter 29.

For two phases, labeled with subscripts 1 and 2, the free energies are:

$$G_1 = U_1 + PV_1 - TS_1$$

and

$$G_2 = U_2 + PV_2 - TS_2.$$

For a phase transition taking place at constant pressure,

$$\left(\frac{\partial G_2}{\partial T}\right)_P - \left(\frac{\partial G_1}{\partial T}\right)_P = -(S_2 - S_1) = -\Delta S$$

$$\left(\frac{\partial G_2}{\partial P}\right)_T - \left(\frac{\partial G_1}{\partial P}\right)_T = (V_2 - V_1) = \Delta V.$$

That is, high temperatures will favor the phase with the higher entropy, and high pressure will favor phases with low volume.

Unfortunately, the relation between entropy and structure type is not a simple one, though a few trends are observed.

(1)   High temperature usually favors higher-symmetry structures.
(2)   High temperature usually favors lower-density phases.

To understand why these trends are observed, it is useful to understand the various mechanisms that contribute to entropy. Entropy is a measure of internal disorder in a material: $S = k \ln W$, where k is Boltzmann's constant, and W are the ways to disorder.

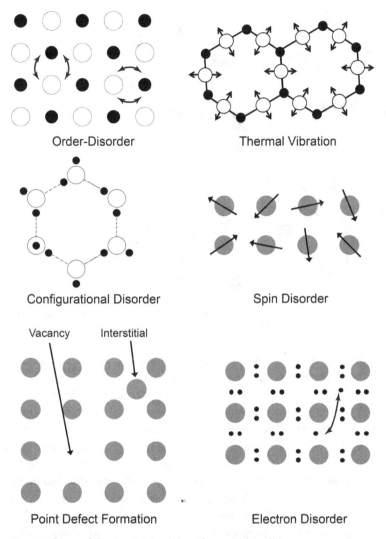

Fig. 20.9 Some of the structural origins of entropy in solids.

Changes in entropy occur at solid-state transformations just as they do during melting, boiling, and sublimation phase changes (Chapter 4).

Some of the mechanisms that contribute to disorder in a solid are illustrated in Fig. 20.9. The total entropy for a given phase is roughly the sum of each of the contributions relevant for that material. Quantitatively, the entropy change ΔS is expressed in entropy units (eu): $\Delta S = \Delta Q/T_t$, where ΔQ is the heat change associated with the transformation and T_t is the transition temperature:

$$[\Delta S] = \frac{[\text{cal}]}{[\text{mole}]} \Big/ [\text{K}] = \left[\frac{\text{cal}}{\text{mole K}}\right] = [\text{eu}].$$

Order–disorder mechanisms have already been discussed for the cases of Cu_3Au and the Al–Si ordering transitions in the feldspars. As is always the case, high temperatures

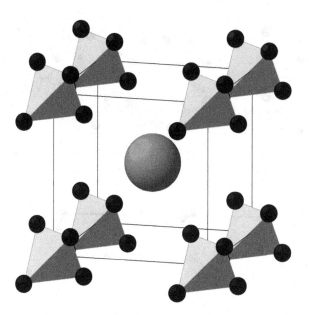

Fig. 20.10 Low-temperature crystal structure of NH_4Cl, with ordered arrangements of the NH_4 tetrahedra. These disorder at higher temperatures.

favor disorder in the atomic arrangements. Ordering at lower temperatures is often driven by differences in size or valence of the atoms.

Configurational disorder (also called *orientational order–disorder*) is important in some molecular crystals, or crystals where non-spherical molecular groupings exist. For NH_4Cl, the basic structure corresponds to the CsCl structure, but with NH_4 groupings replacing the Cs^+ ion. At low temperatures, the NH_4 tetrahedra all have the same orientation, as shown in Fig. 20.10. At higher temperature, however, the tetrahedra disorder in their orientation. A second example of orientational disorder occurs in ice. The crystal structure of ice I was described in Chapter 11; in brief, the O occupy positions in stacked, puckered hexagonal rings like those in tridymite. The hydrogens are arranged around the oxygens such that each O is in tetrahedral coordination with two near H within the water molecule, and two more distant ones to adjacent water molecules. The H positions are partially random, as shown in Fig. 20.11. One hydrogen lies between any two oxygen, with one near O neighbor, and one more distant. The O–O distance is 2.76 Å and the possible short O–H distance is ~0.96 Å. The oxygen positions are completely specified in ice, so that $W_O = 1$. There are many ways of arranging the hydrogens because of disorder. For N molecules of H_2O, each of the $2N$ hydrogens has two possible positions, O–H—O or O—H–O, giving $(2)^{2N}$ arrangements. However, the actual number of arrangements decreases due to imposition of the constraint that each O should have only two near H neighbors within the water molecule. To see why this is so, consider Fig. 20.12, which shows the 16 possible arrangements for the H around one of the oxygen. Of these, one has all four H close to the central O ion; there are also four arrangements with three close H (e.g. a hydronium ion), six with two hydrogens (a water molecule), four with one (a hydroxyl ion), and

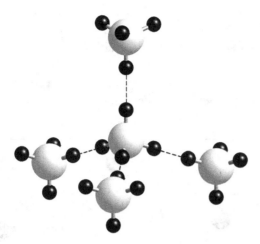

Fig. 20.11 A portion of the ice I structure, showing oxygen positions (large circles) and potential hydrogen sites in black. Only one of the two hydrogen sites along each of the lines between oxygens is occupied. Configurational entropy can be calculated from the Boltzmann relation by enumerating the possible hydrogen configurations. Redrawn from R. E. Newnham, "Phase diagrams and crystal chemistry," in *Phase Diagrams: Materials Science and Technology, Volume V*, Academic Press, New York (1978), with permission.

one with none. Only the six arrangements with two neighbors are probable. Given N oxygens, the number of hydrogen arrangements is $W_H = \left(\frac{6}{16}\right)^N (2)^{2N} = \left(\frac{3}{2}\right)^N$. Experiments have verified this result to within 1%.

Many chain-like molecules and polymers also show considerable configurational disorder, as it is possible to rotate portions of the structure around C–C single bonds, for example. As an example, consider any of the paraffins. These are flexible molecules with a sizable *configurational* contribution to the entropy of fusion. Configurational isomerism in butane is illustrated in Fig. 20.13. The number of possible arrangements increases rapidly with the chain length. When the potential energies between different isomers differ by only a few kT_{melt}, multiple configurations are found in the melt. Crystals generally involve the zig-zag chain conformation, since this allows the most intermolecular bonding. As temperature rises, the enthalpy associated with the better bonding competes with the tendency to increase disorder.

As was described in Chapter 16, *point defects* are thermodynamically stable at finite temperatures due to their contribution to the entropy. A single vacancy in a mole of material has 6.02×10^{23} possible positions, and so provides a significant contribution to disorder at elevated temperatures. Vacancies and interstitials are both possible, with their relative abundance depending on the details of the crystal structure. The extreme in this type of disorder is melting of the solid. The entropy of fusion for Ne, Ar, Kr, and Xe lies close to 14 J/mole °C. For most metallic elements, the entropy of fusion is about 6.3 to 10.5 J/mole °C, and for simple ionic crystals between 16 and 25 J/mole °C, or between 2 and 3 eu/atom.

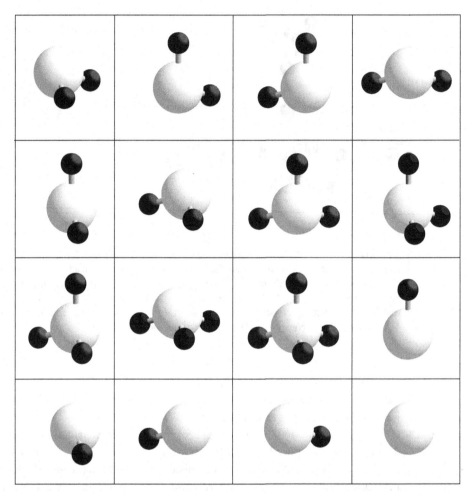

Fig. 20.12 Sixteen possible configurations for the arrangement of hydrogen around any oxygen in H_2O. The first six are equally probable and the remaining ten much less so.

Similar entropy changes are observed for non-melting transitions. When CsCl transitions to the rocksalt structure at elevated temperatures, the entropy change is 2.46 eu. Smaller entropy changes are usually observed for displacive phase transitions. Thus, the α- to β-quartz transformation has an entropy change of 0.18 eu.

Some crystals melt one of the sublattices before the entire crystal melts. In AgI, for example, the Ag atoms disorder at lower temperatures before the I atoms melt. Essentially, the Ag acts like a liquid in the solid I lattice. The Ag^+ have high mobilities in response to applied electric fields, and are extremely good ionic conductors (Chapter 25).

All materials undergo *thermal vibration* as the temperature increases. This introduces disorder, since the absolute position of the atoms at any instant in time is less well defined. The effect of thermal vibration also helps explain many of the observed trends associated with temperature-driven phase transitions. High temperature usually favors

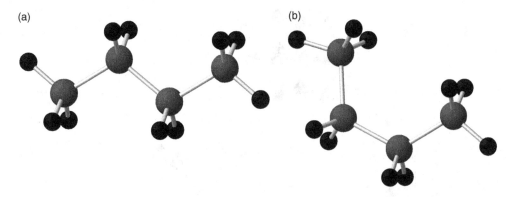

Fig. 20.13 The (a) extended chain and (b) folded chain of butane. Both configurational isomers are found in the melt.

higher-symmetry structures. This is a consequence of the fact that vibrational entropy is maximized when all of the vibrational entropy modes are degenerate.[2] Increased symmetry increases the degeneracy, and hence the entropy.

The observation that more open phases are favored at high temperatures can also be tracked back to the thermal vibration contribution. Consider the volume dependence of entropy: $(\partial S/\partial V)_T = -(\partial V/\partial T)_P/(\partial V/\partial P)_T$. Since most materials expand on heating, $(\partial V/\partial T)_P$ is usually positive. However, $(\partial V/\partial P)_T$ is always negative. Thus, $(\partial S/\partial V)_T$ is generally positive, and the entropy S usually *increases* with volume. Thus, open structures (larger volumes and lower densities) have larger entropies than dense structures.

In magnetic solids, *magnetic spin disorder* is favored at elevated temperatures. Thus, ferromagnetic, ferrimagnetic, and antiferromagnetic ordering is lost at high temperatures, when the tendency to increase entropy overcomes the forces leading to magnetic order.

Multiple mechanisms for *electronic disorder* can also add to the entropy. At low temperatures, electrons in ionically and covalently bonded solids are confined in the bonds, where their position is known. When enough thermal energy is introduced, however, electrons can be liberated from their bonds to occupy the conduction band. As this occurs, there is an addition to the entropy. In some transition metal oxides, structural phase transitions are also associated with changes in the electron localization. This will be described in Chapter 26 for VO_2, which undergoes a semiconducting to metallic transition.

20.3   Entropy Changes in Solids, Liquids and Gases

It is interesting to compare the entropy change in solid state phase transformations with those of melting and evaporation. Transformations between the HCP, FCC and BCC

---

[2] R. G. J. Strens, "Symmetry–entropy–volume relationships on polymorphism," *Mineralogical Magazine,* **36**, 565–577 (1967).

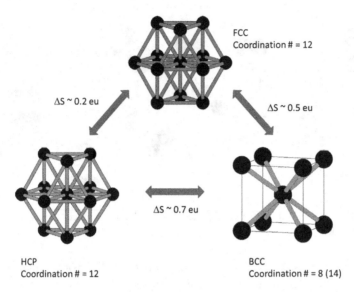

Fig. 20.14 Average entropy changes for solid state phase transformations in pure metals. Data from S. A. Cho, *J. Sol. State Chem.*, 11, 234 (1974).

structures of metallic elements are very common (Fig. 20.14). The entropy changes are smallest in the FCC ⇌ HCP transition. Both structures are close-packed with 12 near neighbors and six next nearest neighbors. The transformation involves a simple shear motion across the close packed plane converting the ABABAB sequence in HCP to the ABCABC sequence in FCC.

Somewhat larger entropy changes accompany the FCC ⇌ BCC and HCP ⇌ BCC transformations observed in a number of metallic elements (Fig 20.12). There are substantial changes in bonding as the coordination number changes from 12 to eight near neighbors, and six of the next-near neighbors move in closer to form a distorted 14-coordination. The packing in the BCC structure is slightly less dense, providing more vibration room, and slightly greater positional disorder. Average entropy changes are largest for the HCP ⇌ BCC phase transition where both the local coordination and the symmetry change. The FCC ⇌ BCC transition does not have a change in symmetry, and there is no coordination change at an HCP ⇌ FCC transition. For all three solid state transformations, the entropy changes are less than 1 eu, far smaller than those in melting and evaporation transitions.

The entropy of melting, S_f, is defined as $S_f = H_f/T_m$, where H_f is the heat of fusion in eu and T_m is the melting point in Kelvin. The entropy of fusion is typically about ten times larger than that of a displacive phase transition. For metals, inert gases, and simple inorganic compounds, the melting entropy is about 1.5 to 3.5 eu/atom. This corresponds to the loss of positional order at the melting point. Typical values are listed in Table 20.1.

The entropy of evaporation is about ten times bigger than the entropy of fusion and about a hundred times larger than that for a displacive phase transition. The amount of disorder (roughly 20 eu) is about the same at all boiling reactions. According to

Table 20.1 Entropy of fusion for several elements and simple compounds

Substance	Melting point (K)	Heat of fusion (cal/mole)	Entropy of fusion (eu/atom)
He	24.6	80.2	3.26
Ar	83.6	280	3.35
Al	923	2570	2.75
Cu	1352	3120	2.3
NaCl	1073	6690	6.2 (3.1)
MgO	3070	18500	6.4 (3.2)

Table 20.2 Evaporation entropies for inert gases, metals, inorganic, and organic compounds

Substance	Boiling point (K)	Heat of evaporation (kcal/mole)	Entropy of evaporation (eu)
Ar	87.4	1.50	17.2
Kr	121.4	2.32	19.1
Al	2714	76.8	25.8
Fe	3148	83.7	26.6
H_2O	373.2	9.71	26.0
NaCl	1680	39.0	23.2
CH_4	111.7	2.23	20.0
C_6H_6	353.3	7.37	20.8

Trouton's rule, the evaporation entropy depends only secondarily on the type of bonding (see Table 20.2).

To summarize, for monatomic materials and simple compounds, the entropy changes associated with displacive phase transitions, melting, and boiling are approximately 0.2, 2, and 20 eu, respectively. As described in the next section, the transitions become more complex in anisotropic molecular structures where orientational and configurational disorder takes place.

20.4 Pressure-Induced Phase Transformations

As described above, the Gibbs free energy ($G = U - TS + PV$) describes the stability of various polymorphs. With increasing pressure, the PV term becomes progressively more important. Since it contributes positively to G, phases with small volumes (that is, high densities) are more stable at high pressures. Thus, pressure *always* favors polymorphs with lower volumes (e.g. higher densities). Increased pressures *often* entail increased coordinations for the atoms, and/or a decrease in symmetry.

Pressure and temperature typically have opposite effects on the stability of a polymorph. As a result, in diagrams with pressure on the vertical axis and temperature on the horizontal one, phase boundaries usually slope up to the right, as in the $CaCO_3$ diagram in Fig. 20.15. High pressures favor aragonite, because its density, 2.93 g/cm^3 exceeds that of calcite, 2.71 g/cm^3. This transition also increases the coordination numbers; Ca is

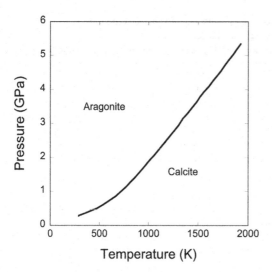

Fig. 20.15 Pressure–temperature diagram for calcium carbonate showing the stability regions of calcite and aragonite. Calcite is the more stable polymorph at low pressures and high temperatures. Its structure is less dense and more symmetric than that of aragonite. Rescaled, by permission, from J. C. Jamieson, "Introductory studies of high-pressure polymorphism to 24,000 bars by X-ray diffraction with some comments on calcite II," *J. Geol.*, **65** (3), 334–343 (1957). Figure 2.

6-coordinated in calcite and 9-coordinated in aragonite, with oxygen coordinations of 3 and 4, respectively.

High-pressure phases are extremely important in geology, as pressure increases with depth in the Earth's crust. Similarly, high pressures resulting from meteorite impacts can stabilize high-pressure polymorphs. To study these under laboratory conditions, pressures can be applied slowly in a diamond anvil cell or rapidly with a shock wave (Fig. 20.16).

For ionically bonded compounds with the stoichiometry AX, the three most widely observed crystal structures are zincblende (4-coordination), rocksalt (6-coordination), and cesium chloride (8-coordination). For a given composition, as pressure increases, there is a tendency to shift to structures with higher coordination numbers. This tendency is related to the compressibility of the ions. In most ionic structures, anions are larger than cations. Anions are also more compressible, because of the low electron density in the outer part of the ion. As a result, the radius ratio of cation to anion increases as pressure increases, and the coordination number increases in accordance with Pauling's rules.

Consider the phase boundary from the rocksalt to the CsCl structure. NaCl undergoes this transition at 26 GPa, KCl at 2 GPa, and RbCl at 0.5 GPa. A comparable transition is predicted for LiCl at ~80 GPa.[3] Why do transition pressures for the NaCl to the CsCl

---

[3] A. Martin Pendas, V. Luna, J. M. Recio, M. Florez, E. Francisco, M. A. Blanco, and L. N. Kantorovich, "Pressure-induced B1–B2 transitions in alkali halides: General aspects from first principles calculations," *Phys. Rev. B*, **49**, 3066–3074 (1994).

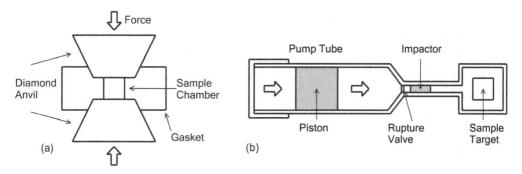

Fig. 20.16 (a) Diamond anvil cell. The diamond anvils apply the pressure to a solid sample confined by a metal gasket. Pressures above 300 GPa (3 Mbar) have been achieved using (b) shock wave apparatus for generating ultrahigh pressures.

structures decrease in the order Li^+, Na^+, K^+, Rb^+, Cs^+? The ionic radii increase in the same order. Larger cations have higher values for $\frac{r_{cation}}{r_{anion}}$, and so require less pressure to reach the point where higher coordination numbers are thermodynamically stable. CsCl, CsBr, and CsI already have the 8-coordinated structure under ambient conditions, and so do not undergo this phase transition at elevated pressures.

In the same way, many III–V and II–VI compounds with the sphalerite or wurtzite structures undergo phase transformations to the rocksalt structure in the 1 to 10 GPa range. This increases the coordination from 4 to 6. AgI, InP, InAs, CdS, CdSe, and CdTe convert to the rocksalt structure, while AlSb, GaSb, and InSb transform to the white tin structure, a distorted NaCl arrangement.

In compounds with three or more atoms, different bonds will compress at different rates, with weaker bonds being more compressible than strong ones. As a result, in layer and chain silicates, the Si–O bond is less compressible than bonds such as Mg–O.

In the Earth's crust, Al and Si shift from tetrahedral to octahedral coordination to 12-coordination with oxygen. For example, $MgSiO_3$ changes from the pyroxene structure to ilmenite, then to perovskite, and finally a newly discovered post-perovskite phase at pressures of 119 GPa and a temperature of 2400 K. The pyroxenes are chain silicates in which the Mg ions are octahedrally coordinated to six oxygens, and tetrahedral silicon to four oxygen atoms. One-third of the oxygens are bonded to two silicons in the silicate chain, the other two-thirds to three Mg and one Si. Pauling's rules are satisfied in the following ways:

$$\text{Chain } O \rightarrow 2(4/4) - 2 = 0.$$

$$\text{Other } O \rightarrow 3(2/6) + (4/4) - 2 = 0.$$

Hazen and Finger report that when the Si–O bond distance is compressed to ~1.59 Å, the Si transforms from tetrahedral to octahedral coordination.[4] Under pressure, $MgSiO_3$

---

[4] For many framework silicates with a room pressure Si–O bond length of 1.61 Å, the tetrahedral to octahedral coordination change occurs at pressures of ~ 10 GPa (100 kbar). The longer the Si–O bond at ambient pressure conditions, the higher the pressure that must be applied to force the coordination change.

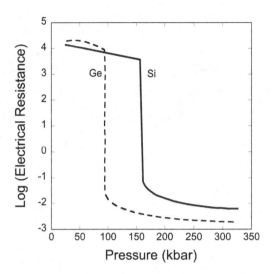

Fig. 20.17 Change in the electrical resistance of Si and Ge as a function of pressure. The sharp drop is associated with a transition from semiconducting to metallic behavior. As a rule of thumb, the degree of metallic character often increases as one moves down the periodic table. Thus, it is not surprising that less pressure is required to convert Ge to the metallic form, than is the case for Si. Figure adapted, by permission, from data in S. Minomura and H. G. Drickamer, "Pressure induced phase transitions in silicon, germanium and some III–V compounds," *J. Phys. Chem. Solids*, 23, 451–456 (1962) (combination of the original Figs. 1 and 2).

undergoes a phase transformation to the ilmenite structure in which both Mg and Si are octahedrally coordinated, and all oxygens are bonded to two Mg and two Si. In this case the charge balance is

$$2(2/6) + 2(4/6) - 2 = 0.$$

The second transformation at still higher pressures is to the perovskite structure where the Mg coordination increases to 12, and the Si coordination remains at six. Each oxygen is bonded to four Mg and two Si leading to

$$4(2/12) + 2(4/6) - 2 = 0.$$

Thus, the cation coordination increases at each transition. Pauling's rules are satisfied in different ways in these three structures.

Because of the changes in structure, many material properties also depend on the applied pressure. As an example, consider the electronic band gap of semiconductors and insulators. It is difficult to predict the effect of pressure on the band structure of materials. As a general rule, as the separation distance between atoms drops, the allowed energy bands broaden, with the net result that band gaps often decrease. In some cases, materials with substantial band gaps are converted to conductors or superconductors as the atoms are pushed closer together. An example of this is shown in Fig. 20.17 for the case of Si and Ge. Si transforms from the familiar diamond structure to the metallic β-Sn structure at 12 GPa and then through several other phases before becoming a face-centered-cubic metal stable to at least 300 GPa.

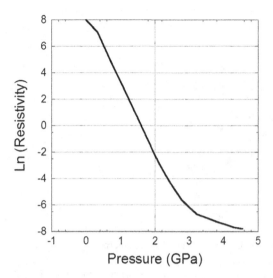

Fig. 20.18 Insulator-to-metal phase transformation induced by pressure in SmSe. Adapted, by permission, from data in E. Bucher, V. Narayanamurti and A. Jayaraman, "Magnetism, metal–insulator transition, and optical properties in Sm- and some other divalent rare-earth monochalco-genides," *J. Appl. Phys.*, 42 (1), 1741–1745 (1971). Figure 2. With permission of AIP Publishing.

Several of the high-pressure Si phases are superconductors at low temperatures. It should be noted, however, that because the energies of different bands in a material may change at different rates with pressure, in other cases, the band gap can *rise*, rather than *decrease*, as pressure increases. Behavior of this type is shown by MgO, stishovite and diamond.

Pressure sometimes alters the electron configuration of transition metal ions, causing a transformation from a high-spin state to a low-spin state. As pressure increases, anion neighbors approach the transition metal ions more closely, and crystal field splitting rises steeply, increasing the likelihood of the low field state. Furthermore, the ionic radius of low-spin Fe^{2+} is about 0.16 Å smaller than that of the high-spin configuration. Hence, there is a tendency for divalent iron compounds to transform to low-spin states under pressure.

Sometimes, materials undergo phase transitions in the properties, even in the absence of structural phase transformations. A good example of this is SmSe. SmSe has the rocksalt crystal structure. The f electrons of the Sm are localized on the Sm at ambient pressure conditions. As pressure increases, the energy gap between these f levels and the empty d band shrinks, so that electrons are promoted into the d band, where they are available to contribute to electrical conduction. As shown in Fig. 20.18, the result is a continuous phase transition between an insulating and a metallically conducting state. Over the entire pressure range, the rocksalt crystal structure is retained.

Many elemental solids undergo phase transitions between polymorphs as a function of pressure. As this happens, some elements take on crystal structures like those of the most stable transition metals. Once this has happened, the properties, including density and compressibility, approach those of the room pressure properties of many transition

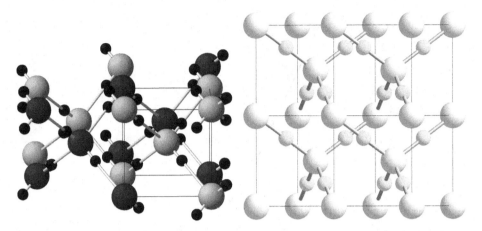

Fig. 20.19 (Left) The crystal structure of Ice VII. The two interpenetrating frameworks are shown using two different colors for the oxygens. All of the darker-color oxygen atoms belong to one framework, the lighter-colored ones to the other. Hydrogens are shown as the smaller ion. (Right) Crystal structure of cuprite, Cu_2O, which also has two interpenetrating networks. Cu are shown as the smaller circles, and O as the large circles.

metals. As a result, relatively low-stiffness materials like Cs become substantially stiffer (much closer to that of tungsten) once these phase transformations have taken place.

Ice VII, the stable form of H_2O at 25 °C and 2.5 GPa, is an example of a high-pressure phase for a molecular solid. Ordinary ice I (see Chapter 11) has a low density due to its open structure, which is why it floats on water. Ice VII has a much higher density, 1.57 g/cm^3, which is why it is observed at elevated pressures. The structure of ice VII consists of *two* interpenetrating frameworks, each of which is similar to the framework found in ice I; this is shown in Fig. 20.19. Cuprite, Cu_2O (Fig. 20.19) is an inorganic analog of this.

20.5   Liquid Crystals and Plastic Crystals

All crystals have long-range order in three dimensions, while liquids are generally ordered only at very local lengths scales, such as the first coordination sphere. These two states should be regarded as extremes, with a number of solids showing states with intermediate order in between. The family of solid-like liquids (e.g. liquids with a higher degree of ordering) are called *liquid crystals*, while the family of the liquid-like solids are called *plastic crystals*.

To understand these intermediate levels of order, consider a liquid crystal such as those shown in Fig. 20.20. In the fully crystalline state, there is three-dimensional periodicity in the molecule alignment, as with other molecular crystals (top left of figure). Many liquid crystals are composed of either rod-, disk-, or banana-shaped molecules, which are anisometric in shape. When rod-shaped molecules like those in *p*-azoxyanisole ($C_2O_3N_2H_6$) come close together, there is a tendency for them to align parallel to a single axis, often with a fairly uniform molecule-to-molecule spacing. Thus,

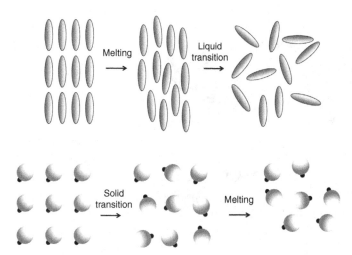

Fig. 20.20 Melting involving the onset of two types of disorder, positional and orientational disorder. In some materials the two commence separately. On heating, liquid crystals lose positional order before orientational order, while the sequence is reversed in plastic crystals. The latter generally consist of near-spherical molecular groups, whereas anisotropic molecules sometimes show liquid crystal behavior. Redrawn from R. E. Newnham, "Phase diagrams and crystal chemistry," in *Phase Diagrams: Materials Science and Technology, Volume V*, Academic Press, New York (1978), Fig. 31, p. 68. Adapted by permission from Elsevier.

even in the liquid crystal state, the alignment of molecules is not truly random. This is shown in the middle of the top row of the figure. The molecule positions are reminiscent of those seen in the crystal, but without translational periodicity. The crystalline to liquid crystalline phase transformation corresponds to the melting temperature. The molecules are not free to rotate until higher temperatures, at which point entropy destroys the remaining order, and the molecule arrangement becomes random. In much the same way, disk-shaped molecules can stack like coins, but with imperfect ordering.

A number of commonalities are observed amongst the known liquid crystals.

(1) The molecules are anisometric in shape, and have a comparatively rigid backbone along the long axis. Several examples of this are illustrated in Fig. 20.21.

(2) Many have aromatic rings along the backbone, which improve the alignment of the molecules.

(3) Polarizable species in the molecule are also common.

There are three types of ordering observed in liquid crystals with rod-like molecules: nematic, smectic, and cholesteric. The arrangement shown in the top row of Fig. 20.20 corresponds to the nematic variety, in which rod shaped molecules are aligned along a given axis, but without translational periodicity. These liquid crystals typically possess Curie group symmetry ∞m. When the molecules align into layers with fixed periodicity between the layers, but some randomness within the layer, the ordering is called smectic. Cholesteric liquid crystals refer to those where the molecular alignment is helical. Transitions between these arrangements occur as a function of temperature.

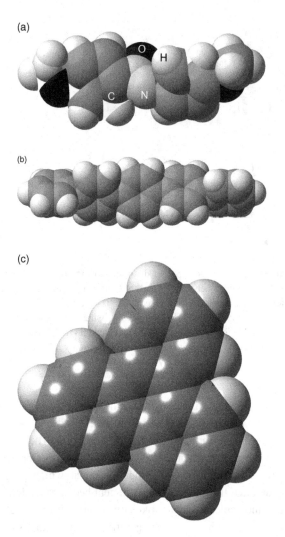

Fig. 20.21 Example liquid crystal molecules with atoms shown at their full van der Waals radii. (Top) 4-4′ Dimethoxyazoxybenzene; (middle) quinquephenyl; and (c) triphenylene. The top two are examples of rod-shaped liquid crystal molecules; the bottom is a disc-shaped one.

Disc-shaped molecules have other preferred stacking sequences. When the discs stack face-to-face, either rectangular or hexagonal arrangements of the resulting columns result. Alternatively, as was described in Chapter 11, the disks can stack into herringbone arrays, as was the case for solid benzene. Of course, in the liquid crystal, the ordering is imperfect.

The key industrial use of liquid crystals is in display applications, where the orientation of the molecules can be changed by an applied electric field. This realignment of the molecules changes the optical properties, and can be used to turn pixels on and off, as is described in Chapter 27.

In contrast to liquid crystals, *plastic crystals* are crystalline, with three-dimensional periodicity of the atomic arrangements. Typical is the case where nearly spherical molecules lose orientational order at a lower temperature than translational periodicity. This can occur either if the molecule is free to rotate around any axis, or if the molecules jump between different possible orientations, so that at any point in time, the distribution is disordered. Further increases in temperature lead to true melting, where the periodicity is lost. Because the material already has orientational disorder below the melting point, it has a high amount of entropy. This extra entropy stabilizes such plastic crystal phases to higher temperatures.

20.6   Phase Change Memories

Non-volatile memories are a mainstay component of many handheld electronic devices. Several mechanisms are employed for non-volatile memory, one of which is based on a phase transformation between crystalline and non-crystalline materials. Many of the phase change memories are made from semi-metals that are easily transformed between amorphous and crystalline states, with a large accompanying change in electrical resistivity. Crystalline materials conduct electricity better than amorphous solids because there is less electron scattering. The permanent storage of information in these memories depends on the phase changes induced by Joule heating with an applied voltage.

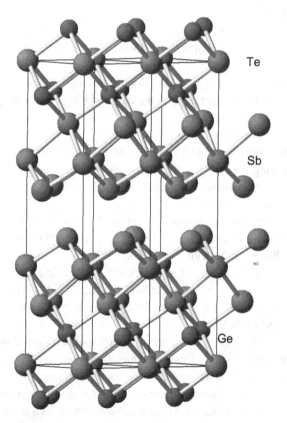

Fig. 20.22 Layer crystal structure of $Ge_2Sb_2Te_5$; c is vertical in the figure.

Beginning in the amorphous state, the alloy crystallizes when heated. Under further heating, the material melts, and then reverts to the amorphous state on rapid cooling.

Antimony–tellurium alloys are used in many of the phase change memories. The low melting points and complex compositions favor the formation of a glassy phase on rapid cooling. Selected doping with germanium leads to reproducible phase changes and faster switching times. An example crystal structure is shown in Fig. 20.22; $Ge_2Sb_2Te_5$ has a layered crystal structure, with Ge and Sb in 6-coordination. Some of the Te has a pyramidal 3-coordination, others have octahedral coordination. Other compositions crystallize to a metastable rocksalt structure. Key material properties for this application are that the material should be easily and reproducibly changed between the amorphous and the crystalline states, and the amorphous phase should not spontaneously crystallize.

20.7   Problems

(1) Cu_3Au undergoes a phase transition. At elevated temperatures it has the structure of Cu, where each position is filled randomly with Cu or gold atoms (i.e. at each position, there is a 1/4 probability that the atom is Au, and a 3/4 probability that the atom is Cu). At lower temperatures, the thermodynamically stable phase is a cubic one with space group $Pm3m$ (#221), with $a = 3.74$ Å. Cu is at (0, 0.5, 0.5) and Au is at (0, 0, 0).

   (a) Compare this structure to that of Cu.

   (b) Do you expect the kinetics of this transition to be fast or slow? Explain your reasoning.

(2) The perovskite crystal structure is known for the variety of transformations it can support. The prototype structure is cubic (e.g. $SrTiO_3$ at room temperature). It has space group $Pm3m$ (#221), with $a = 3.905$ Å. Sr is at (0, 0, 0) and Ti is at (0.5, 0.5, 0.5), and O is at (0.5, 0.5, 0).

   (a) Plot the unit cell of this structure showing the bonds of the atoms involved. Replot it showing the connectivity (and shape) of the Ti polyhedra.

   (b) What are the coordinations of each ion? For the Ti itself, give each of the Ti–O bond lengths.

   (c) $BaTiO_3$ is a ferroelectrically-distorted perovskite (at $T < 130$ °C). The structural distortion is quite small. At room temperature, it is tetragonal with space group $P4mm$ (#99), with $a = 3.99$ Å, $c = 4.03$ Å. Ba is at (0, 0, 0) and Ti is at (0.5, 0.5, 0.514), and O is at (0.5, 0.5, −0.025), and (0, 0.5, 0.485). Compare this structure to that of cubic $SrTiO_3$. (Hint: it helps to draw slightly more than one unit cell.) In your comparison be sure to include the Ti–O distances in tetragonal $BaTiO_3$.

   (d) A ferroelectric, by definition, is a material with a spontaneous dipole which can be reoriented between crystallographically defined states by an applied electric field. In your $BaTiO_3$ unit cell, estimate the positions of the center of positive and negative charge in the unit cell. Also note that there were six possible directions in which the Ti could have displaced; as a result these are the six possible polarization directions in tetragonally distorted $BaTiO_3$.

Note that two additional structural distortions (to orthorhombically and rhombohedrally distorted versions) also occur at lower temperatures.

(e) GdScO$_3$ is an orthorhombically distorted perovskite which is not ferroelectric. The space group is orthorhombic (*Pnma*, space group 62) with $a =$ 5.742 Å, $b = 7.926$ Å, and $c = 5.482$ Å). Gd sits at (0.44058, 0.75, 0.48392), Sc sits at (0, 0, 0.5), and O is at (0.4494, 0.25, 0.1201) and (0.1956, 0.5623, 0.1927). Plot a unit cell of this, and compare the structure to that of cubic SrTiO$_3$. It may be easiest to see this if you show the octahedra around the Sc.

(3) (a) Write the equation for free energy, defining the terms.

(b) Explain why you might expect the transitions in KCl as a function of temperature shown in the diagram below. The figure is adapted, by permission, after C. W. F. T. Pistorius, "Melting curves of the potassium halides at high pressures," *J. Phys. Chem. Sol.*, **26**, 1543–1548 (1965). Figure 3.

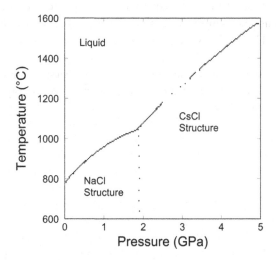

(c) Why is there a transition from the NaCl structure to the CsCl structure as a function of pressure?

(4) Justify why tridymite converts to cristobalite on increasing pressure.

(5) Describe five factors that contribute to entropy in solids. For each factor, give an example of a material showing this contribution. How does entropy affect the stability of polymorphs as a function of temperature?

(6) The stable form of carbon at room temperature and pressure is graphite (space group 186: *P6$_3$mc* with $a = 2.456$ Å, $c = 6.696$ Å) and C at (0, 0, 0) and (1/3, 2/3, 0) and all symmetry-related positions.

(a) Draw a few unit cells of this structure showing the bonds that connect the C ions.

(b) Show the coordination (i.e. the number and geometric arrangement of near neighbors) of C atoms.

(c) Determine the closest C–C distances parallel and perpendicular to the c axis. Based on your knowledge of bonding, describe what you believe to be the predominant bond type(s) present in graphite.

(d) Graphite is an extremely anisotropic material (that is, its properties are very different when measured in different directions). For example, consider the mechanical stiffness of graphite. As a rule of thumb, the mechanical stiffness is high when you are pulling directly against the strong bonds in a structure, and drops as the bond strength decreases. Given this, discuss what you would expect in terms of the anisotropy in the mechanical stiffness in graphite.

(7) At high pressures, graphite undergoes a phase transformation to the diamond crystal structure. Diamond is cubic and has space group 227: $Fd\bar{3}m$ with $a = 3.56$ Å) and C at $(0, 0, 0)$ and all symmetry-related positions. Si and Ge, the two most important elemental semiconductors have this same crystal structure.

(a) Draw a unit cell, showing the bonds that connect the C atoms.

(b) What is the coordination of the C atoms? What is the C–C bond length? Compare the bond length with that determined in Problem (6).

(c) Count the number of dangling bonds per surface area for {100}, {110}, and {111} planes. What does this suggest for the equilibrium morphology of diamond?

(d) Draw the temperature–pressure phase diagram for carbon. Explain why various transitions are observed, based on free energy arguments.

21 Cement and Concrete

21.1 Introduction

Concrete is the most widely used construction material in the world, and yet it remains largely unappreciated. It is made by mixing Portland cement with sand, crushed rock, and water. Total world consumption is several billion tons per year, about one ton for every living human being. No other material except water is consumed in such tremendous quantities. Nevertheless, it is often ignored in books dealing with the physics and chemistry of materials. Comprehensive books on cement and concrete are available, and serve as excellent references for the field.[1]

Despite its familiarity and long history, concrete presents engineers with many interesting questions and uncertainties. The setting and hardening of concrete involves a number of simultaneous chemical reactions and the process often continues for years. Hydration, dehydration, and rehydration are the keys to making cement and concrete. It has proved especially difficult to optimize the principal engineering properties of concrete: strength, durability, dimensional stability, and low cost.

The remote ancestors of modern concrete are mortars made thousands of years ago from limestone. The first use of what can truly be called concrete dates back to 7000 B.C. in Galilee. Lime (CaO) plasters were used in building the Egyptian pyramids and are still in good condition. The Romans developed some of the first cements from mixtures of lime paste and volcanic ash.

In 1824, Joseph Aspdin, an English bricklayer, patented a new cement formed by burning clay and limestone together in a stove. It was named Portland cement because of its resemblance to a rock quarried along the coast of England. The strength and stability of concretes made from Portland cement has led to worldwide acceptance.

Limestone and clay – the same basic ingredients – are still in use today. The raw materials are carefully proportioned and then ground to a fine size to facilitate reaction. After firing at 1450 °C the resulting Portland cement clinker is reground to a fine powder, and mixed with a few percent of gypsum ($CaSO_4 \cdot 2H_2O$) to control the setting rate. The clinker consists of four major phases: tricalcium silicate (Ca_3SiO_5), dicalcium silicate (Ca_2SiO_4), tricalcium aluminate ($Ca_3Al_2O_6$), and tetracalcium aluminoferrite ($Ca_4Al_2Fe_2O_{10}$). Low-cost minerals are the raw materials in common cement leading to some variability in the chemistry.

---

[1] A. M. Neville, *Properties of Concrete*, 4th edn, New York, John Wiley & Sons (1996); H. F. W. Taylor, *Cement Chemistry*, 2nd edn, London: Thomas Telford (1997).

Table 21.1 Abbreviated nomenclature used in cement literature

C = CaO	S = SiO$_2$	A = Al$_2$O$_3$	F = Fe$_2$O$_3$
M = MgO	K = K$_2$O	$\bar{S}$ = SO$_3$	N = Na$_2$O
T = TiO$_2$	P = P$_2$O$_5$	H = H$_2$O	$\bar{C}$ = CO$_2$

C-S-H = amorphous calcium silicate hydrate of unspecified composition.
AFm, AFt = hydrated phases of calcium, alumina and iron formed during
the setting process.

In forming concrete, the Portland cement is mixed with about half its weight in water
to make a mud-like mixture called cement paste. An initial setting process takes place
within a few hours as the cement phases begin to hydrate, but the hardening process
goes on for months. Both fine aggregate (sand) and coarse aggregate (gravel = crushed
rock) are commonly used as fillers in making concrete. It is important that the cement
paste coats all the filler particles and fills all the pores between aggregate particles.
Generally the aggregate is chemically inert and takes no part in the hydration process.

The setting process involves a hydrated calcium silicate gel (C-S-H) which slowly
recrystallizes to a mechanically strong chain silicate resembling the minerals tobermorite and jennite.

Cement scientists have developed a short-hand terminology (Table 21.1) to describe
the many oxide phases encountered in the manufacture of Portland cement clinker and
the subsequent setting and hardening processes. Abbreviations are useful because there
are dozens of different compounds involved as raw materials, and in high-temperature
reactions, and final hydration products. As an example, the hydrated calcium silicate
jennite, Ca$_9$Si$_6$O$_{18}$(OH)$_6 \cdot$ 8H$_2$O, is abbreviated to C$_9$S$_6$H$_{11}$.

21.2   Processing of Cement

Portland cement is by far the most important hydraulic cement used in building construction and civil engineering. Hydraulic cements set and harden as a result of chemical
reactions with water. The surfaces of silica, alumina, lime, and other oxides are all
hydrophilic. Oxides are ionic, so is water, and "like wets like." As water molecules
approach the surface of an oxide, hydrogen bonds are formed to the surface oxygens.
One hydrogen bond links the water molecule to the surface, and two hydrogen bonds leads
to dissociation into two surface hydroxyls, and then to a silanol surface of hydroxyls. The
silanol groups also have a strong affinity for water as they link through additional
hydrogen bonds. The additional water molecules penetrate the surface layer, causing
crevice formation and liberating Si(OH)$_4$ silanol groups into solution. SiO$_2$ dissolves very
slowly in water compared to CaO and most other oxides, which are more ionic than silica.

Portland cement is produced by heating lime, silica, and other minerals to partial
fusion and grinding the cooled "clinker" to a fine powder. Most commonly, the source
of lime is limestone (mostly calcite, CaCO$_3$) and silica and alumina come from clay or
shale. Minor additions of corrective constituents such as iron ore, sand or bauxite are
used to adjust the bulk composition (see Table 21.2).

Table 21.2 Typical Portland cement composition

	Percentages
CaO = C	60–67
SiO_2 = S	17–25
Al_2O_3 = A	3–8
Fe_2O_3 = F	0.5–6.0
MgO = M	0.5–5.5
$Na_2O + K_2O$ = N + K	0.4–1.3
TiO_2 = T	0.1–0.4
$SO_3 = \bar{S}$	1.0–3.0

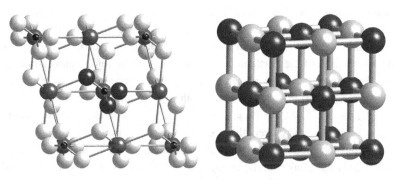

Fig. 21.1 Comparison of the crystal structures of calcite (left) and lime (right). The darker of the two spheres is Ca, the lighter is oxygen. Carbon is shown as the small black sphere. The rocksalt structure is distorted to a rhombohedral one by the introduction of carbonate groups. One of the planar triangular carbonate groups is highlighted in a darker color in the calcite figure.

Calcite ($CaCO_3$), magnesite ($MgCO_3$), and dolomite ($CaMg(CO_3)_2$) are important raw materials for basic refractories and cements. The crystal structures of the three minerals are very similar with triangular carbonate groups and octahedrally coordinated Ca^{2+} or Mg^{2+} ions. Ca–O bonds (~2.3–2.4 Å) are significantly longer than Mg–O bonds (~2 Å), so they do not substitute for each other, but instead form three separate minerals. The Mg and Ca ions occupy alternate octahedral sites in the dolomite structure. Thus, dolomite is an ordered derivative of the calcite structure shown in Fig. 21.1. The calcite structure can be envisaged as a derivative of the rocksalt structure in which the spherical O^{2-} ion in CaO is replaced by the flat planar carbonate $(CO_3)^{2-}$ group.

Limestone is an impure polycrystalline calcite with minor amounts of clay, iron oxide, and other sedimentary minerals. It is used for making lime (CaO), cement, and as a filler in road building.

When heated, calcite and other carbonate minerals decompose into oxides and gaseous carbon dioxide. Decomposition typically occurs at temperatures of 600–950 °C:

$$CaCO_3 \rightarrow CaO + CO_2\uparrow.$$

Some clay-bearing limestones have overall chemical compositions close to those of Portland-cement, making it possible to use a blend of strata from the same quarry.

The main limitation is that the clinker should contain less than 4% MgO and should be low in alkalies such as Na_2O and K_2O.

A distinctive feature of Portland cement in comparison with other natural cements of similar composition is that it is fired to high temperature in a kiln. The minerals lose molecular water below 200 °C, and then begin to decompose. Clay minerals such as kaolin lose hydroxyl ions near 600 °C to form metakaolin. Metakaolin is an amorphous phase ($Al_2Si_2O_7 = AS_2$) that reacts readily with the other raw materials in Portland cement. Partially decomposed clay minerals such as metakaolin retain the silica layers but the octahedral layers have lost much of their crystallinity.

Calcite ($CaCO_3$), the principal mineral phase in limestone begins to decompose to lime (CaO = C) near 800 °C and then to react with partially decomposed layer silicates to give belite ($Ca_2SiO_4 = C_2S$), aluminate ($Ca_3Al_2O_6 = C_3A$), and ferrite ($Ca_4Al_2Fe_2O_{10} = C_4AF$). These three phases together with excess lime (CaO C), are present at 1200 °C.

Further heating to temperatures between 1300 and 1450 °C leads to partial melting. About 20–30% of the Portland cement mix becomes liquid, mainly from the aluminate and ferrite phases. In the presence of this melt, much of the belite phase (dicalcium silicate = $Ca_2SiO_4 = C_2S$) combines with the remaining lime (CaO = C) to form alite (tricalcium silicate = $Ca_3SiO_5 = C_3S$). Alite is the component of Portland cement clinker that provides the early strength during the hydration process.

Cooling the fired product back to room temperature forms nodules as the residual melt bonds the clinker particles together. The various phase changes involved in the processing of Portland cement are illustrated in Fig. 21.2.

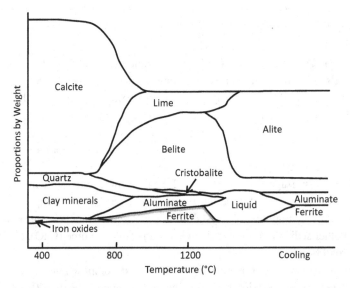

Fig. 21.2 Diagram illustrating the processing of limestone and shale to form Portland cement. The four clinker phases are described in the next section. Adapted, by permission, from article published in *Cement Chemistry*, Academic Press, H. F. W. Taylor, London 1990, Fig. 3.1, p. 61. Copyright Elsevier.

To complete the manufacture of Portland cement, the clinker is milled together with about 2% gypsum ($CaSO_4 \cdot 2H_2O = C\bar{S}H_2$). Gypsum slows down the rapid hydration of the calcium aluminate phase during the setting reaction.

Rapid cooling from the clinkering temperature down to 1100 °C is followed by air-quenching to room temperature. It is important to retain Ca_2SiO_4 as β-Ca_2SiO_4, the polymorph stable near 600 °C. This is partly accomplished by solid solution substitution in β-Ca_2SiO_4, which prevents its conversion to the non-hydraulic polymorph, γ-Ca_2SiO_4.

21.3   Clinker Phases

The four most important components of Portland cement are alite, belite, aluminate, and ferrite. All four are anhydrous oxides.

Four binary phases exist within the CaO–SiO_2 system: $CaSiO_3$, $Ca_3Si_2O_7$, Ca_2SiO_4, and Ca_3SiO_5. The latter two compounds are orthosilicates that interact strongly with water and are the principal components of Portland cement. In cement terminology (Table 21.3) tricalcium silicate = Ca_3SiO_5 = C_3S = alite and dicalcium silicate = Ca_2SiO_4 = C_2S = belite.

Alite

Alite is a highly reactive hydraulic oxide that makes up 50–90% of Portland cements. It is formed from modified limestones heated to temperatures exceeding 1250 °C followed by rapid cooling. C_3S undergoes several phase transformations that are all similar in structure. As shown in Fig. 21.3, the structure of alite consists of SiO_4 tetrahedra linked together by divalent calcium ions. The Ca^{2+} ions are coordinated by eight oxygens. Oxygens are of two types: most are bonded to a single Si^{4+} ion and several calciums, but a few are bonded only to calcium, just as they are in lime. The intense interaction between alite and water has been attributed to these poorly bonded oxygens.

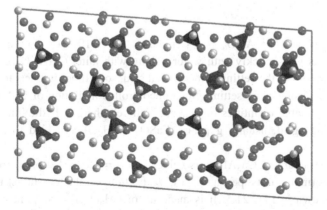

Fig. 21.3 Crystal structure of Ca_3SiO_5 showing the $(SiO_4)^{4-}$ tetrahedra, Ca^{2+} cations (light gray) and isolated O^{2-} anions unbonded to silicon (darker gray atoms). Looking down the *b* axis.

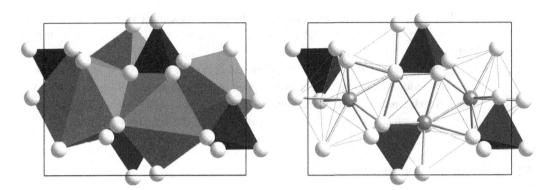

Fig. 21.4 The crystal structure of the β polymorph of belite. Si^{4+} is tetrahedrally coordinated to oxygen. Ca^{2+} cations have seven or eight oxygen neighbors.

Belite

Dicalcium silicate (Ca_2SiO_4 = belite = C_2S) is the second most abundant constituent of Portland cement clinker, about 15–30%. It reacts slowly with water and contributes little to the setting strength in the first month. After a year, however, its contribution is comparable to the more hydraulic alite phase.

Five different polymorphs of belite exist at ordinary pressures. All five consist of SiO_4^{4-} tetrahedra and divalent Ca^{2+} ions. β-C_2S, the polymorph stable around 600 °C, is the preferred constituent for Portland cement. On cooling to room temperature, sintered masses of β-C_2S transform to a powder that enhances the interaction with water.

The structure of β-Ca_2SiO_4 is illustrated in Fig. 21.4. Unlike alite, there are no oxygen atoms unattached to silicon. This may explain why alite is more hydraulic than belite.

Aluminate

Tricalcium aluminate ($Ca_3Al_2O_6$ = C_3A) makes up 5–10% of Portland cement clinkers. It reacts very quickly with water. Along with alite, $Ca_3Al_2O_6$ contributes to the initial strength of the setting cement. That said, sometimes it reacts too quickly with the water, causing rapid setting. Such uncontrolled hydration of C_3A results in the generation of large quantities of heat from the exothermic hydration reactions. A set-controlling agent, usually gypsum ($CaSO_4 \cdot 2H_2O$), is added to counteract the problem.

The crystal structure of $Ca_3Al_2O_6$ is cubic with a = 15.26 Å, space group $Pa3$, and Z = 24 formula units per unit cell. A portion of the crystal structure of $Ca_3Al_2O_6$ is shown in Fig. 21.5. It is made up of $(Al_6O_{18})^{18-}$ rings held together by Ca^{2+} ions. The rings contain tetrahedrally coordinated Al^{3+} ions and two types of oxygen. One type is attached to one aluminum and several calcium ions, the other to two aluminums. The chemical instability of tricalcium aluminate has been

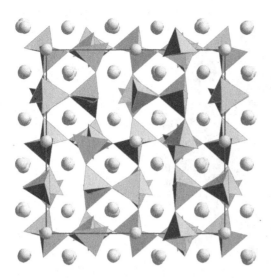

Fig. 21.5 The Al_6O_{18} ring in tricalcium aluminate. AlO_4 tetrahedra share corners in a highly puckered arrangement. Calcium ions are bonded to five or six oxygens in the cubic unit cell. The Ca–O bonds linking the rings together are easily ruptured during the hydration process of cement.

linked to the bridging oxygens bonded to two Al tetrahedra. This configuration violates Pauling's rules and leads to the Al-avoidance principle noted in many aluminosilicate minerals.

Ferrite

The iron-bearing ferrite phase makes up about 5–15% of normal Portland cement clinker. The ideal composition is tetracalcium aluminoferrite, ($Ca_4Al_2Fe_2O_{10}$ = C_4AF), but there is extensive solid solution along the join $Ca_2(Al_xFe_{1-x})_2O_5$.

The crystal structure of the ferrite phase is related to the well-known perovskite structure. Perovskite is the mineral form of calcium titanate ($CaTiO_3$ = CT). As shown in Fig. 21.6, the structure consists of a three-dimensional network of titania octahedra with Ca^{2+} ions located in interstitial sites.

The Ca_2AlFeO_5 structure (Fig. 21.7) is derived by omitting one-sixth of the oxygen atoms in the perovskite structure and converting half of the octahedral sites into tetrahedra in alternate layers. Al^{3+} and Fe^{3+} take the place of Ti^{4+} in both the tetrahedral and octahedral layers while the Ca^{2+} sites are similar in the ferrite and perovskite structures. The smaller Al^{3+} ions show a preference for the tetrahedral sites. Each Ca^{2+} ion in C_4AF has seven oxygen neighbors with bondlengths between 2.3 and 2.6 Å. The comparatively open channels parallel to the silica chains promote hydration of the ferrite phase. The presence of iron in the ferrite phase imparts a brown color to cement.

In Portland cement clinkers the aluminate and ferrite phase often occur together in oriented intergrowths.

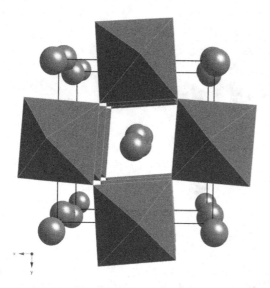

Fig. 21.6 Unit cell of $CaTiO_3$ showing the Ca atoms in light gray, and the Ti octahedra in darker gray. The oxygens are at the corners of the Ti octahedra.

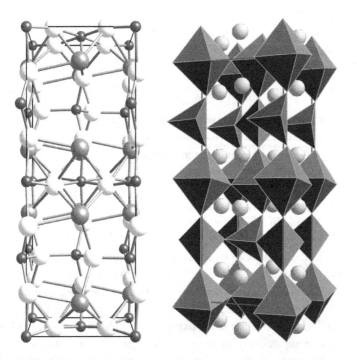

Fig. 21.7 Crystal structure of tetracalcium aluminoferrite (C_4AF) showing octahedral $(Al,Fe)O_6$ and tetrahedral $(Al,Fe)O_4$ groups coordinated with divalent calcium ions. The structure is closely related to the mineral brownmillerite.

21.4   Hydration of Portland Cement

The setting of cement is *not* a drying process. It is just the opposite: a hydration process. When cement is mixed with water, it undergoes a complex dissolution reaction generating various calcium, silicate, and aluminate ions in solution. The hydrated product phases precipitate from solution when their solubility limit is reached. This dissolution – diffusion – precipitation process requires a nucleation time, sometimes called the dormant period. As the hydration reaction proceeds, more and more of the anhydrous components of cement are converted into hydrates, with a decrease in porosity. The volume of the hydrates, including the associated interlayer water, is more than twice that of the initial anhydrous phases. It is the hydrates that given cohesion to hardened cement.

Lime ($CaO = C$) is produced by heating a pure limestone or chalk to 900 °C, the temperature at which carbon dioxide ($CO_2 = \overline{C}$) dissociates from calcium carbonate ($CaCO_3 = C\overline{C}$). To prevent the reverse reaction from setting in, CO_2 must be removed from the kiln on cooling.

Unlike most oxide minerals, pure lime hydrates vigorously in water to form a soft, fine powder of $Ca(OH)_2 = CH$ that mixes with water to give a highly plastic putty which dries to a firmly bonded mass. Cement derives its properties from several calcium silicates and calcium aluminates that behave like lime.

The size of the Ca^{2+} ion is an important feature of cementitious materials. It is a little too large for octahedral coordination with oxygen, but not large enough for the close-packed coordinations favored by Sr^{2+} and Ba^{2+}. Because of the comparatively large size of the Ca^{2+} ion, the oxygen ions in lime are forced apart, opening the structure, and increasing the rate of hydration.

The same is true in some, but not all, calcium compounds. Anorthite feldspar ($CaAl_2Si_2O_8$) and perovskite ($CaTiO_3$) are not hygroscopic. In those minerals, the structures are stabilized by three-dimensional frameworks of $(TiO_6)^{2-}$ octahedra or $(Al_2Si_2O_8)^{2-}$ tetrahedra. The same is true in calcite ($CaCO_3$), where the strong covalent bonds between carbon and oxygen stabilize the structure.

The situation is different in the calcium silicates and calcium aluminates used in cement. The isolated silicate tetrahedra in Ca_3SiO_5 and Ca_2SiO_4 are held together by the Ca^{2+} ions, and in $Ca_3Al_2O_6$ the aluminate rings are also linked by the calcium ions. As a result, these oxides interact rapidly with water to form hydrated structures and impart strength to the setting cement.

As described above, when water is mixed with the cement powder, chemical reactions cause a change in the slurry as anhydrous oxides are converted to hydrates. The hydration reactions are exothermic, with the production of heat reaching a maximum typically in 2 to 3 days. The hydration rates of the four clinker phases are shown in Fig. 21.8. The reactions take place in a water–cement paste which in time produces a firm hard mass – the hydrated cement paste. Grinding the cement to a very fine particle size ensures rapid reaction followed by the development of the required compressive strength.

Calcium silicate hydrates have a very complex crystal chemistry with more than 30 different phases known. Near room temperatures these phases range from

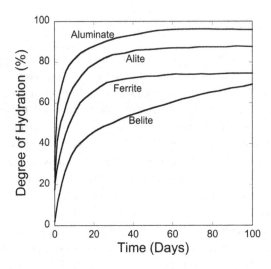

Fig. 21.8 Degree of hydration of phases in Portland cement. Adapted from S. Mindess and J. Francis Young, *Concrete* 1st Ed., Prentice-Hall, Inc. Englewood Cliffs, NJ, © 1981, Fig. 3.3(b), p. 28. With permission of Pearson Education, Inc., New York, New York.

semicrystalline to amorphous and are described by the generic term C-S-H. The C-S-H phases formed from the Ca_3SiO_5 (alite) and Ca_2SiO_4 (belite) of Portland cement are generally amorphous.

As shown in Fig. 21.7, the alite clinker phase hydrates more rapidly than belite. Hydrated alite is responsible for the early strength of cement. At this stage the X-ray patterns of C-S-H develop two-dimensional order resembling the mineral tobermorite, and is sometimes referred to as C-S-H (I).

Tobermorite, $Ca_5Si_6O_{16}(OH)_2 \cdot 8H_2O$, has a structure (Fig. 21.9) based on complex layers of calcium oxide octahedra reinforced by tetrahedral silica chains. The CaO_6 octahedra share edges in the (001) plane with Ca–O bonds ranging between 2.3 and 2.5 Å long. The silica single chains have a peculiar geometry known as *drierketten* that repeats every third tetrahedron. Two of the SiO_4 groups (the paired tetrahedra) share two oxygens with calcium ions in the octahedral layer, and two with other silicons. The third SiO_4 group (the bridging tetrahedron) shares oxygens with two bridging tetrahedra, one calcium in the octahedral layer and one interlayer calcium. Hydrogen bonds from water molecules also contribute to the interlayer bonding. The twisted *drierketten* chains lead to a fibrous microstructure. Interlocking fibers act to strengthen the cement.

The Ca/Si ratio in tobermorite is only 0.83, far smaller than the 3:1 ratio of alite (Ca_3SiO_5). This means that much of the calcium is excluded from the tobermorite portion of C-S-H (I). It remains in the C-S-H gel as portlandite $Ca(OH)_2$ = CH (see Fig. 21.10). This has a layer-like structure that easily intergrows with the tobermorite-like regions in the C-S-H (I) gel.

Changes in the C-S-H gel take place as belite begins to take part in the hydration reaction. Belite (Ca_2SiO_4) becomes a major contributor to the strength after several weeks. The structure of C-S-H begins to crystallize as an imperfect form of the mineral jennite, called C-S-H (II). As the setting process proceeds, the C-S-H gel changes from a

Fig. 21.9 Crystal structure of tobermorite, showing silicate tetrahedra in dark gray and calcium polyhedra in light gray. The space between the Ca–O layers is filled with additional CaO polyhedral and additional water molecules, giving an overall composition close to $Ca_5Si_6O_{16}(OH)_2 \cdot 7H_2O$.

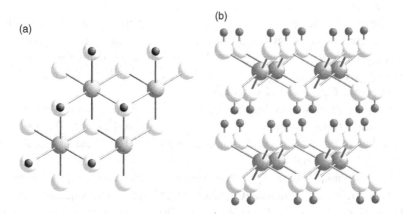

Fig. 21.10 Structure of portlandite, $Ca(OH)_2$: (a) looking down the c axis, perpendicular to the layers, and (b) viewing approximately parallel to the layers. Ca is shown in light gray, oxygen in white, and H is the smallest atom.

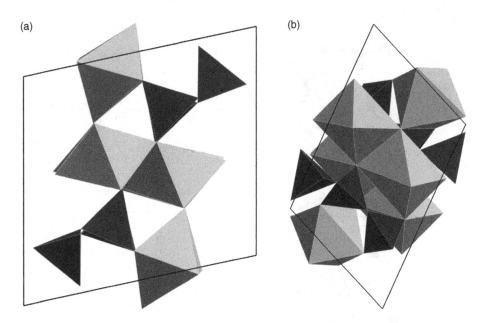

Fig. 21.11 Crystal structure of jennite showing silicate tetrahedra (dark) and Ca octahedra in light gray (a) view along [010], (b) view along [001].

mixture of CH and poorly crystallized tobermorite to a mixture of C-S-H (I) and C-S-H (II), and eventually ages to C-S-H (II) with a structure resembling jennite.

The crystal structure of jennite $Ca_9(Si_6O_{18})(OH)_6 \cdot 8H_2O$, is illustrated in Fig. 21.11 using projections along [010] and [001]. It is built up from three different modules: (1) ribbons of edge-sharing calcium octahedra, (2) chains of corner-sharing silica tetrahedra, and (3) interlayer linkages of calcium octahedra and hydrogen-bonded water molecules. The average Si–O and Ca–O bond lengths are 1.63 and 2.39 Å, respectively.

Like tobermorite, jennite is another crystalline calcium silicate hydrate with drierketten silicate chains, but it has a much higher Ca/Si ratio near 1.5, closer to that of commercial Portland cements (0.7 to 2.3). As with tobermorite, the jennite structure is based on a Ca–O layer flanked on both sides by silicate drierketten and linked together by interlayer Ca^{2+} ions and water molecules.

The principal difference between the tobermorite and jennite structures is that half the drierketten chains in jennite are replaced with hydroxyl groups. This causes a strong puckering of the calcium octahedral layer in jennite. This is the key component of the strong robust material we know as hardened cement.

The three minor phases in Portland cement clinker also play a role in the hydration process. Tricalcium aluminate ($Ca_3Al_2O_6 = C_3A$) hydrates very quickly (Fig. 21.8), per the reaction: $Ca_3Al_2O_6 + 6H_2O \longrightarrow Ca_3Al_2(OH)_{12}$. This leads to rapid crystallization, rapid volume expansion, and a false set with considerable volume expansion, but little improvement in strength. The setting rates of the aluminate and ferrite phases in Portland cement are controlled by adding gypsum to the cement.

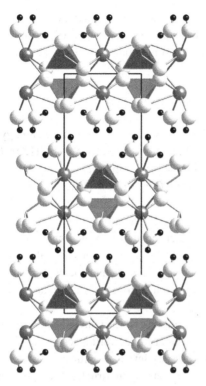

Fig. 21.12 Crystal structure of gypsum showing the CaSO₄ layers held together by water molecules. The sulfate tetrahedra are shown. O is the large light sphere, Ca is the large darker sphere, and H is the smallest darkest sphere.

Large deposits of gypsum are found throughout the world. Gypsum ($CaSO_4 \cdot 2H_2O$ = $C\bar{S}H_2$) and its dehydrated form anhydrite ($CaSO_4$ = $C\bar{S}$) are classified as evaporites, a large family of minerals deposited during the evaporation of seawater. The gypsum added to cement affects the setting time, the strength development, and the volume stability.

Gypsum is an important example of a crystalline hydrate. It is widely used in the building trade as plaster and as an additive in Portland cement to control the setting rate. Gypsum has a rather complex layer structure in which the layers of calcium sulfate are bound together by hydrogen bonds from water molecules (Fig. 21.12). Each water molecule links oxygen atoms belonging to sulfate groups in adjacent layers. O–HO bonds are easily ruptured, leading to easy cleavage parallel to (010) planes. The sulfur in gypsum, like the carbon in calcite, are examples of small, highly charged cations that form soluble complex anions.

When heated to about 120 °C, gypsum undergoes partial dehydration to the hemihydrate $CaSO_4 \cdot 0.5H_2O$ known as plaster of Paris (named for the gypsum deposits in the Montmarte district of Paris). Powdered plaster of Paris rehydrates very quickly when mixed with water. On further heating to 300 °C, the hemihydrate is converted to powdered anhydrite ($CaSO_4$), which also hydrates rapidly in water mixtures. Mixtures of the hemihydrate and anhydrite react with water to form the original dehydrate. Fillers such as wood pulp or sand are added to improve the setting rate and durability.

When added to Portland cement, gypsum reacts with the aluminate and ferrite phases to slow the hydration reaction and allow the alite and belite phases to control the setting and hardening process. The sulfate phases formed during hydration are known as AFt and AFm.

AFt phase

Ettringite is a product of the reaction between the aluminate clinker phase, gypsum, and water. The chemical composition is $[Ca_3Al(OH)_6 \cdot 12H_2O]_2 (SO_4)_3 \cdot 2H_2O = C_6A\bar{S}_3H_{32}$.

The crystal structure of ettringite consists of complex chains of hydrated calcium aluminate link together by sulfate tetrahedra (Fig. 21.13). The chains are arranged in a hexagonal array running parallel to the trigonal c axis. Within the chains, Al^{3+} ions are bonded to six hydroxyl groups. The $Al(OH)_6$ octahedra are in turn bonded together by three calcium ions surrounded by four water molecules located on the outer surface of the calcium aluminate chains. The positively charged columns are balanced by $(SO_4)^{2-}$ tetrahedra. The sulfate tetrahedra are located in the channels between the chains, along with additional water molecules. Hydrogen bonds from the water molecules provide the interchain linkages.

In keeping with the crystal structure, ettringite crystals tend to be long slender prisms, in which elongated hexagonal crystallites of ettringite form needles parallel to calcium aluminate columns. These needle-like crystals interlock with one another to provide initial strength to the setting cement. The ettringite structure accepts a number of substituents. The sulfate ions are sometimes replaced by hydroxyl, carbonate or chlorine anions; Fe^{3+} replaces Al^{3+} in the octahedral sites. The more general designation AFt is used for the impure ettringite phase formed in common cements.

Ettringite coats the particles, retarding the hydration, regulating the temperature, and slowing the setting process.

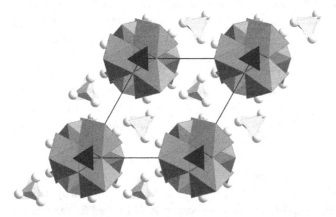

Fig. 21.13 Crystal structure of ettringite: Ca, S and Al polyhedra are shown in various tones of gray. The hydrogens are not shown for simplicity. The central calcium aluminate chains consist of $Al(OH)_6$ octahedra held together by 8-coordinated Ca^{2+} ions. The water molecules in the outer part of the chains (shown here as isolated oxygen atoms) form hydrogen bonds to the sulfate groups between chains.

AFm phase

Reactions between gypsum, water, and the aluminate and ferrite clinker phases also lead to another group of hydrated calcium sulfoaluminates known as the AFm compounds. These are layer structures (Fig. 21.14) with a representative chemical formula of $[Ca_4Al_{2-x}Fe_xO_7](SO_3) \cdot 12H_2O = C_4A\bar{S}H_{12}$.

The structure consists of layers of 7-coordinated Ca^{2+} ions and 6-coordinated Al^{3+} and Fe^{3+} ions. One of the oxygen ions bonded to Ca^{2+} protrudes from the layer and is linked to the water molecules and sulfate groups between layers.

To conclude this section, we consider the structure–property relationships that control the reactivity with water. The ability of a material to act as a hydraulic cement depends on two factors. First, it must react with water at a reasonably rapid rate. Most oxides do not, and have little or no value as a cement. Second, the resulting hydrate phase must have low solubility and adequate mechanical strength. Tricalcium silicate, the principal component of Portland cement clinker, satisfies these criteria, and is a fine hydraulic cement.

An understanding of the hydraulic reaction must be based on our knowledge of the crystal chemistry of the oxide and the molecular nature of water. The key step in hydration is the transfer of protons from the water molecules to the solid. Water is acting as a Brønsted acid in the reaction. The oxygen atoms subject to proton attack are those deficient in strong bonding to nearby cations. Any structural feature that draws electrons away from the oxygen ion in question renders it less reactive with protons.

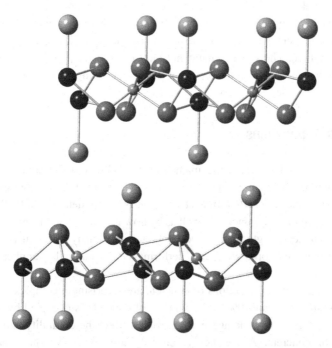

Fig. 21.14 Portion of the crystal structure of an AFm phase. The large dark ions are Ca, the smaller octahedrally coordinated ions are Al^{3+} or Fe^{3+}. Oxygen and OH^- groups are shown in the lighter large spheres. The spaces in between layers are filled by complex anions and water molecules.

This hypothesis explains the following observations regarding the hydraulic behavior of Portland cement.

(1) Tricalcium silicate (Ca_3SiO_5 = C_3S = alite) is more reactive than dicalcium silicate (Ca_2SiO_4 = C_2S = belite). One of every five oxygens in alite is bonded only to calcium, whereas all those in belite are also bonded to one silicon. Lime (CaO = C) is even more reactive than alite because all the oxygens are bonded to calcium, and wollastonite ($CaSiO_3$ = CS) is even less reactive than belite because most of the oxygens are bonded to two silicons.

(2) Oxygen atoms in certain calcium aluminates are more reactive than those in calcium silicates. The oxygen linked to two $(AlO_4)^{3-}$ tetrahedra are especially susceptible to proton attack. The oxygen in the –Al–O–Al– bond is deficient in positive bond strength, making it an attractive target for protons. In feldspars and other aluminum-bearing tetrahedral network structures, this underbonding causes the Al-avoidance principle. $CaAl_2O_4$, $Ca_{12}Al_7O_{13}$, and $Ca_3Al_2O_6$ are all highly reactive.

(3) The silicate chains in hardened cement are one of the reasons for its strength and durability. Alite and belite, the calcium silicate phases in clinker, do not contain silicate chains. Their structures are made up of isolated $(SiO_4)^{4-}$ tetrahedra held together by Ca^{2+} ions. But when mixed with water and recrystallized in the tobermorite and jennite structures they form single-chain $(SiO_3)^{2-}$ silicates like the pyroxene family. It is well known that some of the toughest polycrystalline minerals such as jadeite and nephrite are chain silicates. Other related materials are the "unbreakable" glass-ceramics manufactured commercially. Several of these products contain chain silicate crystallites embedded in a glass matrix. Since hardened cement is formed from an amorphous C-S-H gel, there is a strong similarity here. The silicate chains not only make cement and concrete stronger and tougher but also play an important role in enhancing the chemical durability.

21.5   Engineering Applications

Concrete, asphalt, and steel are the mainstays of highway and bridge construction throughout the world. Concrete is a mixture of Portland cement, sand, aggregate (gravel or crushed stone), and water. Performance characteristics are determined by the proportions and quality of the components as well as by how it is mixed and formed. Steel may be added for tensile strength (reinforced concrete), and many different additives are used to improve the workability and performance in particular applications and conditions.

The underlying chemical reactions of concrete are surprisingly complex and depend on the type of stone. About 70% of the volume of concrete is occupied by aggregate, thus it is not surprising that the aggregate particles affect the durability and structural performance of concrete. Aggregate is generally less expensive than cement, but economy is not the only motivation for its use; common stones confer higher stability and better durability than hydrate cement paste alone, and are less subject to chemical attack.

The size of aggregate used in concrete generally ranges from 0.1 to 10 mm. Many types of rock can be used as fillers, but there are a few deleterious effects that must be avoided. Coal and other organic matter in the aggregate can lead to changes in pH which interfere with the hydration reactions. Clay and other fine-grained filler may form surface coatings that interfere with the bond between the cement paste and the aggregate grains. Salt contamination is a problem with sands dredged from the ocean estuaries and deserts. Apart from the danger of corrosion of steel reinforcement, salt can react with moisture causing unsightly efflorescence. Similar problems can be caused by salts used for deicing. Pyrite (FeS_2) and other sulfide minerals react with water and oxygen to form sulfuric acid that can attack the hydrated cement paste.

Aggregates must have sufficient strength to withstand the volume changes associated with the external loads and with freezing and thawing. Internal pores and excess surface roughness can lead to problems.

Concrete is the most heavily used material in the world. Worldwide, concrete construction annually consumes about 1.6 billion tons of cement, 10 billion tons of sand and crushed stone, and 1 billion tons of water. Given transportation costs, there is a huge financial incentive for using local sources of stone, even if the properties are less than ideal. Concrete is not a homogeneous material and an unlimited number of combinations and permutations are possible.

Compared with other building materials, concrete has the advantages of low cost, castable shapes, and fire resistance, but it lacks tensile strength and ductility. Steel reinforcement greatly improves the mechanical properties. Concrete is about ten times stronger under compression than in tension. Surface flaws leading to fracture or chemical attack are more difficult to form or grow under compression.

Steel reinforcing cables are capable of keeping the concrete in compression. This is achieved by placing the cables under tension while allowing the concrete to set around them. When the concrete is sufficiently hard, the tensile stress is removed to make prestressed concrete. The compressed concrete also protects the steel from corrosion. The chemistry of steel (Chapter 10) is better understood than the chemistry of asphalt and concrete, but it too is a material with different chemical compositions and properties. Toughness, strength, weldability, and corrosion resistance are four of the performance characteristics that vary with the type of steel alloy.

Greater flexibility in highway design and construction have been achieved using new types of specialty concretes. Compressive strengths of 10 000 psi (= 68.9 MPa) are typical in high-performance concretes. Ultra-high-performance concretes with compressive strengths of 30 000 psi (207 MPa) and much greater durability are made from formulations that include silica fume, water reducers, and steel or polymer fibers. Specialty concretes for use in cold weather utilize admixtures that depress the freezing point of water.

Highway construction has a long history of using byproduct materials and industrial waste to reduce costs and improve properties. Recycling construction materials makes economic and environmental sense. Two good examples are the "crumb rubber" fillers made from old automobile tires, and the "fly ash" residue from coal-burning power plants.

21.6   Summary

The common inorganic cements used in constructing roads and buildings are derived from limestone mixtures heated to 1450 °C. These finely divided oxides react with water to form a plastic mass that solidifies at room temperature. The solidification process can be divided into "setting" and "hardening" stages that denote the first appearance of rigidity, and the subsequent increase of hardness and strength that occurs with time.

Several criteria are involved in selecting a good water-based cement. First, the cement powder must be unstable in the presence of water, entering into solution with relative ease, and reprecipitating as a strong insoluble hydrate. The cementing process is generally accompanied by substantial volume changes. Generally the hydrated product has a larger volume than the dehydrated starting material, but a smaller volume than the total volume occupied by the dry powder and the water involved in hydration process. This ensures that the paste will undergo a moderate shrinkage during setting and the increased volume of the hydrated product will be accommodated in the pore space initially occupied by water. If powdered lime is made into a paste, each particle hydrates so quickly that the resulting crystallites are unable to expand into the pore space, but instead press against adjacent crystallites. Consequently there is a rapid expansion during the setting process, opening up flaws, and making strong cohesion impossible.

Therefore, the second criterion for a satisfactory cement is that hydration be sufficiently slow to prevent large expansion in the initial stages. The calcium silicates used in Portland cement hydrate at a modest rate but the tricalcium aluminate phase can be a problem requiring chemical modification.

A third criterion is referred to as "soundness," which means that after setting and hardening takes place, no further hydration can occur, otherwise deleterious internal stress can cause structural flaws and disintegration. It is important that unreacted phases be properly sealed inside the hardened hydrated cement. Some engineers consider this property to be the main requirement of a good hydraulic cement. Related to this criterion is a fourth condition, which states that the final hardened cement must be highly insoluble in water. This ensures long-term stability and enhances soundness by reducing permeability. Insolubility in these hydrating systems is enhanced by incorporating aluminum in the gels. C-A-S-H is less soluble than C-S-H.

The sources of strength and toughness in cement and other inorganic solids have led to considerable discussion and controversy. There are several mechanisms that may improve the mechanical properties of cement. Plaster of Paris derives its strength from a felted mass of needle-like gypsum crystals. The toughness of hydrated Portland cement is similar to other chain silicates, like the two varieties of jade. The silica chains in tobermorite and jennite resemble those in nephrite and jadeite (see Chapter 30).

The main sources of weakness in cement are internal cracks and cavities that develop during improper processing and it is here that calcium silicate gels play a role. The colloidal C-S-H gels coat the original clinker particles, filling pore space, sealing off unreacted material, and creating a dense hydrate with very small crystallite size. Contemporary scientists might refer to the colloidal structure as a natural nanocomposite.

21.7   Problems

(1) The crystal structure of portlandite has space group $P\bar{3}m$, with $a = b = 3.5918$ Å, and $c = 4.9063$ Å. The Ca are at positions $(0, 0, 0)$; O are at $(1/3, 2/3, 0.2341)$; H are at $(1/3, 2/3, 0.4248)$.

    (a) Draw the crystal structure of this compound, showing the Ca–O and O–H bonds.

    (b) Compare the volume occupied per calcium atom for CaO and portlandite. Recall that CaO has the rocksalt crystal structure, with $a = 4.8105$ Å.

    (c) Explain the importance of your answer in part (b) to the setting of cement.

(2) Describe the progression of phases that develop in Portland cement.

22 Surfaces and Surface Properties

22.0 Introduction

The chemical and structural nature of surfaces is described in this chapter. Special emphasis is placed on the surface chemistry of ore minerals, and the use of surfaces for catalysis. Very perfect surface structures are required for growth of epitaxial thin films, where the crystal structure of the growing films templates from the crystal structure of the underlying substrate. Surfaces can also degrade over time on exposure to the atmosphere, through the process of weathering.

22.1 Surface Structure and Surface Energy

Surface phenomena are important in a variety of fields, including heterogeneous catalysis, corrosion, surface functionalization, biological systems, and many solid state devices. Crystallographic relationships between substrate structure and surface structure can be determined by techniques such as low-energy electron diffraction (LEED). Ideally, surfaces are periodic in two directions. Because the surface may not be actually two dimensional (e.g. the surface can be corrugated), this surface translational symmetry is *diperiodic*. The symmetry conforms to one of the 10 diperiodic point groups.

Surfaces differ in numerous ways from the bulk of the material. Because surface atoms are undercoordinated (i.e. they have fewer neighbors than they would in the bulk of the material), the atoms often displace from their expected positions. Sometimes "extra" bonds are formed with other surface atoms in order to reduce the number of dangling bonds. A good example of this is given by the reconstruction of clean Si (100) surfaces. In the ideal case, each Si atom on a (100) surface has only two out of four of its neighbors. This is energetically expensive. To reduce the number of dangling bonds, the surface atoms move closer together in rows, so that they can form bonds with each other, as shown in Fig. 22.1. This reduces the energy of the surface atoms, and simultaneously changes the surface periodicity. The reconstructions are named for the periodicity that is formed on the surface layer, relative to the periodicity of the original lattice. The simplest surface unit cell should be periodic with characteristic repeat distances related to the bulk unit cell. In this simple case, for a (001) surface, the periodicity is a single unit cell distance of a and b along the x and y axes respectively. This termination is referred to as a 1×1 reconstruction. If, however, the surface structure changed to a repeat pattern of $2a$ and $2b$, in the principle surface directions, this is a 2×2 reconstruction. Figure 22.1 shows the 2×1 reconstruction of the Si (100) surface.

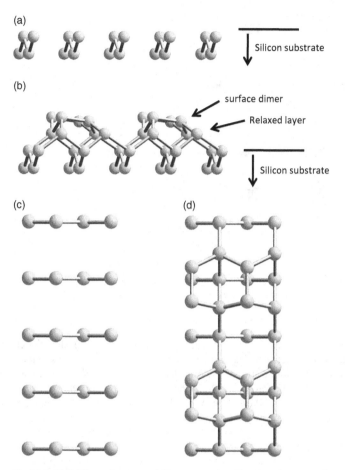

Fig. 22.1 Si (100) surfaces would have two dangling bonds per surface atom if no atomic rearrangement occurred. This is show from the side in part (a), and from the top in part (c). Electron diffraction experiments indicate, however, that the true surface is partially reconstructed with double bond formation – parts (b) and (d). Locally, the surface symmetry is 2*mm* rather than 4*mm*. Figures drawn from data in P. R. Watson, M. A. Van Hove, and K. Hermann, *Journal of Physical and Chemical Reference Data, Monograph No. 5 Atlas of Surface Structures, Vol. 1A*, American Chemical Society and American Institute of Physics, Woodbury, New York (1995).

Another way in which surfaces can differ from bulk materials is in chemical reactivity. The III–V semiconductor compounds show interesting electrochemical behavior on the {111} faces. There are two types of these faces: (111)A faces in which the trivalent ions are on the outermost surface, and (111)B faces with group V elements on the outermost surface. On (111)A faces, each surface atom forms three pair bonds to neighboring atoms. The surface atoms for (111)B also form three pair bonds but have two "dangling" unshared electrons remaining. Experimentally, it is found that the (111)B surfaces have higher reactivity in oxidizing solutions, a fact attributed to the dangling electron pair.

Surfaces can also have different compositions from the bulk of the material. As shown in Fig. 22.2, the (111) surface terminated by In reconstructs by eliminating 1/4 of the In atoms. This drops the coordination of the surface atoms from four to three.

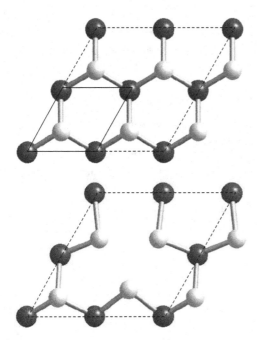

Fig. 22.2 Top views of the indium (111) surface in InSb. (Top) The ideal structure. (Bottom) The reconstructed surface involves removal of some In atoms, with movement of other atoms. In the figure, In is shown in dark spheres, and Sb in the lighter spheres. Figure redrawn from R. E. Newnham, *Properties of Materials: Anisotropy, Symmetry, Structure*, Oxford University Press, New York (2005). Fig. 32.5, p. 360. Adapted with permission by Oxford University Press.

Surface terminations depend greatly on the environment. Most surfaces left exposed to air will adsorb gaseous species, reducing the number of dangling bonds. Indeed, removal of adsorbed layers can be quite challenging, and may require high temperature heat treatments under ultrahigh vacuum.

In many cases, chemisorbed layers on single crystal surfaces form ordered structures. The surface structure depends on the symmetry of the substrate and on the size and chemical nature of the adsorbed gas species. Chemisorbed molecules tend to form close-packed surface structures with small unit cells showing the same rotational symmetry as the substrate.[1]

Surface reconstructions on clean metals are much more subtle than those shown for Si. This is probably a function of the non-directional metallic bonding, as well as the ability to redistribute the electron "sea" near the surface. Cu (100) surfaces, for example, are reported to relax in the outer two layers, without undergoing major changes in bonding. This is shown in Fig 22.3. The outermost surface layer shows a slightly smaller interplanar distance than in the bulk of the material, while the layer

[1] G. A. Somorjai and F. J. Szalkowski, *J. Phys. Chem.*, 54, 389 (1971).

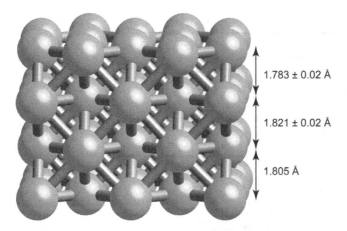

1.783 ± 0.02 Å

1.821 ± 0.02 Å

1.805 Å

Fig. 22.3 Side view of the relaxation of the outer two layers of the Cu (100) surface. The separation distance between adjacent planes is 1.805 Å in the bulk of the crystal.

immediately underneath it shows a slightly larger separation distance than the bulk of the crystal. The perturbations are modest, and are on the order of a percent of the normal interplanar spacings. Similar small contractions of the outermost layer atoms towards the bulk of the crystal have been reported for Cu (111) surfaces. The atom displacements can change as atoms are adsorbed onto the surface but, again, the changes are modest.

In other metals, such as W, the (100) surface has been reported to show a lower symmetry under certain surface preparation conditions, in which the atoms in the top layer may displace in the plane, as well as produce a buckled out-of-plane distortion.

In ionically bonded compounds, surface reconstructions and surface relaxations depend on both how well bonded the surface atoms are, and the charge neutrality on the surface. For example, in the case of the MgO (100) surface, the Mg atoms may displace in towards the bulk by a fraction of an angstrom (various reports list this as no displacement to 0.105 ± 0.05 Å). This leaves the more polarizable O^{2-} ion more exposed. As shown in Fig. 22.4, the measured changes in surface structure occur within the top one or two atomic planes, and produce a very small change in the surface. Numerous different surface reconstructions are reported for $SrTiO_3$, depending on the temperature, the surface termination, and the surface preparation conditions.

The crystallography of numerous surfaces, with and without adsorbed layers, is reported by Watson, Van Hove, and Hermann.[2]

Surface atoms are missing about half the chemical bonds, so that the *surface energy* is about one-half of the energy required to break all bonds. This is equivalent to half the latent heat of vaporization for one molecular layer, or about 1 J/m^2 for many engineering materials. The observed surface energies for several metals are listed in Table 22.1. Note that the energies are comparable to those of non metals, and correlate well with the melting points.

[2] P. R. Watson, M. A. Van Hove, and J. Hermann, *Atlas of Surface Structures, Vol. 1A and 1B*, American Chemical Society and American Institute of Physics, Woodbury, New York (1995).

Table 22.1 Surface energies for metals

Metal	Melting point (K)	Surface energy (J/m^2)
In	429	0.45
Pb	600	0.5
Au	1331	1.3
Fe	1810	2.0
Pt	2042	2.1
W	3653	2.9

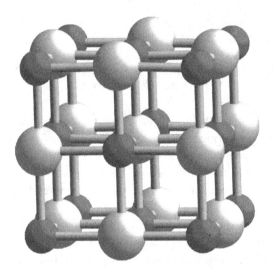

Fig. 22.4 One of the reported surface reconstructions of MgO (100). The surface is at the top of the figure, and the Mg atoms are shown displaced slightly towards the bulk in the top layer, and slightly towards the surface in the second.

The surface of a solid can be considered a defect in that the bulk structure is altered in the outermost layers. There are several types of changes that lower the surface energy: (1) polarization of surface ions, (2) rearrangement of surface atoms, (3) stoichiometry changes at the surface, and (4) chemisorption of foreign atoms and molecules.

In inorganic solids, anion polarizabilities generally exceed those of the cations. Since surface anions are subject to asymmetric electric fields, the electron clouds are pulled toward the interior of the crystal, producing a dipole on the anion and thereby decreasing its electrostatic energy. Surface energy decreases as polarizability increases. The measured surface energy of CaF_2 is 2.5 J/m^2. Replacing Ca^{2+} with the more polarizable Pb^{2+} ion reduces the energy to 0.9 J/m^2 for PbF_2. PbI_2 has one of the lowest surface energies known, 0.13 J/m^2.

Surface energy is important in nucleation. When it first forms, a crystal nucleus is entirely surface, and the environment must provide the surface energy. Supersaturation increases with surface energy, and therefore decreases with polarizability. Iodides, bromides, sulfides, and compounds containing heavy metals do not form supersaturated solutions, whereas solutions of CaF_2 in water often acquire a supersaturation of 500% before forming crystals.

Electrical neutrality is ordinarily taken for granted, but small departures may reduce the surface energy. In addition, it is common for the concentration of point defects (see Chapter 16) to be higher near a surface. When the outer surface consists of highly polarizable anions, highly charged cations are screened, lowering the electrostatic energy. Nonstoichiometry is common in tiny crystals and of great importance in colloid chemistry.

As was described above, the structure of a surface sometimes differs from that of the bulk. Surface energy is lowered when more polarizable ions move towards the surface, and less polarizable ions towards the bulk; this can give rise to a surface dipole layer. Many oxide surfaces are good examples of this, as the surface properties are often dominated by the more polarizable oxygen ion. Surface energy can also be lowered through the attraction of foreign substances, especially anions. The surface of freshly fractured quartz chemisorbs oxygen molecules, producing ozone in the process. Partly because of its reactivity, quartz dust causes silicosis of the lungs.

22.2   Wetting and Adhesion

Among the properties that are strongly dependent on the surface is the wetting of a liquid on a solid, or the adhesion between dissimilar materials. Wetting describes how readily a liquid spreads out on a solid surface. Wettability is dependent on a number of factors, the most critical being the surface tension of the liquid and the surface energy of the solid. Surface wetting is measured quantitatively from the contact angle θ, as shown in Fig. 22.5.

A useful rule of thumb is that "like wets like." That is, the more similar the bonding within the liquid is to the bonds in the solid, the more likely that the liquid will spread out on the solid surface. Thus, ionic liquids like water wet the surface of most oxide minerals and ceramics, while covalent liquids such as kerosene wet covalent solids like sulfides. Water molecules are attracted to the surface oxygens through the formation of hydrogen bonds. Water wets the surface of silicates very well because of the similarity in Si–O and H–O bonds. This difference in surface wetting has been used for many years in recovering sulfide ore minerals from silicates. Other good examples of

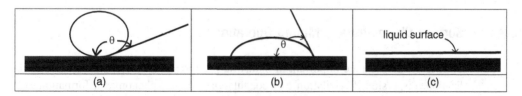

Fig. 22.5 Surface wetting is quantified by the contact angle θ. A drop of liquid is placed on the surface and θ is measured tangentially with a microscope. Schematic illustrations of wetting, (a) non-wetting, (b) partial wetting, and (c) wetting are shown. Ionic liquids such as water wet ionically bonded solids, whereas covalent liquids tend to wet covalent solids.

non-wetting situations include mercury droplets on glass, water on a waxed car, or water on a lotus leaf; note that in this latter example, physical structure at the nanoscale plays an important role.

It is critical to note that it is the bonding *at the surface* of the solid that will govern wetting. Thus, for example, a clean Si surface will not be wet by water, while a Si surface that has been oxidized to SiO_2 will be wet by water. Many metal surfaces also oxidize in air, and so will be wet by water. This is critical in corrosion of metals, as the corrosion process can be slowed down if water is repelled by the surface.

Adhesion is seldom a problem in bonding to clean metal surfaces, because the surface energies are high. These high energies can in fact be problematic in terms of keeping the metal surfaces clean.

In much the same way that the "like wets like" rule describes the wetting between a metal and a liquid, a propensity for strong bonding across solid interfaces leads to good adhesion between two different materials. It is imperative that adhesion promoters also form strong bonds to both surfaces. There are numerous cases in which electronic components require metal electrodes in intimate contact with oxide layers – this situation is promoted when the two materials wet. Several criteria have been identified which encourage good bonding between the surfaces in this case.

(1)   Wetting is favored when the metal has low electron density and the oxide has a small band gap.

(2)   Trace impurities promote wetting when the impurity reacts with the oxide substrate. An excellent example of this is the use of metals like titanium or chrome to promote adhesion of non-reactive metals like gold to glass. Ti is an "oxygen-friendly" metal, meaning that it readily oxidizes, even if it has to remove oxygen from the underlying oxide to do so. Thus, when Ti is deposited from the vapor phase (for example by evaporation or sputtering) onto an oxide surface it can scavenge a small amount of O from the substrate, forming a thin suboxide that bonds well to the underlying material. If Pt is deposited sequentially, without breaking vacuum, onto this Ti-coated oxide surface, it will have significantly better adhesion than a single layer of Pt on the same surface since Pt bonds to the Ti very well. In a similar manner, the wetting of oxides is further enhanced by adding small amounts of oxygen friendly metals like Ti to the liquid metal.

(3)   Liquid metals tend to wet non-metal substrates with a refractive index exceeding 2.2. Many metals wet the surface of diamond ($n = 2.4$) and rutile ($n = 2.7$).

22.3   Surface Chemistry and Mining Operations

Surface phenomena are important in industrial processes such as precipitation, crystallization, dispersion, flocculation and coagulation, wetting phenomena, formation and breaking of emulsions and foams, detergency, lubrication, and corrosion inhibition. Examples in the petroleum industry include oil recovery and catalytic cracking, and in metallurgy, reduction of iron ores in blast furnaces and stress corrosion. All heterogeneous reactions involve adsorption and desorption steps that can be controlled electrochemically or by surfactants (organic molecules containing both polar and non-polar groups).

Surfactants are best known as soaps, detergents, and insecticides, but they are also utilized in separating minerals by the froth flotation process. The idea behind this approach is relatively simple: if air bubbles are blown through oil, to create a froth in water, then the bubbles will adhere to hydrophobic surfaces. Consider an ore that contains a covalently bonded mineral (such as diamond or sulfide minerals like chalcopyrite) and other ionically bonded rocks (including most oxides such as quartz or calcite). Minerals with ionic bonding are generally hydrophilic; those with covalent or metallic bonding are hydrophobic. After a sequence of crushing and grinding operations, the fine-grained ore minerals are transferred to a series of flotation cells for separation. Hydrophobic materials attach themselves to air bubbles, and can be floated to the surface. In contrast, hydrophilic materials are incapable of attachment since they are fully water wetted.

In some cases, regulating agents (H^+, OH^-, HS^-, CN^-, HCO_3^-) can be used to prepare mineral-particle surfaces for subsequent absorption of a collector (C_2–C_6 dithicarbon-ates, C_8–C_{18} carboxylates or amines), making the mineral hydrophobic. Using different regulating agents and collectors, selective separation of many different minerals becomes possible. Even wurtzite and sphalerite, the polymorphic forms of ZnS, can be separated by flotation. Frothers are added to the water to aid in bubble formation and in the attachment of solid particles. Surfactants such as C_5–C_{10} alcohols or polyoxyethylenes are excellent frothing agents. Mineral particles are attached to bubbles in milliseconds through the strong molecular interactions between the film and the frother on the air bubbles and the collector absorbed on the solid particle. The buoyancy of the bubble lifts the particle to the top of the flotation cell, where it is skimmed off and laundered to form the concentrated ore.

Mining engineers are well aware of the intense activity in chemical reagents for treating a wide variety of ores and recycled products. The billion-dollar markets are served by hundreds of chemical firms. Because of the depletion of high-grade ores, there is an increased need for developing selective chemicals to economically concentrate the ore minerals. Industries based on copper, aluminum, iron, coal, and precious metals have undergone remarkable growth during the past few decades. Some of the important reagents used as collectors, frothers, flocculants, solvent extraction reagents, depressants, and dispersants are discussed in the following paragraphs. All involve surface reactions that promote the separation of minerals, and the cleansing of effluents.

Sulfide ores such as chalcopyrite ($CuFeS_2$) make use of *thio collectors* such as ethyl or propyl xanthate:

$$R-O-C\underset{\diagdown S\ M^+}{\overset{\diagup\!\diagup S}{}}$$

These reagents consist of a polar functional group, hydrophilic in character, and a sulfur atom bonded to carbon. *Cation collectors* are used to float off minerals that have a negative surface charge. They are commonly used to process the oxide ores of iron, potash, and phosphates. These cation reagents generally contain a nitrogen group with unpaired electrons. An ether amine

$$R-O-C-C-C-NH_2$$

in liquid form is preferred when the flotation feed is very fine and good selectivity for silica is required. *Anionic collectors* are modified alkyl sulfonates developed originally to selectively concentrate iron-bearing minerals in low-grade iron ores. They are effective collectors for goethite (FeOOH), magnetite (Fe_3O_4), hematite (Fe_2O_3), siderite ($FeCO_3$), barite ($BaSO_4$), and chromite ($FeCr_2O_4$). *Fatty acid collectors* are typically oleic or linoleic acids in which the recovery rate of industrial minerals is adjusted by the length of the carbon chain. Non-polar oils are used as *hydrocarbon collectors* in the flotation of many types of minerals. Napthalene (C_nH_{2n}) and paraffins (C_nH_{2n+2}) act by coating hydrophobic particles and improving the adhesion to air bubbles. Hydrocarbon collectors are made from crude oil. Polyglycols, $R–(O–C_3H_6)_n–OH$, are strong surface-active *frothers* that maximize load support with coarse grinds in almost all pH ranges.

In addition to frothers and several types of collectors, there are a number of modifying reagents used in the processing of engineering materials. The most important of these chemicals are the *pH regulators* such as lime (CaO), brucite ($Mg(OH)_2$), sulfuric acid (H_2SO_4), and hydrochloric acid (HCl). These acidity regulators are cheaper than most of the processing chemicals, but are used in very high volumes.

Liquids containing suspended solids are encountered in the processing of most minerals and recycled products. *Dispersants* and *depressants* are two of the classes of reagents used to control the suspension process. *Dispersants* are used to prevent agglomeration and control slurry viscosity. *Depressants* are chemical additives that produce hydrophilic surfaces that cause particles to remain in the liquid, rather than attaching to air bubbles. Lime (CaO) is a depressant for pyrite (FeS_2) in copper plants.

High efficiencies in metallurgical processing are often achieved with polymer *flocculants* that cause coagulation. High molecular weight polyacrylamides are used to thicken ore pulps and concentrates.

Solvent *extraction* reagents are designed to selectively remove a metal species from one solution to another. Chelation extractants are organic molecules that bond to an individual metal ion and release hydrogen ions into the aqueous solution. Chelating surfactants are widely used for the recovery of copper from sulfuric acid leach solutions.

As the foregoing discussion makes clear, a number of surface phenomena are involved in meeting the world-wide demand for engineering materials.

22.4   Oxidation of Metals

The formation of oxide coatings on metals involves molecular dissociation, chemical reaction, and diffusion (Fig. 22.6). For example, at the surface of metallic iron, the metal's atoms dissociate into cation cores and electrons that diffuse separately through the oxide layer to the outer surface. At the oxide surface, the electrons interact with O_2 molecules to form O^{2-} anions, which in turn react with Fe^{2+} to form FeO. As the oxide film thickness grows, the diffusion distances increase, increasing the resistance to ionic

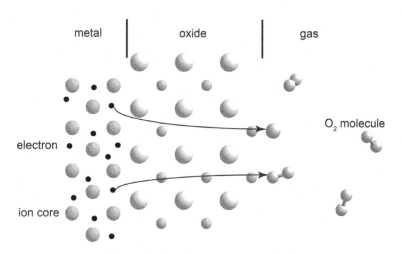

metal · oxide · gas

electron

ion core

O_2 molecule

Fig. 22.6 Schematic illustration of growth of an oxide on a metal surface.

and electric current flow, slowing the growth rate. The result is that the oxide layer thickness often grows parabolically with time.

The oxidation of iron is of great practical interest. Iron forms four oxides: FeO (rocksalt structure), Fe_3O_4 (spinel structure), γ-Fe_2O_3 (defect spinel), and α-Fe_2O_3 (corundum structure). Below 200 °C, the oxide film formed on Fe consists of a solid solution of Fe_3O_4 and γ-Fe_2O_3, with a partially ordered distribution of ferric (Fe^{3+}) and ferrous (Fe^{2+}) cations. Comparatively thin oxide layers are formed. At higher temperatures, α-Fe_2O_3 films form and the oxidation rates increase rapidly. Thick oxide scales limit the usefulness of iron and common carbon steels to about 550 °C. Above 600 °C, a very thick layer of FeO begins to grow.

The oxidation of metals sometimes involves epitaxial relationships between the metal and the oxide coating. *Epitaxy* occurs when there is a three-dimensional registry of the crystal structure of an underlying substate with the crystal structure of the film on top. In many cases, this is achieved when one or more of the sublattices of the two structures are well matched in size and geometric arrangement. For example, the (100) surface of BCC α-Fe and wustite (FeO) are compared in Fig. 22.7. The lattice parameter in FeO (4.28 Å) is similar to the Fe–Fe separation distance in the [011] direction of α-Fe (4.05 Å). This leads to an oriented overgrowth relationship between the unit cells, as shown in Fig. 22.7(c). The cube edges in FeO are rotated by 45° from those of the underlying metal. When the epitaxial layer is very thin, it can be strained coherently to the substrate. As its thickness grows, however, when there is a mismatch between the size of the film and the substrate, dislocations often nucleate to relieve the strain. This was discussed in Chapter 16.

The surface structures of oxide coating sometimes transform with temperature. For iron oxide, there is a change in oxidation state near 570 °C as the structure changes from wustite (FeO) to magnetite (Fe_3O_4). Magnetite has the spinel structure discussed in Chapter 29. The transformation at 570 °C is accomplished by omitting 25% of the Fe^{2+} ions and converting 50% of them to Fe^{3+}. The remaining 25% are shifted to tetrahedral sites. A similar epitaxial arrangement is possible with the underlying metal (Fig. 22.7(d)).

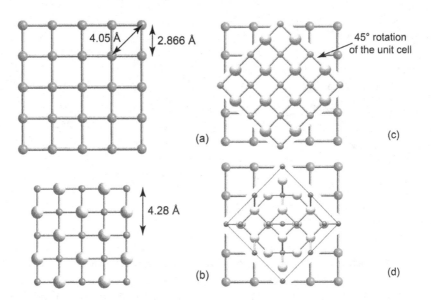

Fig. 22.7 (a) The (100) face of BCC Fe, (b) the (100) face of FeO, with the oxygen shown as the larger circles, (c) the epitaxial arrangement between BCC Fe and an oxide layer of FeO, and (d) the epitaxial arrangement of magnetite on Fe.

The ability of a metal to withstand high-temperature oxidation is of great engineering importance. A key feature of oxidation resistance is the formation of a stable oxide barrier. Spalling, a type of delamination phenomenon, may result from mechanical stresses in the oxide layer that develop during growth or as a result of differential thermal expansion between the oxide scale and the metal substrate.

When an oxide forms on the surface of a metal, the volume change is given by the Pilling–Bedworth ratio (PBR):

$$PBR = \frac{\dfrac{\text{Volume of the oxide}}{\text{Metal atom}}}{\dfrac{\text{Volume of the metal}}{\text{Metal atom}}}.$$

The specific volume of the oxide is rarely the same as the metal consumed by the oxidation process. The PBR value is used to predict the sign and magnitude of the growth stress. A compressive stress develops on the oxide when PBR > 1, and a tensile stress when PBR < 1. The larger the deviation from 1, the larger the growth stress.

Aluminum and sodium provide interesting examples. Aluminum is a representative of the metals that form a protective oxide layer. The volume, V, of one mole of the oxide is equal to the molecular weight, M, divided by the density, ρ. For Al_2O_3, $V = \frac{102 \text{ g/mole}}{3.987 \text{ g/cm}^3} = 25.58 \text{ cm}^3$. For Al metal, $V = \frac{27 \text{ g/mole}}{2.70 \text{ g/cm}^3} = 10.0 \text{ cm}^3$. Remembering that there are two moles of aluminum atoms in a mole of Al_2O_3, the PBR value is $\frac{25.58 \text{ cm}^3}{2 \times 10.0 \text{ cm}^3} = 1.28$. The oxide layer is therefore under compression, providing a non-porous protective coating. Such is not the case for alkali and alkaline-earth metals.

For sodium, the Na_2O/Na layer has a PBR of $\dfrac{\dfrac{61.98\ \mathrm{g/mole}}{2.27\ \mathrm{g/cm^3}}}{2 \times \dfrac{22.99\ \mathrm{g/mole}}{0.97\ \mathrm{g/cm^3}}} = 0.57$. The sodium monoxide layer is under severe tension, making it highly porous. As a result, the alkali and alkaline earth metals do not form a protective coating, and so oxidize easily.

From a practical point of view, adherence of the oxide scale is as important as imperviousness. The protective nature of aluminum oxide is enhanced by the continuous nature of the Al/Al_2O_3 interface, in which some Al atoms are part of *both* the oxide and the metal.

The rate of heating and cooling and the relative thermal expansion coefficients are also important. To be protective, the oxide coating must also be non-volatile and non-reactive with the chemical environment. Heat- and corrosion-resistant steels resist attack by air, water, and corrosive chemical solutions at moderate temperatures by virtue of their formulations. Chromium enhances passivity by developing a thin oxide film of Cr_2O_3 (corundum structure). Fe–Cr alloys containing about 11% Cr are used to make *stainless steel*.

22.5   Epitaxy and Topotaxy

The word "epitaxy" is Greek for "on-arrangement." This word is used by scientists and engineers to refer to the growth of a crystalline layer on a crystalline substrate where there is a three-dimensional registry of the crystal structures of the two layers. Perhaps the most industrially important use of epitaxy is in the growth of the multilayer stacks used for semiconductor lasers, such as $Al_{1-x}Ga_xAs$ on GaAs (see Fig. 22.8). In this case, both materials have the same structure (zinblende), but different refractive indices and band gaps. Stacks of this type can be used to confine both the electrons and the photons for high-efficiency, high-brightness light sources.

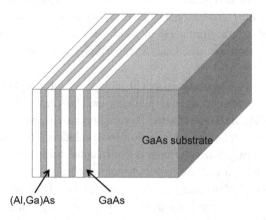

Fig. 22.8 A schematic of a multiple quantum well laser, showing a stack of epitaxial layers on a GaAs substrate.

A classic mineral example of epitaxy is the oriented overgrowth of staurolite on kyanite, in which the (010) plane of staurolite is parallel to (100) of kyanite, and the *c* axes of the crystals are collinear. A large number of factors affect epitaxy – temperature, adatom mobility, deposition rate, film thickness, atmosphere – some success has been achieved in relating crystallochemical factors to the oriented overgrowths. Lattice mismatch is one important criterion, and the tolerance for epitaxy has been established for some of the simpler structure types. For example, the allowed mismatch for alkali halides is about 15%, so that LiBr, NaCl, NaBr, LiI, NaI, KBr, and RbCl are in the right size range to form epitaxial layers on KCl, whereas LiCl and KF are too small, and KI and RbI are too large.

It is important to point out that it is not really lattice parameters that are important, but interatomic distances. In the epitaxial growth of some metal oxides, the oxygen sublattice is continuous across the film–substrate interface. For metallic films on alkali halides, epitaxy occurs when the metal atoms in the overgrowth have close registry to the halogen atoms in the substrate. High polarizabilities and low ionization potential also aid epitaxy in this case, so that metallic films are more likely to form epitaxial layers on NaCl than LiF.

Other epitaxial systems of considerable industrial importance include the oriented overgrowth of γ-Al_2O_3 on Al, which protects the metal from corrosion. In the semiconductor arena, silicon on sapphire and strained Si–Ge epitaxial layers are of current interest.

Epitaxy has been used to modify crystal habit. Glycine (CH_2NH_2COOH) matches the {110} faces of rocksalt, slowing their growth rate, and altering the stable form from cubic to dodecahedral. Epitaxial chemical conversion occurs in matlockite, PbFCl, and can be altered to epitaxial layers of PbF_2 or $PbCl_2$. Sulfate contamination of cryolite is a problem in aluminum manufacture. The contaminant proved to be Na_2SO_4 epitaxial overgrowths on Na_3AlF_6.

Topotaxy is the transformation of one crystal into another of different chemical composition without any major changes in axial directions. That is, the initial crystal acts as a template for the new one. The definition of topotaxy is sometimes broadened to include reactions in which the composition does not change: polymorphic transitions and exsolution phenomena may also be topotactic. Topotaxy appears to be widespread, and is a common method of transformation in solid state reactions. Many mineral polymorphs result from topotactic reactions. Pyrite (FeS_2) transforms to goethite (FeOOH) topotactically.

The practical significance of topotaxy is that it allows crystal morphology to be controlled. As an example, platelet $SrTiO_3$ particles can be prepared by reacting $Sr_3Ti_2O_7$ platelets with TiO_2. $Sr_3Ti_2O_7$ has the layered structure shown in Fig. 22.9, so it tends to adopt a platy morphology when grown from solution. This shape can be retained on conversion to $SrTiO_3$. Since the latter is from a cubic point group at elevated temperatures, platelet-like particles are difficult to generate directly.

A more beautiful example of topotaxy is the exsolution of TiO_2 from sapphire crystals. The host crystal structure is corundum, α-Al_2O_3. At high temperatures, Ti can substitute on the Al site, but as the temperature decreases, TiO_2 needles precipitate out with defined crystallographic orientations with respect to the sapphire. Because

Fig. 22.9 Crystal structure of $Sr_3Ti_2O_7$ showing the Ti octahedra, Sr (darker circles), and O (lighter circles). The octahedra form layers (horizontal in this diagram) that favor a platy morphology. This material can be topotactically converted to cubic $SrTiO_3$, retaining the platelet morphology.

the refractive index of TiO_2 is very high, light scatters effectively at the sappire-rutile interface. This produces the optical effect known as a star sapphire.

22.6   Catalysis

A catalyst is a substance that increases the rate of a chemical reaction without being consumed by the reaction. Since catalysts must come into contact with the constituent chemicals to promote reaction, their surface properties are of paramount performance. Catalysts are required for a wide range of reactions in the chemical, petroleum, and polymer industries. Catalysts are often prepared with very high surface areas, as the key reactions take place on internal or external surfaces. Catalysis is not a simple phenomenon, since it involves at least five important steps:

- diffusion to the active surface,
- chemisorption of the reactant molecules,
- surface reaction,
- desorption of the reactant molecules,
- diffusion of the products away from the surface.

The reaction rate can be controlled by any of these.

Catalysts can be used both to break apart and to synthesize molecules. Finely divided Ni, Pt, and W are all used as cracking catalysts. In a cracking reaction, C–C bonds of a hydrocarbon are ruptured, yielding products with a lower molecular weight. Catalysis is important in polymerization as well as decomposition. In the Fischer–Tropsch reaction, CO and H_2 react at an iron surface to produce hydrocarbons.

Clean metal surfaces are often highly reactive, so that chemisorbed gas molecules may decompose on forming chemical bonds to the surface. When molecular hydrogen

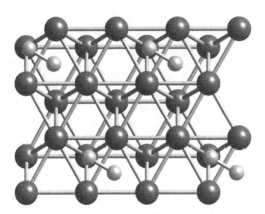

Fig. 22.10 Arrangement of C_2H_2 molecules (only C shown as small circles) on an Ni (111) surface. The view is straight down on the (111) surface.

is chemisorbed on a clean nickel surface, the molecule dissociates to yield a pair of hydrogen atoms bonded to the metal. In much the same way, nickel also catalyzes the decomposition of formic acid vapor, HCO_2H, to H_2 and CO_2. Experiments show that the reaction rate is independent of vapor pressure, from which it is concluded that rapid chemisorption maintains a monolayer of reactant species on the surface. Figure 22.10 shows that the adsorbed species (shown in the figure for ethylene) can form ordered arrays on the surface of the metal. Platinum group metals are superior to nickel and gold as catalysts for many reactions, due in part to their inertness. Precious metals like platinum, palladium, and rhodium are used as the catalysts in automobile catalytic converters. The mechanisms which make some materials better catalysts than others are complicated, and not always well understood.

Another important family of materials used both as catalysts and adsorbent hosts by the chemical and petrochemical industry are the zeolites. The name zeolite is Greek for "boiling stone," in reference to the fact that the mineral releases water as it is heated. Zeolites are a large family of low density three dimensional structures usually built from corner-shared $(SiO_4)^{4-}$ tetrahedra. Silicon can be replaced by a number of different elements, including Al, Ga, Ge, and P. The simplest of the zeolites is the mineral sodalite, shown in Fig. 22.11. With over 100 different structure types, the zeolites possess an interconnected interior pore space which is accessible through openings larger than 3 Å.

The pores and cavities within the zeolite structures have dimensions similar in size to many organic molecules such as paraffins, olefins, and aromatics. Its high selectivity for Ca^{2+} ions makes Linde A (Fig. 22.12) an excellent water softener. In this application, the cages are impregnated with Na^+ ions. When hard water, which is rich in Ca^{2+}, passes by the zeolite, ion exchange occurs due to the composition gradient. This reduces the Ca^{2+} concentration in the water, softening it. The eight-fold rings in Linde A have a diameter of 4.2 Å, through which molecules the size of CF_4 and CH_3I may pass, as well as smaller molecules like water.

The faujasite framework pictured in Fig. 22.13 is also industrially important. Faujasite zeolites are widely used as cracking and hydrocracking catalysts, including conversion of

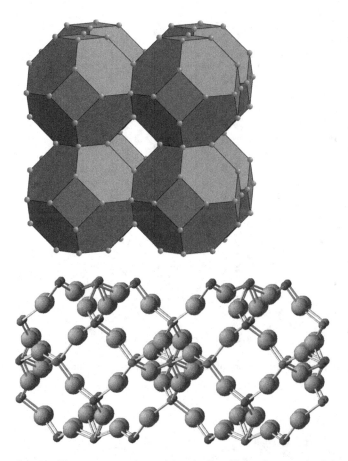

Fig. 22.11 Two representations of the aluminosilicate framework of the crystal structure of the sodalite zeolite. Si and Al atoms are in tetrahedral coordination with O to form a series of interconnected four and six rings. Large cavities are observed in between the framework units. These cavities can be filled by alkali or alkaline earth ions, or by small molecules such as water.

over a billion gallons of crude oil per day into fuels and chemicals. Faujasite has extremely large cages interconnected by 12-fold rings with a pore diameter of more than 9 Å, allowing passage of cyclohexane, naphthalene, and carbon tetrabromide.

The pentasil synthetic zeolites have revolutionized many petrochemical processes. The structure (Fig. 22.14) has an intermediate pore size that allows molecules with a diameter below about 5.5 Å to diffuse through the channels. This corresponds to the size of a benzene molecule. Long-chain hydrocarbons in petroleum can diffuse through the zeolites, but those with bulky side groups cannot. This allows the zeolites to be used as molecular sieves. Substitution of Al^{3+} for Si^{4+} provides control of the acidity over a wide range. A few of the many commercial applications of the pentasil zeolite ZSM-5 are listed in Table 22.2.

Because the aluminosilicate framework is inherently flexible in terms of connectivity, it is possible to design zeolites for particular applications. For the $NaAlSi_xO_{x+2}$ family zeolites, x can vary from 2 to infinity. Key engineering variables are as follows:

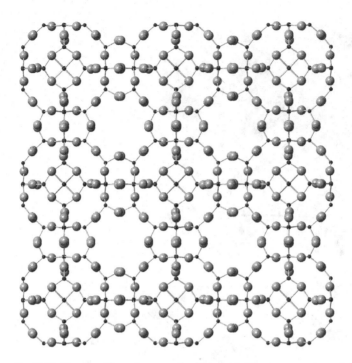

Fig. 22.12 Aluminosilicate framework of Linde A. The cages are bounded by 4-, 6-, and 8-rings.

Fig. 22.13 Tetrahedral framework structure of a faujasite zeolite. Large, interconnected cavities are visible.

(1)   The *framework structure* controls molecular sieving by setting geometric constraints on the size of the molecules that can pass through the interstices. The framework also controls catalytic transition states.

(2)   The *framework composition* is largely responsible for the stability at elevated temperatures or under hydrothermal conditions (Al > Ga > Fe > B), acid site

Table 22.2 ZSM-5 petrochemical catalytic processes[3]

Process	Product	Commercial use
Xylene isomerization	*p*-xylene	Polyesters
Toluene disproportionation	Benzene and xylenes	Polyesters
Benzene alkylation	Ethylbenzene to cumene	Styrene
Alcohols to olefins	Ethylene, etc.	Polyethylene, etc.
Alcohols to aromatics	Substituted benzene	High octane gasoline
C_2H_6, C_3H_8 aromatization	Benzene, toluene, etc.	Gasoline, chemicals
Aldehyde isomerization	Ketones	Solvents, chemicals
Toluene alkylation	*p*-ethyltoluene	High temperature polymers
Lube wax (C_{20+}) reduction	Lower C number paraffins	Diesel, jet fuel

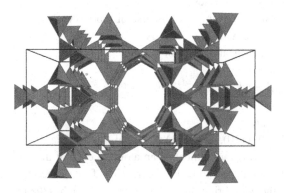

Fig. 22.14 Aluminosilicate framework of a pentasil zeolite.

strength, and active site density. Al can be partially replaced by other trivalent ions in the framework.

(3)   Sodium and other *exchangeable cations* control selectivity and the activity as catalysts and sorbents, including the acidity and basicity of the system.

(4)   *Transition metal additions* are important in bifunctional activity in which both the metal and the zeolite framework play important roles in the catalysis of chemical reactions.

As purification media, zeolites are used to remove sulfur compounds from hydrocarbon streams, as dessicants in the drying of natural gas, and in small-scale applications such as home refrigerators and multipane windows, where they scavenge harmful solvents.

A third type of catalyst utilizes transition metal oxides with multiple valence states. Catalytically active sites in the spinel family are associated with disorder or defects in the structure. Transfer of an electron from the solid to a chemisorbed molecule allows molecules like nitrous oxide, NO, to split, ultimately enabling production of N_2 and O_2 reaction products. The electron is returned back to the transition metal oxide.

---

[3] D. E. W. Vaughan, "Crystal design in zeolite synthesis," in *Crystal Engineering*, ed. K. Seddon and M. Zawarotko, Kluwer Academic Publishers, The Netherlands (1999).

22.7   Problems

(1) Calculate a Pilling–Bedworth ratio for silicon and its oxide and comment on its importance in solid state electronics.

(2) Iron forms several oxides: FeO, Fe_3O_4, and Fe_2O_3, with densities 5.70, 5.18, and 5.24 g/cm^3, respectively. Fe metal has a body-centered cubic unit cell with lattice parameter $a = 2.87$ Å. Calculate the PBR values. The changes in crystal structure and in oxidation states (Fe^{2+}/Fe^{3+} ratio) are very complex.

(3) Define what is meant by a catalyst. Give an example.

(4) Explain why zeolites can be used as molecular sieves and as ion exchange materials.

(5) What is meant by surface reconstruction? Why might covalent materials be more susceptible to it than metallic ones?

(6) What is topotaxy? Give an example of the use of topotaxy.

(7) Ni has the FCC structure, with $a = 3.524$ Å.

   (a) Calculate the theoretical density of Ni:

   $M = 58.7$ g/mole
   $V = (3.524 \text{ Å})^3 = 43.76 \times 10^{-24} \text{ cm}^3$
   $Z = 4$
   density $= 8.91$ g/cc.

   (b) Sketch the (100), (110), and (111) planes of this material.

   (c) What is the coordination number of an atom on each of these planes?

   (d) Which would you expect to be the equilibrium face of the material?

   (111) should be the equilibrium face of the compound. It has the fewest number of dangling bonds per surface atom.

23 Neumann's Law and Tensor Properties

23.0 Introduction

A change in external environment elicits a response from a material, and physical properties express the relationship between a response and the applied force. Properties can be classified as equilibrium, steady-state, hysteretic or irreversible. The material response depends explicitly on the bonding that holds a material together, as well as the geometric arrangement of those bonds in space. As a result, there is a strong dependence of many physical properties on symmetry and crystal structure. This chapter outlines these ideas, as well as the mathematical representation of the dependence of properties on direction.

23.1 Anisotropic Properties

Anisotropic properties vary with the direction in which they are measured. It makes sense that physical properties of solids are often anisotropic, given that most physical properties depend on the geometric arrangement of the atoms and bonds in the solid. The way in which a property varies spatially depends on the symmetry of the solid and the tensor nature of the property.

To illustrate this, consider the case of measurement of the electrical conductivity of various materials. Such measurements can be made by applying an electric field, E, across the specimen, and measuring the current density, J, that flows. The conductivity, σ, is then

$$J = \sigma E.$$

For a material such as NaCl, when a voltage is applied in the [100] direction, and the induced current is measured in the same direction, the $\sigma_{[100]}$ can be determined. Recall that the atomic arrangements in NaCl are identical along the a, b, and c directions, so it stands to reason that $\sigma_{[100]} = \sigma_{[010]} = \sigma_{[001]}$ in halite.

In contrast, in the layered material graphite, electrons can move more readily parallel to the layers than perpendicular to the layers. For an electric field applied in-plane, the current density and electrical conductivity will be comparatively high in the plane. However, when the electric field is applied perpendicular to the layers, the current density will be smaller, and the electrical conductivity lower. Clearly, the material is anisotropic. So, what will happen when the electric field is applied at an angle to the

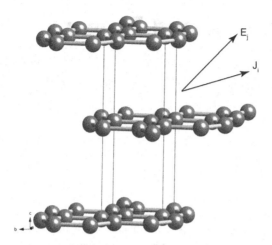

Fig. 23.1 Illustration of the crystal structure of graphite, showing an applied electric field, E, and the resulting direction of current density, J.

planes? In this case, one can describe the electric field as a vector that has components in three orthogonal directions x, y, and z. Using Einstein notation, this is written as E_i, where i is an integer that runs from 1 to 3. E_1, for example, is parallel to the x direction, E_2 is parallel to y, and E_3 is parallel to z. The resulting current flow will also be a vector, *and that vector will not necessarily be parallel to the applied electric field*. In the case of graphite, because the conductivity is higher in the plane, the net current flow will be closer to the plane than the electric field vector, as shown schematically in Fig. 23.1.

In order for the conductivity to relate a vector (the electric field) to another non-parallel vector (the current density), a 3×3 matrix is required. That is

$$\begin{pmatrix} J_1 \\ J_2 \\ J_3 \end{pmatrix} = \begin{pmatrix} \sigma_{11} & \sigma_{12} & \sigma_{13} \\ \sigma_{21} & \sigma_{22} & \sigma_{23} \\ \sigma_{31} & \sigma_{32} & \rho\sigma_{33} \end{pmatrix} \begin{pmatrix} E_1 \\ E_2 \\ E_3 \end{pmatrix}.$$

Thus, $J_1 = \sigma_{11}E_1 + \sigma_{12}E_2 + \sigma_{13}E_3$, etc.

The short-hand notation for this is $J_i = \sigma_{ij}E_j$.

23.2   Tensors of Rank 0, 1, 2, 3, and 4

Tensors are named for the number of directions that must be specified to completely describe the anisotropy of the property. Zero-rank tensor properties are scalars that do not depend on direction. *Density* $= \frac{mass}{volume}$ is a scalar property. It does not matter how a crystal is oriented when measuring its density because neither mass nor volume are directional properties. In the same way, the specific heat, C, of a solid is a scalar quantity, since $C = \frac{dQ}{dT}$. Neither the heat, Q, nor the temperature, T, is associated with any directionality.

But most properties depend on direction. The "rank" of the tensor quantity is a measure of how many directions need to be specified to define the property.

Pyroelectricity is an example of a first rank tensor quantity. As will be described in detail in Chapter 28, pyroelectricity refers to a change in polarization, ΔP_i induced by a change in temperature: $\Delta P_i = \pi_i \Delta T$. The material property that describes the change is the pyroelectric coefficient, π_i; polarization is a dipole moment per unit volume. Dipoles have a positive end and a negative end, and hence a direction is needed to describe the vector from the negative charge to the positive charge. Thus, the polarization is a vector, and is described by a single subscript. Temperature, however, is a scalar. In order to relate a scalar excitation to a vector response, the pyroelectric coefficient is a vector that describes how the magnitude, and potentially the orientation, of the polarization change with temperature.

Many properties are second-rank tensor properties, including electrical conductivity, dielectric permittivity, thermal expansion coefficient, thermal conductivity, and the diffusion coefficient, among others. To describe just two of these, electrical conductivity is a second rank tensor, since the current density (a vector quantity) is not necessarily parallel to the applied electric field (another vector quantity). Relating one vector to a non-parallel one requires a 3×3 matrix as shown above. Alternatively, a second-rank tensor property can also relate another second-rank tensor to a scalar. A good example of this is the thermal expansion coefficient: $\varepsilon_{ij} = \alpha_{ij}\Delta T$, where ε_{ij} is the strain, ΔT is the change in temperature, and α_{ij} is the linear thermal expansion coefficient. Two subscripts are required to describe the strain, since in addition to elongating or contracting, the material can deform on heating or cooling; thus, the two directions refer to the direction of the displacement and the orientation of the measurement axis. Thermal expansion will be discussed in detail in Chapter 24.

Piezoelectricity is a third-rank tensor property: $\Delta P_i = d_{ijk}\sigma_{jk}$. In this equation, there is a change in polarization, ΔP_i, induced by an applied stress, σ_{jk}. The materials constant that links these two is the piezoelectric charge constant, d_{ijk}. This property can also be written in terms of applied strain, rather than applied stress. In this case, the equation is $\Delta P_i = e_{ijk}\varepsilon_{jk}$.

Elasticity is a fourth-rank tensor quantity. Written generally, Hooke's law is $\sigma_{ij} = c_{ijkl}\varepsilon_{kl}$, where σ_{ij} is stress, ε_{kl} is strain, and c_{ijkl} is the elastic stiffness. Two directions are needed to specify stress (the direction of the force and the direction normal to the face on which the force acts); thus, stress is a second-rank tensor.

23.3 Neumann's Principle and Application of Symmetry to Physical Properties

Neumann's principle states that the geometrical representation of any physical property will contain the symmetry of the point group of the material. Breaking this down, the geometric representation of the physical property can be thought of as a surface in three-dimensional space, where the distance from the origin to the surface represents the magnitude of the physical property in that direction. Neumann's principle says that this representation will contain all of the symmetry of the point group (and possibly more). That is, *symmetry acts on physical properties*. Point groups, rather than space groups, appear in Neumann's law, since the dimensions at which physical properties are measured are vastly larger than the scale of translations within the unit cell.

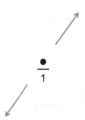

Fig. 23.2 Effect of a center of symmetry, $\bar{1}$, on a vector property. The sum of the vector and its opposite is zero, leading to a null property.

Consider the application of a center of symmetry to a first-rank tensor property. First-rank properties such as pyroelectricity are vector properties. A center of symmetry says that the vector will be symmetry-related to its equal and opposite, as shown in Fig. 23.2, and thus will be equal to zero. That is, first-rank tensor properties are constrained to have zero magnitude (e.g. to be *null properties*) when a center of symmetry is present. In the same way, all odd-ranked tensor properties are zero when the material has a center of symmetry.

Sometimes, combinations of other symmetry elements can also lead to properties being null, or to properties being equivalent along different directions. Consider, for example, a material with the tetragonal crystal system. All tetragonal materials have the minimum symmetry of a four-fold rotation axis. Thus, the a and b crystallographic axes are related by this four-fold axis. Any properties in the a axis are thus symmetry-related to those along b, $-a$, and $-b$. Thus, if we consider a polarization pointed in an arbitrary direction, as shown in Fig. 23.3, then the component along the a axis is cancelled by the one along $-a$; in the same way, the components along b are cancelled by those in the $-b$. However, the components along the c axis are allowed to exist by symmetry. Thus, the pyroelectric coefficient for a material with the point group symmetry 4 allows one non-zero component along the crystallographic c axis.

Symmetry also acts on even-rank tensor properties. Even-rank tensor properties (e.g. second, fourth, sixth, . . . rank tensors) have at least one non-zero coefficient. For second, third, and fourth rank tensor properties, the four- fold rotation axis will simply mean that the properties are the same along the a and b axes.

If the point group symmetry is $4/m$, then there is a mirror plane perpendicular to the crystallographic c axis. Thus, the polar component along the c axis is symmetry-equivalent to the one along $-c$. In this case, there are no remaining components of the pyroelectric vector.

If a material belongs to an orthorhombic crystal class, then the minimum symmetry for the system is three mutually perpendicular two-fold rotation axes along a, b, and c. Since there is nothing relating any of the axes to each other, the physical properties along these axes can all differ from each other. As a result, the thermal expansion coefficients $\alpha_a \neq \alpha_b \neq \alpha_c$.

The convention for where symmetry elements are located was described in Chapter 7. Table 23.1 shows this for the crystal physics convention where Z_1, Z_2, and Z_3 refer to three perpendicular axes. Where possible, these axes are often arranged to be parallel to one or more of the crystallographic axes.

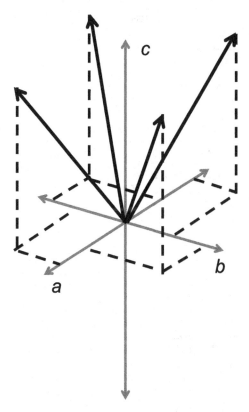

Fig. 23.3 Operation of a four-fold rotation axis on a vector. Components along the a and b crystallographic axes cancel out.

In the *triclinic system*, Z_3 is chosen to be parallel to c, and Z_2 is normal to the a–c plane (010). The Z_1 axis lies in the (010) plane and is perpendicular to Z_1 and Z_3, forming a right-handed coordinate system.

For *monoclinic crystals*, Z_2 is chosen to be parallel to the b [010] axis. It is parallel to the two-fold symmetry axis or perpendicular to the mirror plane, as is customary in crystallography. Axis Z_3 is aligned parallel to c and Z_1 is perpendicular to Z_2 and Z_3, again forming a right-handed system.

In the *orthorhombic system*, crystallographers generally choose a unit cell in which $c < a < b$. The Z_1, Z_2, and Z_3 crystal physics axes are taken as a, b, and c, respectively.

Trigonal crystals have a minimum symmetry of a three-fold rotation parallel to the crystallographic c axis. This axis is chosen to be the Z_3 direction. The Z_1 axis is then assigned to the a crystallographic direction, with Z_2 completing the right-handed orthogonal set.

For the *tetragonal system*, Z_3 and c are parallel to the four-fold axis; Z_1 and Z_2 are parallel to a and b, respectively.

Table 23.1 Orientation of the crystal symmetry axes with respect to the symmetry elements of the 32 point groups

Crystal class	Symmetry elements
1	1
$\bar{1}$	$\bar{1}$
2	$2 \parallel Z_2$
m	$m \perp Z_2$
$2/m$	$2 \parallel Z_2, m \perp Z_2$
222	$2 \parallel Z_1, 2 \parallel Z_2, 2 \parallel Z_3$
$mm2$	$m \perp Z_1, m \perp Z_2, 2 \parallel Z_3$
mmm	$m \perp Z_1, m \perp Z_2, m \perp Z_3$
3	$3 \parallel Z_3$
$\bar{3}$	$\bar{3} \parallel Z_3$
32	$3 \parallel Z_3, 2 \parallel Z_1$
$3m$	$3 \parallel Z_3, m \perp Z_1$
$\bar{3}m$	$\bar{3} \parallel Z_3, m \perp Z_1$
4	$4 \parallel Z_3$
$\bar{4}$	$\bar{4} \parallel Z_3$
$4/m$	$4 \parallel Z_3, m \perp Z_3$
$6/m$	$6 \parallel Z_3, m \perp Z_3$
622	$6 \parallel Z_3, 2 \parallel Z_1$
622	$6 \parallel Z_3, 2 \parallel Z_1$
$6mm$	$6 \parallel Z_3, m \perp Z_1$
$\bar{6}m2$	$\bar{6} \parallel Z_3, m \perp Z_1$
$6/mmm$	$6 \parallel Z_3, m \perp Z_3, m \perp Z_1$
23	$2 \parallel Z_1, 3 \parallel [111]$
$m3$	$m \perp Z_1, 3 \parallel [111]$
$\bar{4}3m$	$\bar{4} \parallel Z_3, 3 \parallel [111]$
$m3m$	$m \perp Z_1, 3 \parallel [111], m \perp [110]$

Among *hexagonal crystals*, Z_3 is chosen to be along c, the six-fold symmetry axis. As with trigonal crystals, Z_1 is parallel to a, and Z_2 is chosen to be perpendicular to Z_1, and Z_3.

Cubic crystals have Z_1, Z_2, and Z_3 parallel to a, b, and c, respectively.

Table 23.2 shows the number of non-zero coefficients, and the number of independent coefficients (in parentheses) for first- through fourth-rank tensors for all of the point groups. The details of how to calculate which coefficients are zero and non-zero are described in books on crystal anisotropy, such as Nye's *Physical Properties of Crystals*, or Newnham's book *Crystal Anisotropy*.

23.4   Geometric Drawings

The geometrical representations of the physical properties are surfaces in three dimensions, where the distance from the origin to the surface represents the magnitude of the physical property in that direction. As an example, consider the directional dependence of these four tensor properties of aluminium nitride. AlN has the hexagonal wurtzite

Table 23.2 Listing of the number of non-zero coefficients, and the number of independent coefficients (in parentheses) for first- through fourth- rank tensors for all of the point groups

Crystal class	Tensor rank			
	1	2	3	4
1	3 (3)	9 (6)	18 (18)	36 (21)
$\bar{1}$	0	9 (6)	0	36 (21)
2	1 (1)	5 (4)	8 (8)	20 (13)
m	2 (2)	5 (4)	10 (10)	20 (13)
2/m	0	5 (4)	0	20 (13)
222	0	3 (3)	3 (3)	12 (9)
mm2	1 (1)	3 (3)	5 (5)	12 (9)
mmm	0	3 (3)	0	12 (9)
3	1 (1)	3 (2)	13 (6)	24 (7)
$\bar{3}$	0	3 (2)	0	24 (7)
32	0	3 (2)	5 (2)	18 (6)
3m	1 (1)	3 (2)	8 (4)	18 (6)
$\bar{3}m$	0	3 (2)	0	18 (6)
4	1 (1)	3 (2)	7 (4)	16 (7)
$\bar{4}$	0	3 (2)	7 (4)	16 (7)
4 /m	0	3 (2)	0	16 (7)
422	0	3 (2)	2 (1)	12 (6)
4mm	1 (1)	3 (2)	5 (3)	12 (6)
$\bar{4}2m$	0	3 (2)	3 (2)	12 (6)
4/mmm	0	3 (2)	0	12 (6)
6	1 (1)	3 (2)	7 (4)	12 (5)
6	0	3 (2)	6 (2)	12 (5)
6/m	0	3 (2)	0	12 (5)
622	0	3 (2)	2 (1)	12 (5)
6mm	1 (1)	3 (2)	5 (3)	12 (5)
$\bar{6}m2$	0	3 (2)	3 (1)	12 (5)
6/mmm	0	3 (2)	0	12 (5)
23	0	3 (1)	3 (1)	12 (3)
m3	0	3 (1)	0	12 (3)
432	0	3 (1)	0	12 (3)
$\bar{4}3m$	0	3 (1)	3 (1)	12 (3)
m3m	0	3 (1)	0	12 (3)

crystal structure, as shown in Fig. 23.4. The physical properties are referred to a set of orthogonal axes, Z_1, Z_2, and Z_3. Figures 23.5–23.8 show how the physical properties of AlN depend on direction. Pyroelectricity is a first rank tensor, and the coefficient follows a simple linear dependence on $\cos\theta$, where θ is the angle from the polar axis. The maximum response is observed along the polar axis $x_3 = [001] = c$. Directions perpendicular to [001] are not pyroelectric.

Second-rank tensor properties are proportional to the second power of the trigonometric functions. The dielectric constant of aluminum nitride is slightly larger along [001] than in directions perpendicular to the six-fold rotation axis. For hexagonal

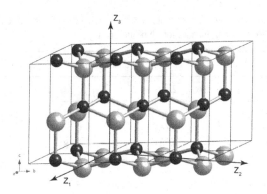

Fig. 23.4 Hexagonal crystal of AlN showing the orientation of the crystallographic axes and the orthogonal axes used to describe the physical properties: Z_1, Z_2, and Z_3. Physical property measurements are made on plates cut at different θ angles.

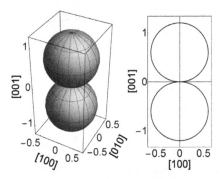

Fig. 23.5 Pyroelectric coefficient of AlN plotted as a function of direction: $\pi_3 = 1.2\ \mu\text{C/cm}^2$. Both three-dimensional representations and a two-dimensional cross section are shown.

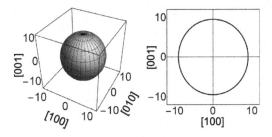

Fig. 23.6 Dielectric constant of AlN plotted as a function of direction: $\varepsilon_{r,11} = 9, \varepsilon_{r,33} = 9.5$.

crystals, the maximum values are always parallel or perpendicular to the symmetry axis, see Fig. 23.6.

Piezoelectricity is a third-rank tensor property. The coefficients can be either positive or negative depending on the sign of the polarization charge produced by an applied tensile stress. This is illustrated for the case of AlN in Fig. 23.7.

The elastic compliance coefficients follow an equation involving the fourth power of trigonometric terms. Maximum values can occur in directions other than symmetry

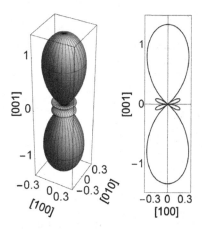

Fig. 23.7 Longitudinal piezoelectric coefficient of AlN plotted as a function of direction:
$e_{333} = 1.34 \frac{C}{m^2}$; $e_{311} = -0.6 \frac{C}{m^2}$; $e_{113} = -0.32 \frac{C}{m^2}$.

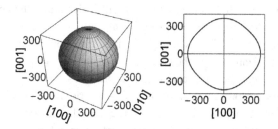

Fig. 23.8 Elastic stiffness of AlN plotted as a function of direction: $c_{1111} = 402.5$ GPa;
$c_{1122} = 135.6$ GPa; $c_{1133} = 101$ GPa; $c_{3333} = 387.6$ GPa; $c_{2323} = 122.9$ GPa.

axes, as shown in Fig. 23.8. The [001] direction is the stiffest in AlN, while the most
compliant coefficient is about 50° from [001].

For AlN, all four types of tensor properties are cylindrically symmetric about [001].
In lower-symmetry crystals, this cylindrical symmetry can be lost.

23.5   Continuous (Curie) Groups

Neumann's law is still useful even when the material is not a single crystal. Polycrystalline materials are made up of many grains (essentially single crystals that are randomly
oriented with respect to each other). In this case, although each grain has the characteristic point group symmetry of the crystal structure, the polycrystalline sample has higher
effective symmetry when the macroscopic properties are measured, due to averaging of
the responses for different grains. In other cases, all or part of the sample is amorphous,
so that there is no starting point group. Nonetheless, an effective, non-crystallographic
symmetry can be defined.

For both of these cases, *Curie group symmetries*, as were discussed in Chapter 7, can
be applied, and used in Neumann's law. Polycrystalline materials with randomly

Table 23.3 Minimum symmetry operations for textured materials in limiting groups

Curie group	Symmetry operators
∞	$\infty \parallel Z_3$
∞m	$\infty \parallel Z_3$, $m \perp Z_1$
$\infty 2$	$\infty \parallel Z_3$, $m \parallel Z_1$
∞/m	$\infty \parallel Z_3$, $m \perp Z_3$
∞/mm	$\infty \parallel Z_3$, $m \perp Z_3$, $m \perp Z_1$
$\infty\infty$	$\infty \parallel Z_3$, $\infty \parallel Z_1$
$\infty\infty m$	$\infty \parallel Z_3$, $\infty \parallel Z_1$, $m \perp Z_1$

oriented grains and most amorphous materials have effective symmetries of $\infty\infty m$: spherical symmetry. The mirror symmetry element is lost if the material has handedness or chirality (e.g. a sugar solution). In this case, the Curie group symmetry is $\infty\infty$. If the material is processed under conditions where a net texture appears along a single axis (e.g. the c axes of the grains are aligned parallel to a specific orientation) then one of the infinite fold rotation axes is lost, yielding symmetry $\infty m/m$: cylindrical symmetry. When there is a net vector property in a polycrystalline material, for example in a polycrystalline material with a net polarization, then the Curie group symmetry is ∞m, since the top and bottom of the sample can be differentiated; this produces a conical symmetry. Magnetized polycrystalline ferromagnets are in ∞/m. Cholesteric liquid crystals adopt a helical patterning with Curie group symmetry $\infty 2$. Ferroelectric liquid crystals possess a polar axis. They belong to ∞m or ∞, depending on whether or not the molecules are chiral. A poled right- or left-handed polycrystalline substance belongs to the lowest Curie group, ∞. Table 23.3 shows the location of the symmetry elements for the crystal physics axes for the Curie groups.

23.6   Problems

(1) Draw a stereographic projection for the point group of α-quartz. The space group is $P3_221$. On your diagram, show the location of all of the symmetry elements. Could such a material be pyroelectric? Explain your reasoning.

(2) Write the equation for a second-rank tensor property, defining all of the terms.

(3) Explain why piezoelectricity is a third-rank tensor property.

(4) For each of the crystal systems, explain whether the electrical resistivity should be the same or different along the crystallographic a and b axes. Along a and c? Explain your reasoning in terms of the minimum symmetry of the crystal systems.

(5) The crystal system of rutile is $P4_2/mnm$. Could rutile be piezoelectric? Explain why or why not.

(6) In poled ferroelectric crystals, the Curie group symmetry is ∞m, whereas before poling the symmetry is $\infty m/m$. On the basis of crystal physics, explain why poling is needed to observe a net pyroelectric coefficient.

(7) MoS_2 is being researched as a semiconductor. Its space group is $P6_3/mmc$, $a = 3.1604$ Å, $c = 12.295$ Å. Mo is located at positions (1/3, 2/3, 1/4), S is located at (1/3, 2/3, 0.629).

(a) Draw several unit cells of this crystal.

(b) Explain whether you believe this material will have the same electrical resistivities along a, b, and c.

24   Thermal Properties

24.0   Introduction

Rigid models are often used to represent crystal structures, but the atoms in a real solid oscillate rapidly around their equilibrium positions. As Dame Kathleen Lonsdale wrote, "A crystal is like a class of children arranged for drill, but standing at ease, so that while the class as a whole has regularity both in time and space, each individual child is a little fidgety."[1] This atomic motion around the equilibrium lattice positions (thermal motion) influences a wide range of thermal properties, including the specific heat, the thermal expansion coefficients, and the thermal conductivities.

This chapter introduces the temperature dependence of atomic motion, and then describes the fundamental mechanisms that contribute to several key thermal properties. The link between bonding, structure, and the magnitude of the property coefficients is then made for a wide variety of materials.

24.1   Thermal Vibrations

Solids consist of billions/trillions of vibrating atoms in thermodynamic equilibrium with a Boltzmann energy distribution. As a consequence, at a given temperature, the vibration amplitudes will be large for low-frequency oscillations and small for high-frequency oscillations. Low frequencies correspond to low force constants and heavy masses, which implies a correlation between large vibration amplitudes and weak interatomic bonds. This is illustrated in Fig. 24.1, where the thermal amplitudes of the group IV elements are shown.[2] As the size of the atom increases and the bond length decreases, the vibration amplitude in the elemental solid increases.

The root-mean-square (rms) displacements of atoms are determined by X-ray structure analysis. Vibration amplitudes are generally about 3–7% of the nearest neighbor distances near room temperature. As shown in Fig. 24.2, the rms values increase with temperature, often with a slow, nearly linear rise. Motion continues to very low temperatures.

It is worth noting that the amplitude of the atomic vibration is direction-dependent in all crystals, so that the locus of motion is described using a thermal ellipsoid. For example,

[1] K. Lonsdale, *Crystals and X-rays*, D. Van Nostrand Co., Princeton, New Jersey (1949).
[2] V. V. Levitin, "Atomic vibrations in solids: Amplitudes and frequencies," *Phys. Rev.*, **21**, 1–162 (2004).

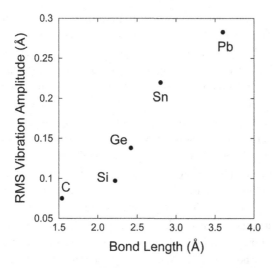

Fig. 24.1 Relationship between bond length and vibration amplitude for group IV elements. Redrawn from V. V. Levitin, "Atomic Vibrations in Solids: Amplitudes and Frequencies," *Phys. Rev.* **21**, 1–162 (2004). Adapted with permission from Cambridge Scientific Publishers.

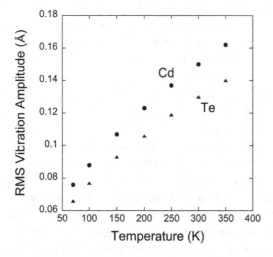

Fig. 24.2 Vibration amplitude of the Cd and Te atoms in CdTe. Data from R. D. Horning and J.-L. Staudenmann, "X-ray vibrational studies on (100)-oriented CdTe crystals as a function of the temperature (8–350 K)," *Phys. Rev. B*, **34** (6), 3970–3979 (1984).

Mg and Zn crystallize in the hexagonal close-packed structure with 12 neighbors around each atom. The packing in magnesium is almost ideal, with nearly equal interatomic distances for all neighbors. As a result, the thermal vibrations are nearly isotropic. In contrast, in zinc, the *c/a* ratio exceeds the ideal value, so that the Zn–Zn distances within a close-packed plane are shorter than those between atoms in adjacent layers. As a result, atoms can vibrate with larger amplitude along *c* than perpendicular to *c*.

The amplitude of motion depends on the atomic masses, the strength of the bonding and the amount of room available for motion. For solids made out of atoms with the same mass, those having stronger bonds are harder to perturb as a function of temperature, and as a result, the amplitudes of motion tend to be smaller. This can be readily visualized by recognizing that bonds provide spring-like restoring forces to the displaced atoms. Stronger bonds act like stiffer springs, with higher spring constants.

A variety of illustrative phenomena are observed in metallic solid solutions. For example, when a metal is alloyed with atoms that increase the average bond strength, the vibration amplitude decreases at a given temperature, because stronger bonds are more difficult to perturb thermally. Secondly, when a solid solution is made by adding a heavier atom to a lighter metal, the vibrational amplitude is observed to decrease more than would be expected based on higher masses alone. It is believed that this is a result of the fact that the heavier atoms act rather like "nodes" in the string of vibrating atoms. This, in turn, favors higher-frequency, lower-amplitude vibration modes.

In group IV semiconductors, the rms vibration amplitudes increase linearly with temperature. Because the bonds in diamond are stronger than those in Si or Ge, the vibration amplitude of the carbon atoms is smaller at a given temperature. In compound semiconductors, the bonds connect dissimilar atoms. In this case, the lighter atoms in the compound displace more than the heavy atoms. This is indicative of a general trend that in diatomic compounds, lighter atoms generally have larger amplitudes of vibration than do heavy atoms. As might be expected, for III–V compounds with a given group III element (e.g. GaP, GaAs, and GaSb), as the mass of the group V element increases, the bond strength decreases. As a result, the rms Ga displacement rises.

In ionic compounds the amplitudes of vibration are generally smallest along the directions of the bonds. Instead, atoms tend to vibrate towards available open space in the crystal structure. Thus, in compounds with the fluorite structure, the F atoms vibrate towards the empty center of the unit cell; in silicates, the O^{2-} ions vibrate towards the large interstices at the center of ring structures.

In molecular solids, the largest vibration amplitudes develop where there is the most room in the structure to accommodate the displacement. In organic compounds, since C, N, and O all have similar atomic weights, atoms near the edge of molecules usually vibrate with larger amplitude than those near the center of the molecule.

Surface vibrations With fewer chemical bonds to hold them in place, surface atoms vibrate with greater amplitude than those inside the crystal. As temperatures begin to approach the melting point, surface melting can take place, leading to reduced friction. The slipperiness of ice is thought to originate in this way. Diffraction measurements offer convincing evidence for the liquid-like layer on the surface of ice. The surface layer exhibits rotational disorder with intact long-range positional order in between −13.5 and 0 °C. Liquid-like surface layers have also been reported to form on metals, semiconductors, molecular solids, and inert gas crystals.

24.2   Heat Capacity

The heat capacity of a material is a measure of how much heat, Q, must be added to increase the temperature, T, of a mole of the material by 1 °C. It is important to note the precise definition of "heat capacity" in any context, as it can be defined in terms of a

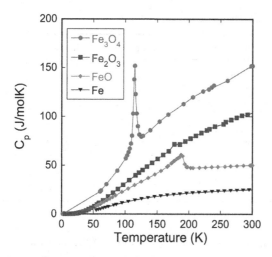

Fig. 24.3 Temperature dependence of the specific heats of Fe and a number of iron oxides. The greater the number of atoms in the chemical formula unit, the higher the saturation value for the specific heat. For some compounds, anomalies due to phase transitions are also observed. Data from Y. S. Touloukian, *Thermophysical Properties of Matter, Volume 4: Specific Heat, Metallic Elements and Alloys*, eds. Y. S. Touloukian and C. Y. Ho, IFI/Plenum, New York, Washington (1970); and *Thermophysical Properties of Matter, Volume 5: Specific Heat: Nonmetallic Solids*, ed. Y. S Touloukian, IFI/Plenum (1970).

mole, a gram-atom, or a given mass of the material (the latter is referred to as the specific heat). Experimentally, the heat capacity is usually measured at constant pressure (denoted by the subscript P), where $C_P = \left(\frac{dQ}{dT}\right)_P$. A related quantity often used in physics and thermodynamics is the specific heat at constant volume, $C_V = \left(\frac{dQ}{dT}\right)_V$. Note that $C_P \geq C_V$ because work is required when the material expands against the fixed pressure.

As a solid absorbs heat, atom vibration amplitudes increase, electron energy levels rise, atomic defects form, and in general the entropy of the material increases. All of these are mechanisms by which the material can absorb energy without changing its temperature. Thus, outside of phase transition regions, these mechanisms account for the heat capacity of the solid. In many temperature regimes of practical interest, the contribution due to atomic vibrations is dominant. Consequently, within an elemental solid, where each atom can vibrate in three dimensions, the heat capacity approaches $3R$ at high temperatures (accounting for the degrees of freedom for the vibration modes), where $R = 8.314$ J/(mole K). This is shown in Fig. 24.3. More generally, $C_V = 3nR$, where n is the number of atoms in the formula unit. In a diatomic solid such as NaCl, the saturation value for C_V is 6R, while in Al_2O_3, where there are five atoms per formula unit, the saturation value for C_V is 15R.

At 0 K, atomic vibrations are very limited and the heat capacity is zero; with increasing temperature, the heat capacity curve varies smoothly up to the saturation value. Near absolute zero, the energy separation between the quantized energy levels for the atomic vibrations is large compared with the available thermal energy. As a result, the system remains in its ground state, and can't absorb thermal energy. As the

Table 24.1 Debye temperatures for representative materials[3]

Material	Θ_D (K)	Material	Θ_D (K)	Material	Θ_D (K)
Al	385	C (diamond)	2200	LiF	670
B	1220	Si	635	LiCl	420
Ca	230	Ge	360	LiBr	340
Cu	310	Sn (grey)	260	LiI	280
Fe	460	Sn (white)	170	CsF	245
Mg	330	Pb	85	CaF$_2$	470
Ni	440			MgO	800
Pt	225			SiO$_2$ (quartz)	255
W	315			TiO$_2$ (rutile)	450

temperature rises, the thermal energy ($k_B T$) allows more of the states to be populated. The rate at which the heat capacity curve approaches the high-temperature bound depends on the bonding in the material. Weakly bonded solids like Pb can easily be excited into vibration, and so reach the saturation value at low temperatures. In strongly bonded solids like diamond, on the other hand, much more thermal energy is required to activate all of the vibration modes, and the heat capacity approaches the asymptotic value more slowly.

The Debye temperature (Θ_D) is an indicator of how rapidly the specific heat approaches the saturation value. The Debye temperature arises from a simplified model of the vibrations in solids in which it is presumed that the number of phonons increases quadratically with frequency until a cutoff frequency is reached at which all phonons are active. In reality, this model describes the low-temperature specific heat relatively well, but fails at higher temperatures. For substances with small Θ_D, C_V increases rapidly with temperature, leveling off at the classical value well below room temperature. To illustrate basic trends between the Debye temperature, periodic properties of the elements, and bonding, Θ_D values for a set of common materials are listed in Table 24.1. It can be seen that solids made up of heavy elements with long, weak bonds have low Debye temperatures, and hence high specific heats at low temperatures.

At high temperatures, the specific heat can rise above the saturation value. There are several means by which this occurs. First, when there are free electrons in the solid, there is also a contribution to C_V and C_P from electrons with energies near the Fermi energy. Second, the material can absorb energy through the creation of point defects.

Nakamura[4] examined the specific heat data for ionic and covalently bonded solids. For alkali halides, the Debye temperatures, average mass $\overline{m}$, and bond lengths d, follow the relation $\Theta_D^2 \overline{m} d^3 = constant$. For the covalently bonded column IV elements with the diamond structure, a better fit was obtained with $\Theta_D^2 \overline{m} d^4 = constant$.

Anomalies in the heat capacity curve appear when a material goes through a phase transition. This can be easily understood as a consequence of the heat absorbed or

---

[3] From G. Burns, *Solid State Physics*, Academic Press, Orlando, Florida (1985).

[4] T. Nakamura, "Influences of mass and bond length on the Debye temperatures of ionic and covalent substances," *Jpn J. Appl. Phys.*, **20** (9), L653–L656 (1981).

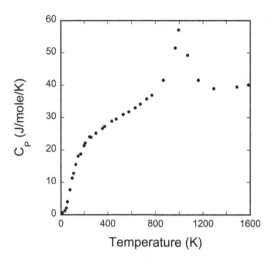

Fig. 24.4 Specific heat as a function of temperature for Fe showing the anomaly resulting from the BCC-to-FCC phase transition. Data from Y. S. Touloukian, *Thermophysical Properties of Matter, Volume 4: Specific Heat, Metallic Elements and Alloys*, eds. Y. S. Touloukian and C. Y. Ho, IFI/Plenum, New York, Washington (1970).

dissipated at the transition temperature. This is illustrated, for example, in Fig. 24.4, where a peak in the thermal conductivity of iron is observed where the material undergoes a phase transition from a body-centered cubic (BCC) to a face-centered cubic (FCC) crystal structure. Note that the heat capacity anomaly at phase transitions can be quite large. For this reason, phase transitions in solid phases are of interest for thermal energy harvesting. A large quantity of heat energy can be stored over a small temperature transient.

The temperature dependence of the specific heat of polymers is also a function of the rotation and vibrational motion within the sample. This is often described using the language of phonons, even in the case where the polymer is amorphous. In polymers, a variety of molecular motions are excited with increasing temperature, leading to changes in the specific heat; the case of polypropylene is shown in Fig. 24.5 as an example. At low temperatures, e.g. <60 K, specific heat is described in terms of vibrations of a three-dimensional continuum. From ~70–200 K, vibrations of linear chains dominate (bending and stretching). As temperature increases, phonons with decreasing wavelengths are excited, raising the specific heat further. Most C_P values in polymers are relatively large compared to metals and fall in the range of 1–2 kJ/kg K. Replacement of H with heavier elements such as F or Cl lowers C_P.

It is notable that at the lowest temperatures, the specific heat of amorphous polymers exceeds that of a crystalline polymer of the same composition due to a combination of localized vibrations and decreased sound velocity due to the lower packing density of the amorphous material. Above the glass transition temperature, T_g, the specific heat rises because additional non-vibrational contributions (e.g. chain rearrangement) add to the specific heat.

The high specific heat of water is also important in keeping food warm. Indeed, if you consider the shape of a pizza, its large radiating surface is not well suited to retaining

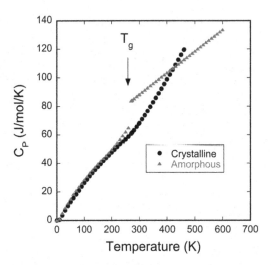

Fig. 24.5 C_P as a function of temperature for amorphous and crystalline polypropylene, where the glass transition temperature, T_g, is marked. At T_g, the specific heat rises due to the ability to re-arrange the chain configuration. Data from Gaur and Wunderlich, "Heat capacity and other thermodynamic properties of linear macromolecules: IV. Polypropylene," *J. Phys. Chem. Ref. Data*, **10** (4), 1051 (1981).

heat. Thus, pizza delivery is made possible, in part, by the large specific heat of tomato sauce, which helps keep the pizza warm from the oven to your kitchen table.

24.3   Thermal Expansion

The previous section established that adding heat to a material increases the population and magnitude of atomic/ionic vibrations in solids about their equilibrium positions. An inescapable consequence of this activity is a net increase in interatomic distance, and thus macroscopic physical dimension in most materials. We refer to this property as thermal expansion and quantify it with the second-rank property tensor α_{ij}.

The macroscopic linear thermal expansion coefficient, α_{ij}, is defined based on the strain ε_{ij} induced by a change in temperature, ΔT: $\varepsilon_{ij} = \alpha_{ij}\,\Delta T$. From a microscopic perspective, $\alpha_{ij} = \frac{1}{length}\left(\frac{\partial length}{\partial T}\right)_P$, the derivative of the sample dimension with respect to temperature, measured at a constant pressure, P, normalized to the dimension. The thermal expansion coefficient is a second-rank tensor quantity, and so can be different in different directions in some crystals. For simplicity, this anisotropy is often ignored, and the subscripts are dropped from the equation. However, it is important to note that in some materials, particularly those with asymmetric structures like graphite, thermal expansion anisotropy can be very large. The volume thermal expansion coefficient $\beta = \frac{1}{V}\left(\frac{\partial V}{\partial T}\right)_P$, is the sum of the linear thermal expansion coefficients along the three principal axes.

Many properties of solids associated with atomic vibrations can be calculated by assuming that atoms only undergo small displacements from their equilibrium positions,

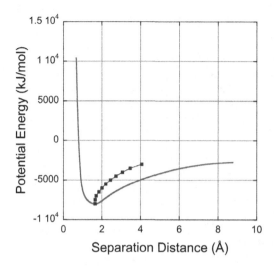

Fig. 24.6 The solid line shows the potential energy as a function of separation distance for the Si–O bond. The symbols show the average separation distance as a function of energy (and hence temperature).

so that only the leading (harmonic) term of the energy as a function of displacement needs to be retained when the energy is expanded around the equilibrium value. This approximation fails, however, to describe thermal expansion. This is illustrated in Fig. 24.6, which shows the potential energy as a function of separation distance for the Si–O bond at 0 K, the minimum in the curve corresponds to the equilibrium bond length. As temperature rises, the solid gains thermal energy $k_B T$, and can begin to sample higher-energy states, i.e., the atoms can oscillate about the bottom of the potential minimum. For a fully symmetrical potential energy profile, the mean separation distance between atoms is *not* a function of temperature. That is, as the temperature rises, the amplitude of the thermal vibrations will increase, but the average atom separation distance will not change. On the other hand, when the energy well is asymmetric, the average separation distance does depend on the energy available, and hence on temperature. Since we understand the potential well vs. separation distance as a balance between attractive and repulsive forces that have different functional dependencies on distance, we expect an asymmetry that favors net expansion with increasing $k_B T$. The squares in Fig. 24.6 show the average separation distance as a function of energy.

Thermal motion underlies both thermal expansion and specific heat, so it is not surprising that the linear thermal expansion coefficient α is proportional to the molar specific heat, C_P.

A number of key trends illustrate the links between the thermal expansion coefficient and the structure and bonding in a solid.

(1)   Most materials expand a few percent from 0 K to their melting temperature. For example, many metals show ~8% volume expansion (which corresponds to a linear expansion of between 2% and 3%) as illustrated in Fig. 24.7. Thus, of necessity, solids with high melting temperatures have low thermal expansion coefficients. The same general trend is also true for other bonding types.

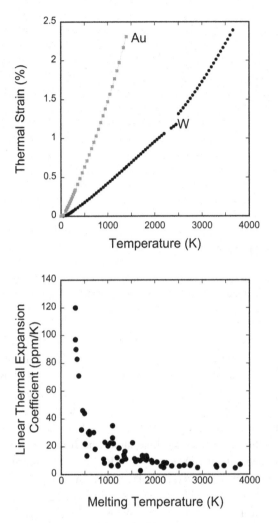

Fig. 24.7 (Top) Expansion of Au and W as a function of temperature from 0 K to their melting temperatures. It can be seen that although the melting temperatures are very different, the total expansion before melting is comparable. (Bottom) As a result of this, materials with high melting temperatures tend to have low thermal expansion coefficients, and vice versa.

Typical thermal expansion coefficients for polymers are an order of magnitude larger than those of most metals and ceramics, in keeping with their lower melting points, and with the weak secondary bonding that holds the chains together. Thermal expansion coefficients for polymers (Table 24.2) often range from 50 to 200 × 10^{-6}/K (which is also written as 50 to 200 ppm/K).

(2)   The thermal expansion coefficient drops as the bond strength rises. Megaw has shown that α is inversely proportional to the square of Pauling's electrostatic bond strength. In the same way, the thermal expansion coefficient decreases with bond valence. A corollary to this is evident in chain structures where the strongest bonds are typically arranged parallel to the chain. Consequently, the thermal

Table 24.2 Typical thermal expansion coefficients for polymers

Polymer	Thermal expansion coefficient $(10^{-6}/K)$
Polyurethane	150
Nylon	80
Polymethyl methacrylate	70
Polytetrafluoroethylene	100
Polystyrene	75
Polypropylene	90
Polyethylene	110

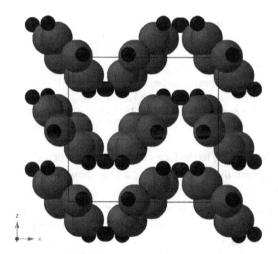

Fig. 24.8 Crystal structure of benzene. The large atoms are C, the smaller ones are H.

expansion coefficient parallel to the chain is smaller than that perpendicular to the chain. Indeed, in many crystalline polymers, large thermal expansion coefficients are observed perpendicular to the chains, while *contraction* parallel to the chains is sometimes observed due to the large transverse vibrations of the atoms on heating. Likewise, in layered structures, lower thermal expansion coefficients are observed parallel to the layers rather than perpendicular to them.

For crystals structures containing aligned planar molecules, the thermal expansion coefficient perpendicular to the molecules is often the largest, as weaker secondary bonds typically provide the attractive force in those directions. As an example, for benzene (see Fig. 24.8), the C_6H_6 molecules lie closer to the (001) plane of the orthorhombic crystal. The thermal expansion coefficients along the three crystallographic directions are $\alpha_a = 119$ ppm/°C, $\alpha_b = 106$ ppm/°C, and $\alpha_c = 221$ ppm/°C.

(3)   Anomalies in thermal expansion occur near phase transitions, as atoms displace or re-arrange into different structures. For example, when iron converts from the

Table 24.3 Average linear expansion coefficients for various cation polyhedra

Bond	Cation polyhedron mean linear expansion coefficient (ppm/°C)
Mg–O (VI coordinated)	14
Ca–O (VIII coordinated)	14
Ni–O (VI coordinated)	14
Ba–O (IX coordinated)	15
Na–O (VI coordinated)	17
K–O (VI coordinated)	21
K–Cl (VI coordinated)	46
K–Br (VI coordinated)	49
Al–O (VI coordinated)	9
Bi^{3+}–O (VI coordinated)	9
Si–O (IV coordinated)	0
Ti^{4+}–O (VI coordinated)	8
Zr^{4+}–O (VI coordinated)	8

From Robert M. Hazen and Larry W. Finger, *Comparative Crystal Chemistry: Temperature, Pressure, Composition and the Variation of Crystal Structure*, John Wiley and Sons, New York (1982).

body-centered crystal structure to the face-centered cubic structure, it contracts by several tenths of a percent due to the higher packing density of the FCC phase.

(4) Comparative studies demonstrate that a given bond tends to expand by approximately the same amount as a function of temperature in many different structures. Hazen and Finger tabulated the expansion coefficients of many bonds; a subset of these is shown in Table 24.3.

(5) In materials with open crystal structures, prediction of the thermal expansion coefficients is complicated by interaction between two factors: the increased amount of "vibration room" available in the structure, and the possibility of cooperative rotation of polyhedra.

Consider, for example, a silicate framework structure. In such open structures, atoms vibrate anisotropically; the amplitudes of vibration tend to be larger in directions that correspond to more open spaces in the structure. Transverse vibrations of the oxygen atoms allow the time average of the Si–O bond distance to increase with temperature without the Si – Si distances changing much. In more generic terms, atomic vibrations are favored in directions toward open spaces. In this way, the thermal energy can be accommodated with minimal changes in overall bond length.

As a result, the thermal expansion coefficient of some open silicate structures is very small. A good illustration of this is cordierite, $Mg_2Al_4Si_5O_{18}$, which has the crystal structure of beryl, with open silicate 6-rings, as shown in Fig. 24.9. In the hexagonal form of cordierite, the structure *contracts* slightly ($\alpha_c = -1.7$ ppm/°C) on heating along the c direction, and *expands* in the a–b plane ($\alpha_a = \alpha_b = 2.9$ ppm/°C). The thermal expansion coefficient of cordierite is low over a wide

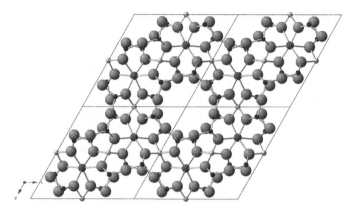

Fig. 24.9 Crystal structure of hexagonal $Mg_2Al_4Si_5O_{18}$ cordierite. The larger atoms are oxygen. The small tetrahedral atoms are Si and Al, and the octahedrally coordinated atoms are Mg.

temperature range, which has led to its use in applications that require resistance to thermal shock, such as substrates for catalytic converters. *Thermal shock* is a mechanical failure mechanism induced by the stresses which develop due a temperature differential across a part. Since the strains which develop are proportional to the thermal expansion coefficient, α, materials with low thermal expansion coefficients are inherently robust against thermal shock.

In open structures, the polyhedra can rotate, causing unusually large or unusually small expansion coefficients. An excellent example of this occurs in quartz. Near the α–β phase transition in quartz, the silicate tetrahedra rotate from a configuration in which they have three-fold symmetry to one with six-fold symmetry (see Fig. 24.10). In this case, a large thermal expansion coefficient is associated with the change in the Si–O–Si bond angle.

In contrast, polycrystalline sodium zirconium phosphate $NaZr_2(PO_4)_3$ (see Fig. 24.11) shows a modest net thermal expansion coefficient due to cooperative rotation. As the phosphate tetrahedra rotate with increasing temperature, the material expands along the c axis, such that $\alpha_c = 25.5$ ppm/°C. However, since the oxygens at the apices of the tetrahedra are also part of the Zr octahedra, the octahedra must also rotate. This, in turn, induces a *contraction* in the a–b plane: $\alpha_a = -6.4$ ppm/°C. In a polycrystalline ceramics, expansion along one axis is partially counterbalanced by contraction along others.

(6) As the concentration of vacancies within a solid increases, its thermal expansion coefficient also tends to increase. This often produces an increase in α near the melting point of the solid, where the concentration of vacancies increases exponentially. In addition, chemical expansion in a lattice can occur when a change in temperature introduces significant changes in the defect chemistry. This can occur, for example, when oxygen loss causes reduction of a multivalent cation in a material (such as CeO_{2-x}), where Ce^{4+} can be reduced to Ce^{3+}. The larger size of the lower-valent ion leads to swelling of the lattice. At least in some systems, this chemical expansion exceeds the lattice contraction that might otherwise be expected due to relaxation around the oxygen vacancy.

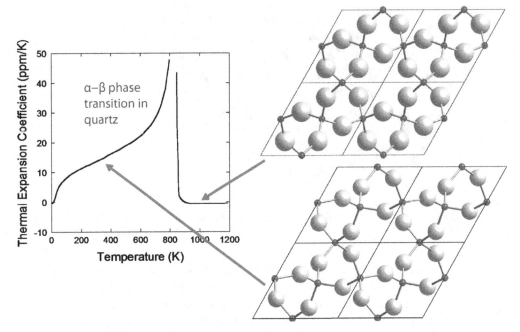

Fig. 24.10 Thermal expansion coefficient of quartz across the α–β phase transition. In the α phase, cooperative rotation of polyhedra produces a large thermal expansion coefficient. On heating into the β phase, cooperative rotation ceases as a result of the higher-symmetry six-fold rotation axis. The thermal expansion coefficient then drops significantly, as would be expected for a strongly bonded material with a large amount of room for transverse vibrations of the O atoms into the interstices. Quartz expansion data redrawn by permission, from R. K. Kirby, "Thermal expansion of ceramics," in *Mechanical and Thermal Properties of Ceramics*, Proceedings of a Symposium Held at Gaithersburg, MD April 1–2, 1968, National Bureau of Standards Special Publication 303 issued May 1969.

(7) For metal alloys, the thermal expansion coefficient of the alloy usually decreases when an end member with a lower thermal expansion coefficient is added. Exceptions are known to occur in the transition metals.

(8) Among the factors that usually do not have a strong effect on the thermal expansion coefficient are density, grain size, impurities (at <1% level), dislocations, and grain boundaries.

The metal alloy Invar (which has compositions near 0.64 Fe–0.36 Ni) has a low thermal expansion coefficient over a wide temperature range below the ferromagnetic Curie point. The key to understanding this is the realization that the ferromagnetic ordering eliminates the d electron component of the bonding. As a result, at low temperatures, where the ferromagnetic ordering is complete, the interatomic distances are large. As temperature rises, the magnetic ordering is gradually lost. This, in turn, increases the d electron contribution to the bonding, and hence the bond strength. The resulting contraction due to the change in bond strength largely counterbalances normal thermal expansion, yielding a low thermal expansion coefficient of ~1.2 ppm/K.

Fig. 24.11 NZP crystal structure. Na (or K) spheres; P = tetrahedral coordination with O; Zr = octahedral coordination with O. The c axis is vertical in this diagram.

Most materials expand on heating, but a few do not. Negative thermal expansion coefficients are observed in several molybdates and tungstates with the general formula $A_2M_3O_{12}$, where M is Mo or W, and A is a trivalent cation such as Sc or Y. The crystal structure is shown in Fig. 24.12. All three unit cell dimensions of orthorhombic $Y_2W_3O_{12}$ decrease continually between room temperature and 1100 °C. Over this temperature range, the average thermal expansion coefficient is -7.0×10^{-6}/K.[5] In the structure, the Y is octahedrally bonded to six oxygens, and tungsten tetrahedrally coordinates to four oxygens. Each oxygen is in a bent linear coordination with one yttrium and one tungsten neighbor. Low thermal expansion coefficients are expected in keeping with the universal relationship between α and bond strength, but this does not explain the *shrinkage* in cell size. This effect has been ascribed to rocking motions between the interconnected polyhedra. As temperature increases, the oxygen atoms vibrate transversely, causing a decrease in the W–O–Y bond length and a compaction of the unit cell.

24.4   Thermal Conductivity

Thermal transport is defined by the expression $\frac{Q}{At} = -\kappa \frac{dT}{dx}$, where Q is the heat carried over a cross-sectional area A, in time t, in response to a temperature gradient $\frac{dT}{dx}$. The

[5] P. M. Forster and A. W. Sleight, *Int. J. Inorg. Mat.*, **1**, 123–127 (1999).

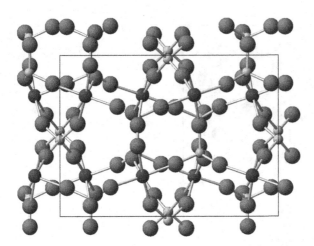

Fig. 24.12 Crystal structure of $Y_2W_3O_{12}$, a negative thermal expansion material. O shown in gray, Y shown in lighter gray, and W shown in dark gray.

material response is embedded in the thermal conductivity, κ, which is a second-rank tensor property. The negative sign in the equation results from the fact that heat is transferred from the hot side to the cold side of the material, i.e. in a direction opposite to the driving force.

There are four key mechanisms that contribute to the thermal conductivity, as shown in Fig. 24.13: *phonons, free electrons, photons,* and *convection*. Convection is important in solids only if they contain fluid or gaseous inclusions.

Phonons are collective vibrations of atoms within solids (e.g. quantized lattice vibrations), caused by thermal excitation. An intuitive grasp of how phonons transport heat can be gleaned by returning to the model of a solid being composed of atoms held together by bonds that act like springs. Strong bonds behave as stiff springs, while weak bonds correspond to compliant springs. Stiff springs are efficient in transmitting vibrations (and hence heat) from the hot side to the cold side, because when one atom is vibrating, the motion is strongly coupled to that of the neighboring atoms. In contrast, for very weak springs, one end can be shaken hard without coupling the vibration to the other end. In much the same way, weakly bonded solids show smaller phonon contributions to the thermal conductivity over a wide range of temperatures.

The impact of bond strength on thermal conductivity is illustrated in Table 24.4. Notice the decrease in thermal conductivity as you move down the column four elements, from diamond to silicon to germanium in the direction of longer, weaker bonds. Similarly, because of their short strong bonds, diamond, aluminum nitride, and beryllium oxide are excellent thermal conductors. All three have been proposed as replacements for alumina in electronic packaging where thermal conductivity is needed to withdraw heat from silicon chips. For binary compounds like AlN and BeO, the thermal conductivity is particularly high due to strong bonding in addition to the relatively similar atomic masses of the constituent atoms. As a general trend, vibrations propagate more efficiently along a lattice when the mass contrast is small.

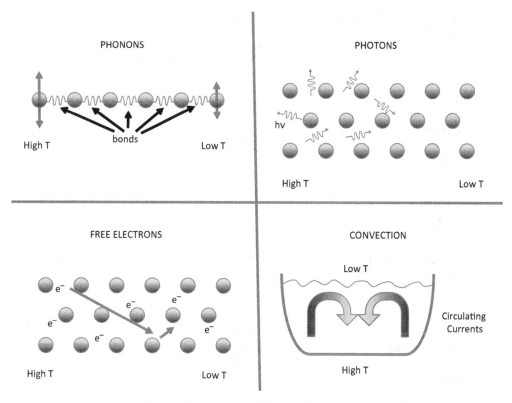

Fig. 24.13 Schematics of the primary mechanisms for thermal conductivity. The circles denote atoms.

Graphite illustrates the relationship between thermal conductivity and structure. Graphite conducts heat much better parallel to the carbon layers than in the perpendicular direction. In graphite, phonons, and to a lesser extent electrons, control thermal conduction. As was described above, strong springs, like the bonds *within* the graphite layer, pass along thermal energy more efficiently than do weak springs, like the long bonds *between* graphite layers. More generally, good thermal insulators have plenty of open spaces (think of the insulators used in buildings, such as fiberglass mats); good thermal conductors do not. Silica aerogels are good thermal insulators because of the trapped porosity and the weak links between particles.

The phonon contribution to the thermal conductivity is $\kappa = 1/3\rho C_V \lambda$, where ρ is the density, C_V is the specific heat at constant volume, and λ is the mean free path for the phonon. The mean free path describes the average distance that a phonon travels before it is scattered. This relationship helps explain the strong temperature dependence of the thermal conductivity as solids are heated from 0 K (see Fig. 24.14). Since C_V approaches zero at 0 K, so does κ. If no vibrational modes are active, phonon transport cannot occur. At very low temperatures, phonon generation leads to a rapid rise in the thermal conductivity, and the lowest frequency phonons dominate. However, phonons can be scattered by any perturbation in the periodic potential. Important contributions to phonon scattering include sample surfaces, defects such as grain boundaries, other

Table 24.4 Typical thermal conductivity values for several inorganic materials measured at room temperature

Material	Thermal conductivity (W/cm · K)
Diamond	9
Silicon	1.48
Germanium	0.599
Aluminum	2.37
Copper	3.98
Gold	3.15
Silver	4.27
Iron	0.803
Platinum	0.714
Tungsten	1.78
Graphite	
parallel to layers	20
perpendicular to layers	0.095
Alumina	
Single crystal	0.46
Polycrystal	0.36
Beryllia	2.72
Magnesia	0.60
Fire clay refractory	~0.011
Quartz	0.104
Silica glass	0.0138
Borosilicate glass	0.011

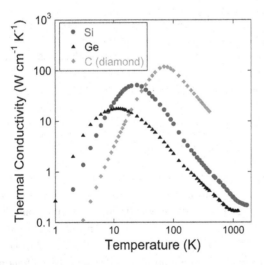

Fig. 24.14 Comparison between the thermal conductivities of diamond, silicon, and germanium. As the bond strength increases, the maximum in the thermal conductivity rises. Data from *J. Phys. Chem. Ref. Data 3 Supp. 1, Thermal Conductivity of the Elements: A Comprehensive Review*, eds. C. Y Ho, R. W. Powell, and P. E. Liley, Am. Chem. Soc. & Am. Inst. Phys. & Nat. Bur. Standards (1975).

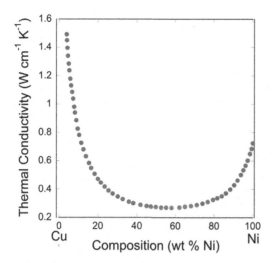

Fig. 24.15 Lattice thermal conductivity across the Cu–Ni alloy system. Data from M. W. Ackerman, K. Y. Wu, and C. Y. Ho, "Thermal conductivity and Lorenz function of Cu–Ni and Ag–Pd alloy systems," from *Thermal Conductivity 14*, eds. P. G. Klemens and T. K. Chu, Springer, New York (1976).

phonons, impurities, and even the presence of other isotopes of the constituent atoms. Phonon conductivity experiences a maximum as a function of temperature as illustrated in Fig. 24.14. This behavior can be explained by phonon–phonon scattering. As more phonons become active in the material and their vibration amplitude increases, they begin to scatter each other. For most solids a maximum value in the thermal conductivity occurs at $T < 50$ K, at which point the increase in scattering reduces thermal transport more than additional phonons boost it; the maximum is higher in diamond because the Debye temperature is high.

It is important to remember that anything that interrupts the perfect periodicity of a lattice can scatter a phonon, and hence reduce the thermal conductivity. Typical scattering sites include impurity atoms, grain boundaries, surfaces, or different isotopes of the constituent atoms. An example of the importance of scattering is shown in Fig. 24.15. Recall that Cu and Ni form a complete substitutional solid solution, so that the samples are single phase all of the way across the phase diagram. However, the thermal conductivity goes through a sharp decrease as the sample composition is changed from either pure Cu or pure Ni. This is because the phonons are scattered when they encounter the small changes in the periodic potential associated with substitution of Cu for Ni or vice versa.

Glass and other amorphous solids have smaller thermal conductivities than crystals because the phonon waves are scattered very strongly in aperiodic materials. Near room temperature, the mean free path for a phonon in vitreous silica is of the order of 10 Å, about the size of the silicate rings in the disordered structure of glass. When the mean free path is approximately constant, the thermal conductivity of amorphous solids increases with T as the specific heat increases. This accounts for the slowly rising thermal conductivity of glasses on increasing temperature. The thermal conductivity of

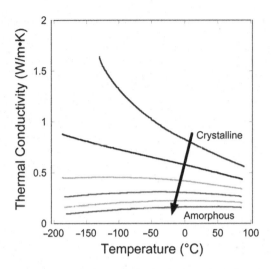

Fig. 24.16 Typical changes in thermal conductivity of a polymer as a function of the degree of crystallinity. Curves for 100%, 80%, 60%, 40%, 20%, and 0% crystallinity are shown. Adapted, by permission, from D. W. van Krevelen and K. te Nijenhuis, *Properties of Polymers*, Elsevier (2009). Fig. 17.3, p. 649.

glass increases rapidly above 1000 °C where photons help in transporting energy, as will be discussed below. The same is true with other transparent materials.

The trends associated with thermal conductivity and structure in crystalline inorganic materials are also expressed in polymeric materials. For example, most crystalline polymers, such as high-density polyethylene (HDPE), polypropylene (PP), and polytetrafluoroethylene (PTFE) possess higher κ values than amorphous polymers like low-density polyethylene (LDPE), atactic polystyrene (PS), and atactic polyvinylchloride (PVC). The latter group of polymers in this set features lower density and lower crystallinity which lead to more phonon scattering and less efficient thermal transport. This is illustrated in Fig. 24.16.

Stretched polymers have (partially) aligned chains, and conduct heat best along the direction of elongation, parallel to the covalently bonded backbone. Thermal conductivity values are smaller in transverse directions where weak van der Waals bonds are found. The anisotropy increases with chain length. The addition of small molecules as plasticizers generally reduces the thermal conductivity. Foamed polymers are excellent insulators because static air is a poor conductor. Thus, some coffee cups and house insulation are made from foamed polymers. Typical κ values of polymers are listed in Table 24.5.

A second key contribution to the thermal conductivity arises from *mobile charge carriers, such as electrons* in the solid. Free electrons transport heat, as excited electrons from the hot portion of the sample undergo collisions with the lattice in the colder region of the sample, transferring their energy to it. Thus, solids which have high electrical conductivities, σ (so that electrons move well in response to an applied electric field) also have high thermal conductivities. This correlation results in the Weidemann–Franz law: $\frac{\kappa}{\sigma T} = constant$, illustrated in Fig. 24.17.

Table 24.5 Thermal conductivities of selected polymers in W/mK

	κ		κ
High-density polyethylene	0.44	Low-density polyethylene	0.35
Polypropylene	0.24	PTFE	0.27
Polyurethane	0.31	Polyurethane (foamed)	0.03
Polyvinylchloride	0.16	Polyvinylchloride (foamed)	0.03
Polystyrene	0.16	Polystyrene (foamed)	0.04

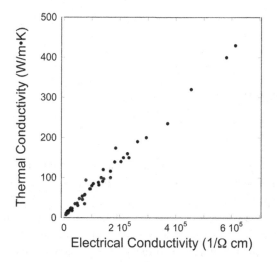

Fig. 24.17 Comparison between thermal conductivity and electrical conductivity for elemental metals, showing the linear correlation between the two.

In many metals near room temperature, the contributions to the thermal conductivity from phonons and photons are of the same order of magnitude. However, the electron contribution to the thermal conductivity is less temperature-dependent than the phonon contribution over a wide temperature range. Since the total thermal conductivity is the sum of all of the contributions, the thermal conductivity of metals often plateaus at temperatures above ~100 K, as shown in Fig. 24.18. This is in contrast to the temperature dependence of the thermal conductivity for intrinsic semiconductors, as seen in Fig. 24.14. In those materials, the electron contribution to the thermal conductivity is quite small, and the thermal conductivity continues to drop at elevated temperatures.

The third contribution to the thermal conductivity arises from *photons*. It is easily observed that objects heated to a high temperature glow, and that the hotter the object, the shorter the wavelength of the peak in the emitted radiation. Thus, photons emitted from the hot section of the material can be absorbed in a cold section of the material, transferring heat. If the material is too opaque, then the contribution from this mechanism is small relative to phonon conduction. Conversely, if the material is completely transparent to the radiation, then the photon will not be reabsorbed in the cold region, and no heat transfer takes place. An important consideration is the optical opacity of the material

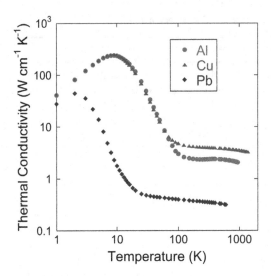

Fig. 24.18 Thermal conductivities as a function of temperature for selected metals.

in the spectral range of peak emission. Consider the temperature range above ~400 °C where strong emission exists between the near-IR and visible spectra. For example, if the material is a wide band gap insulator (e.g., Al_2O_3) with very few defects to absorb light, photons emitted at the hot side will exit the material without absorption and there is no thermal transport. Alternatively, if the hot material is a shallow direct band gap semiconductor (e.g., Ge) visible light photons are absorbed strongly and photon conduction occurs locally. Finally, if the material is an intermediate indirect band gap material with a low optical absorption coefficient (e.g., CdO) thermal energy can be transported over comparatively long distances. It is important to keep in mind that photon heat conductivity must always be considered in the context of the hot-side temperature, which establishes peak emission, and the optical properties in that spectral range of interest.

Finally, *convection* can contribute to the thermal conductivity in cases where circulating currents in liquids or gases transport heat.

24.5   Thermoelectricity

Thermoelectricity refers to the conversion of heat energy into electrical energy. Consider a conductor to which a thermal gradient is applied. The electronic carriers on the hot side are excited, and can migrate towards the colder end, producing a voltage across the material. This will continue until the density gradient of the charge carriers provides a counterbalancing potential that causes the system to reach a steady state. The resulting voltage across the conductor depends on the temperature gradient, defining the Seebeck coefficient in V/K. The Seebeck coefficient is widely used in thermocouples to measure temperature.

The thermoelectric effect can also be used to harvest electrical energy from systems that produce waste heat. Think, for example, of an exhaust manifold from a combustion

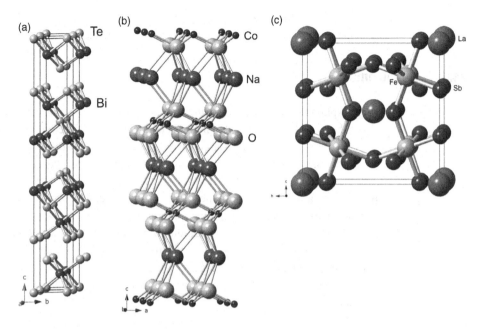

Fig. 24.19 Crystal structures of some important thermoelectric materials: (a) Bi_2Te_3, (b) β-Na_xCoO_2, and (c) skutterudite.

engine, where a gradient of several hundred degrees Celsius exists to ambient air. The voltage generated by this gradient across an efficient thermoelectric harvester could be used to charge a battery. The material's figure-of-merit for thermoelectric energy harvesting is $ZT = \frac{S^2\sigma}{\kappa}$, where S is the Seebeck coefficient, σ is the electrical conductivity, and κ is the thermal conductivity. Optimizing the energy harvesting thus requires a high Seebeck coefficient, a high electrical conductivity, and a low thermal conductivity. The latter two points are particularly challenging, since usually materials with a high electrical conductivity also have a high thermal conductivity. Thus, the challenge in designing new thermoelectric devices is to make a material that is simultaneously *an electron crystal and a phonon glass.* That is, the electronic carriers such as electrons or holes should be transported readily, while the phonon contribution to the thermal conductivity should be minimized by increasing phonon scattering (as in a glass).

Good candidates for thermoelectric applications exhibit a number of interesting crystal structures. These include the layer structure of Bi_2Te_3, the cobaltites, and the network structure of the skutterudites.

Consider for example, the skutterudite shown in Fig. 24.19. A typical composition would be $LaFe_3CoSb_{12}$. In these materials, the rare earth ion is too small for the cage in which it is found. The net result is that it "rattles" in its cage structure with a large amplitude. This localized vibration mode significantly reduces the thermal conductivity of the network, without greatly reducing the electronic conductivity. Bi_2Te_3 provides a second example where local structure can produce a favorable ZT. Bi_2Te_3 is a layered compound in which the atoms stack in a Te–Bi–Te–Bi–Te arrangement. The large atomic masses of Bi and Te, their large atomic mass contrast, and the layered structure produce a particularly low thermal conductivity, especially along the *c* axis. The partially metallic bonding, however, preserves a comparatively large electronic conductivity.

24.6   Problems

(1) **(a)** Explain the factors that contribute to the thermal conductivity of dense solids.

   (b) Shown in Fig. 24.15 is the thermal conductivity as a function of composition for the Cu–Ni system. Based on your answer to part (a), and your understanding of the phase diagram for this system, why do you think the curve has the shape it does?

(2) Based on what you know about the structures involved, why does silica glass have a lower thermal expansion coefficient than α-quartz?

(3) Describe the mechanisms that contribute to thermal conductivity. Rank order the following in terms of the thermal conductivity, and explain your reasoning: Cu, polystyrene, corundum, and rocksalt.

(4) For each pair, which material should have a higher thermal expansion coefficient? Explain why.

   (a) Na and W.

   (b) SiO_2 glass and quartz.

   (c) Polyethylene and NaCl.

(5) Based on what you know about the crystal structures, explain why you think the following materials have the thermal expansion coefficients that they do.

   (a) Graphite: normal to *c* axis, $\alpha = 1$ ppm/°C,

   parallel to *c* axis, $\alpha = 27$ ppm/°C.

   (b) MgO $\alpha = 13.5$ ppm/°C.

   (c) Quartz (see graph). (Figure adapted, by permission, from R. K. Kirby, "Thermal expansion of ceramics," in *Mechanical and Thermal Properties of Ceramics*, Proceedings of a Symposium Held at Gaithersburg, MD April 1–2, 1968, National Bureau of Standards Special Publication 303, issued May 1969. Fig. 6, p. 44.)

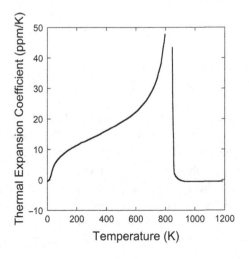

(6) The graph shows the thermal conductivity of several materials as a function of temperature.

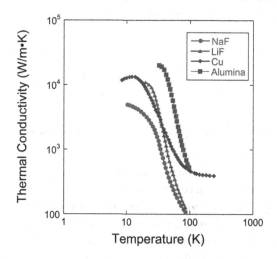

(a) Why do most curves show a decreasing thermal conductivity with temperature over the range shown?

(b) Why does alumina have a high thermal conductivity?

(c) Why does the curve for copper flatten out at ~100 K?

(7) Zircon ($ZrSiO_4$) is tetragonal (space group $I4_1/amd$, #141) with $a = 6.6$ Å and $c = 5.88$ Å. Zr is at (0,0,0,), Si is at (0,0,0.5), and O is at (0, 0.2, 0.34).

(a) Plot a unit cell of zircon showing the Si polyhedra.

(b) Below 1100 K, the a lattice parameter follows the temperature dependence: $a(Å) = 6.6003 + 126 \times 10^{-7}T + 82 \times 10^{-10}T^2$. Plot the lattice parameter and the instantaneous thermal expansion coefficient from 300 K to 1100 K.

(c) What is the value of the linear thermal expansion coefficient along the a axis at 300 K?

(8) Consider a rutile single crystal.

(a) Along which axis would you expect a higher thermal conductivity? Explain why.

(b) Compare the thermal conductivity for a polycrystalline ceramic to that of a single crystal.

(9) Na has a thermal expansion coefficient of ~70 ppm/K, while W has a thermal expansion coefficient of ~ 5 ppm/K. Explain this.

(10) Write an expression for the thermal expansion coefficient, defining all of the variables. Rank order the following materials in terms of their average thermal expansion coefficients and explain your reasoning in detail: Cu, SiO_2 glass, Ar, Pb, water ice.

(11) Draw a stereographic projection for point group $6/m$. Write an equation for thermal expansion, defining all of the terms. Would you expect the thermal expansion coefficients in the a, b, and c directions to be the same or different in a material with this point group? Explain your reasoning.

(12) The following table gives values for the average thermal conductivity of a number of materials.

Material	Thermal conductivity (cal/s · K · cm)	
	100 °C	**1000 °C**
Al_2O_3	0.072	0.015
BeO	0.525	0.049
MgO	0.09	0.017
Graphite	0.43	0.15
Fused silica	0.0048	0.006

(a) Explain the physical mechanisms that contribute to the thermal conductivity in solids, using data from the chart (or elsewhere) to illustrate your points.

(b) Why does the thermal conductivity of most materials decrease with increasing temperature, and yet the thermal conductivity of fused silica rises?

(c) Describe the structure of a material that should have an anisotropic thermal conductivity, detailing the origin of the anisotropy.

(13) The following graph shows the thermal conductivity of KCl, some of which has been doped. (Figure from J. B. Wachtman, Jr., *Mechanical and Thermal Properties of Ceramics*, National Bureau of Standards Special Publication 303 (1969), used with permission.)

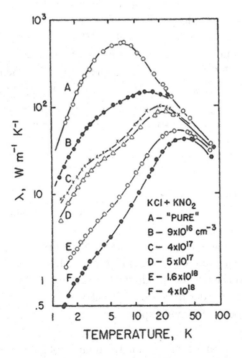

(a) Explain why the pure material has the observed temperature dependence of the thermal conductivity.

(b) On doping, the thermal conductivity is reduced primarily at low temperatures, but nearly converges with the data for the pure material above 50 K. Why is this?

(c) Would you expect the thermal expansion coefficient to be the same or different measured along [100] and [001]? Explain your reasoning.

(14) β-Silicon nitride is hexagonal with space group $P6_3/m$, and $a = 7.606$ Å, $c = 2.909$ Å. The Si atoms are at (0.174, –0.234, 0.25), and N atoms are at (1/3, 2/3, 1/4) and (0.321, 0.025, 1/4).

(a) Draw at least four unit cells looking down the c axis.

(b) The thermal expansion coefficient is approximately 1 ppm/°C. Explain why, based on the structure and bonding.

(15) Compare the thermal conductivity for diamond, copper, and polypropylene. Be specific about which mechanisms you believe dominate the response in each case. Which of these could be made anisotropic?

(16) Spinel, $MgAl_2O_4$ is cubic with space group $Fd3m$ (#227), with $a = 8.09$ Å. Al is at (0.625, 0.625, 0.625), Mg at (0,0,0) and O at (0.39, 0.39, 0.39).

(a) Plot a unit cell of spinel.

(b) Determine the coordination of each of the ions.

(c) Discuss the oxygen packing arrangement in this structure. (Hint: it may be easiest to do this by hiding the cations in your plot.)

(d) The lattice parameter of spinel as a function of temperature is given by: $a(Å) = 8.0739 + 3.565 \times 10^{-8}T^2 + 9.90 \times 10^{-11}T^3$ for T in K over the temperature range from 77–300 K. Plot the lattice parameter and the instantaneous thermal expansion coefficient as a function of temperature over this range.

(e) What is the value of the thermal expansion coefficient at room temperature? Based on the structure and bonding, why do you think the material has this thermal expansion coefficient?

(f) The structure plotted in part (a) is close to the structure shown by Fe_3O_4 at room temperature, with the exception that Fe_3O_4 is an inverse spinel, so that the Fe^{3+} is at (0,0,0), while the other Fe^{3+} and the Fe^{2+} are distributed at (5/8, 5/8, 5/8), and O is at (0.379, 0.379, 0.379). Determine the tetrahedral cation–O–octahedral cation angles in this compound. These angles are critical in the superexchange process that governs the magnetic ordering in the spinel ferrites.

(17) Describe the crystal structure of a material with strong bonds *and* a high thermal expansion coefficient. Explain the structure–property relations that produce this result.

(18) Consider C (diamond) and Ge. Both are group IV elements.

(a) Draw curves of the specific heat of these two materials as a function of temperature on the same graph. Justify why they have the shape that they do.

(b) Which should have a larger thermal conductivity at room temperature, and why?

(19) Consider the following graph (from I. David Brown, *The Chemical Bond in Inorganic Chemistry: The Bond Valence Model*, Oxford University Press, Oxford (2002). Fig. 9.4, p. 115. Used by permission.)

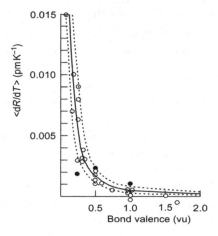

(a) In the crystal structure of α-Al_2O_3, the average Al–O bond length is 190 pm (1.9 Å); $d_{ij} = R_O - b\{\ln(v_{ij})\}$. For the Al–O bond, $R_O = 1.644$ Å, and b is 0.38. Determine the coordinations of the Al and O.

(b) Estimate the linear thermal expansion coefficient, based on the temperature dependence of the bond length (dR/dT). Does this make sense based on your knowledge of the structure and bonding?

25   Diffusion and Ionic Conductivity

25.0   Introduction

Diffusion is the movement of atoms (or less frequently molecules) through materials in response to a concentration gradient. Since most solids are comparatively dense, it can be difficult to move atoms through the small available interstices in solids. While heat capacity and many other bulk properties are relatively insensitive to defect concentration, this is not true for transport properties such as diffusion. In general, as the concentration of defects such as vacancies rises, the space available for diffusion through a lattice increases. Diffusion often controls the rate of solid state reactions, including densification.

A closely related property is that of *ionic conductivity* – the movement of ions through a material in response to an applied electric field. This property is critically important in batteries, some gas sensors, and fuel cells. This chapter describes the structure–property relations that govern the magnitude of diffusion coefficients and ionic conductivities.

25.1   Diffusion – Moving Atoms Through Solids

The diffusion coefficient, D, relates the flux of diffusing atoms to their concentration gradient: $\frac{dn}{dt} = -D\frac{dc}{dx}$, where n is the number of diffusing species passing through a unit area, t is time, c is the concentration, and x is position. Thus $\frac{dc}{dx}$ is the concentration gradient, and $\frac{dn}{dt}$ is the atom flux, the number of diffusing species passing through a unit area per second. The materials constant D is a second-rank tensor, and is specific to both the diffusing particle and the lattice through which it moves.

Figure 25.1 shows four mechanisms that enable diffusion within solids. In the *ring* mechanism, several atoms rotate their positions collectively. This produces comparatively little strain in the surrounding lattice. The *ring* and *interchange* mechanisms are more important in metals than in ionically bonded solids, because an exchange of neighboring cations and anions requires a great deal of energy. In the *vacancy* mechanism, an atom moves from its lattice position into an adjacent vacancy. Because the space available for this motion is comparatively large, atom diffusion occurs readily by this mechanism, with the limiting factor being the concentration of available vacancies. *Interstitial* diffusion occurs when an atom moves from one interstitial location to an adjacent one, without permanently perturbing the equilibrium position of surrounding

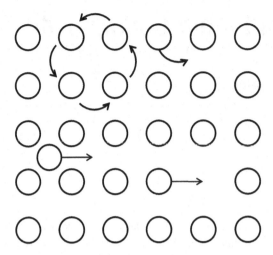

Fig. 25.1 Various diffusion mechanisms: ring, vacancy, interstitial, and dissociative.

atoms. This mechanism is limited by the space available in the channel between interstitial sites. *Vacancy* and *interstitial* processes require less energy than *interchange* because only one atom is displaced at a time, rather than two or more. *Interstitialcy* diffusion can occur when a comparatively large atom diffuses by displacing an atom on a lattice site to an interstitial position, so that it can take the lattice site. *Dissociative* diffusion involves moving an atom from its lattice site into an interstice. More complete information on diffusion is available in the book by Shewmon.[1]

Physical intuition suggests that diffusion should be strongly temperature-dependent. First, at elevated temperatures, thermal vibration increases in amplitude, and the material expands. Thus, at higher temperatures atoms have more energy available to overcome barriers. Thermal expansion slightly increases the size of the interstices. In principle, interstices between atoms can become instantaneously larger due to thermal fluctuations, facilitating squeezing atoms through small interstices. Finally, point defect concentrations rise as temperature rises towards T_{melt}, as discussed in Chapter 16. Thus, conditions for vacancy-based diffusion processes become more favorable with increasing temperature as well. A useful rule of thumb is that diffusion coefficients for an isostructural series of materials at a given temperatures are approximately inversely proportional to their melting points.

In most solids, the diffusion coefficient is $\sim10^{-15}$ to 10^{-18} cm^2/s at low temperatures such as room temperature. In contrast, for ions in solution, the diffusion coefficient at room temperature is on order 10^{-5} cm^2/s. To obtain practically useful solid state chemical reaction rates, the diffusion coefficient needs to rise considerably, to values on order 10^{-7} cm^2/s. This can be achieved by providing thermal energy to the material; as a rough rule of thumb, increasing the temperature to $\sim2/3$ T_{melt} raises the diffusion coefficients adequately. Diffusion is a thermally activated process; its temperature dependence can be written as $D = D_o e^{-\frac{E_A}{k_B T}}$, where D_o is a constant, T is temperature,

[1] Paul G. Shewmon, *Diffusion in Solids*, J. Williams Book Company, Jenks, Oklahoma (1983).

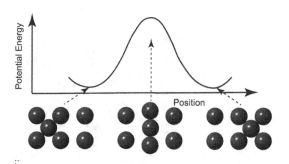

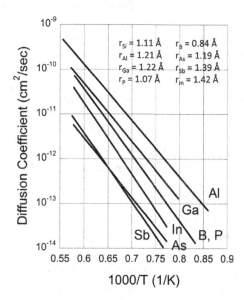

Fig. 25.2 (Top) Potential energy as a function of position showing the energy barrier associated with squeezing atoms through interstices. (Bottom) Diffusion coefficients for various atoms in silicon, showing that the diffusion is thermally activated. Also shown are the covalent radii for the atoms. This figure was drawn with permission from data in *CRC Handbook of Chemistry and Physics*, David R. Life ed. 81st edition, CRC Press, 2000, Boca Raton, FL, American Chemical Society.

k_B is Boltzmann's constant, and E_A is the activation energy associated with the potential barrier associated with moving the atom through the intermediate position to its new location. A characteristic energy barrier for diffusion is illustrated in Fig. 25.2, along with the temperature dependence of the diffusion coefficient of various atoms in silicon.

The size of migrating atoms is very important to the magnitude of their diffusion coefficients, with small atoms like carbon and hydrogen diffusing rapidly via interstitial sites in many metals. In general, smaller atoms have higher diffusion coefficients than larger atoms, *unless they are bound too strongly in position*. Thus, D depends on the nature of the diffusing atom and the host lattice. For rapid diffusion, the host lattice must provide suitable interstitial sites and the channels between those sites must not be too small. For example, carbon atoms in α-Fe and γ-Fe have different diffusion coefficients. The solubilities are also different: α-Fe (BCC) cannot dissolve as much carbon as FCC

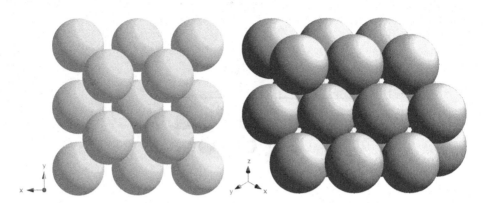

Fig. 25.3 Crystal structures of (left) BCC α-Fe and (right) FCC γ-Fe showing that there are smaller channels between interstices in FCC Fe, slowing interstitial diffusion.

γ-Fe, because the size of the interstices is larger in the FCC structure, as was shown in Chapter 10. However, the channels between those interstices are significantly smaller in the denser FCC structure. The very small opening between the close-packed atoms decreases the diffusion coefficient in γ-Fe, because the C must squeeze through a smaller channel to reach the next interstitial position. In α-Fe, the larger channels between the interstitial sites provide more space for the C to squeeze through, and higher C diffusion coefficients (see Fig. 25.3).

There are some exceptions to this general trend of smaller atoms having higher diffusion coefficients. Swalin's rule states that in metal alloys which are substitutional solid solutions, large A atoms in a matrix of B atoms tend to diffuse faster than the B atoms. This is a consequence of the fact that the lattice distortion caused by size misfit increases the abundance of vacancies near the A atoms. The higher local vacancy concentration supports the higher diffusion rates for the larger A atoms.

Anisotropy in the diffusion coefficients is related to the anisotropy of the crystal structure. For example, Mg has the hexagonally close-packed crystal structure. In hexagonal cells, the c axis is not related to a or b by symmetry. Thus, the diffusion coefficients along the a and c axes should be expected to differ. Experimentally, at 300 K, $D_x = D_y = 4.0 \times 10^{-24}$ cm^2/s, and $D_z = 4.5 \times 10^{-24}$ cm^2/s. In a more anisotropic hexagonal crystal, PbI$_2$ (see Fig. 25.4) at 316 K, $D_x = D_y = 5.02 \times 10^{-11}$ cm^2/s, and $D_z = 2.9 \times 10^{-11}$cm^2/s. This anisotropy is due to the layered structure of PbI$_2$. The larger diffusion coefficients parallel to the layers arise from the shorter jump distances required to move from one atomic position to the next.

In ionically bonded materials having open structures with well-connected channels, such as the zeolites, diffusion coefficients are large. In the same way, quartz has a relatively open crystal structure, with channels along the trigonal c axis. In directions normal to c, quartz is a good insulator, but parallel to c the diffusion coefficients and ionic conductivity are higher. At 250 °C, Na$^+$ ions are easily transported through the quartz crystal, but K$^+$ ions pass through less readily because of their larger size.

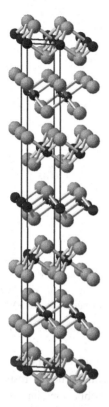

Fig. 25.4 PbI$_2$ crystal structure with the Pb atoms (darker color) in any layer sandwiched between iodine atoms (lighter color). The long axis is the c axis.

Activation energies for diffusion increase steadily with ionic radius, ranging from ~75 kJ/mol for Li$^+$ (~ 0.6 Å radius) to ~138 kJ/mole for Cs$^+$ (1.7 Å). Passage through the c axis channel becomes increasingly difficult for ions exceeding the channel diameter in size.

In ionically bonded solids, different ionic species often diffuse at vastly different rates, although electroneutrality requires that the diffusion coefficients be coupled. Diffusion coefficients for various ions in simple oxides are shown in Fig. 25.5. The diffusion coefficients of the large oxygen ions are small, except for special structures such as stabilized zirconia, which contains a large concentration of oxygen vacancies.

25.2   Ionic Conductivity in Oxides

If diffusing atoms are ionized, then in addition to ordinary diffusion mechanisms, they can also be driven through the lattice by application of an electric field. This is the origin of *ionic conductivity*, the contribution to the total electrical conductivity due to the

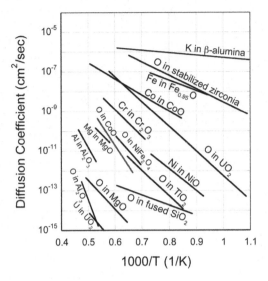

Fig. 25.5 Diffusion coefficients in a few oxide materials. Adapted, by permission, from W. D. Kingery, H. K. Bowen, and D. R. Uhlmann, *Introduction to Ceramics*, John Wiley & Sons, New York (1976).

movement of ions, rather than electrons or holes. Ions are much larger than electrons or holes, and therefore they usually have very much lower mobilities in response to the applied electric field. When similar concentrations of mobile ions and electrons are present, the electrons will dominate the overall conductivity. However, in some materials, where the concentration of mobile electrons or holes is low, the ionic conductivity can exceed the electronic conductivity by orders of magnitude.

Because diffusion and ionic conductivity both involve motion of atoms through the lattice, the ion mobility is thermally activated. The two quantities are directly related through the Einstein equation:

$$\sigma_{ion} = \frac{Dnq^2}{k_B T}.$$

Here, n is the concentration of charge carriers, q is the charge per ion, k_B is Boltzmann's constant, and T is the absolute temperature. Because of this thermal activation, ionic conductivity increases as temperature increases. In many cases, this mandates high operation temperatures (often $> 400\ °C$) for solid state ion conductors.

Ionic diffusion generally occurs by the movement of ions to neighboring vacancies. For pure stoichiometric compounds, the vacancy concentration and the ionic conductivity are generally small. In NaCl, σ_{ion} is on order 10^{-8} $1/\Omega m$ at room temperature, and 10^{-5} $1/\Omega m$ right below the melting temperature. At the melting temperature, σ_{ion} increases by about five orders of magnitude, since ions are freed from the bonds holding them in place. Impurities and other defects often play a decisive role in ionic conduction. In a salt crystal, the mobile charges may be interstitial ions (Frenkel defects) or vacancies (Schottky defects). The position of a vacancy changes when a neighboring ion moves to fill it. Schottky defects predominate in KCl, where both cation and anion vacancies occur. In AgCl, some Ag^+ ions occupy interstitial sites, increasing the ionic conductivity.

There are a number of common structural features of materials with high ionic mobility.

(1)   Since σ_{ion} is proportional to the carrier concentration, a large concentration of one type of ion should be mobile.

(2)   There must be empty space in the structure through which ions can move in order to increase the mobility of the ions. Thus, high levels of non-stoichiometry often increase ionic conductivity.

(3)   Open structures are favorable, because they provide lower activation energies for the ion jump from one position to another. Thus, glasses usually have higher σ_{ion} than crystals of the same composition, because there is more open space in the glass.

(4)   In crystalline solids, the network of ion channels responsible for ionic conductivity should be connected in at least two dimensions to avoid blockage of the channels. Many equivalent conduction paths are possible in high-symmetry directions in highly symmetric crystal structures.

(5)   There should be a framework (preferably a three-dimensional one) permeated by open channels through which the ions can move. This framework helps keep the structure from collapsing as ions are moved. It is also helpful if the framework contains one ion that is somewhat polarizable; the distortion allows the migrating ion to move more readily.

(6)   Small ions diffuse more readily than large ones, since they fit through the available interstices more readily. Thus, in many structures, the cation contribution to σ_{ion} is larger than the anion contribution to σ_{ion}. This is the case, for example, in solid NaCl, where the Na^+ carries most of the current.

(7)   Ions should not be too strongly bound into structures. That is, σ_{ion} requires that the bonds holding the migrating ion in place be broken and reformed as it moves. If the bond strength is too high, this becomes difficult, and the ionic conductivity drops. Thus, lower-charged cations may move more readily in multicomponent materials than highly charged cations. In many silicate glasses, for example, the network-forming cations are too strongly bonded in place to be mobile. An excellent example of this is observed in Na_2O–SiO_2 glasses, where the Na^+ ion is far more mobile than the Si^{4+}, even though it is nearly four times larger in size. Therefore, when low ion mobility is required (e.g. in display glasses for many cell phones and computers) alkaline-earth-modified, rather than alkali, glasses are typically preferred due to the stronger bonding of the 2+ cations. It is noted that the ionic conductivity of glasses depends on the heat-treatment conditions. The more open the glass structure, the higher the ionic conductivity. Annealing of a glass can change the ionic conductivity by an order of magnitude.

Materials that exhibit very high ionic conductivities with negligibly small electronic conductivity are sometimes referred to as superionic conductors. The ionic conductivity as a function of temperature is shown for a number of these superionic conductors in Fig. 25.7. The crystal structures of some of these compounds are discussed below.

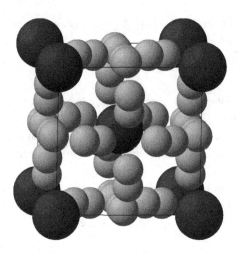

Fig. 25.6 Crystal structure of α-AgI. The iodines are on a BCC lattice, and the Ag are disordered amongst many positions (shown in gray).

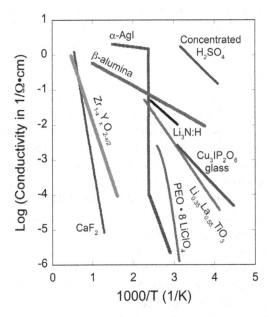

Fig. 25.7 Ionic conductivity as a function of temperature for a number of compounds. Adapted, by permission, from A. R. West, *Basic Solid State Chemistry*, 2nd Edn, John Wiley & Sons, New York (1999).

Very high ionic conductivity is observed in the α-form of AgI stable at temperature above 420 K, shown in Fig. 25.6. In this structure, the iodine atoms are arranged on a body-centered cubic lattice. The Ag ions are highly disordered amongst many possible, closely spaced sites – primarily tetrahedral sites on the faces of the unit cell. Stoichiometry demonstrates that only two of these sites will be filled per unit cell, but the large number of possible sites, and the very short distances required to move to an adjacent site means that the Ag conductivity is unusually high. Indeed, one can think of the Ag ions as being essentially melted long before the iodine lattice. The very sharp drop in conductivity at lower temperatures seen in Fig. 25.7 is associated with a phase transformation to either a wurtzite or zincblende crystal structure, where the Ag atoms are bound in position. In order to retain the higher ionic conductivity at lower temperatures, the material is often modified compositionally. $RbAg_4I_5$, for example, has a room temperature ionic conductivity about 17 orders of magnitude higher than that of NaCl. Its conductivity at higher temperatures is very similar to that of AgI, but it does not undergo the phase transition that locks the Ag^+ in place. In both AgI and $RbAg_4I_5$, the conductivity is improved because the Ag^+ ion has a modest radius (~1 Å); the $4d^{10}$ electron configuration also lends it a degree of covalency that helps polarize the I^- lattice, increasing the ion mobility.

Li^+ conductors are of considerable interest for Li-ion batteries. Among the important crystal structures for this application are ordered rocksalts such as $LiCoO_2$ (see Fig. 25.8). This compound has Li^+ and Co^{3+} ions alternating in adjacent (111)

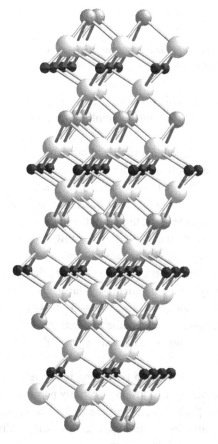

Fig. 25.8 Crystal structure of $LiCoO_2$, with Co as the smallest ion, Li as the darker, medium-sized ion, and oxygen as the largest, lightest ion. This is a derivative of the rocksalt structure, in which the cations are ordered along the (111) planes of the original structure.

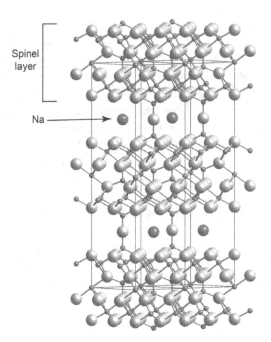

Fig. 25.9 Crystal structure of β-alumina. The spinel layers consist of Al and O. The O pillars separate the spinel layers, leaving considerable open space for Na^+ motion.

planes of the original rocksalt structure. This ordered arrangement is favored, since the two cations differ in size and charge. The high Li^+ mobility occurs parallel to the Li planes.

High Li^+ ionic conductivity is also observed in a variety of other structures, including some Li-containing spinels, Li-containing olivine structures, and Li_3N.

β-Alumina is a misnomer for compounds with the approximate composition $Na_2O \cdot 11Al_2O_3$. The crystal structure, shown in Fig. 25.9, has a hexagonal unit cell with cation-deficient spinel blocks separated by O pillars. In each of the spinel layers, the O is in a nearly close-packed arrangement with the stacking sequence ABCA. However, every fifth layer, 3/4 of the oxygens are missing. The Na ions are too large for the tetrahedral or octahedral sites in the spinel layers. As a result, they are expelled into the space between the layers. High Na^+ ionic conductivity occurs parallel to the layers. β″-Alumina has a closely related structure, and a typical composition close to $Na_2O \cdot 6Al_2O_3$. In these two structures, Ag^+ and Li^+ are close to the optimum size for the conducting ion. K^+ is too large to have a high mobility; H^+ is too small and bonds too strongly to the pillar oxygens, lowering its mobility.

It is not surprising that the ionic conductivity in β-alumina is strongly anisotropic, with high conductivity parallel to the planes, and low conductivity perpendicular to the planes. As a result, in polycrystalline β-alumina, where the individual crystallites are randomly oriented, the conduction paths do not line up, and the total conductivity is reduced below the in-plane value of the crystalline form.

In many samples of β-alumina, there is up to 25% more Na than would be expected for a stoichiometric compound. The extra positive charge is balanced by oxygen

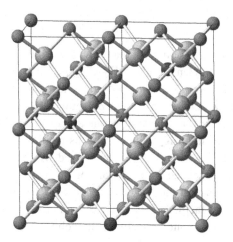

Fig. 25.10 Stabilized zirconia structure. The Zr atoms have 8-coordination, and are partially replaced by Ca atoms, shown in a different color. Both are on the face-centered positions of the cubic cell. For charge balance, oxygens are eliminated, leaving behind oxygen vacancies. In this set of four unit cells, three oxygen vacancies are shown.

interstitials in the conduction plane. The diffusion of Na^+ is slowed by these interstitial oxygens. To prevent this, some of the aluminum atoms can be replaced in the structure with lower-valent ions such as Mg^{2+} or Ni^{2+}.

Like many glassy ionic conductors, β-alumina shows a mixed alkali effect. That is, when two different alkali ions are present in a solid solution, the ionic conductivity is lower than for either end member.

Ionic conductivity can also occur on the anion sublattice. An excellent example of this is observed in the fluorite structure shown in Fig. 25.10. This is a low-density crystal structure, with large interstices at the center of the unit cell. Furthermore, the structure is unusually tolerant of large levels of nonstoichiometry. Consider, for example, stabilized zirconia: $Zr_{1-x}Ca_xO_{2-x}\square_x$. For each Ca^{2+} that is substituted for Zr^{4+}, one oxygen must be removed for charge balance. This introduces a large concentration of oxygen vacancies. The larger Ca^{2+} ion helps stabilize the fluorite structure, and prevents the transformation to the 7-coordinated Zr in baddelyite. The high ionic conductivity is then a result of several factors: (a) the high vacancy concentration, (b) the large interstitial spaces through which atoms can move, and (c) the short O–O distance (~2.6 Å) that facilitates motion.

Stabilized zirconia is widely used industrially as an oxygen-ion conductor for oxygen sensors and oxygen pumps.

There are related compounds with the pyrochlore crystal structure that also show ionic conductivity via the anion sublattice. Pyrochlores have the general chemical formula $A_2B_2O_7$. They can be regarded as derivatives of the cubic fluorite structure in which one out of every eight O atoms is missing. The vacancies are ordered in a regular fashion, reducing the coordination of the "A" cation to a distorted octahedral coordination, while "B" remains in cubic coordination.

25.3   Ionic Conductivity in Polymers

Some polymer materials also have high ionic conductivity when mixed with ionic salts. Good examples include polyethylene oxide $(CH_2-CH_2O)_n$ and polypropylene oxide $(CHCH_3-CH_2O)_n$. Consider, for example, polyethylene oxide mixed with $LiClO_4$. A simplified picture is shown in Fig. 25.11. The polyethylene oxide backbone has a zig-zag configuration; the chains are randomly coiled and twisted. The Li^+ is attracted to the electron density of the O in the chain, and its mobility results from hopping from site to nearby site. Such motions are facilitated by local structural relaxations in the polymer chains associated with rotations of chain segments around single bonds in the backbone. Segmental motion of the chains with high amplitude and high frequency also help. This is facilitated above the glass transition temperature of the polymer, so that the chains are not effectively immobile. Moreover, it is necessary to avoid crystalline phases in the polymers, as the Li-ion mobility drops in this case. The preparation of branched or cross-linked chains helps to minimize crystallization, as does incorporation of a plasticizer, or introduction of ceramic particles. When crystallization is avoided and the glass transition temperature is low, room temperature conductivities on the order of 10^{-5} $1/\Omega \cdot cm$ can be achieved. As is true in many crystalline ionic conductors, higher free volumes increase the ionic conductivity of polymers.

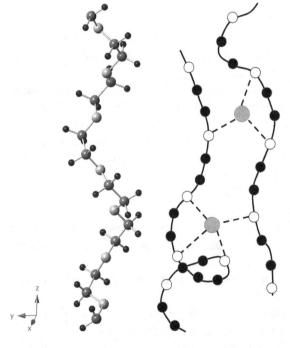

Fig. 25.11 (Left) One chain of polyethylene oxide. Along the backbone, the repeat unit is C–C–O. The side groups are H. (Right) Schematic showing bonding of the Li ions to the polymer backbone. Dark circles are C, open circles are O, larger gray circles are Li ions. It can be seen that the polymer chains coil around the mobile Li^+ ions.

It is notable that in amorphous structures like glasses and ion-conducting polymers, the energy barrier to ion movement varies as a function of position within the solid. The actual value for the ionic conductivity thus depends on the details of the structure, and hence the specific processing conditions. Generally, the more open the structure is, the higher the ionic conductivity.

25.4   Batteries

There are many different types of batteries. A few will be discussed here.

The nickel–metal-hydride (NiMH) battery is used in hybrid automobiles, consumer products, and stationary power supplies. It has significant advantages over other rechargeable batteries including cycle life, safety, and non-toxic materials. All commercial NiMH batteries use negative electrodes made of metal hydrides immersed in an aqueous electrolyte of potassium hydroxide, with nickel hydroxide as the positive electrode. NiMH batteries can be manufactured in almost any size from 10^{-2} A · hr to 10^{2} A · hr. Steel is compatible with the KOH electrolyte, so the batteries can be packaged in rugged steel cans.

The half-cell reactions are:

$$MH_x + xOH^- \rightleftharpoons M + xH_2O + xe^-$$

at the negative electrode and

$$NiOOH + H_2O + e^- \rightleftharpoons Ni(OH)_2 + OH^-$$

at the positive electrode.

During the discharge cycle, protons are transferred from the metal hydride to the nickel oxyhydroxide electrode, causing electrons to pass through the external load.

Nickel hydroxides have been used as electrodes in alkaline batteries for more than a century. There are two closely related crystal structures for nickel hydroxide. β-Ni(OH)$_2$ has the same crystal structure as portlandite Ca(OH)$_2$ and brucite Mg(OH)$_2$ (see Chapter 19). Each Ni^{2+} ion is bonded to six hydroxyl ions with a Ni–O bondlength of 2.07 Å. The O–H distance in the hydroxyl ion is 0.97 Å. A less-well crystallized form, α-Ni(OH)$_2$, is partially hydrated with water molecules between the hydroxide layers.

During the electrochemical cycle, Ni(OH)$_2$ is oxidized to NiOOH, nickel oxyhydroxide. The oxidation process involves the liberation of protons and a change of valence state from Ni^{2+} to Ni^{3+}. The basic brucite structure is retained throughout the reaction, but the changes in unit cell dimensions indicate changes in the internal structure involving interlayer water molecules.

At the negative electrode, hydrogen is the active material, moving in and out of the metal electrode. Hydrogen atoms occupy interstitial positions in the metal structure. Intermetallic compounds such as LaNi$_5$ are capable of holding between 1% and 7% H by weight. As a storage medium, the metal hydride has a better volumetric efficiency than liquid H$_2$. The crystal structure of LaNi$_5$ is illustrated in Fig. 25.12.

Lithium batteries are available in a wide variety of shapes and sizes. The electrode potential of lithium is the lowest among all metals. The electrochemical series

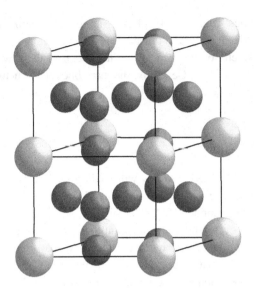

Fig. 25.12 The crystal structure of LaNi$_5$. La atoms are at the corners of the unit cell. The c direction is vertical.

shown in Table 25.1 indicates how easily a metal is oxidized or its ions are reduced. It serves as a measure of the comparative utility of different metals as battery materials.

In addition to its excellent electrochemical potential, lithium has the lowest density (0.54 g/cm^3) and electrochemical equivalent (0.26 gA/hr) of all solids. As a result, lithium batteries offer the possibility of high energy density and high voltage. There are many kinds of lithium batteries with a variety of cathode-active materials.

Many disposable lithium batteries are lithium–manganese dioxide batteries. Figure 25.13 shows a schematic view of the reaction during the discharge of the manganese dioxide crystals. Tetravalent manganese ions are reduced to trivalent manganese as lithium ions enter the lattice from a nonaqueous electrolyte such as lithium chlorate (LiClO$_4$). The reaction at the anode is

$$MnO_2 + Li \longrightarrow MnO_2^-(Li+).$$

The crystal chemistry of MnO$_2$ is rather complex, with several different polymorphs. Of special interest are those with tunnels that permit rapid movements of interstitial ions such as Li$^+$ or H$^+$. Pyrolusite (β-MnO$_2$) has the rutile structure (Fig. 25.14) with Mn^{4+} ions located in edge- and corner-sharing octahedra. A chain of edge-sharing polyhedra lies parallel to the tetragonal c axis. Interstitial Li$^+$ ions are accommodated in a chain of empty octahedra running parallel to the Mn-filled chain.

Larger tunnels are found in other polymorphs of manganese dioxide. β-MnO$_2$ is the stable form of stoichiometric manganese dioxide, but other polymorphs are observed in chemical reactions. Figure 25.14 compares the structure of ramsdellite and hollandite with pyrolusite. Double octahedral chains are observed in these structures, with correspondingly larger tunnels. The 1×1 tunnels in pyrolusite are enlarged to 2×1 tunnels in ramsdellite and an intergrowth of 1×1 and 2×1 tunnels in γ-MnO$_2$. Still larger

Table 25.1 Electrochemical series

Electrode reaction	Standard electrode potentials at 25 °C, in V
$Li^+ + e^- \leftrightarrow Li(s)$	−3.045
$K^+ + e^- \leftrightarrow K(s)$	−2.931
$Rb^+ + e^- \leftrightarrow Rb(s)$	−2.925
$Ba^{2+} + 2e^- \leftrightarrow Ba(s)$	−2.912
$Sr^{2+} + 2e^- \leftrightarrow Sr(s)$	−2.899
$Ca^{2+} + 2e^- \leftrightarrow Ca(s)$	−2.76
$Na^+ + e^- \leftrightarrow Na(s)$	−2.712
$Mg^{2+} + 2e^- \leftrightarrow Mg(s)$	−2.372
$Be^{2+} + 2e^- \leftrightarrow Be(s)$	−1.847
$Al^{3+} + 3e^- \leftrightarrow Al(s)$	−1.676
$Zn^{2+} + 2e^- \leftrightarrow Zn(s)$	−0.7618
$Fe^{2+} + 2e^- \leftrightarrow Fe(s)$	−0.447
$Cd^{2+} + 2e^- \leftrightarrow Cd(s)$	−0.403
$Tl^+ + e^- \leftrightarrow Tl(s)$	−0.336
$Ni^{2+} + 2e^- \leftrightarrow Ni(s)$	−0.257
$Sn^{2+} + 2e^- \leftrightarrow Sn(s)$	−0.1375
$Pb^{2+} + 2e^- \leftrightarrow Pb(s)$	−0.1262
$H^+ + e^- \leftrightarrow H_2(g)$.	0
$AgBr(s) + e^- \leftrightarrow Ag(s) + Br^-$	0.071
$Sn^{4+} + 2e^- \leftrightarrow Sn^{2+}$	0.151
$Cu^{2+} + e^- \leftrightarrow Cu^+$	0.158
$Bi^{3+} + 3e^- \leftrightarrow Bi$	0.308
$Cu^+ + e^- \leftrightarrow Cu(s)$	0.521
$Fe^{3+} + e^- \leftrightarrow Fe^{2+}$	0.770
$Hg^{2+} + 2e^- \leftrightarrow Hg(l)$	0.851
$O_2(g) + 4H^+ + 2e^- \leftrightarrow 2H_2O$	1.229
$Cl_2(g) + e^- \leftrightarrow 2Cl^-$	1.3583
$Au^{3+} + 3e^- \leftrightarrow Au(s)$	1.498
$Mn^{3+} + e^- \leftrightarrow Mn^{2+}$	1.5415

Data from *CRC Handbook of Chemistry and Physics*, 96th edition.

2×2 tunnels are found in hollandite. In electrochemical cells, these tunnels provide pathways for motion of charged ions. The manganese oxide in batteries is highly defective, with mixed structures and mixed valences. Mn^{4+}, Mn^{3+}, and Mn^{2+} (the three common valences of manganese) all occupy the type of octahedral sites found in defective pyrolucites like ϵ-MnO_2.

25.5   Fuel Cells

Fuel cells are electrochemical conversion systems that convert fuel directly, efficiently, and cleanly to electrical and thermal energy without the noxious emissions associated with combustion engines.

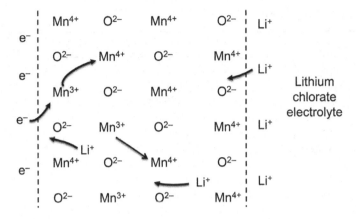

Fig. 25.13 Primary Li–MnO$_2$ battery undergoing discharge. Arrows indicate movements of electrons and Li ions.

Figure 25.15 shows the basic principles of a high-temperature *solid oxide fuel cell* (SOFC) operating on hydrogen fuel and using oxygen as an oxidant. Solid oxide fuel cells operate at high temperatures, typically 800 to 1000 °C. They use a ceramic electrolyte such as yttria-stabilized zirconia. At the cathode, oxygen reacts with incoming electrons from the external circuit, forming O^{2-} ions which migrate through the zirconia electrolyte to the anode.

$$\text{Cathode: } O_2 + 4e^- \rightarrow 2O^{2-}$$

At the anode, the oxide ions react with hydrogen to produce water vapor. The reaction is accompanied by the liberation of electrons to the external circuit.

$$\text{Anode: } H_2 + O^{2-} \rightarrow H_2O + 2e^-$$

The overall process is simply the reaction of hydrogen with oxygen to produce water and electrical energy.

$$\text{Cell: } 2H_2 + O_2 \rightarrow H_2O + \text{energy}$$

The open circuit voltage of a solid oxide fuel cell is around 1 V.

Alternatively, the SOFC can oxidize fuel gas, which is typically a mixture of H$_2$ and CO. SOFCs can reach efficiencies of 60% and are resistant to poisoning by carbon monoxide and sulfur contamination.

The electrolyte in a *proton-exchange membrane fuel cell* is a thin polymer membrane such as polyperfluorosulfonic acid (Nafion). The chemical structure for Nafion is shown in Fig. 25.16. The SO$_3$H sulfonic acid group is highly acidic because of the strong electronegativity of the nearby fluorine atoms. Nafion is representative of a class of polymers called *ionomers*. Typical applications for ionomers are thermoplastics, coatings, fuel cell membranes, ion exchange membranes, and permselective membranes capable of separating cations from anions. Fluorocarbon ionomers like Nafion consist of a linear perfluorinated backbone with side chains capable of bonding to other ions.

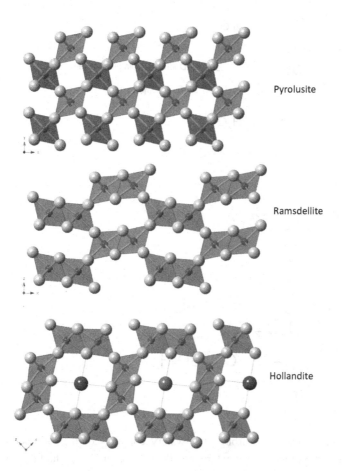

Pyrolusite

Ramsdellite

Hollandite

Fig 25.14 The crystal structures of three polymorphs of MnO_2: pyrolusite with 1×1 channels in between the MnO_6 octahedra, ramsdellite with 2×1 channels, and hollandite with 2×2 channels. The very large channels in hollandite need to be propped open with water molecules or foreign cations. In all of these structures, the oxygens (light ions) are in hexagonal close-packed layers. The Mn ions are octahedrally coordinated. Small ions such as lithium or protons move about in interstitial sites.

Ionic polymers are quite different from hydrocarbon polymers without ions. Ionomers interact strongly with water, forming hydrogen bonds between the water molecules and the fluorine atoms of the polymer chain. The absorbed water plays a significant role in the ion selectivity and ion transport. Proton conductivity and gas permeability are two properties that increase with water content within the membrane.

In a proton-exchange membrane fuel cell, the electrochemical reactions take place on the surface of a thin (typically ~0.2 mm) ionomer membrane. The membrane surfaces are coated with thin catalyst layers of platinum. The oxidation half reaction at the anode is initiated at the Pt surface by the dissociation of the hydrogen into protons and electrons:

$$H_2 \longrightarrow 2H^+ + 2e^-.$$

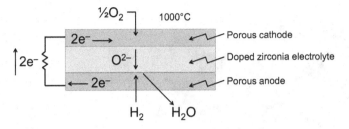

Fig. 25.15 High-temperature solid oxide fuel cell.

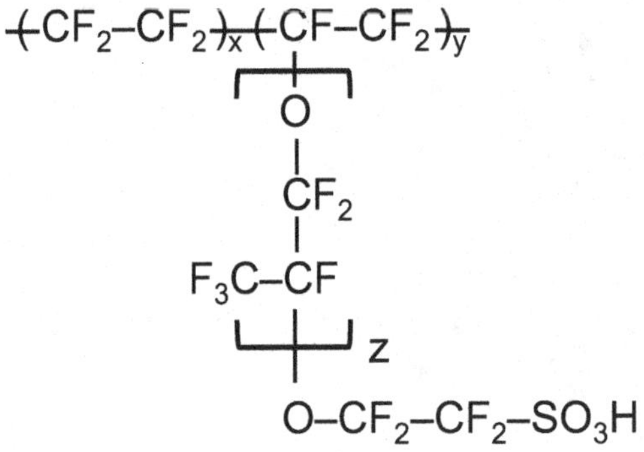

$$+CF_2-CF_2\!\!\rightarrow_{\!\!x}\!(CF-CF_2\!\!\rightarrow_{\!\!y}$$

O

CF$_2$

F$_3$C–CF

z

O–CF$_2$–CF$_2$–SO$_3$H

Fig. 25.16 Structure of Nafion.

The protons are conducted through the watery channels of the ionomer membrane, while electrons pass through the external load. At the cathode, the half reaction of O_2 is also catalyzed by platinum. Electrons from the external circuit combine with protons passing through the membrane:

$$O_2 + 4H^+ + 4e^- \longrightarrow 2H_2O.$$

The overall reaction is

$$2H_2 + O_2 \longrightarrow 2H_2O + \text{energy}.$$

Proton-exchange membrane fuel cells have a number of attributes that make them ideal candidates for automotive and domestic applications. They operate near room temperature, which allows them to start up rapidly from cold conditions, and they are compact in size because of their high power density.

25.6   Problems

(1)   What is meant by the vacancy solid solution in stabilized zirconia?

    (a)   Explain the structure–property relations that make stabilized zirconia a good ionic conductor.

(b) Draw the crystal structure for cubic zirconia (or a projection).

(c) The lattice parameter a for zirconia is 5.462 Å. The mass of Zr is 91.22 g/mole and the mass of O is 16 g/mole. Avogadro's number is 6.02×10^{23}/mole. Calculate the theoretical density of cubic ZrO_2.

(2) (a) Describe the crystal structure of one material with high ionic conductivity.

(b) Describe which ion is responsible for the conduction, and why the structure is helpful in achieving high ionic conductivity.

(c) For the compound that you chose, write the defect formation reactions for
   – a Schottky defect
   – a cation vacancy.

(3) (a) List four factors that influence the magnitude of the ionic conductivity of solids.

(b) Describe the structures of an anion and a cation conductor, explaining why high conductivity is favored in the two different structures.

(4) An idealized form of β-Al_2O_3 (a misnomer that has stuck for the composition $Na_2O \cdot 11Al_2O_3$), is hexagonal (space group $P6_3/mmc$) with $a = 5.595$ Å, $c = 22.49$ Å. Na is at (2/3, 1/3, 1/4), Al at (0, 0, 0); (1/3, 2/3, 0.022); (1/6, 1/3, –0.106); and (1/3, 2/3, 0.178), and O is at (1/6, 1/3, 0.05); (1/3, 2/3, –0.05); (0,0,0.144); (0.5, 0, 0.144); and (1/3, 2/3, 0.25).

(a) Plot a unit cell of this material.

(b) Compare the structure to that of spinel.

(c) What is the Na coordination in this compound? There is a considerable amount of open space available in this structure for the Na to move. This is the origin of the high ionic Na^+ conductivity of the material.

(d) Along which direction(s) would you expect the thermal conductivity to be higher? Justify your reasoning.

(5) How could insulating ZrO_2 be made into an electronic conductor? An ionic conductor? Be specific concerning the mechanisms and equations involved.

(6) $LiCoO_2$ has space group $R\bar{3}m$, with $a = 4.96$ Å, $\alpha = 32.97°$: Co is at (0, 0, 0), Li is at (0.5, 0.5, 0.5), and O is at (0.262, 0.262, 0.262).

(a) Draw a unit cell of the crystal structure.

(b) What crystal structure is this one a derivative of? Explain the structural relationship.

(c) This material is critical for Li battery applications. Why?

(7) Figure 25.7 shows the ionic conductivity of a number of materials as a function of temperature.

(a) Describe why ionic conductivity is strongly dependent on temperature.

(b) Explain why yttria-stabilized zirconia (YSZ) has a high O^{2-} ion conductivity, based on its structure.

(8) If it were possible to make stoichiometric MgO at 1600 °C, it can be shown that the concentration of Schottky defects would be ~1.4×10^{11}/cm^3, and the concentration of thermally promoted electrons and holes would be ~3.6×10^9/cm^3. Assume all of the ionic conductivity is due to the Mg. Given $D(V_{Mg}) = 0.38 \exp(-2.29 \text{ eV}/k_B T)$ in cm^2/s and $\mu_e = 24$ cm^2/V $\cdot$ s, and $\mu_h = 7$ cm^2/V $\cdot$ s, what fraction of the conductivity is ionic at this temperature?

(9) CuCl is an electronic p-type conductor at high chlorine pressures. As the chlorine pressure decreases, ionic conductivity takes over. Suggest a mechanism to explain this behavior.

(10) Consider a glass of composition $x\text{Na}_2\text{O} \cdot (1-x)\text{SiO}_2$. What is the principal electrical conduction mechanism in this glass? How would you expect the conductivity to change as a function of (a) composition (i.e. x) and (b) temperature?

26 Electrical Conductivity

26.0 Introduction

Electrical conductivity refers to the movement of electrically charged species in response to an applied electrical field. Metallic materials can be used to carry an electrical current or to create equipotential surfaces. Electrical insulators are used to block currents, and are technically important to separate signal-carrying lines, to prevent electrical shorting, and sometimes to store electrical energy. Semiconductors have conductivities in an intermediate range, due to their smaller band gaps. Semiconductors are essential to numerous technologies, including integrated circuits, and some sensor applications. A comparison of the electrical resistivities of typical undoped metals, semiconductors, and insulators is shown in Fig. 26.1, where the resistivity is the inverse of the conductivity. The origins of these curves, as well as modifications that can lead to changes in the observed behavior are discussed in this chapter.

Electrical conduction can occur via the motion of any charge carrier, including electrons and holes, ions, or charged vacancies. The electrical conductivity, σ_{total}, is the sum of all of the mechanisms that contribute to the response. That is:

$$\sigma_{total} = \sum_i n_i q_i \mu_i,$$

where n_i is the concentration of carriers of species i, q_i is the charge per carrier, and μ_i is the mobility of the carrier. The Einstein notation of Chapter 23 is employed in describing the electrical conductivity as a second-rank tensor. For metals, the temperature dependence of the conductivity is a function of the mobility term; in band-type semiconductors, the factor dominating the temperature dependence is the carrier concentration. This chapter will focus on *electronic conductivity*, due to electrons or holes, including both band conduction and charges hopping from one localized state to another location. Ionic conductivity is discussed in detail in Chapter 25.

26.1 Energy Bands

In order to understand electrical resistivity, it is imperative to have some grasp on how electrons behave in solids. Quantum mechanics teaches us that electrons have both particle-like and wave-like characteristics. Thus, the electron can be described using a wavefunction, Ψ. The allowed energy levels of electrons in isolated atoms are defined

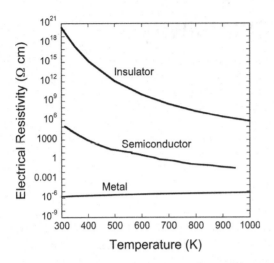

Fig. 26.1 Typical electrical resistivities of an insulator (SiO_2), a semiconductor (Si), and a metal (Cu).

by quantum mechanics, and are familiar to us in terms of characteristic wavelengths for absorption of electromagnetic radiation.

In order to develop some understanding of how band structure emerges in solids, consider the simple case of the interaction of two atoms with one electron to share between them, e.g. the molecule H_2^+. One approach that can be employed to describe this case is the coupled-mode approach adopted by Feynman.[1] Schrodinger's equation for the electron can be written as:

$$\left(\frac{-\hbar^2}{2m}\nabla^2 + V\right)\Psi = i\hbar\frac{\partial\Psi}{\partial t}, \qquad (26.1)$$

where $\hbar$ is Planck's constant/2π, m is the electron mass, V is the potential (typically written as a function of the position vector $\vec{r}$), t is time, and Ψ is the wavefunction. This can be re-written as:

$$H\Psi = i\hbar\frac{\partial\Psi}{\partial t}, \qquad (26.2)$$

where H is the Hamiltonian. Via separation of variables, the solution Ψ can be written as:

$$\Psi = \sum_j w_j(t)\Psi_j(\vec{r}), \qquad (26.3)$$

where Ψ_j are different possible states for the electron. Sensible states for the electron would be around one or the other nucleus, as shown in Fig. 26.2. It would be useful to

---

[1] R. P. Feynman, R. B. Leighton, and M. Sands, *The Feynman Lectures on Physics*, Addison-Wesley Publishing Co., Reading, MA (1963).

State 1   State 2

Fig. 26.2 Two possible states for an electron in a molecule H_2^+. Atomic nuclei are shown as dots; the electron is shown orbiting one of the nuclei. Adapted from *The Feynman Lectures on Physics, Vol. 2: Mainly Electromagnetism and Matter* by Richard P. Feynman, Robert B. Leighton, and Matthew Sands, Copyright © 2011. Fig. 10-1, p. 10-2. With permission of Basic Books, an imprint of Perseus Books, LLC, a subsidiary of Hachette Book Group, Inc.

know the probability that the electron is in a particular state at a particular time. This allows one to ignore the spatial variation, and concentrate just on the time dependence.

$$\sum_j w_j H \Psi_j = i\hbar \sum_j \Psi_j \frac{dw_j}{dt}. \tag{26.4}$$

As outlined in Solymar and Walsh's text,[2] this ultimately leads to the following situation:

$$i\hbar \frac{dw_k}{dt} = \sum_j H_{kj} w_j, \tag{26.5}$$

where $H_{kj} = \int \Psi_k H \Psi_j dv$ and dv is the volume element.

For simplicity, we can consider only two states

$$i\hbar \frac{dw_1}{dt} = H_{11} w_1 + H_{12} w_2$$
$$i\hbar \frac{dw2}{dt} = H_{21} w_1 + H_{22} w_2. \tag{26.6}$$

If H_{12} and H_{21} are both zero, then the two states are not coupled. Thus, once the electron is in state 1 (i.e. around nucleus #1), it stays there. This might make sense if the two nuclei are very far apart from each other. In this case, there is little incentive for the electron to leave the attractive potential of one nucleus, to go off in search of the other. However, if the two nuclei are close together, there is a finite probability that the electron can tunnel through the potential barrier and get to the other proton. In this case, the states are coupled.

One simple case is obtained when $H_{11} = H_{22} = E_o$ and $H_{12} = H_{21} = -A$, where A describes the coupling between the states. Larger A values correspond to higher tunneling probabilities. Since A is related to tunneling, it must drop exponentially with distance. E_o is the energy of the states, and might also be expected to rise quickly if the protons get too close together. In this case, as the energy is plotted as a function of separation distance, a graph like Fig. 26.3 is obtained. Thus, this approach predicts both the bonded ground state of the molecule, as well as an excited state.

[2] L. Solymar and D. Walsh, *Electrical Properties of Materials*, 6th edn, Oxford Science Publications, New York (1998).

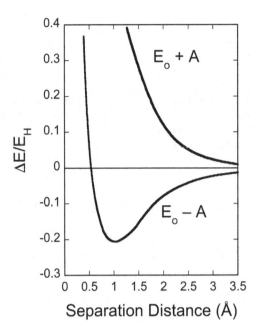

Fig. 26.3 Ground ($E_o - A$) and excited ($E_o + A$) states as a function of atomic separation distance between the two nuclei in H_2^+. E_H is the energy of the electron in an isolated H atom. Adapted from *The Feynman Lectures on Physics, Vol. 2: Mainly Electromagnetism and Matter* by Richard P. Feynman, Robert B. Leighton, and Matthew Sands, copyright © 2011. Fig. 10-3, p. 10-3. With permission of Basic Books, an imprint of Perseus Books, LLC, a subsidiary of Hachette Book Group, Inc.

Unfortunately, as the number of electrons in the system increases, it becomes progressively more difficult to completely describe the wavefunctions of the system. However, many of the same ideas associated with ground states and excited states carry over to crystals. The key point to recognize in thinking about this is that the more electrons that are in the material, the more the energy levels split slightly, to avoid violating the Pauli exclusion principle. The net result is that the discrete energy levels of the electrons in the atom form a series of very closely spaced energy levels in bands, separated by energy levels which are forbidden (the so-called band gaps). This ultimately produces the band structure for the solid. A schematic illustration for the case of Na is provided in Fig. 26.4.

26.2   Metals

Conventional metals can be envisaged as an array of ion cores that donate electrons to a sea of electrons holding the atom cores together. These electrons are then available at all temperatures for conduction. In the parlance of quantum mechanics, metals have partially filled conduction bands or overlapping filled and unfilled bands at all temperatures. More explicitly, materials with only a single valence electron will have partially filled valence bands; many metals have this characteristic. When a material has a partially filled valence band (that is, there are more electronic states than valence electrons) the material will be a good conductor. Partially filled valence bands are also

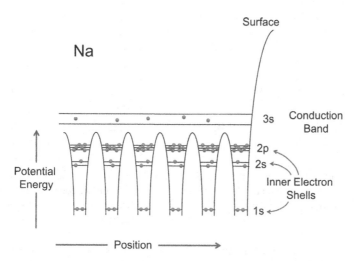

Fig. 26.4 Schematic illustration of the development of bands and band gaps based on the atomic orbitals of the atoms, for the case of Na. The energy levels are shown by lines, the electrons by small dots. Each potential well is associated with an atom core. Note that the energy levels of orbitals which are deeply buried in the atom are very little changed by near neighbors. However, orbitals that interact with the neighbors split into closely spaced energy levels that become bands of allowed energy, separated by energy gaps. In the case of Na, the outermost band, originating from the 3s orbitals, is partially filled. Adapted, by permission, from figure in M. S. Tyagi, *Introduction to Semiconductor Materials and Devices*, John Wiley and Sons (1991).

often called conduction bands. A material may also be a good conductor if it has two (or more) valence electrons if the resulting filled band overlaps with an empty (or partially filled) band. In either case, the result is that the valence electrons are free within a metal, and within the metal, the potential is approximately constant. Furthermore, the electrons in the metal are confined in the metal (i.e. they don't float around in free space above the metal), which means that there must be an energy barrier for electrons at the material's surface. This energy barrier is the *work function*, E_w, of the metal.

Electrons prefer regions of lower potential, while holes are attracted to regions of high potential, or as my first quantum mechanics instructor put it "Electrons like to ski, holes like to mountain climb."[3] Many of the inner electrons are still bound to the atom core; thus, the potential is locally lower at those points, as shown in Fig. 26.5. Nonetheless, to a very good approximation, in the metal, the potential can be treated in the simplified way shown in the same figure. For anyone that has taken a class in quantum mechanics, this looks very like the classic "electron in a box" problem, where the size of the box is now the size of the metal. For macroscopic samples (of order 1 cm on a side), this means that the energy levels are quasi-continuous, with energy spacings between allowed levels on the order of 10^{-19} eV. In practice, even the largest energy spacings in the conduction band are still several orders of magnitude lower than thermal energy at room temperature.

When an electric field, E_j is applied to a metal, the electrons are accelerated opposite to the electric field. In practice, the electrons cannot be accelerated indefinitely due to

[3] Professor Wayne Huebner.

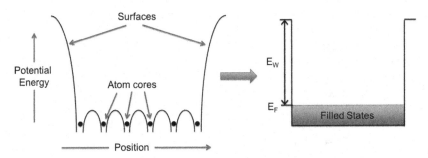

Fig. 26.5 Schematic illustration of the potential energy of a metal as a function of position. The figure at the left shows potential wells for each atom core, and a higher potential at the surfaces. In the figure at the right, the potential profile is simplified to treat the metal as having a constant potential inside. E_F is the Fermi energy, E_w is the work function.

collisions of the carriers with impurity atoms, phonons, or other electrons. Instead, electrons attain an average drift velocity, $\langle v_d \rangle = \frac{-|e|E\tau}{m}$, where e is the fundamental charge of the electron (1.602×10^{-19} C), m is the free electron mass (9.11×10^{-28} g), and τ is the average time between collisions. The current density, J, is related to the applied field via the electrical conductivity of the material:

$$J_i = \sigma_{ij}E_j.$$

Thus, the electrical conductivity, σ, is given by

$$\sigma = \frac{ne^2\tau}{m} = nq\mu,$$

where n is the number of carriers per unit volume, and μ is the carrier mobility. The *electrical resistivity* is $1/\sigma$.

The key concept to remember in thinking about scattering is that carriers can be scattered by any discontinuity in the periodic potential shown in Fig. 26.5. A schematic illustrating common scattering mechanisms is shown in Fig. 26.6. As was shown in Chapter 24, at finite temperatures, atoms oscillate around their equilibrium lattice positions, as phonons are excited. These phonons scatter carriers more at elevated temperatures, thus, the electrical conductivity of metals decreases as temperature increases. For many metals, the electrical conductivity is approximately proportional to T^{-1} near room temperature. Scattering can also occur owing to ionized impurities. Because there is some Coulombic interaction (attraction or repulsion) between the electronic carrier and an ionized species, the path of the carrier is deflected (scattered) by the charged impurity. At low temperatures, where the carriers have lower kinetic energy, the carrier spends more time close to the ionized impurity, and the effect on the electron path is larger. At higher temperatures, the carrier is influenced by the impurity for less time, and thus this mechanism is less effective as a scattering mechanism at higher temperatures. Neutral impurity scattering is caused by the variations in the periodic potential associated with replacing a host atom with a different element. This, too, leads to scattering. In principle, carriers can also scatter each other, or scatter from vacancies, but in most cases, these are a minor contribution to the total scattering. Finally, large imperfections in crystals, such as dislocations, stacking faults, grain

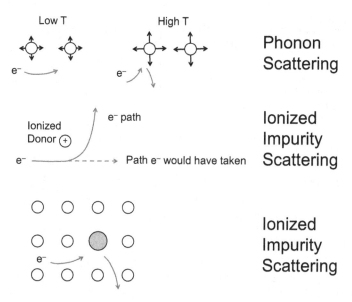

Phonon Scattering

Ionized Impurity Scattering

Ionized Impurity Scattering

Fig. 26.6 Typical electronic carrier scattering mechanisms.

boundaries, and surfaces also scatter carriers. This can radically reduce carrier mean mobility. The net mobility is given by:

$$\frac{1}{\mu} = \frac{1}{\mu_{phonon}} + \frac{1}{\mu_{ionized\ impurities}} + \frac{1}{\mu_{neutral\ impurities}} + \frac{1}{\mu_{defects}} + \frac{1}{\mu_{carrier-carrier}} + \cdots$$

The mobility can vary markedly as more mechanisms are introduced for scattering. For example, single crystal Si with few impurities has an electron mobility of about 1400 cm^2/V · s at room temperature. In contrast, in amorphous Si, where there are variations in the atomic positions and no long range periodicity to the potential, the room-temperature mobility is ~1 cm^2/V · s.

Near 0 K, where there are fewer phonons, and the amplitude of the phonons is smaller, the minimum electrical resistivity of a metal decreases as the impurity concentration drops (see Fig. 26.7). The more impurities disrupt the perfect periodicity (e.g. due to a different atomic size), the higher the electrical resistivity becomes. *Matthiessen's rule* says that for small concentrations of impurity atoms, the increase in resistivity caused by the impurity atoms is independent of temperature. Thus, the impurity concentration controls the saturation value for the electrical resistance in metals below ~10K. The higher the dopant concentration, the larger is the background level of scattering.

The composition dependent conductivity in many metal alloys is described by Nordheim's rules. For binary alloys, *Nordheim's rule* states that the resistivity is proportional to $x \times (1 - x)$, where x and $1 - x$ are the mole fractions of the two constituents. Thus, in many metallic substitutional solid solutions, the electrical resistivity is highest near the middle of the solid solution region, where the material is most disordered.

Electrical conductivities of metals are compared in Table 26.1. Silver is slightly more conductive than copper but it is far more expensive and tarnishes easily. The electrical conductivity of copper is sensitive to impurities (Fig. 26.7). Solid-solution elements will

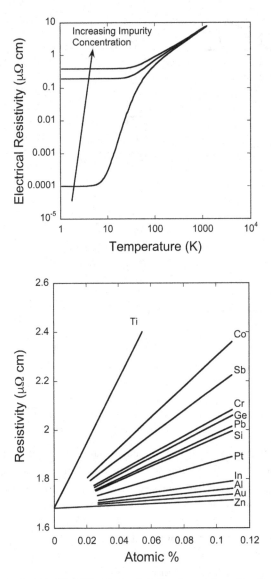

Fig. 26.7 (Top) Electrical resistivity of a metal as a function of temperature for various impurity concentrations. (Bottom) The electrical resistivity of Cu increases as a function of the amount of substitutional solid solution. The more similar the doping element is in terms of size and electronegativity, the less the electrical resistivity is increased. Figure redrawn from data in *ASM Specialty Handbook: Copper and Copper Alloys*, ed. J. R. Davis, ASM International (2001).

strengthen copper but lower the electrical conductivity, calling for alloys that balance the two effects. Note that silver, zinc, and cadmium increase the resistivity only slightly while higher-valent metals such as phosphorus (not shown) and titanium degrade the conductivity enormously. The addition of 1% Cd doubles the tensile strength of Cu while raising the resistivity by only 5%.

The presence of order–disorder phenomena has a profound effect on the electrical resistivity of metals. This is illustrated in Fig. 26.8, which shows the resistivity of ordered and disordered Cu–Au alloys. At low temperatures, the ordered structure is favored, as described in Chapter 20, and the electrical resistivity is low at compositions

Table 26.1 Room temperature electrical resistivities of several metals

Material	Electrical resistivity ($\mu\Omega$ cm)
Li	9.47
Na	4.77
K	7.39
Rb	13.1
Cs	20.8
Be	3.70
Mg	4.48
Ca	3.42
Sr	13.4
Ba	33.2
Al	2.709
99 wt% Al–1 wt % Cu	2.82
90 wt% Al–10 wt % Cu	3.67
80 wt% Al–20 wt % Cu	4.67
70 wt% Al–30 wt % Cu	5.41
60 wt% Al–40 wt % Cu	5.99
V	20.1
Cr	12.6
Mn	144
Fe	9.87
Ni	7.12
Cu	1.678
Zn	6.01
Ag	1.617
Au	2.255
Ta	13.4
W	5.39
Pb	21.1
Pd	10.73
Pt	10.7

where intermediate compounds form. For alloys processed at elevated temperatures, entropy is favored, and the Cu and Au are disordered on the sites of the FCC lattice. This disorder can be quenched in at room temperature if samples are cooled quickly enough. Electrons are scattered more effectively by the quenched disorder in the solid solution, raising the resistance compared with the ordered compound. Thus, the resistance rises at the order–disorder transition temperature.

The Fermi energy, E_F, describes the energy position of the highest filled state at a temperature of 0 K (whether or not a state exists at that energy position). At other temperatures, it is the energy at which the number of empty states below the Fermi energy is the same as the number of filled states above E_F. The probability that a given state at energy E_i will be filled by an electron is given by the Fermi–Dirac distribution:

$$f(E_i) = \frac{1}{\exp\left(\dfrac{E_i - E_F}{k_B T}\right) + 1}.$$

Because the Fermi probability function is symmetric, the Fermi energy, E_F, is also the energy where $f(E_F) = 1/2$. Figure 26.9 shows the Fermi–Dirac distribution at several

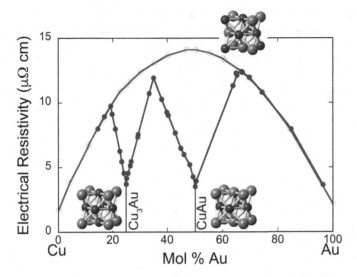

Fig. 26.8 Electrical resistivity of Cu_3Au as a function of compositions. Lower electrical resistivity is observed in the low-temperature ordered compounds relative to the disordered phase (gray curve).

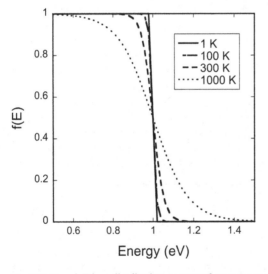

Fig. 26.9 Fermi–Dirac distribution at several temperatures, calculated for $E_F = 1$ eV.

temperatures. For temperatures exceeding 0 K, the electrons can absorb energy and have an energy value $E > E_F$. As temperature increases, the Fermi-Dirac distribution broadens progressively.

From a thermodynamics perspective, E_F describes the chemical potential of the electrons in the system at $T = 0$ K, and a reasonable approximation of the chemical potential of the electrons at other temperatures. The latter point is particularly useful in describing contacts between dissimilar materials. The Fermi energy must be constant across the junction in the absence of applied fields.

Whether or not an allowed energy level exists at a given energy for a given material is described by the density of states, $g(E)$. Real densities of states in materials are

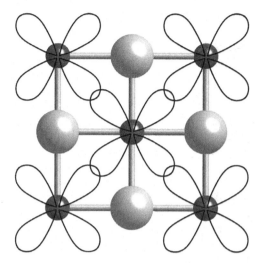

Fig. 26.10 Crystal structure of TiO. The d_{xy} orbitals overlap, producing a partially filled band and metallic conductivity. Adapted, by permission, from A. R. West, *Solid State Chemistry and Its Applications*, John Wiley and Sons, New York (1984).

complicated, and not easily predicted. That said, to a first approximation, the density of states rises approximately as $g(E) = C_1\sqrt{E - E_o}$, if E_o is the edge of a band.

In metals, the Fermi energy either lies in a partially filled band of allowed energies, or in overlapping filled and unfilled bands. Furthermore, E_F is a very weak function of temperature – it may shift only a few thousandths of eV as the temperature rises 1000 K. Thus, conduction electrons are available at all temperatures.

This relatively simplistic model describes most of the properties of elemental metals, including the high thermal and electrical conductivities, as well as the properties of photoemission, thermionic emission, and specific heat.

Not all metallic conductors are made from metallic elements or alloys. In addition to band conduction due to s and p bands, it is possible to get narrow bands that arise from the overlap of d orbitals. Consider, for example, TiO, which has a rocksalt crystal structure. As shown in Fig. 26.10, if the separation distance between Ti^{2+} ions is below a critical distance, the d orbitals overlap, and hence can form narrow d bands. For Ti^{2+} there are two electrons in the d shell, and they preferentially occupy the d_{xy}, d_{yz}, and d_{xz} orbitals due to crystal field splitting for the octahedral coordination of rocksalt (as shown in Chapter 14). Because the d band is partially filled, the material is a metallic conductor.

Metallic conductivity of this type is likely when the transition metal cations approach closely; e.g. in structures that have edge- or face-sharing of the cation polyhedra, rather than corner sharing. This is the origin of the metallic electrical conductivity in VO_2 (rutile structure) and Ti_2O_3 (corundum structure). The 4d and 5d transition metal series, where the d orbitals extend to a larger fraction of the ionic radius, have increased probability of d orbital overlap, and hence increased likelihood of metallic conductivity.

Returning to Fig. 26.10, two other scenarios exist. The first occurs if the transition metal cations are far apart relative to the spatial extent of the d orbitals. In this case, the d electrons are bound to the transition metal nucleus, so that crystal field theory will

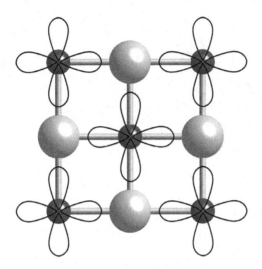

Fig. 26.11 Crystal structure of NiO. The $d_{x^2-y^2}$ and d_{z^2} orbitals point directly at the oxygen atoms. No band is formed, and NiO is an electrical insulator. Adapted, by permission, from A. R. West, *Solid State Chemistry and Its Applications*, John Wiley and Sons, New York (1984).

apply. MnO is an excellent example of this; it is an electrical insulator with a light pink color associated with the splitting of the Mn d electron energy levels.

A second possibility is illustrated by NiO. Ni^{2+} is a $3d^8$ ion, so that the electron configuration is shown in Fig. 26.11. In this case, the band that would form from the d_{xy}, d_{yz}, and d_{xz} orbitals is full, and hence does not contribute to conduction. The $d_{x^2-y^2}$ and d_{z^2} point directly towards the oxygen neighbors, blocking the formation of d-only bands. Thus, this material is also a band electrical insulator (though some conductivity can occur due to electron hopping as described in this chapter).

In a limited number of cases, metallic conductivity can develop due to mixing of the d orbital from the transition metal with the 2p orbital of the anion (e.g. oxygen). Examples of this include the excellent metallic conductivity appearing in ReO_3 – a perovskite derivative structure in which all of the A sites are vacant.

A. R. West[4] summarizes key trends for metallic conduction in transition metal oxides in the following way:

(1)   Low cation charges are helpful, as this both leaves some electrons on the cation, and increases the cation size.

(2)   For transition metals in octahedral coordination, electron overlap is easier for the d_{xy}, d_{yz}, and d_{xz} orbitals. Thus, cations near the beginning of the transition metal series are favorable for metallic conductivity.

(3)   4d and 5d atoms are involved.

(4)   The anion has a comparatively low electronegativity. This tends to decrease the band gap. As a result, NiO is much more insulating than NiS, NiSe, or NiTe.

---

[4] A. R. West, *Solid State Chemistry and its Applications*, John Wiley and Sons, New York (1984).

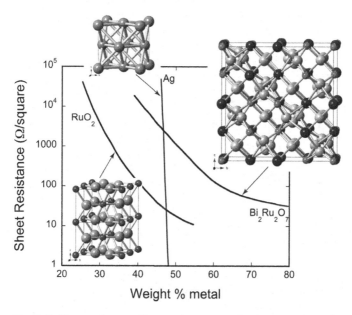

Fig. 26.12 Sheet resistance (sheet resistance $= \frac{\rho}{thickness}$) of a number of metallic composites as a function of composition. Many elemental metals have very steep drops in resistance once a percolation path for the metal particles has been achieved. This makes control of the resistance difficult. In contrast, some oxide electrodes allow finer tunability of resistance via formulation.

Such conducting transition metal oxides are used in a variety of applications, including thick film resistors, and transparent conductors. Oxide conductors used in thick-film resistors include RuO_2 (rutile structure) and $Bi_2Ru_2O_7$ (pyrochlore structure). The structures of these compounds are shown in Fig. 26.12. Powders of $Bi_2Ru_2O_7$ are typically mixed with a low-melting borosilicate glass and an organic vehicle to make a thick paste that can be screen-printed onto a substrate. Following drying and firing to remove the organics and densify the layer, the result is a network of metallic particles separated by thin insulating glassy regions. The resistance value and its temperature dependence can be adjusted by the relative amount of the conducting particles.

Another important metallic conductor is LaB_6. B is smaller than most other metallic atoms, and has a strong tendency to self-associate. In LaB_6, the B atoms are arranged in octahedral clusters that link together into a three-dimensional network, with the La atoms interspersed into large interstices in the structure (Fig. 26.13). Because LaB_6 has a low work function, compared to metals such as tungsten, it can be used in high brightness electron sources in electron microscopes.

26.3   Semiconductors and Insulators

In both semiconductors and electrical insulators, there is a finite band gap, E_g, in the material; the distinction between the two is somewhat arbitrary. Many early developments in semiconductor devices used semiconductors with band gaps <2 eV. For

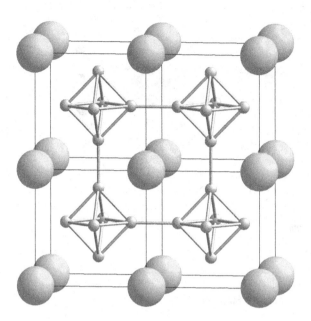

Fig. 26.13 Crystal structure of LaB_6. B_6 octahedra are interconnected in three dimensions with B–B bonds. La fits into the large interstices.

example, the band gaps of Ge, Si, and GaAs are 0.67 eV, 1.12 eV, and 1.424 eV, respectively. More recently, semiconductors with much larger band gaps have found application, including SiC, GaN, and ZnO with band gaps of 2.86 eV (6H SiC), 3.2 eV, and 3.37 eV, respectively. Diamond is also a semiconductor, with a band gap of approximately 5.6 eV; undoped diamond is an excellent electrical insulator. The band gap arises because the outmost electrons are confined in the bonds. Thus, the assumption of a constant potential inside the material (as was done for a metal) is inappropriate. Instead, the properties of semiconductors and insulators can be thought of as resulting from a periodically varying potential. The resulting periodic potential produces regions of allowed energy levels separated by band gaps, as shown schematically in Fig. 26.14 (see also the discussion on this subject in Chapter 8). That is, electrons experiencing a periodic potential can have only certain energies. The more strongly bound the electrons are, the narrower is the range of allowed energy zones.

As was described in Chapter 8, for a given structure type, the band gap increases as either the bond strength or the ionicity increases. Thus, band gaps are larger for compounds with shorter bond lengths.

A key to understanding conductivity in band semiconductors is the observation that only partially filled bands contribute to conduction. Therefore, at 0 K, where all of the states below the Fermi energy are filled, and all the states above the Fermi energy are empty, the electrical resistivity of an undoped semiconductor is very high.

Electrons promoted over the band gap to the conduction band leave behind an empty state, called a hole, in the valence band. In a pure material without defects, the concentration of electrons in the conduction band, n, is identical to the concentration of holes in

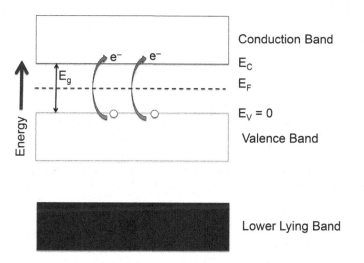

Fig. 26.14 Schematic energy band diagram of a semiconductor or an insulator.

the valence band, p. The total current is then the sum of the currents carried by free electrons (e^-) and free holes ($h^\bullet$):

where:
$$\sigma = n(e)\mu_{e^-} + p(e)\mu_{h^\bullet},$$

$$\mu_{e^-} = \frac{e\tau_{e^-}}{m_e^*} \text{ and } \mu_{h^\bullet} = \frac{e\tau_{h^\bullet}}{m_h^*}$$

and m_e^* and m_h^* are the effective masses of the electrons and the holes, respectively. The effective masses are the masses that a free electron (or hole) would need to have in order for the change in electron (hole) velocity due to an applied electric field to equal the actual velocity of the conduction electron (hole) under the same field. It is a shorthand way of describing the fact that the conduction electrons and holes in semiconductors experience not only the electric field that is applied, but also the periodic potential. Effective masses can be less than or greater than the electron rest mass, m_o.

In a semiconductor, the number of conduction electrons per unit volume, n, is given by

$$n = \int_{E_C}^{\infty} g(e)f(E)dE,$$

where g is the density of states and f is the Fermi–Dirac distribution; the integration to infinite energy is an approximation, but causes little error due to the exponentially decreasing $f(E)$. In the case where $E - E_F \gg k_B T$, then the Fermi–Dirac distribution simplifies to the Boltzmann approximation

$$f(E) \cong \exp\left(\frac{E - E_F}{k_B T}\right).$$

Inserting this equation and a parabolic approximation of the density of states produces

$$n = 2\left(\frac{2\pi m_e^* k_B T}{h^2}\right)^{3/2} \exp\left(-\frac{(E_C - E_F)}{k_B T}\right)$$

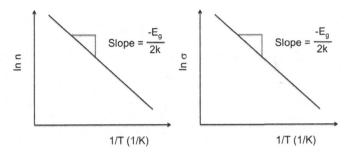

Fig. 26.15 Carrier concentration and electrical conductivity as a function of temperature in an intrinsic semiconductor or insulator.

and likewise
$$p = 2\left(\frac{2\pi m_h^* k_B T}{h^2}\right)^{3/2} \exp\left(\frac{-(E_F - E_V)}{k_B T}\right).$$

Thus, in intrinsic semiconductors, the carrier concentration rises somewhat more rapidly than exponentially with temperature $\left(\text{as } T^{3/2} \exp\left(\frac{-E_g}{k_B T}\right)\right)$. This is shown schematically in Fig. 26.15. Since the carrier concentration is much more strongly temperature dependent than is scattering, for intrinsic semiconductors the electrical conductivity also increases exponentially with temperature due to carrier promotion over the band gap.

Furthermore, since $n = p$ in an intrinsic semiconductor
$$2\left(\frac{2\pi m_h^* k_B T}{h^2}\right)^{3/2} \exp\left(\frac{-E_F}{k_B T}\right) = 2\left(\frac{2\pi m_e^* k_B T}{h^2}\right)^{3/2} \exp\left(-\frac{(E_g - E_F)}{k_B T}\right).$$

On solving for the Fermi energy,
$$E_F = \frac{E_g}{2} + \frac{3}{4} k_B T \left(\frac{m_h^*}{m_e^*}\right).$$

At 0 K, then, the Fermi energy is simply half the band gap. In cases where the effective masses of electrons and holes are not identical, the Fermi energy is a function of temperature, but at 300 K the intrinsic Fermi level position differs from half the band gap by less than $k_B T$ even when the effective masses differ by a factor of 10.

Dopant atoms introduce states into the band gap of the semiconductor or insulator; doped semiconductors are also called extrinsic semiconductors. A substitutional dopant is a *donor* if it has more valence electrons than the atom it is replacing and donates extra electrons to the material making it an *n-type* extrinsic semiconductor. A substitutional dopant is an *acceptor* if it brings fewer valence electrons than the atom is it replacing, making the material a *p-type* extrinsic semiconductor. Group V elements such as P, As, and Sb can substitute for silicon on the diamond lattice. Since these elements have five valence electrons, and only four are needed for the covalent bonding, the extra electron can contribute to the conductivity. Trivalent group III dopants (B, Al, Ga, In), on the other hand, require an additional electron to complete their tetrahedral bonding in silicon. This creates a mobile hole.

The Fermi energy describes the energy point at which the number of filled states above E_F is equivalent to the number of empty states below E_F. Thus, as donors are

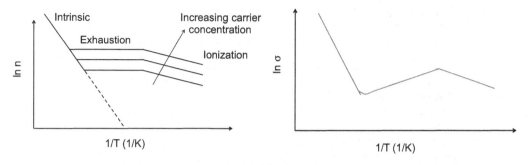

Fig. 26.16 Carrier concentration and electrical conductivity as a function of temperature in an extrinsic semiconductor. At the lowest temperatures, the dopants are ionized. In the exhaustion regime, all of the dopants are ionized, and the carrier concentration significantly exceeds the intrinsic carrier concentrations. At high temperatures, the number of electrons promoted over the band gap exceeds that due to the dopant, and the material acts like an intrinsic semiconductor.

added, the Fermi energy moves up towards the conduction band; the Fermi energy moves down towards the valence band in an acceptor-doped material. Consider, for example, the case of Ge that has been doped with $10^{15} \frac{donors}{cm^3}$ and $10^{14} \frac{acceptors}{cm^3}$. If one uses $m_e^* = 0.55 m_o$ and $m_h^* = 0.37 m_o$, then the calculated $E_F - E_V = 0.42$ eV at 300K. This is not in the middle of the band gap ($E_g = 0.661$ eV for Ge). Indeed, it is generally true that in order to increase the Fermi energy of a material, either more electrons can be added so that holes are removed, or vice versa.

The room temperature conductivity of silicon and other semiconductors is controlled by chemical doping. There are approximately $5 \times 10^{22} \frac{silicon\ atoms}{cm^3}$. If only one part per million of these atoms is replaced by donors, then there will be $5 \times 10^{16} \frac{donors}{cm^3}$. Since there will be ~1 free electron per donor atom, $n \sim 5 \times 10^{16} \frac{e^-}{cm^3}$. Given that the *intrinsic* electron concentration would have been about $n_i \sim 1 \times 10^{10} \frac{e^-}{cm^3}$ at 300K, $\frac{\sigma_{doped}}{\sigma_{intrinsic}} \approx \frac{5 \times 10^{16}}{10^{10}} \approx 5 \times 10^6$. Since the product of n and p is a constant at a given temperature, the hole concentration drops about six orders of magnitude on doping, and those holes are negligible in terms of their contribution to the total conductivity. Stop and think about the consequences of this for a moment. Very small levels of impurities produce profound changes in the electrical conductivity of semiconductors. Thus, it is imperative that the purity of the starting materials be very high. Intrinsic silicon has about 10^{10} electrons and holes per cm³. Adding 10^{12} donors to intrinsic silicon would increase the electron concentration by about a factor of 100 and decrease the hole concentration by the same factor. Adding 10^{12} donors to Si is about the same concentration as adding one grain of salt to 30 tons of sugar. While it is possible to purify some semiconductors, including Si and Ge, to this level, it is also true that for most materials, intrinsic semiconductors are rarely or never encountered.

In a doped semiconductor, the carrier concentration from the donors or acceptors thus exceeds the intrinsic concentration of electrons or holes over a wide temperature range, as illustrated in Fig. 26.16. At very low temperatures, the dopants are neutral. However, as the temperature rises, the dopants begin to ionize, freeing conduction electrons or holes. Typically, by ~100 K the dopants are all ionized, and the carrier concentration is

Table 26.2 Typical ranges for room temperature electron mobilities for a variety of different classes of materials

Material family	Mobilities $(cm^2/V \cdot s)$
Inorganic semiconductors	1 to a few ten thousands
Elemental metals	~10
Transparent conductors	$\lesssim 10$
Graphene	up to 200 000
Polymer semiconductors	$\lesssim 10$
Small molecule organic semiconductors	$\lesssim 15$

constant. The higher the dopant concentration, the higher is the baseline carrier concentration in the extrinsic region (sometimes called the exhaustion regime). The exhaustion regime is the range in which semiconductor circuits are designed to operate. If the device overheats, the promotion of electrons over the band gap will exceed the dopant concentration, and the material then acts like an intrinsic semiconductor. It should be noted that because the temperature dependence of the scattering is much weaker than the temperature dependence due to thermal promotion of carrier concentrations, an exponential increase in conductivity is observed in both the ionization and intrinsic temperature regimes. In the exhaustion regime, the electrical conductivity is not constant, but drops slightly with increasing temperature, due to phonon scattering.

Table 26.2 compares typical ranges for electron mobilities for a variety of different classes of band-type metals and semiconductors. Because mobilities depend specifically on crystal quality as well as the sample purity, the results depend heavily on the preparation methods. In III–V semiconductors, the electron mobilities generally rise for larger atoms. Thus, compounds like GaAs, InAs, InSb and PbTe have unusually high electron mobilities exceeding 5000 $cm^2/V \cdot s$ at room temperature. More ionic bonding tends to lower electron mobilities.

Not all semiconductors entail electrons or holes that are in the conduction band or valence band, respectively. In electron hopping (or hole-hopping) semiconductors, the carriers hop from one atom core to the next, as shown in Fig. 26.17. Because atom sizes depend on the number of electrons present (see Chapter 9), when the carrier moves, so does a lattice distortion. The combination of an electronic carrier and an associated strain is called a *polaron*. The probability of a hopping event rises the closer any two potential sites are, and the closer they are in energy. In crystals, this is often accomplished by doping a transition metal oxide to create multiple oxidation states at equivalent crystallographic sites. As an example, consider magnetite: $Fe_3O_4 \equiv FeO \cdot Fe_2O_3$. In the spinel lattice, both tetrahedral and octahedral cation sites exist. At low temperatures, the tetrahedral sites are occupied by one of the Fe^{3+}; the Fe^{2+} and other Fe^{3+} ion per formula unit are ordered on the octahedral sites. The electrical resistivity under these conditions is comparatively high, as the probability of electron hopping is modest. However, at the Verwey transition at 123 K, the Fe^{2+} and Fe^{3+} ions disorder on the octahedral sites, the probability of electron hopping rises, and the electrical resistivity falls dramatically (see Fig. 26.18).

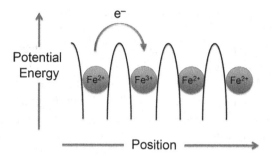

Fig. 26.17 Schematic of electron hopping from atom core to atom core.

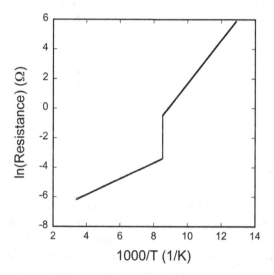

Fig. 26.18 Electrical resistance as a function of temperature in magnetite.

In electron-hopping semiconductors, the mobility is thermally activated:

$$\mu_{hopping} \propto e^{-\frac{\Delta E}{k_B T}}.$$

In general, the mobility of electrons or holes in bands exceeds the mobility due to electron hopping. Both of these tend to be vastly larger than the mobility of ions.

The fraction of cations with a particular oxidation state will depend on the defect chemistry of the compound. For example, if a transition metal oxide such as Fe_2O_3 is reduced, so that some of the oxygen is lost, the material will compensate electronically by reducing some of the Fe^{3+} to Fe^{2+}: $Fe^{3+}_{2-2x}Fe^{2+}_{2x}O_{3-x}$. As a result, the electrical conductivity of multivalent transition metal oxides depends strongly on oxygen pressure. The Friedrich–Mayer rules state that the electronic conductivity of oxides derived from the *highest valence state* is increased when the partial pressure of the oxygen in contact with the oxide is decreased. When the partial pressure is increased, the conductivity decreases. The reverse holds true for the *lowest valence state*. When oxygen pressure is increased, the conductivity increases, and when the pressure is decreased, the oxide is more resistive.

Controlled valence semiconductors are polaron-hopping semiconductors in which the valence state is controlled by doping. Consider, for example, Fe_2O_3 that has been doped with TiO_2. As Ti^{4+} is substituted for Fe^{3+}, the material charge compensates by shifting some of the Fe^{3+} to Fe^{2+}. That is, the chemical formula becomes $Fe^{3+}_{2-2x}Ti^{4+}_x Fe^{2+}_x O_3$. This is an n-type semiconductor in which electrons are transferred between iron atoms of different valence:

$$Fe^{3+} + e^- \rightleftharpoons Fe^{2+}.$$

It is also possible to make p-type polaron hopping semiconductors by substitutional solid solution. As an example, consider NiO doped with Li_2O. The lower-valent Li^+ ion converts some of the Ni^{2+} to Ni^{3+} to achieve electroneutrality: $Ni^{2+}_{1-2x}Li^{1+}_x Ni^{3+}_x O$. In this case, conduction occurs as $Ni^{2+} + h^{\bullet} \rightleftharpoons Ni^{3+}$.

It is important to remember, though that doping in compounds will not always result in electronic compensation, so that electrons and holes are produced. Instead, compensation can also occur by formation of ionic defects like interstitials and vacancies. Thus, aliovalent doping in compounds will not always produce significant changes in the electrical conductivity.

26.4   Conducting Polymers

There is a very wide range in electrical resistivities of polymers. This is illustrated by a comparison of polyethylene and polyacetylene. Both have carbon backbones. The two differ in hydrogen content and chemical bonding. Polyethylene, $-(CH_2-CH_2)_n-$, has covalent single bonds within the chain. When all of the electrons participate in strong sigma covalent bonds, the electrons are confined, and large energies are required to liberate an electron from the bond to reach the conduction band. In contrast, electrons that are participating in multiple bonds via π bonding are much less strongly confined. As a result, the band gap decreases in conjugated systems with alternating single and double bonds. Consider, for example, polyacetylene, $-(CH-CH)_n-$ (Fig. 26.19). The base polymer is a semiconductor due to electron transport along the backbone. In polyacetylene, the π bonding orbital is full, while the π^* antibonding orbital is empty. Because of the narrow band gap between π and π^* orbitals, polyacetylene becomes a conducting polymer when doped with charged species such as Na^+ or I_3^-. The doping process differs from that in silicon and other inorganic semiconductors. The conductivity of Si is controlled by substituting atoms such as trivalent boron or pentavalent phosphorus for silicon. In polyacetylene and other polymeric semiconductors with narrow band gaps, doping is often carried out by inserting charged atoms or molecules between polymer chains, or by attaching charged sidegroups. Ions such as Na^+ donate an electron to the polymer, making it an n-type semiconductor. The iodine molecule I_3^- captures an electron from polyacetylene, making it a p-type conductor.

Conjugation, where alternating single and double bonds, is also common in 5- and 6-membered ring structures based on C (sometimes with some replacement of C atoms

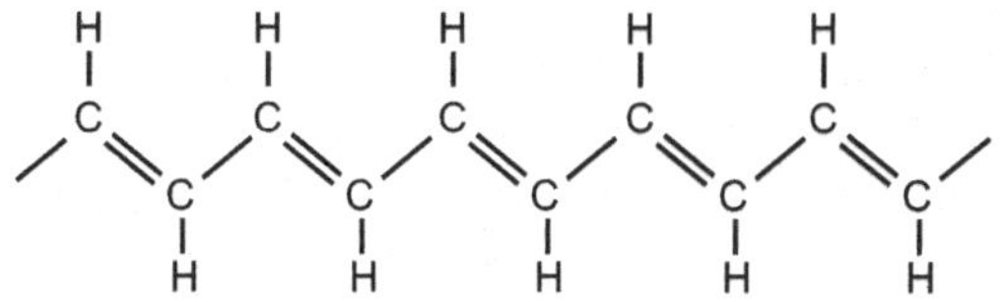

Fig. 26.19 Schematic illustration of structure of polyacetylene showing the single and double bonds alternating along the backbone.

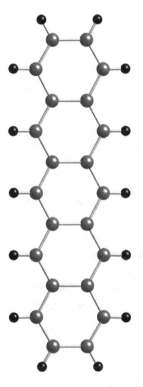

Fig. 26.20 Molecule of pentacene, $C_{22}H_{14}$.

with O, N, or S). Here also, the conductivity can be further increased by doping with electron donors (like the alkali metals), or electron acceptors (like Br_2, SbF_5, WF_6, etc.).

Organic semiconductors include materials such as pentacene, polypyrrole, polythiophene, polyfuran, and the polyanilines. These materials are becoming increasingly important in electronic applications, including light-emitting diodes for displays. Pentacene, for example, is composed of five benzene rings linked along a single axis, as shown in Fig. 26.20. The resonating double covalent bonds with the interconnected carbon 6-rings produce the modest band gap of E_g ~0.7 eV. Because this is a comparatively small molecule, it is relatively easy to crystallize pentacene, although defects are common due to the low energies associated with the van der Waals bonding connecting one molecule to the next. It should also be noted that it is important not only to provide a conduction path within a molecule. For a bulk sample to conduct well, conduction

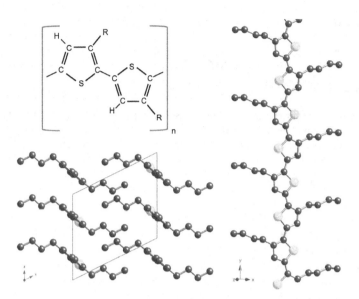

Fig. 26.21 Poly(3-hexyl thiophene). The S is the largest sphere. C is the darker one. H are not shown.

also needs to occur between molecules. Pentacene is sometimes functionalized by replacing some of the hydrogens with larger atomic groups to improve the carrier transport from one molecule to the next.

As a representative example of the conducting polymers, consider poly(3-hexyl thiophene), abbreviated as P3HT. P3HT is a p-type semiconductor with hole mobilities of ~0.2 cm^2/V · s in samples with 30–60% crystallinity. Its chain structure and chain alignment are shown in Fig. 26.21 for the crystalline portion regioregular structure with a planar backbone. If the backbone twists around its length, the density at which chains can be packed decreases, and the conductivity drops.

26.5   Sensors

Changes in the electrical resistance of materials as a function of some change in the environment, such as temperature or chemistry, can be exploited to make sensors. For example, humidity sensors are often used in air conditioning systems, or in microwave ovens, as well as for electronic spark timing in automobile engines. Such devices often rely on a change in the surface resistance by several orders of magnitude. High-surface-area oxide substrates with salt coatings are sensitive to small humidity changes. The mechanism for surface conduction involves the absorption of water vapor, followed by dissociation into hydronium and hydroxyl ions:

$$2H_2O \rightleftharpoons (H_3O)^+ + (OH)^-.$$

Conduction occurs as the protons are transferred from one water molecule to another absorbed on the oxide surface, via a *Grotthus reaction*. Oxide surfaces are useful in this case because of the strong attraction between surface hydroxyl groups and adjacent water molecules.

Chemically driven valence changes can also produce resistance changes useful for sensing. As an example, in RuO_2 and IrO_2, both metallically conducting isomorphs of rutile, protons can react with surface oxygens, reducing the transition metal cation, and changing the conductivity.

The temperature dependence of the electrical conductivity of semiconductors is much stronger than that of metals, and is sometimes employed in *thermistors* to measure temperature or as inrush current limiters. Thermistors are temperature dependent resistors. For many semiconductors, the temperature dependence of the electrical conductivity can be approximated as: $\sigma(T) \approx T^{-b} \exp\left(\frac{-E_A}{k_B T}\right)$. As described above, the exponential temperature dependence tends to dominate, so that the electrical resistance can be described reasonably well as $R = A \exp\left(\frac{\beta}{k_B T}\right)$. The temperature coefficient of resistance, α, is then $\alpha = \frac{1}{R}\frac{dR}{dT} = \frac{-\beta}{T^2}$. Amorphous silicon, vanadium oxide, and numerous mixed-valence oxides with the spinel crystal structure are used as thermistors.

26.6   Metal–Insulator Transitions

Some materials undergo phase transitions at which the electrical properties change from metallic to semiconducting characteristics, typically accompanied by decreases in the electrical conductivity by several orders of magnitude. Metal-to-semiconductor transitions are usually associated with a change in crystal structure, as in the transition of gray tin (semiconductor) to white tin (metallic). The cases in which a change in electrical properties are not associated with a major change in structure are rare, and will only occur if the atomic positions are sufficiently variable to accommodate a major change in bonding. The lower oxides of vanadium, V_2O_3 and VO_2 are examples. The V^{4+} ion has a single remaining d electron. At room temperature, VO_2 has a monoclinic crystal structure, which can be regarded as a distorted rutile structure, as shown in Fig. 26.22. Down the axis of the diamondback chain, the diamonds alternate between short and long separation distances between the V atoms. The longer of these two distances

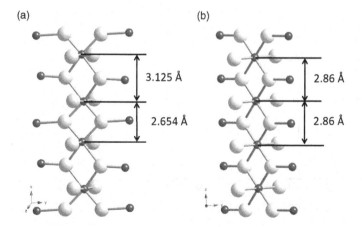

(a)   (b)

3.125 Å   2.86 Å

2.654 Å   2.86 Å

Fig. 26.22 VO_2 crystal structures: (a) semiconducting monoclinic room temperature phase, and (b) metallic tetragonal phase stable above 68 °C.

exceeds the critical separation distance for orbital overlap. Thus, the distortion leads to the d electrons being confined on the V atoms, producing semiconducting characteristics. At 68 °C, a higher-symmetry tetragonal phase is recovered. As the phase transition occurs, the V–V distances along the diamond chain become uniform, and are short enough that band conduction becomes possible. This produces metallic conductivity.

26.7   Superconductivity

Some materials, when cooled, lose their electrical resistivity, as shown in Fig. 26.23. Typically, above the critical temperature, such materials are normal metallic conductors, and below this temperature, they are *superconductors*. Superconductivity is found in elemental metals, intermetallic compounds, as well as in some copper oxide, iron oxide, or boride compounds among others. Table 26.3 gives critical temperatures for a number of superconductors.

The characteristic properties of superconductors are as follows:

(1)   Zero dc electrical resistance below a critical temperature, T_c. In practice, it has been demonstrated that eddy currents in superconductors persist for at least three years. In contrast, those in a normal metal decay in around 10^{-12} s as faster electrons are scattered into lower energy states due to the finite electrical resistivity.

(2)   The Meissner effect, in which a material cooled through the semiconducting transition in the presence of a magnetic field expels the magnetic flux. The magnetic fields cannot penetrate large distances into the solid because the applied magnetic field, $\vec{H}$, sets up superconducting currents which oppose $\vec{H}$. That is, the material behaves as a perfect diamagnet. If, however, the magnetic field is

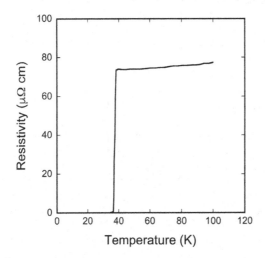

Fig. 26.23 Resistivity as a function of temperature for MgB_2 showing a superconducting transition at 39 K.

Table 26.3 Phase transition temperatures for a number of superconductors

Material	T_c (K)
Hg	4.15
Pb	7
Nb_3Sn	18
$(La,Ba)_2CuO_{4-x}$	38
MgB_2	39
$YBa_2Cu_3O_{7-x}$	95
$Bi_2Sr_2Ca_2Cu_3O_{10}$	110
$Tl_2Ba_2Ca_2Cu_3O_{10}$	~120
$HgBa_2Ca_2Cu_3O_{8+x}$	134

increased above a critical field, H_c, the superconductor is driven back to a normal state.

(3)   The superconducting state can also be lost if the current density in the material exceeds a critical current density, J_c.

In superconducting materials, there is a small population of electron pairs, called Cooper pairs, which cooperate to avoid being scattered. The strength of the interaction between these two electrons is largest when the two electrons have opposite spin and opposite momentum. In metallic superconductors, a Cooper pair of electrons moves in a coupled way through the lattice. As the first electron moves through the lattice, it attracts or repels atoms, generating a phonon. The phonon provides an attractive force for the second electron in the Cooper pair. Their mobility is effectively infinite. In short, Cooper pairs move coherently, and so do not scatter. Therefore, even though their concentration is small, the Cooper pairs dominate electrical conduction. Interestingly, however, Cooper pairs do not carry heat well, so the thermal conductivity can still be quite low, even when the electrical conductivity is very high. The energy required to break the binding of the Cooper pairs is the superconducting energy gap.

So-called Type I superconductors become normal once the magnetic field exceeds H_c. Above this field, the magnetic flux lines can penetrate into the bulk of the sample. Al, Cd, Mo, and Hg are all Type I superconductors.

In contrast, a Type II superconductor is a perfect diamagnet for fields below H_{c1}. Between H_{c1} and H_{c2}, the magnetic field begins to penetrate the material as discrete flux lines called fluxoids. The core of each fluxoid is a normal conductor, rather than a superconductor. Typically, these fluxoids are arranged periodically in the solid. Assuming sufficient pinning by defects, as the magnetic field rises, the fluxoids get closer and closer together, until the normal cores overlap at H_{c2}, at which field the entire solid is a normal conductor. Type II superconductors are critical for generation of high magnetic fields, such as those used in magnetic resonance imaging applications.

A key crystal structure adopted by superconducting intermetallic phases such as Nb_3Sn, Nb_3Al, and V_3Si, is shown in Fig. 26.24. Many of these compounds have superconducting transition temperatures between 15 and 25 K. The structure is characterized by very short separation distances between the Nb atoms. This family of materials is widely adopted for generation of very large magnetic fields in

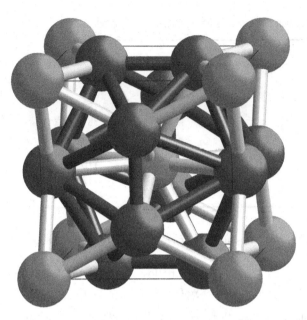

Fig. 26.24 Crystal structure of the Nb_3Sn superconductors. The tin atom in the body-centered positions is bonded to 12 niobium atoms, and niobium is bonded to four tin atoms and ten niobiums.

superconducting magnets. They are able to remain superconducting in a magnetic field and when carrying a current.

The recently discovered superconductor MgB_2 ($T_c = 39$ K) also has an excellent critical current at T_c. The boron atoms in MgB_2 form a hexagonal honeycomb pattern, resembling the carbon layers in graphite. Close-packed layers of magnesium atoms separate these layers. The electrons responsible for superconductivity are associated with the boron layers where there is very strong bonding within the layer, and much weaker bonding between layers. Boron–boron distances are 1.78 Å within the honeycomb and 3.52 Å along the c axis to the next boron layer. The crystal structure is illustrated in Fig. 26.25. It is believed that the superconducting electrons in the boron are strongly coupled to each other through in-plane lattice vibrations.

In late 1986, superconductivity was discovered in an electronically conducting oxide based on La, Ba, Cu, and O. Other copper-oxide-based high-T_c superconductors were discovered in the following years. Many of these are layered compounds. The structures of three representative cuprate superconductors are shown in Fig. 26.26.

Structural analysis leads to three significant insights into the nature of superconductivity in the copper-oxide family.[5] First, a nearly planar layer of CuO_2 transports the superconducting current. Second, as the distance between the CuO_2 layers increases, the critical current decreases as the magnetic flux lines become unpinned. Third, the lattice

[5] R. J. Cava, "Oxide superconductors," *J. Am. Ceram. Soc.*, **83** (1), 5–28 (2000).

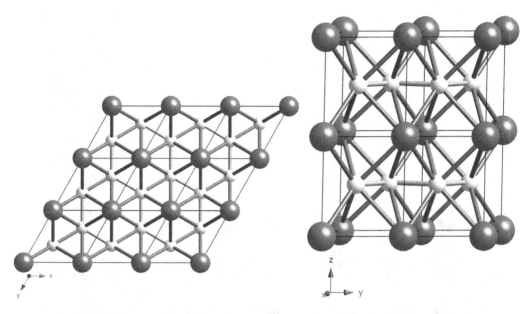

Fig. 26.25 The crystal structure of hexagonal magnesium diboride. Each Mg (larger atom) is bonded to 12 boron atoms (smaller atom) at a distance of 2.50 Å. The shortest Mg–Mg contacts are at 3.08 Å in the (001) plane and at 3.52 Å along [001]. Each B is bonded to three borons at 1.78 Å and six Mg at 2.50 Å.

mismatch between the CuO_2 layers and the non-superconducting layers controls the oxidation state of the copper ions. An average valence of about +2.2 optimizes superconductivity.

There are a number of important chemical, structural, and electronic aspects of copper oxides that are associated with superconductivity. First, both Cu and O are highly involved in the conduction process. Secondly, the Cu^{2+} ions have a d^9 electron configuration that favors Jahn–Teller distortions. The octahedral, pyramidal, and square planar polyhedra in the superconducting cuprates all have four short Cu–O bonds in the (001) planes. One or two longer bonds are formed along [001] with pyramids or octahedra. As described in Chapter 14, the filled d_{xy}, d_{xz}, and d_{yz} orbitals point between O neighbors. The repulsion between the d electrons and the neighboring atoms is largest for the $d_{x^2-y^2}$ orbital. As a result, that is the highest energy orbital. The material lowers the energy of the d_{z^2} orbital by pushing those neighbors further away. The partially occupied $d_{x^2-y^2}$ orbital means that electrons moving in the (001) plane can form Cooper pairs.

For isolated Cu^{2+} atoms, the orbital energy diagram consists of five discrete energy levels, four filled and one half-filled. In solids, the discrete energy levels merge together with the oxygen 2p orbitals to form bands (Fig. 26.27). The upper band formed from the $d_{x^2-y^2}$ orbitals is further split into upper and lower Hubbard bands because of electron–electron repulsion. This is the so-called "on-site repulsion energy" characteristic of Mott insulators. The lower Hubbard band is embedded in the valence band formed by the filled d states of the copper ions and the 2p states of the oxygen ions. The upper Hubbard band is empty and the Fermi level of a Mott insulator like this lies in the forbidden energy gap.

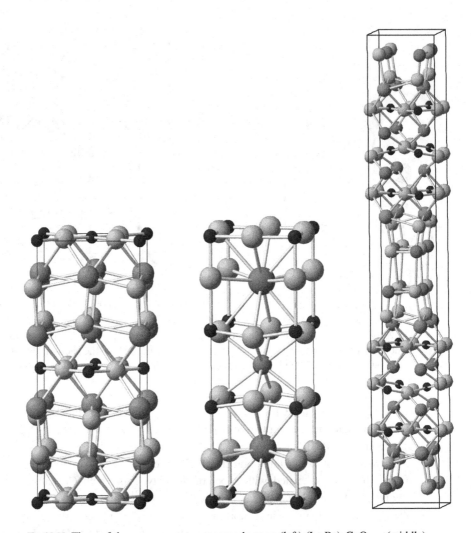

Fig. 26.26 Three of the many cuprate superconductors: (left) $(La,Ba)_2CuO_{4-x}$, (middle) $YBa_2Cu_3O_7$, and (right) $(Bi_{1.7}Pb_{0.3})Sr_2(Ca_{1.7}Bi_{0.3})Cu_3O_{10}$. Most are based on a family of perovskite-related structures known as the Ruddlesden–Popper family, in which perovskite layers are interleaved with rocksalt-like layers. All possess copper-oxide layers separated by insulating metal-oxide layers that act as charge reservoirs for the Cooper pairs in the CuO_2 layers. While the transition temperatures for these materials are excellent, the J_c and H_c values are rather modest. Cu is the small dark sphere; O is the lightest sphere.

So far, only the CuO_2 conducting layer has been considered. What happens to this picture when the interleaved metal oxide layers are included? Like any insulator or semiconductor, n- or p-type doping will take place, depending on the charge balance. These so-called "charge-reservoir layers" control the number of electrons in the CuO_2 planes, and that control is crucial to the existence of superconductivity. The amount of excess charge is controlled by substitution of Sr^{2+} for La^{3+} in a compound such as $La_{2-x}Sr_xCuO_4$, or by oxygen intercalation in a compound like $YBa_2Cu_3O_{6+x}$ (Fig. 26.28).

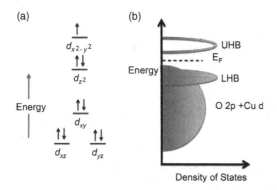

Fig. 26.27 (a) The $3d^9$ electron configuration of Cu^{2+}. (b) Split $d_{x^2-y^2}$ band for interacting electrons in cuprate superconductors showing upper (UHB) and lower (LHB) Hubbard bands. Adapted, by permission, from R. J. Cava, "Oxide superconductors," *J. Am. Ceram. Soc.*, **83** (1), 5–28 (2000).

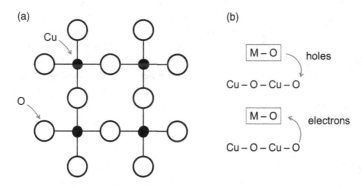

Fig. 26.28 (a) Square planar layers of copper and oxygen in the superconducting cuprates. (b) The doping is controlled by the transfer of electrons and holes between the CuO_2 layers and the metal oxide layers that make up the charge reservoir. Adapted, by permission, from R. J. Cava, "Oxide superconductors," *J. Am. Ceram. Soc.*, **83** (1), 5–28 (2000).

In the cuprate oxides, superconductivity is generally caused by p-type doping, with holes created in the oxygen 2p states of the valence band. The holes interact with the magnetic spins of the copper atoms. Neutron diffraction experiments indicate that antiferromagnetic alignment of the copper spins continues, even as the number of charge carriers increases and the $3d^9$–sp^6–$3d^9$ superexchange coupling begins to break down. This gradual breakdown of magnetic order involves the formation of Zhang–Rice singlets in which a copper spin and an oxygen hole combine to form a spin-neutral charge carrier. Many physicists believe that this rather exotic conduction process is the key to understanding the appearance of superconductivity in this class of materials.

The final stage (See Fig. 26.29) is the association of these singlet states into Cooper pairs to generate superconductivity. Coupling of charge carriers through magnetic spins is analogous to the phonon coupling in the theory of conventional superconductors.

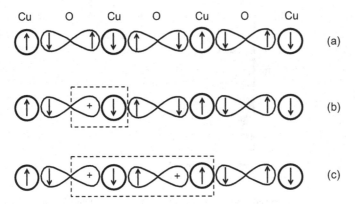

Cu   O   Cu   O   Cu   O   Cu

(a)

(b)

(c)

Fig. 26.29 Copper oxygen interactions in CuO_2 layers. (a) In the undoped Mott insulator state, the unpaired spins of the Cu^{2+} ions couple antiferromagnetically through superexchange interactions that will be discussed in Chapter 29. Two electrons are transferred out from each CuO_2 group, leaving a filled valence band and an empty conduction band. The Cu^{2+} spins are coupled through the 2p orbitals of the oxygens. (b) Holes are created in the valence band when less than two electrons are transferred. At low doping levels, the holes in the O 2p states cause p-type conductivity. The associated Cu^{2+} spin and the empty oxygen 2p state are sometimes referred to as a Zhang–Rice singlet. (c) Cooper pairs and superconductivity develop at moderate doping levels, where about 1.8 electrons are transferred to the charge reservoir. At this stage, the effective valence of copper is +2.2. A substantial number of holes are present in the oxygen 2p states and they begin to interact with the aligned spins of the copper ions. The aligned spins couple the holes in the Cooper pairs through the magnetic spins.

26.8   Problems

(1) Consider Si doped with 10^{17} $\frac{\text{As donors}}{\text{cm}^3}$. If the number of intrinsic carriers in Si is 1.5×10^{10} $\frac{1}{\text{cm}^3}$ at room temperature, find the concentration of holes, and the new Fermi level.

(2) **(a)** Draw a stereographic projection for point group 422.

   (b) Write an equation for electrical conductivity, defining all of the terms. Would you expect the electrical conductivity in the a, b, and c directions to be the same or different in a material with this point group? Explain your reasoning.

(3) For FeO, the $r_{Fe} = 0.85$ Å and $r_O = 1.33$ Å.

   (a) Predict the cation and anion coordinations of FeO using Pauling's rules. What crystal structure is expected for this compound?

   (b) FeS has the NiAs structure. Contrast this crystal structure in terms of edge-, corner-, and face-sharing to the one you determined in part (a).

   (c) Which material would you expect to be a better electrical conductor? Why?

(4) Describe a use for a metallically conducting ceramic and name an appropriate material for the application. Why is a ceramic used in this application?

(5) Draw schematics showing the band structures of a metal and a semiconductor, with the Fermi energy shown.

(6) Draw a curve showing the temperature dependence of the conductivity for an acceptor-doped semiconductor. Explain in detail the factors responsible for the observed behavior.

(7) Fermi–Dirac statistics.

   (a) Plot (using a computer) the Fermi–Dirac distribution for $E_F = 1$ eV, at temperatures of 300, 1000, and 2000 K.

   (b) Show that the probability that a state ΔE above E_F is filled equals the probability that a state ΔE below E_F is empty.

   (c) Calculate the temperature for a solid at which there is a 1% probability that an electron in the solid will have an energy 0.3 eV above the Fermi energy of 4 eV.

(8) If a lot of Si wafers has only 1 part per billion of P (by weight), will this material act like intrinsic Si?

   (Density of Si: 2.33 g/cc; conductivity of intrinsic Si at room temperature ~5 × 10^{-6} 1/Ω cm; mobility of electrons in Si = 0.19 m²/V·s.)

(9) You need to design a thick-film resistor that can carry 20 mA of current continuously and maintain a resistance of 10 kΩ, with an accuracy on the resistance value of 0.5%. The sheet resistance is 5 × 10^3 Ω/square, and temperature coefficient of resistance = 300 ppm/°C.

   (a) What is the required length to width ratio?

   (b) What is the maximum temperature rise allowed?

   (c) Suppose that for these specifications the rated power density for the thick film resistor is 5 W/cm² (i.e. a higher power than this would lead to an unacceptable temperature rise). The area of the resistor required to maintain this performance is: $A \frac{P_{max}}{P_{rated}}$, where P_{max} = maximum power to be dissipated (in W), and P_{rated} is the rated power density (in W/cm²). Given this, what are the minimum dimensions required for the resistor?

(10) Explain how electron hopping differs from electronic conduction in a normal semiconductor. Give an example of a material that shows hopping conductivity, and describe a use for it.

(11) If 10% of Cu atoms are replaced by gold in a solid solution, the resistivity *increases* by a factor of about 3.8. If you add 10 parts per billion arsenic to pure germanium, the resistivity *decreases* by a factor of 16. Why is the change in resistance opposite in sign and vastly different in magnitude for these two cases? How would you expect the conductivity to change as a function of temperature in each case?

(12) The density of aluminum at 20 °C is 2.70 g/cm³. Its atomic weight is 26.98 g/mole, its electrical conductivity is 35 × 10^6 (Ωm)$^{-1}$, and its Fermi energy is 11.70 eV.

   (a) What is the density of free electrons calculated from the Fermi level? In a metal, the Fermi energy is related to the density of free electrons, N, as:
   $E_F = \frac{\hbar^2}{2m} \left(\frac{3N}{8\pi}\right)^{2/3}$.

   (b) What is the density of Al atoms?

   (c) By comparing answers in (a) and (b), how many conduction electrons are donated by each Al atom?

(d) What is the drift velocity of Al when it is exposed to an applied electric field of 120 V/cm?

(13) Draw a figure showing the mean free path of an electron (i.e. the average distance traveled between collisions) in a metal as a function of temperature.

(a) What is the origin of this behavior?

(b) What is the consequence on the temperature dependence of the electrical resistivity?

(c) How would you expect the curves of mean free path and electrical resistivity as a function of temperature to change if the impurity content of the metal increased?

(14) Experimental measurements on the hole concentration and electrical conductivity of a semiconductor are given below.

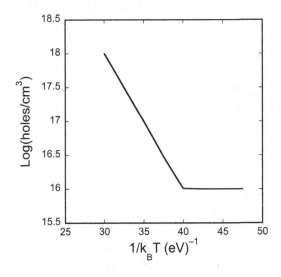

(a) What is the approximate net acceptor impurity concentration?

(b) Estimate the electron concentration at $1/kT = 45$ (eV)$^{-1}$.

(c) Estimate E_g.

(d) Assuming that the mobility does not vary with temperature in the range 250 to 350K, estimate the electron and hole mobilities.

(e) Apply an electric field of 10 kV/cm to the specimen. What is the electron drift speed?

(f) What would you expect to happen to the conductivity at $1/kT = 45$ (eV)$^{-1}$ if the sample were doped with 5×10^{15} donors?

$1/kT$ (eV)$^{-1}$	σ (Ω cm)$^{-1}$
35	30
40	1.6
45	1.0

(15) In intrinsic Si: $E_g = 1.1$ eV, $m_e^* = 0.26m_o$, $m_h^* = 0.39m_o$, $\mu_e = 0.15$ m^2/Vs, $\mu_h = 0.05$ m^2/V $\cdot$ s.

    (a)   Calculate the room-temperature resistivity.

    (b)   Calculate the resistivity at 350 °C.

    (c)   If the resistance of this sample of Si is R, find the temperature coefficient of resistance $(1/R)(dR/dT)$ at room temperature.

(16) Al is a metal. The next element on the periodic table, Si, is a semiconductor. What are the fundamental differences in the bonding of the materials, and why does this result in the development of a band gap in Si?

(17) Contrast the temperature dependence of the conductivity for

    (a)   Cu

    (b)   intrinsic Si

    (c) $Li_xNi_{1-x}O$.

    Explain the mechanisms responsible for controlling the behavior in each case.

(18) Figure 26.7 shows the temperature dependence of the electrical resistivity of copper. Explain in detail why the curves have the shape that they do.

(19) The curve below represents the carrier concentration of electrons as a function of temperature in a particular semiconductor. Explain the origin of the different regions on this curve.

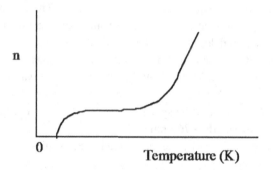

(20) (a) Give a possible explanation for why TiO_2 is a wide band gap semiconductor, and RuO_2 is a metallic conductor. The two materials have the same crystal structure (rutile structure).

    (b)   Describe a use for RuO_2 and explain why it is used in preference to Cu in the application you choose.

Property	Titanium	Ruthenium
Electron configuration of element	$1s^22s^22p^63s^23p^64s^23d^2$	$1s^22s^22p^63s^23p^64s^23d^{10}4p^6$ $5s^24d^6$
Oxidation state of ion	4+	4+
Ionic size	0.605 Å	0.62 Å

(21) The relative resistivity ratio is the ratio of the electrical resistivity of a metal at room temperature to its value at 4 K.

    (a)   Explain the mechanisms that would control the relative resistivity ratio.

    (b)   Suppose that you prepared two samples of Cu with different grain sizes. How would you expect the resistivity ratio of the two samples to compare?

27 Optical Properties

27.0 Introduction

In describing the factors that control the optical properties of solids, it is important to recall that light is an electromagnetic wave. For a light wave in free space, the electric, $\vec{E}$, and magnetic, $\vec{H}$, fields are orthogonal to the wavenormal (see Fig. 27.1). It is the interaction of these oscillating fields with the solid material that underlies the optical properties. As most materials are not magnetic, the optical properties are dominated by the interaction of the electric field vector of the light with the charges in the solid. In the visible frequency range, the only charges that are small and light enough to follow the rapidly oscillating electric field of the light waves are electrons.

The electromagnetic spectrum is sometimes divided, for convenience, into different ranges, as shown in Fig. 27.2. Visible light corresponds to the wavelength range from around 400 to 750 nm. Higher frequencies (wavelengths from ~10 to 390 nm) correspond to the ultraviolet portion of the spectrum, while the infrared range corresponds to lower frequencies (wavelengths ~760 nm to 1 mm).

The polarization (P_i) induced by the electric field (E_j) is given by:

$$P_i = \varepsilon_0 \chi_{ij} E_j + \varepsilon_0 \chi_{ijk}^{(2)} E_j E_k + \varepsilon_0 \chi_{ijkl}^{(3)} E_j E_k E_l + \cdots,$$

where χ_{ij} is the dielectric susceptibility; $\chi_{ijk}^{(2)}$ and $\chi_{ijkl}^{(3)}$ are the higher-order susceptibilities, and ε_0 is the permittivity of free space.

For most light sources, the electric fields are small, so that only the leading term needs to be retained. In this case,

$$P_i = \varepsilon_0 \chi_{ij} E_j = \varepsilon_0 (\varepsilon_{ij} - 1) E_j,$$

where $\varepsilon_r = (\tilde{n}^2) = (n + ik)^2$. Here $\tilde{n}$ is the complex index of refraction, n is the real part

$$n = \frac{speed\ of\ light\ in\ a\ vacuum}{speed\ of\ light\ in\ the\ material},$$

and k is the extinction coefficient. The complex behavior of $\tilde{n}$ results from the phase differences between the polarization response and the driving electric field. This is manifested as absorption of some of the light wave. Recalling Beer's law:

$$I = I_0 \exp(-\alpha d),$$

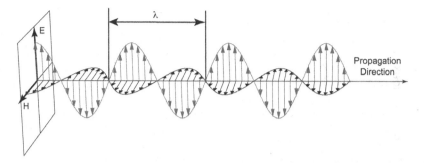

Fig. 27.1 Depiction of the oscillating electric and magnetic field vectors of a light wave moving from left to right through free space (or an isotropic nonmagnetic solid). The wavelength, λ, is also illustrated.

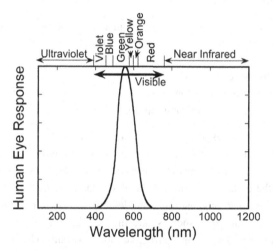

Fig. 27.2 The electromagnetic spectrum.

where I is the intensity of a light beam a distance d away from the point at which the intensity is I_0. The material constant that describes how rapidly light is absorbed as a function of propagation distance is the absorption coefficient, α. The absorption and extinction coefficients are related by $\alpha = \frac{4\pi k}{\lambda}$, where λ is the wavelength of the light.

27.1   Refractive Index and Absorption Coefficients and Their Dispersion

Refractive Indices

We will begin this discussion with the optical properties of a transparent non-magnetic material, so that the absorption coefficient can be ignored. In this case, the refractive index describes the material's response to the light wave. The refractive index in the visible frequency range is then governed by:

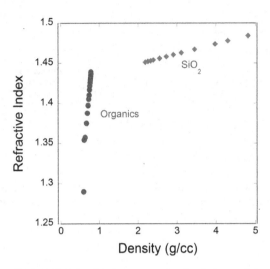

Fig. 27.3 Relationship between refractive index and density in a variety of molecular solids (organics) and SiO_2. In both cases, higher densities promote higher refractive indices.

(1)   the number of electrons per volume,
(2)   the polarizability of the electron cloud, and
(3)   the energy match between photons and band gap.

For the first of these factors, it is intuitively obvious that as the electron density increases, there are more charged particles available to interact with the light wave, and hence the light is slowed further. Consider, for example, the case where the material composition (and hence the electronic polarizability) is held constant, but the material density is changed. Figure 27.3 shows that the atom density and the refractive index are linearly related in both molecular solids such as paraffins, and in ionically bonded solids such as SiO_2. This link between density and refractive index explains why the refractive index of amorphous polymers falls below that of their crystalline counterparts – the density is lower in the amorphous materials.

The electronic polarizability describes how much the electron cloud distorts when exposed to the electric field vector of the light wave. That is, in transparent solids, the electron clouds around the atoms are bound to the nucleus, but undergo localized motion due to the light wave. In thinking about this, it is worthwhile considering the electric field, $\vec{E}$, experienced due to the attraction of electrons to the nucleus as compared to the electric field vector of the light wave. In cases where the electron cloud (and hence the radius) is very small, i.e. the electrons are close to the nucleus, the attraction between the electron cloud and the nucleus is very strong. As a result, the light wave does not perturb the electron cloud much, and the polarizability is small. However, as the atom (or ion) size increases, and the number of inner electron shells increases, the outermost electrons are shielded from the full nuclear charge. Thus, the effective electric field from the nucleus drops, and the polarizability rises. Thus, the polarizability typically increases with atom size. This is shown in Fig. 27.4, for a series of different types of ions. The plot shows that within a family, cation polarizability rises with the cube of the effective radius. Anion polarizabilities are larger, but vary much

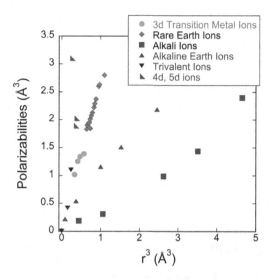

Fig. 27.4 Polarizabilities of several different families of cations, plotted as a function of the cube of the effective radius. The polarizability rises with the ion size.

more from one compound to another, as a result of changes in coordination number and cation–anion separation distance. Extensive tables for electronic polarizabilities are reported for ions by Shannon,[1] and for organic compounds by van Krevelen and te Nijenhuis.[2]

Thirdly, it is important to recognize that the refractive index increases as the band gap is approached. This is the origin of the dispersion in the refractive index as shown in Fig. 27.5. The figure illustrates several important points. First, note that the refractive index of GaAs exceeds that of GaN. This makes sense because As is larger and has more electrons than N. Secondly, it is clear that the GaN has a higher band gap than GaAs, which could be predicted based on the smaller bond length and higher ionicity (see Chapter 8). Moreover, as the energy of the photons approaches the band gap, the refractive index rises. Simple oscillator models describe this frequency dependence. From a classical physics viewpoint, this can be envisioned in the following way. The electron cloud will undergo the largest distortion when the applied electric field forcing the oscillation approaches the natural resonance frequency of the system. Mathematically, this can be approximated as:

$$\tilde{n}^2 \cong 1 + \frac{S\lambda^2}{\lambda^2 - \lambda_0^2 + i\gamma\Gamma},$$

where $\tilde{n}$ is the complex index of refraction, S is a constant, λ_0 is the wavelength corresponding to the band gap, and Γ is a damping term accounting for absorption.

[1] R. D. Shannon and R. X. Fischer, "Empirical electronic polarizabilities in oxides, hydroxides, oxyfluorides, and oxychlorides," *Phys. Rev. B,* 73, 235111 (2006).

[2] D. W. van Krevelen and K. te Nijenhuis, *Properties of Polymers: Their Correlation with Chemical Structure; Their Numerical Estimation and Prediction for Additive Group Contributions,* Elsevier (Singapore) (2009).

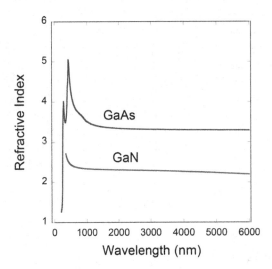

Fig. 27.5 Dispersion in the refractive indices of GaN and GaAs.

From a quantum mechanical perspective, the resonance frequency corresponds to the energy required to promote an electron from the valence band to the conduction band. This can be treated in terms of transition probabilities associated with there being an electron available in the lower-lying state, a hole available in the higher state, and a photon with the correct energy and momentum to cause the transition.

Full description of the optical properties over a wide frequency range typically requires multiple oscillators. The exact shape of the frequency dependence depends on the details of the band structure.

Figure 27.6 shows a comparison of the refractive index and dispersion for a variety of insulating materials. There are a number of points that are apparent from an inspection of the figure:

(1)   In the visible portion of the frequency range, most solids undergo normal dispersion, where the refractive index decreases as the wavelength increases.

(2)   For a given crystal structure, it is important to consider the balance of polarizability and density. In many cases, larger atoms will be more polarizable, and so will produce higher refractive indices. This is often true when the anion size is increased, since the anions contribute much of the polarizability of the compound. In some cases, however, the unit cell size increases more than the polarizability does, and the refractive index drops. This can be seen in Fig. 27.6, where the refractive index of NaCl exceeds that of KCl over the visible frequency range.

(3)   Denser crystal structures have higher refractive indices than lower-density structures in cases where the polarizabilities are similar. Consider, for example, a comparison of the refractive indices of α-Al_2O_3 and α-quartz. Corundum has a structure in which the O lattice is approximately hexagonally close-packed, and the cation polyhedra have face-, corner-, and edge-sharing. In contrast, the crystal structure of quartz is much less dense, with corner-sharing only. This leads to a higher refractive index in alumina than quartz.

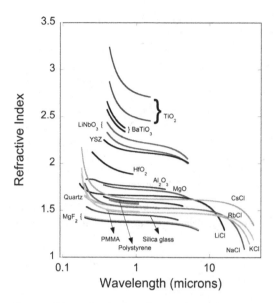

Fig. 27.6 Dispersion in the refractive index for a number of transparent materials.

(4)   A few cations have unusually large polarizabilities. The high polarizability of Ti^{4+} produces very high refractive indices in materials such as rutile TiO_2 and $BaTiO_3$. In addition, TiO_2 has a band gap near 3.5 eV, in the near ultraviolet. As was described earlier, proximity to the band gap increases the refractive index.

Free electrons can also contribute to the optical response of solids. The optical properties of metals in the visible frequency range, for example, are governed by the free electrons. They respond readily to the electric field vector of the light wave, and there is no restoring force for the valence electrons. The net result is that the light is readily reflected from the surface of smooth metal, i.e. the metal appears shiny. The appearance is silvery if all of the visible wavelengths are reflected well. Good examples of this are aluminum and silver. In contrast, there is an absorption associated with the d electrons near 2 eV in Cu, in the red portion of the visible spectrum. As a result, the reflectance of Cu is highest in the red portion of the visible range, producing the reddish hue of Cu.

Absorption Coefficients

For a non-metallic solid, there are two principal causes for absorption of light: lattice vibrations and electronic transitions. In most materials, these two mechanisms respond at different frequencies, so that there is a region of transparency, where the absorption coefficient is low.

For the electronic transitions, it is important to recognize that light carries large energies, but comparatively little momentum. Therefore, light is associated with nearly vertical transitions on energy–wavevector, E–k diagrams. Transitions which entail significant momentum changes require participation of a phonon. Consider, for example, the direct and indirect band gaps in Fig. 27.7. In the case of a direct band

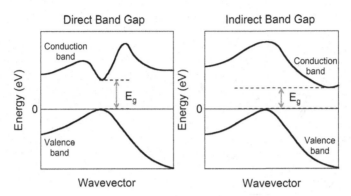

Fig. 27.7 Illustration of band structures for direct and indirect band gaps. Adapted, with permission, from B.E.A. Saleh and M.C. Teich, *Fundamentals of Photonics*, John Wiley & Sons, New York, 1991, Fig. 15.1–3.

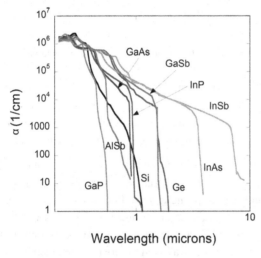

Fig. 27.8 Absorption coefficient as a function of wavelength for a number of semiconductors. The absorption coefficient due to electronic processes is nearly zero until the band gap is reached, then rises for higher energies.

gap material such as GaAs, a photon with an energy $\geq E_g$ can promote an electron from the valence band to the conduction band. That is, absorption occurs when there is an electron in the valence band, a hole in the conduction band, and a photon of the correct energy. The result is that the absorption coefficient rises rapidly once the photon energy exceeds the band gap (see Fig. 27.8). In contrast, in indirect band gap materials, a photon with energy E_g will not necessarily be absorbed unless a phonon with the correct momentum is available simultaneously. This significantly lowers the probability of an absorption event, so the absorption coefficient rises more slowly with photon energy. This can clearly be seen in the figure for the cases of Si and Ge, for example, where the absorption coefficient rises much more slowly once the band gap is exceeded.

It should also be noted that when the absorption coefficient reaches values of $\sim 10^5$–10^6 1/cm, around 2/3 of the light is absorbed in the top 10–100 nm of material. That is,

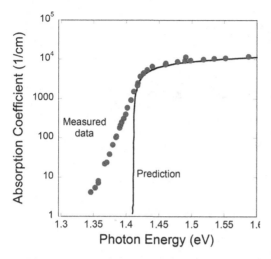

Fig. 27.9 Absorption edge of GaAs. The points show measured data, illustrating the Urbach tail, while the prediction shows the behaviour that would be predicted without the Urbach tail.

the penetration depth of the light is really quite small once the photon energy exceeds the direct band gap.

Once an electron is excited into the conduction band, its lifetime in the excited state will depend on the details of the band structure. When the material has a direct band gap, radiative transitions in which photons are emitted are more likely, and the carrier lifetimes in the excited state are often on the order of nanoseconds. In contrast, for indirect band gap materials, indirect recombination often occurs, in which localized energy levels in the band gap act as stepping stones for the excited carrier to reach the ground state. In this case, the lifetimes in the excited states can be much longer, often from the microsecond to the millisecond range.

In practice, solids don't have an abrupt "on–off" change in the absorption coefficient at the optical band gap: in many cases, there is an exponential tail to the absorption edge known as the Urbach tail (see Fig. 27.9). This occurs because the density of states around the electronic band edges have finite tails to their distribution as a function of energy. This can occur due to perturbations in the periodic potential of a crystal caused by lattice vibrations. For fast processes such as optical absorption, these perturbations act as static irregularities in the lattice. As the temperature increases, the Urbach tail tends to shift to lower energies. The net result is that the absorption coefficient often increases exponentially as a function of energy in a temperature dependent fashion: $\alpha = \alpha_o \exp\left(\frac{h\nu - E_f}{k_B (T - T_o)}\right)$, where E_f is the Urbach focus, $h\nu$ is the photon energy, k_B is Boltzmann's constant, T is temperature, and T_o is a constant for a given material. The existence of the Urbach tail can make precise determination of the band gap difficult.

If the wavelength range is expanded to include the infrared regime, an additional absorption mechanism develops in materials with some degree of ionic bonding. As the frequency decreases, ionic polarizability begins to contribute to the measured properties. In cases where the excitation frequency of the light corresponds to the

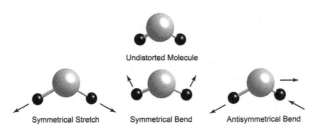

Fig. 27.10 Infrared vibration modes of a water molecule.

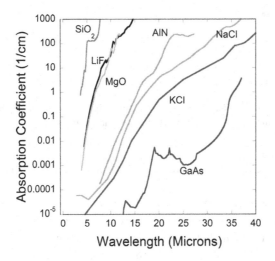

Fig. 27.11 Absorption edges in the infrared for a number of ionic crystals.

characteristic frequency of a particular bond, the light is strongly absorbed. It is possible to describe the mechanics of the system by treating the atoms as masses and the bonds that hold them together as springs. Strong bonds are difficult to perturb, and correspond to stiff springs, with a large spring constant k, while weak bonds act like compliant springs. Recall from basic physics that the resonance frequency, f_r, of such a spring–mass system is given by $f_r \propto \sqrt{\frac{2k}{m^*}}$, where k is the spring constant and m^* is the reduced mass:

$$\frac{1}{m^*} = \frac{1}{m_{cation}} + \frac{1}{m_{anion}}.$$

The observed vibration frequencies also depend on the vibration mode that is being excited. Figure 27.10 shows the modes that are characteristic of a vibrating water molecule. For all vibration modes, heavy atoms (high m^*) held together by long weak bonds (low k) have low resonant frequencies in the deep infrared. In contrast, for bonds between small light atoms such as O and H, the vibrational frequency rises for both the fundamental resonance and the overtones, pushing the absorption into the near infrared. This is illustrated in Fig. 27.11, where the infrared absorption edges of a number of ionic crystals are shown. Note though, that as the mass of the ionic species rises and the bond strength drops, resistance to attack by moisture also drops as well,

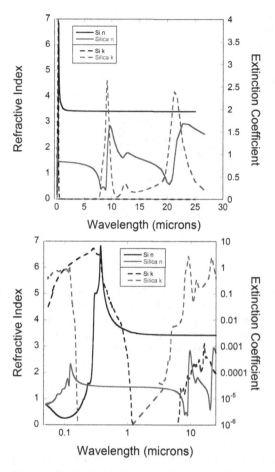

Fig. 27.12 Refractive index and extinction coefficient of Si and amorphous SiO$_2$ as a function of wavelength. Peaks in the extinction coefficient are apparent at each resonance. Both Si and SiO$_2$ have electronic transitions associated with the band gap. SiO$_2$ has pronounced absorptions due to ionic polarizability. The top shows wavelength on a linear scale, the bottom on a log scale.

and the mechanical durability also decreases. This can be problematic in designing robust materials that are transparent deep into the infrared regime, for use as infrared windows.

In cases where this is problematic, one can shift to a covalently bonded solid, where the ionic polarization mechanism (and its attendant absorption) is eliminated. This is illustrated in Fig. 27.12 comparing the refractive index and absorption coefficients of Si and SiO$_2$. In both cases, absorption in the visible and ultraviolet portion of the frequency range is caused by electronic transitions. The higher band gap of SiO$_2$ pushes the main electronic transitions in to the deep ultraviolet. However, in the infrared portion of the frequency range, the absorption coefficient of SiO$_2$ shows resonant frequencies associated with ionic polarization, whereas the absorption coefficients for Si are substantially smaller. The observed absorption is due to second-order effects, in which the light induces a dipole, which in turn couples to the light.

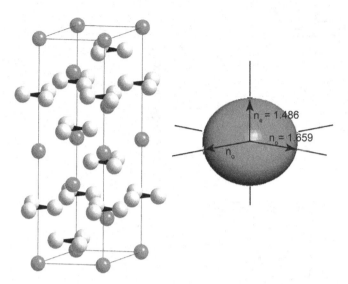

Fig. 27.13 Structure of calcite drawn with a hexagonal cell. The Ca atoms are shown as the darker shaded circles; the O are the lighter shaded circles. The triangles denote the carbonate groups with a C at the center bonded to three O atoms (a few oxygen atoms have been deleted from the unit cell for clarity). The optical indicatrix shown on the right mimics the flattened shape of the carbonate groups. n_o = ordinary index, n_e = extraordinary index.

27.2   Birefringence

In cubic crystals, as well as those with the Curie group symmetry $\infty\infty m$, the refractive index and absorption coefficients are isotropic. However, as the symmetry decreases, the refractive index will depend on the polarization direction of the light; such materials are described as *birefringent*. The optical anisotropy can be visualized using the optical indicatrix, as shown in Fig. 27.13. The plot depicts an ellipsoid given by $\frac{x^2}{n_x^2} + \frac{y^2}{n_y^2} + \frac{z^2}{n_z^2} = 1$, where x, y, and z are orthogonal axes, which are chosen to be coincident with the a, b, and c crystal axes in rectilinear crystal systems.

The origin of the birefringence can be thought of in terms of the anisotropy of the local electric fields experienced by the atoms. The electrons in the solid respond to the net electric field which is the sum of the applied electric field of the light wave as well as internal electric fields associated with the attraction to the nucleus as well as other surrounding dipoles. The electrons respond by displacing to produce local dipoles. In cubic crystals, the forces acting on the dipoles are isotropic, so that the polarization response is the same in all directions. In contrast, for lower symmetries, the forces acting on the dipoles are different in different directions. The consequence is that the magnitude of the polarization becomes a function of direction. Since the local polarization is ultimately responsible for the refractive index, the refractive index is now a function of direction.

The net result is that the optical indicatrix *mimics the shape of the molecular groupings in the crystal.* Consider, for example, the crystal structure of calcite shown

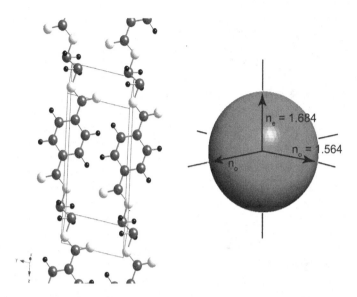

Fig. 27.14 A piece of the structure of polyethylene terephthalate showing the alignment of the chains. The C atoms are shown as the darker shaded circles; the O are the lighter shaded circles. The H is the small dark circle. The optical indicatrix shown on the right mimics the elongated shape of the polymer chains. n_o = ordinary index, n_e = extraordinary index.

in Fig. 27.13. One of the key structural features for this mineral is the planar CO_3^{2-} carbonate group. In calcite, all of the carbonate groups are aligned parallel to a common plane. This leads to the refractive index being largest parallel to this plane, as is illustrated by the shape of the optical indicatrix. Figure 27.14 shows the optical indicatrix for a partially aligned sample of polyethylene terephthalate, along with the polymer chains of the crystal structure. In this case, the optical indicatrix is elongated parallel to the chains.

In many anisotropic materials, the birefringence becomes particularly large near the band gap. The birefringence of most silicates is small because the silicate group itself is quite symmetrical. Most layer silicates have larger refractive indices in the layer than perpendicular to it. In the network silicate quartz, the birefringence is quite small, on the order of 0.01.

27.3   Optical Activity

Optical activity is a special type of birefringence in which the refractive indices for right and left circularly polarized light differ. This kind of behaviour is observed in chiral crystals such as quartz, the quartz derivative $AlPO_4$, Se, cinnabar HgS, and many liquid crystals. The crystal structure of Se is shown in Fig. 27.15, for two different orientations. Se is trigonal, with a 3_1 screw axis parallel to the crystallographic c axis. The Se atoms make spiral chains of a single handedness (either right or left handed). Light with different circular polarization states travels through the optically active material with different speeds, with the net result that on recombination the plane of polarization of

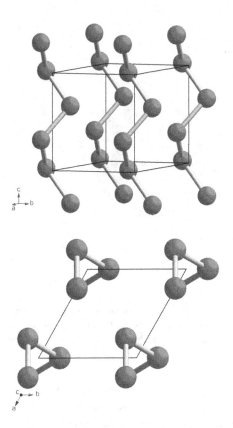

Fig. 27.15 The crystal structure of selenium shown (top) viewed orthogonal to the c axis (bottom) viewed along the c axis.

the light has rotated. The optical rotatory power, p, is defined as $p = \frac{\pi}{\lambda}\left(n_{left} - n_{right}\right)$, where n_{left} and n_{right} are the refractive indices for the left and right circularly polarized light.

In helical inorganic crystals, the most polarizable ions in the helix dominate the optical activity. In contrast, in organic crystals, the sign of p is a result of a competition between the chirality of the molecule and the chirality of the molecular arrangements. Near the band gap, where materials are especially polarizable, the optical activity rises.

One of the most technologically important uses of optical activity is in the liquid crystal displays used in portable electronics. In this application, a liquid crystal is confined between two closely spaced glass plates. The alignment of the liquid crystal molecules is typically constrained to rotate through the thickness of the part using grooves on the glass surfaces to mechanically align the liquid crystal. If the grooves on the two surfaces are oriented perpendicular to each other, then the optic axis of the liquid crystal twists across the display, forming a helix. The material is then optically active, and the polarization of the light rotates between the two plates.

When an electric field is applied to a given pixel of the display, the liquid crystal molecules rotate so that their long axis is parallel to the electric field. When this occurs,

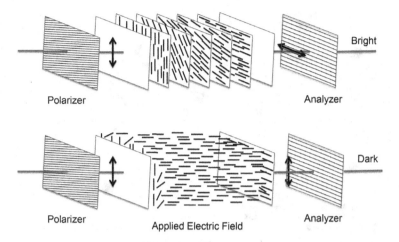

Fig. 27.16 Operation of a liquid crystal display. (Top) With no applied field, a twisted nematic liquid crystal rotates the polarization direction of light due to its optical activity. (Bottom) An applied electric field reorients the liquid crystal molecules, eliminating the optical activity, and turning the pixel off. Adapted, with permission, from B.E.A. Saleh and M.C. Teich, *Fundamentals of Photonics*, John Wiley & Sons, New York, 1991, Fig. 18.3–6.

the helical arrangement of molecules is destroyed, and the optical activity is essentially eliminated. This can be used to electrically control whether light will pass through polarizers, turning the cell on or off, as shown in Fig. 27.16.

27.4   Paint

Paints are used to cover surfaces for decorative and protective purposes. In thinking about the optical functions of paint, it is worthwhile recalling that light is (partially) reflected when it encounters a discontinuity in the refractive index. The bigger the contrast in the optical properties, the more light is reflected.

The three principal components of *paints* are a binder, a pigment, and a thinner. A *binder* adheres to a surface, holding the pigment in place and acting as a moisture barrier. A *pigment* provides covering power and color for the paint, and *thinners* are solvents that help in applying the paint, and then evaporate.

For many years, linseed oil has been the most common binder. It contains linolenic acid, which has three carbon–carbon double bonds (see Fig. 27.17). On contact with air, linolenic acid reacts with oxygen to form a hard, resinous material. The double bonds are polymerized via a complex reaction cycle. In a typical oil-based paint, an alkyd resin is prepared by beating linseed oil with glycerol. The reaction product is then mixed with phthalic anhydride to obtain a low-molecular-weight resin. When the paint is spread in a thin film, the organic thinner evaporates and the resin polymerizes to form a highly cross-linked polymer.

Water-based latex paints use a binder that forms a stable emulsion in water. Generally this is a copolymer prepared from an acrylate ester and vinyl acetate.

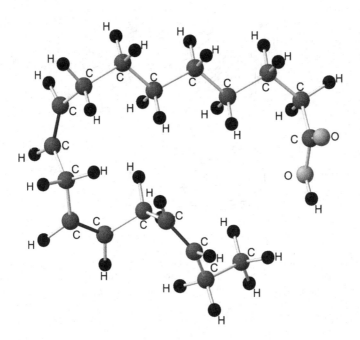

Fig. 27.17 Structure of linolenic acid. The C–C double bonds are shown as darker bonds.

Paint pigments are fine powders with high refractive indices, leading to high reflectivity and good covering power. The high refractive index means that light is reflected well both at the air–paint interface, and at any interfaces between the binder and the pigment particles. Typical white pigments include rutile (TiO_2), antimony oxide (Sb_2O_3), zinc oxide (ZnO) and barium sulphate ($BaSO_4$). Commonly used colored pigments include iron oxide (red), chromium oxide, or a variety of organic solids.

27.5   Color

Color can be introduced into solids by the presence of absorption bands in the visible spectrum (extending from ~400 to 700 nm in wavelength). Color arises through many mechanisms, all of them are associated in some way with electronic transitions between states. Among the most common origins of color are the following:

(1)   *Electronic transitions within partially filled d or f electron shells.* As was described in Chapter 14, when transition metals or rare earths are incorporated into a matrix, the energy levels of the d or f shells split due to crystal field theory. That is, the electrons in orbitals that are pointing more directly at their anion neighbors have higher energy levels than those pointing in between neighbors. Particularly in the case of the 3d and 4f orbitals, the typical energy splits correspond to visible wavelengths. Thus, a photon can cause a transition amongst the orbitals, resulting in an absorption at specific wavelengths. An example of this

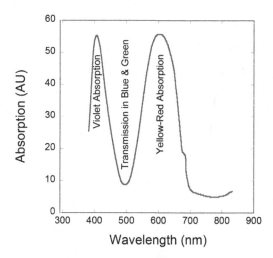

Fig. 27.18 Absorption (in arbitrary units) as a function of wavelength in emerald. The green color is a result of the transmission window in the green/blue wavelength range.

is shown in Fig. 27.18 for the case of the gemstone emerald. In emerald, the Cr^{3+} substitutes in small amounts for Al^{3+} in the crystal structure of beryl. Beryl, of itself, is transparent over the visible wavelength range. However, when Cr is added, the d electrons in Cr^{3+} split due to the octahedral crystal field. This results in absorption lines in the violet and orange/yellow/red portions of the spectrum. There is a strong transmission window in the green. The eye detects the transmitted green light, and perceives the gemstone as an intense green color.

(2). *Charge-transfer processes.* Charge transfer occurs when an electron is transferred from one atom to another. This mechanism is particularly important when multivalent ions are located on crystallographically equivalent sites. For example, magnetite (Fe_3O_4) is jet black, because visible light causes electrons to hop from one iron atom to another in the spinel structure:

$$Fe^{3+} + e^- \rightleftharpoons Fe^{2+}.$$

(3) *Electronic transitions associated with imperfections and defects in solids.* When colorless alkali halides are either irradiated, or heated in an alkali vapor, halogen vacancies are created. If electrons are trapped at this defect site, then the allowed energy levels for the electrons are quantized. Light can cause transitions among the allowed levels, producing absorptions in the visible wavelength range. For example, NaCl will turn yellow, and potassium bromide will turn blue, as a result of these color centers.

(4) *Band gap transitions.* Band gap transitions can also cause color in a material when visible light is sufficiently energetic to promote electrons from the valence band to the conduction band. Cadmium sulfide (CdS) is yellow because blue and violet light is absorbed by the crystal.

(5) The colors observed in organic solids can be explained by molecular orbital theory. In this case, color is often associated with resonating double bonds in a

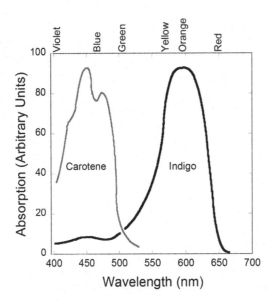

Fig. 27.19 Absorption spectrum for the organic dyes carotene and indigo.

molecule. In some cases, the molecule also has electron donor and acceptor groups that participate in the visible light absorption. Examples of colored organic solids include the yellow dye saffron, the orange color from carotene, the green of chlorophyll, and the blue dye indigo. Figure 27.19 illustrates the absorption spectrum for carotene and indigo. Carotene produces the orange color of carrots as a result of absorbing light with the blue and green wavelengths. The eye takes the transmitted yellow, orange, and red, and sees yellow. The deep blue color of indigo is a result of absorption of the yellow and orange portion of the visible wavelength range.

Fluorescence in phosphors and lasers arises in the same way. The neodymium glass laser, for example, uses transitions in the 4f levels of the Nd^{3+} modifiers in silica glass. Similarly, synthetic fluorapatite is used as a phosphor in fluorescent lamps when activated with antimony and manganese. Mn^{2+} fluoresces red and Sb^{3+} blue. Both are broad-band emitters, so the proportions of the two activators can be adjusted to give white light. The color rendition is further controlled by substituting Cl for F in the $Ca_5(PO_4)_3F$ host lattice. This tunes the crystal field about the Ca^{2+} sites occupied by the Sb^{3+} and Mn^{2+} activators, shifting energy levels.

X-rays are not efficiently captured by photographic film. This led to a search for phosphors which absorb X-rays and emit visible light. Calcium tungstate ($CaWO_4$ = scheelite) was used in X-ray-intensifying screens for many years, but has now been replaced by more efficient X-ray phosphors. With a density of 6.06 g /cm^3, $CaWO_4$ has only mediocre absorption in the 20–100 keV range used in X-ray medical radiography. The optical emission is centered around 430 nm, but the X-ray-to-visible-light conversion efficiency is only 5%. Examining the crystal structure (Fig. 27.20), it is apparent that only the tungsten atom is capable of absorbing

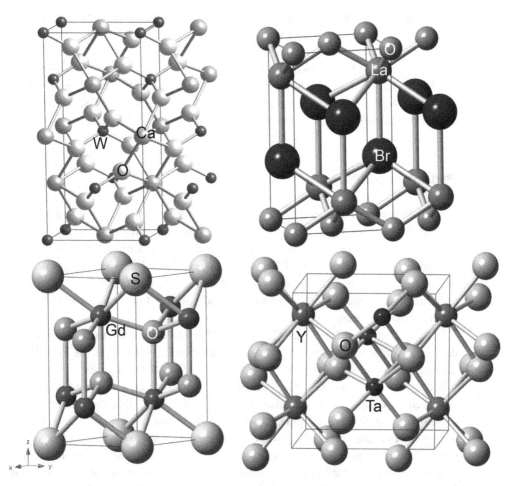

Fig. 27.20 Four of the host lattices used for X-ray phosphors. Heavy elements are important for absorbing the X-rays, and dopants such as Tm, Tb, and Nb convert the absorbed energy into visible light. The resulting fluorescent light is recorded. Clockwise from top left: $CaWO_4$ – the Ca has an unusual 8-coordination geometry that is approximately a square antiprism. LaOBr – the La has 9-coordination. $YTaO_4$ – Y is in 8-coordination. Gd_2O_2S – The Gd has 7-coordination.

50 keV X-rays – the oxygen and calcium atoms are too light. Only about 35% of the X-rays are absorbed in a 200 μm thick screen.

The introduction of rare-earth phosphors overcame some of these problems. Lanthanum oxybromide (LaOBr) doped with thulium (Tm^{3+}) has twice the absorption power of $CaWO_4$ and strong emission in the ultraviolet and visible range. LaOBr:Tm has the BaFCl structure with thulium and lanthanum bonded to nine anions, five bromines, and four oxygens. The compound crystallizes in a plate-like morphology, which is a distinct disadvantage because of the lower packing density.

Gadolinium oxysulfide (Gd_2O_2S) doped with terbium (Tb^{3+}) also exhibits improved absorption and emission over $CaWO_4$. The predominant emission is in the green portion of the optical spectrum near 530 nm. The crystal structure (Fig. 27.20) contains rare-earth ions in unusual 7-coordination, four oxygens, and three sulfurs. It has a density of 7.34 g/cm^3.

Yttrium tantalate has a structure closely related to $CaWO_4$, with Y in 8-coordination and Ta in tetrahedral sites. When doped with 2–5% Nb, this phosphor gives excellent absorption, higher speeds, and better resolution than other X-ray phosphors. Further improvements were made with Tm additions.[3]

27.6   Photoconductivity

When light impinges on a semiconductor, electrons are promoted from the valence band to the conduction band if the energy of the light is greater than the band gap. The mobile holes and electrons generated by light energy will produce a current in response to a voltage or built-in electric field. Because the magnitude of the current is a measure of the light intensity, photoconductors are useful in detecting and measuring the intensity of light. Other applications for photocells include devices to turn on street lights and automatic door openers.

Semiconductors such as CdS, CdSe, and CdTe are commonly used as photoconductors. Photoconductivity is plotted as a function of wavelength in Fig. 27.21. Starting from long wavelengths in the infrared, CdTe is the first to show a response, followed by CdSe and CdS. This is the order of their band gaps and bond lengths, as described in Chapter 8. It can also be seen in Fig. 27.21 that the photoresponse decreases at wavelengths shorter than the forbidden energy gap. These higher-energy photons are so readily absorbed by the crystals that only the surface of the semiconductor is affected. Since most of the crystal is inactive, the photoconductivity effect is small. CdS is most sensitive to green light (corresponding to the peak intensity in daylight), CdSe to red colors, and CdTe to the near infrared. Infrared detectors are commonly used in burglar alarms and for military systems for night vision or for heat-seeking missiles. Semiconductors such as InSb and (Hg,Cd)Te with very narrow band gaps are very sensitive to infrared wavelengths. It is often necessary to cool these infrared detectors to lower the background carrier concentration.

27.7   Nonlinear Optics

In cases where large electric fields are applied to a material, or where the incident light is of particularly high intensity, higher-order terms need to be considered in the polarizability equation:

---

[3] L.H. Brixner, "New X-ray phosphors," *Mat. Chem. Phys.*, 16, 253–281 (1987).

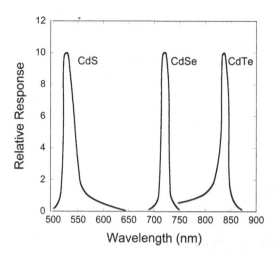

Fig. 27.21 Photoconductivity of the Cd chalcogenide semiconductors. The maximum photoconductivity is measured at the wavelength that corresponds to the band gap of the semiconductor.

$$P_i = \varepsilon_0 \chi_{ij} E_j + \varepsilon_0 \chi_{ijk}^{(2)} E_j E_k + \varepsilon_0 \chi_{ijkl}^{(3)} E_j E_k E_l + \cdots.$$

At heart, the origin of nonlinear optical effects is associated with the fact that no material is infinitely polarizable. Thus, when larger and larger electric fields are applied, the polarization response begins to saturate. An intense laser beam can produce electric fields up to 10^8–10^{12} V/m, vastly higher than the dc dielectric breakdown strengths of solids. Indeed, the field strength can be comparable to that of the atomic fields. The higher-order susceptibilities, $\chi_{ijk}^{(2)}$ and $\chi_{ijkl}^{(3)}$ which describe this produce a variety of interesting optical effects, including dependence of the refractive indices on dc electric fields, intensity dependence of the refractive index, second-harmonic generation, parametric amplification, and the photorefractive effect.

Symmetry acts on the higher-order susceptibilities, as on other properties. Important materials that have a useful $\chi_{ijk}^{(2)}$ include $LiNbO_3$, $LiTaO_3$, $NH_4H_2PO_4$, KH_2PO_4, β-BaB_2O_4, $KTiOPO_4$, $Sr_{5-x}Ba_xNb_{10}O_{30}$, and CdTe. Many of these materials are ferroelectric; their structures will be discussed in detail in Chapter 28. Because $\chi_{ijk}^{(2)}$ is a third-rank tensor, it disappears when there is a center of symmetry, as do all odd-ranked tensors. In this case, the lowest-order nonlinearity is $\chi_{ijkl}^{(3)}$.

From the standpoint of structure–property relations, the higher-order susceptibilities generally rise as the refractive index itself increases. Thus, more polarizable ions (e.g. large anions and some cations) increase the optical nonlinearity. In addition, as the coordination number for oxygen rises in glasses (e.g. as a result of the addition of modifier ions), the volume available for the oxygen increases. As a result, the nonlinearity rises.

27.8   Problems

(1) Below is a picture of some text viewed through a crystal of calcite. Explain why the text is doubled.

(2) Suppose you were able to grow a single crystal of polyethylene. Explain what factors would affect the refractive index and its anisotropy.

(3) Describe the contributions to the refractive index of solids. Draw the crystal structures for two different materials, one with a high refractive index, and one with a low one. Explain why they have high or low refractive indices on the basis of the structure and the mechanisms involved.

(4) Draw a graph showing the optical absorption coefficient versus wavelength curve for an ionically bonded material which is transparent to visible light. Why is the absorption coefficient high for some wavelengths? Explain the mechanisms.

(5) Consider a layered mineral like mica. Explain how the crystal structure affects:
(a) the shape of the optical indicatrix,
(b) the morphology of the crystal, and
(c) the anisotropy of the thermal expansion coefficient.
For each case, be sure to detail your reasoning.

(6) Draw a figure showing the refractive index as a function of wavelength from the near ultraviolet to the near infrared for a transparent material.
(a) Explain the physical phenomena that underlie the behavior.
(b) Suppose that you made a substitutional solid solution between this material and a compound with a larger (more massive) end member. How would this change the curve that you drew and why?
(c) Give examples of organic and inorganic materials (one example each) which are optically anisotropic. In each case explain why the anisotropy is observed based on the crystal structures and the bonding.

(7) $CaCO_3$ (calcite) is a rhombohedral crystal (space group 167: $R\bar{3}c$, with $a = 6.36$ Å and $\alpha = 46.08°$) and Ca at (0, 0, 0), C at (0.25, 0.25, 0.25), O at (0.506, −0.007, 0.25) and all symmetry-related positions).

(a). Calcite is known for its large birefringence (i.e. the refractive index is different for different polarization directions). The optic axis is parallel to the $\langle 111 \rangle$. By looking at your crystal parallel and perpendicular to this axis, speculate why the refractive indices are different.

(b) Suppose you were to substitute a small amount of Fe^{2+} for some of the Ca^{2+}. What sort of crystal field would you expect for the d electrons?

(8) The following graph shows the refractive indices of a number of transparent materials. Based on your knowledge of structure and polarizability, explain the relative positions of CsBr, NaCl, and LiF on the diagram. Figure reproduced, by permission, from W. D. Kingery, H. K. Bowen, and D. R. Uhlmann, *Introduction to Ceramics*, 2nd edition, John Wiley and Sons, New York (1976), Fig. 13.3(b), p. 652.

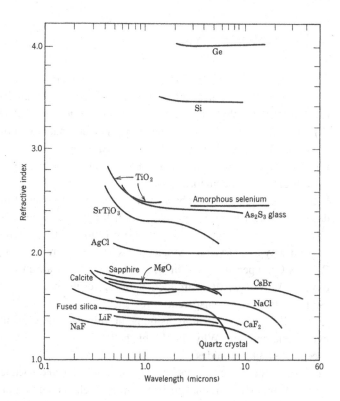

(9) Shown below are two views of the crystal structure of cinnabar (HgS). Its space group is $P3_121$, where the three-fold axis is directed along the c (or z) axis, and the two-fold axis is directed along a (or x). Only bonds to the near neighbors are shown.

(a) Describe what you would expect for the bonding characteristics of this material.

(b) For the Hg–S bond, $v_{ij} \approx \exp\left(\frac{2.308 - r_{ij}}{0.37}\right)$. The nearest-neighbor Hg–S distance in this material is 2.359 Å. How many bonds would you expect to next-near neighbors at an average distance of 3.23 Å?

(c) Draw the stereographic projection for the point group.

(d) Would you expect the material to show optical activity? Why?

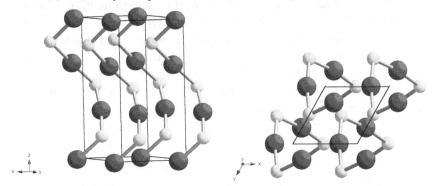

(10) Draw a figure showing the refractive index as a function of wavelength from the near ultraviolet to the near infrared for a material of your choice.

(a) Explain the physical phenomena that underlie the behavior.

(b) Illustrate the types of changes in your figure that would be expected as the bond strength of the material increased. Explain.

(c) Describe the structure of a material that is optically active, detailing the origin of the behavior.

(11) Figure 27.8 shows optical absorption curves for a number of different semiconductors. Explain the fundamental reason for the dependence of absorption on photon energy. Rationalize the relative position of the materials on the graph. Explain why the curves for Si and GaAs are so different in shape.

(12) Ruby is Cr-doped α-Al$_2$O$_3$, in which the Al and Cr are in 6-coordination with oxygen. Why does the introduction of Cr result in color in ruby? Note that α-Al$_2$O$_3$ itself is a transparent, wide band gap insulator. It is experimentally observed that the optical transitions in ruby are dependent on the applied hydrostatic pressure. Why should this be?

(13) F-centers in alkali halides.

(a) Write the defect formation reaction for a chlorine vacancy in NaCl. How will this reaction change if you trap an electronic carrier at the vacancy? (When this happens, the defect is referred to as an F-center.)

(b) Shown is a figure of the optical absorption coefficient as a function of wavelength for several alkali halides with F-centers. Explain why the different alkali halides show absorption at different wavelengths. Data from the figure were replotted from D. A. Robinson, A Study of the F-Center of Several Alkali Halides, Ph.D. thesis, Iowa State University, 1967.

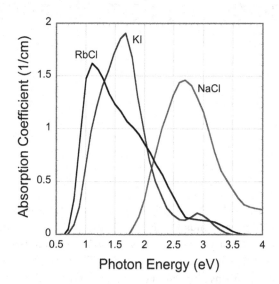

(14) Give an example of a material with a needle or fibrous morphology.

    (a) Explain on the basis of the crystal structure why this shape is adopted.

    (b) What would you expect concerning the birefringence of this material? Justify your reasoning.

28 Dielectrics and Ferroelectrics

28.0 Introduction

When an electric field is applied to an insulating material, little current flows. Instead, the material responds by displacing charges over distances ranging typically from fractions of an angstrom to a few microns. Numerous polymers, glasses, and crystalline materials are used as dielectrics. This chapter describes the different polarization mechanisms that produce the measured dielectric response and polarization, the frequency dependence, as well as representative crystal structures of key materials.

While all single-phase materials possess a dielectric response to an applied electric field, in materials where it is permitted by symmetry, polarization can be induced by an applied stress, producing *piezoelectricity*. The main mechanisms responsible for the electromechanical coupling are described from a crystallographic perspective.

Pyroelectric materials possess a unique polar axis that is not related to any other direction by symmetry. Because the magnitude and/or the orientation of the polarization are a function of temperature, an electrical charge will develop on the sample surface when the temperature is changed. Finally, *ferroelectric* materials are ones in which the spontaneous polarization can be re-oriented between crystallographically defined states by application of an electric field. In this chapter, the structural origins of a spontaneous polarization and its temperature and field dependence are illustrated for a series of ceramics and polymers.

28.1 Relative Permittivity and the Polarization Mechanisms

As an electric field is applied to an insulator, the material responds by displacing charges as far as it can, to produce a polarization. The induced polarization of the solid can be quantified by the relative dielectric response. Consider, for example, the parallel plate capacitor shown in Fig. 28.1, with electrode area A and a separation distance between the electrodes of d. In this case, the capacitance is given by:

$$C = \frac{\varepsilon_o \varepsilon_r A}{d},$$

where ε_o is the permittivity of free space (8.85×10^{-12} F/m), and ε_r is the relative permittivity, sometimes called the dielectric constant. It should be noted that in many

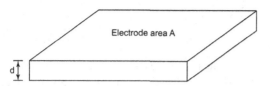

Fig. 28.1 Parallel plate capacitor.

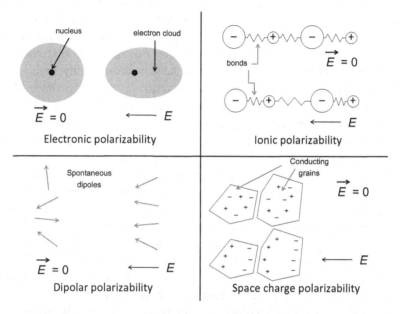

Fig. 28.2 Polarization mechanisms in dielectric materials. Adapted from A R. von Hippel ed., *Dielectrics Materials and Applications, student edition*; jointly published by the Technology Press of MIT and John Wiley and Sons, INC, NY (Chapman and Hall, London) 1954, Fig. 1.1, p. 19. © 1966 Massachusetts Institute of Technology, by permission of The MIT Press.

materials "constant" is a misnomer, due to the dependence of ε_r on frequency and temperature, for example.

Dielectric materials are electrical insulators that can be used to isolate signal-carrying wires from each other, or to store and deliver electrical charge from a capacitor. Dielectric solids are electrically insulating; a useful rule of thumb is that such materials will have band gaps exceeding approximately 3 eV. As a result, when an electric field is applied, the material will respond by polarizing. That is, the charges will be displaced over relatively small distances. There are multiple polarization mechanisms that contribute to the observed response, as illustrated in Fig. 28.2.

Electronic polarizability refers to the displacement of the electron cloud of an atom with respect to the nucleus. Because electrons are small and light, the electron cloud can follow the applied electric field up to frequencies on the order of 10^{15} Hz. As described in Chapter 27, this is the same mechanism that is responsible for the optical properties in the visible frequency range. This contribution to the dielectric response typically varies in the range of ~1.5 to 10 in solids. In most ionic materials, this will account for ~1/3–1/2 of the low-frequency dielectric constant. In covalent insulators, it accounts for all of the low-frequency dielectric constant. Larger atoms, with more inner electron shells, generally have more polarizable electron clouds that produce higher electronic polarizabilities.

Ionic polarizability results from the relative displacements of cations and anions from their equilibrium positions due to the applied electric field. As shown in Fig. 28.2, this leads to alternating compression and expansion of the bonds holding the solid together. No bonds are broken during this process, and the atom displacements are typically a small fraction of an ångström. When the electric field is removed, the atoms will recover to their equilibrium separation distances. The bonds in the solid serve as the restoring force for this. The more elastically compliant the material is, the more readily bonds can be compressed/extended, and the higher is the ionic polarizability. In most solids, ionic polarizability contributes from ~0–10 to the relative permittivity. Obviously, the lower limit is observed when there are no ions in the material (e.g. for covalently bonded solids). In a limited number of solids, ionic polarizability can contribute up to a few hundred to ε_r. This mechanism contributes to the total relative permittivity from dc to ~10^{12} Hz.

Dipolar polarizability refers to the reorientation of pre-existing dipoles due to the applied electric field. Consider, for example, a comparison between CO_2 and water molecules. CO_2 is a linear molecule, and any dipole present on one side of the molecule is cancelled by an equal and opposite dipole on the other side. In contrast, for water, the oxygen ion is the center of negative charge, while the center of positive charge is located at the midpoint between the two hydrogen atoms in the bent molecule (see Chapter 11). When an electric field is applied, this molecule will attempt to reorient, such that its permanent dipole moment is parallel to the applied electric field, within the constraints of any bonds in which it is participating. Typically, this dipole reorientation can follow the applied field up to frequencies of ~10^9–10^{10} Hz. High relative permittivities can be achieved if either the relative orientation or magnitude of the dipole can be changed by an electric field. Such materials are said to have high polarizability. It is important not to confuse *polarization* (the $\frac{static\ dipole\ moment}{unit\ volume}$) with *polarizability* (which describes how polarization is induced or modulated by an applied field).

Space charge polarizability refers to the motion of charge carriers across macroscopic distances, until they are blocked by an insulating interface. Obviously, the magnitude of the contribution depends on the mobility of the charge carriers, as well as the distance traveled. In well-prepared samples, this mechanism can produce effective permittivities of 10^5–10^6. In most cases, though, the response is comparatively slow. While in principle this mechanism can contribute into the GHz frequency range, it is usually lost by ~1 MHz.

The measured dielectric constant is the sum of the responses from all of the polarization mechanisms that are available at the frequency and temperature of the measurement. A schematic of this is shown in Fig. 28.3. It is critical to note that in real materials there can be multiple contributions to any mechanism (e.g. many phonons contribute to the total ionic polarizability); the schematic shows all of these lumped together. Table 28.1 lists typical relative permittivities of a range of dielectrics and semiconductors measured at 1 MHz at room temperature.

Table 28.1 Representative relative dielectric permittivities at 1 MHz

Material	ε_r	Material	ε_r	Material	ε_r
C (Diamond)	5.68	Amorphous SiO_2	3.9	LiF	9
Si	11.7	Borosilicate glass	~4	NaCl	5.90
Ge	16	Amorphous polyethylene	2.3	Al_2O_3	~10
GaP	10.2	Amorphous polypropylene	2.2	MgO	9.65
GaAs	13.1	Polystyrene	2.55	Water	~80
InP	12.1	Polyethylene terephthalate	~3	TiO_2	85.8 ∥ a
					179 ∥ c
InAs	12.5			$BaTiO_3$	~1000–14 000
InSb	17.7				

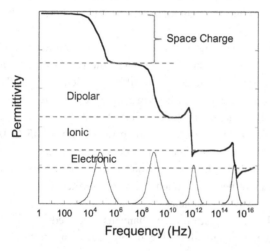

Fig. 28.3 Schematic illustration of the frequency dependence of different contributions to the permittivity. The darker line shows the real part of the dielectric permittivity. The thinner lines show the imaginary part.

28.2   Charge Storage and Loss

When an electric field is imposed on a capacitor, it leads to charge flow that goes towards zero as the capacitor reaches its maximum energy storage level during the charging process. In real materials, it is not possible to recover all of the dc current flow when the capacitor discharges. That is, real materials have finite electrical resistivities, so there will be some energy loss in the capacitor. Properly speaking, to account for the energy loss, the relative permittivity is better described as a complex quantity:

$$\tilde{\varepsilon}_r = \varepsilon' - i\varepsilon''.$$

Note the analogy to the complex dielectric function described in the chapter on optical properties. The dielectric loss tangent, $\tan\delta$, is defined as $\frac{\varepsilon''}{\varepsilon'}$, and can be used to account for the phase lag between the induced polarization and the applied electric field.

The power dissipation in the capacitor exposed to an alternating electric field is $power = I_{rms}V_{rms} \tan \delta$. Here I_{rms} is the root mean square current and V_{rms} is the root mean square voltage. This dissipated power rises with the loss tangent and with the frequency of the applied electric field. Thus, it is interesting to consider the mechanisms that contribute to the loss tangent at different frequencies.

First, it is useful to note that at frequencies where a polarization mechanism is no longer able to keep up with the electric field, the loss in permittivity is coupled to an increase in loss tangent. This is shown schematically in Fig. 28.3.

To illustrate in more detail, let us consider *microwave absorption*: Microwaves are electromagnetic waves ranging from 1 μm to 1mm in wavelength, at frequencies of 0.3 to 300 GHz. Applications include cell phones, military hardware, and microwave ovens. The interactions between microwaves and materials takes place through the electric field vector $\vec{E}$ (V/m) and the magnetic field vector $\vec{H}$ (A/m) belonging to the microwave.

Most of the low-frequency dielectric loss phenomena originate from conduction or space charge effects (Fig. 28.4). In minerals like magnetite (Fe_3O_4), this may involve electron hopping between transition metals with different valence states. This causes conduction and localized heating which further lowers the electrical resistance and may lead to thermal runaway. It is interesting that ferrites absorb microwave energy more efficiently than metallic Fe. When heated in a microwave oven (2.45 GHz, 1 kW), mixed-valence oxides heat faster than metal powders despite the higher electrical conductivity of metals. Metals do not absorb microwaves effectively because high frequency electric fields cannot penetrate much below the surface. For copper, the skin depth is only 1 μm at microwave frequencies.

Several other molecular and electronic mechanisms involved in the interaction between microwaves and matter are illustrated in Fig. 28.5. In addition to hopping

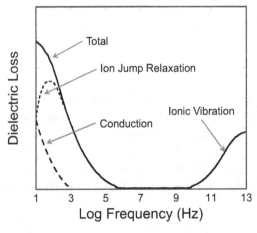

Fig. 28.4 In common oxide insulators, the low frequency losses are dominated by conduction and space charge effects. Far infrared vibrations become important in the microwave region.

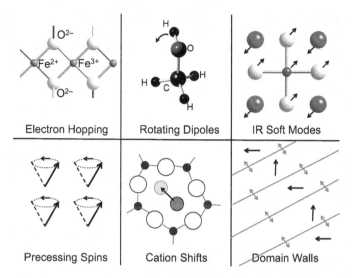

Fig. 28.5 Six ways in which microwaves are absorbed by solids: electron motion, molecular dipole rotation, infrared vibration, magnetic spin resonance, localized ionic motion, and domain wall movement.

electrons and other conduction phenomena, electromagnetic waves in the GHz range are absorbed by rotating dipoles, precessing spins, soft infrared modes, cation shifts, and shivering domain walls. Localized ionic motions promote losses in ceramics and glasses. Fused silica has low dielectric loss over wide frequency and temperature ranges, but many commercial glasses contain significant amounts of Na^+ and Ca^{2+} ions, which promote dielectric loss. At low frequencies, the loss is associated with ion transport through the silicate network, giving a space charge contribution to the dielectric permittivity. Vibration losses become important at higher frequencies. Alkali ion motions in silicate cages lead to sizeable absorption in the microwave range. Cooking-ware made of $NaAlSiO_4$ glass-ceramics is unsuitable for use in microwave ovens because of the rapid heating rate.

Dipole reorientation is the principal cause of microwave absorption in water, polymers, and most organic compounds. The loss spectrum of water follows a classic relaxation curve (Fig. 28.6). Water molecules have large dipole moments because of their bent linear geometry and partially ionic nature. Under an electric field, the dipoles reorient easily at room temperature, producing a large dielectric constant of nearly 80. At microwave frequencies, there is a rapid decrease in the dielectric constant, accompanied by a large loss peak that is utilized in cooking food. The presence of water in meat and vegetables couples the microwave energy into the food.

Other polar liquids behave in a manner similar to that of water. Most contain electronegative oxygen atoms or hydroxyl groups that enhance the electric dipole moment of the molecule. Table 28.2 lists the microwave properties of several polar and non-polar liquids.

The microwave absorption peaks of polar liquids are strong functions of temperature. Lowering the temperature makes the liquids more viscous and reduces the relaxation

Table 28.2 Microwave dielectric properties of liquids at 3 GHz and 25 °C. Polar liquids have higher permittivities and higher losses

Polar liquids	ε'_r	$\tan \delta = \frac{\varepsilon''_r}{\varepsilon'_r}$
Water (H_2O)	77	0.16
Methyl alcohol (CH_3OH)	24	0.64
Ethyl alcohol (CH_3CH_2OH)	6.5	0.25
Non-polar liquids		
Heptane ($CH_3(CH_2)_5CH_3$)	2.0	0.0001
Carbon tetrachloride (CCl_4)	2.0	0.0004

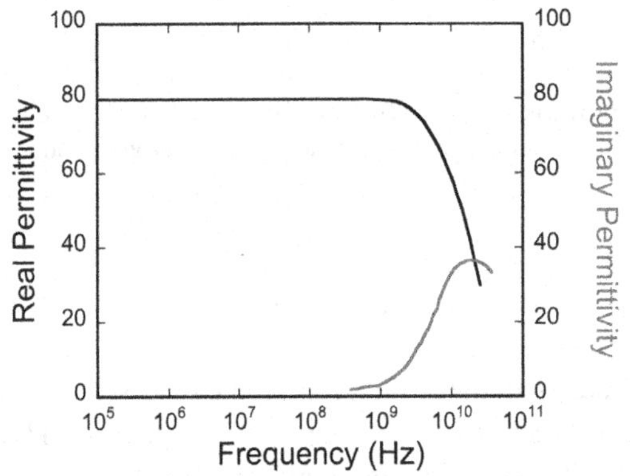

Fig. 28.6 Water has a high dielectric constant (ε') and sizeable losses (ε'') in the microwave region. Adapted from A R. von Hippel ed., *Dielectrics Materials and Applications, student edition*; jointly published by the Technology Press of MIT and John Wiley and Sons, INC, NY (Chapman and Hall, London) 1954, Chapter 1, Theory, part B, A. Von Hippel, Fig. 4.2, page 38. © 1966 Massachusetts Institute of Technology, by permission of The MIT Press.

frequency. It is the "tunability" of energy absorption that makes microwave processing especially interesting for materials engineers. Unlike normal furnaces with broad-band heat sources, monochromatic heat sources such as microwave ovens or high-power lasers offer the possibility of shifting heating from one constituent to another as reaction and densification take place.

The dielectric properties at microwave frequencies often arise from pendulum-like motions of polar side groups. Generally speaking, the smaller segments of polymer chains resonate at higher frequencies and are the last to be frozen out at low temperatures.

Low-frequency infrared modes (see Chapter 27) sometimes extend into the microwave range. Among ferroelectric oxides, these "soft modes" often involve vibrations of heavy cations such as Ba or Pb against a network of oxygen octahedra. The mode softens in high-permittivity perovskites and eventually condenses at the Curie

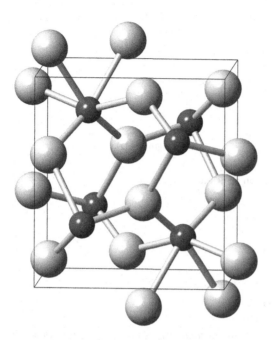

Fig. 28.7 Crystal structure of high-temperature zirconium titanate viewed a few degrees off of [100]. Oxygen and metal ion positions are represented by the light and dark colored circles, respectively. Oxygens form a distorted hexagonal close-packed array with disordered Zr and Ti ions occupying half of the octahedral interstices. The average bond length is 2.08 Å.

temperatures to give spontaneous polarization. This helps explain why high-permittivity dielectrics are generally more absorbing than low-permittivity dielectrics in the microwave region.

The dielectric resonators used in cellular communication networks, marine satellite communications, and microstripline filters are sometimes made from soft-mode materials. Dielectric resonance occurs when the specimen size corresponds to an integral number of microwave wavelengths. The resonant frequency is given by:

$$f = \frac{c}{\lambda\sqrt{\varepsilon_r\mu_r}},$$

where c is the speed of light in vacuum, λ is the wavelength, ε_r is the relative permittivity, and μ_r is the relative magnetic permeability. High-permittivity dielectrics reduce the wavelengths, and miniaturize the microwave filters, but the losses must be low, eliminating most ferroelectrics from consideration. Dielectric resonators made of modified zirconium titanate ($\varepsilon_r \sim 40$) and modified barium zirconate ($\varepsilon_r \sim 30$) are widely used in the 1–20 GHz range. The crystal structure of $ZrTiO_4$ is shown in Fig. 28.7.

Finally, losses associated with motion of domain walls will be discussed below, after domain structures are introduced.

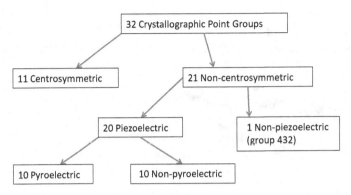

Fig. 28.8 Hierarchy of symmetry.

28.3   Hierarchy of Symmetry

A subset of dielectric materials are *piezoelectric*, so that application of a stress induces a polarization that is proportional to the applied stress. *Pyroelectric* materials possess a unique polar axis that is not related to any other direction by symmetry. Because the magnitude and/or the orientation of the polarization will generally be a function of temperature, an electrical charge will develop on the sample surfaces when the temperature is changed. Finally *ferroelectric* materials are ones in which the orientation of the spontaneous polarization can be switched between crystallographically defined states by application of an electric field.

Figure 28.8 shows the symmetry hierarchy of these effects. Of the 32 point groups, 11 are centrosymmetric, so that all odd-rank-tensor properties, including pyroelectricity and piezoelectricity, are symmetry-forbidden. Of the 21 non-centrosymmetric point groups, point group 432 has enough symmetry that piezoelectricity is impossible. The remaining 20 point groups are potentially piezoelectric. Of these 20 point groups, 10 have a unique symmetry axis, and so are polar. These are the 10 pyroelectric crystal classes. All ferroelectric materials are a subgroup of the pyroelectric classes in which it is possible to reorient the spontaneous polarization by applying an electrical field.

It is particularly easy to visualize the link between symmetry and the possible directions for the spontaneous polarization and pyroelectric coefficient. Consider, for example, the point groups and associated stereographic projections shown in Fig. 28.9. One can imagine drawing an arrow from the origin to each of circles on the stereographic projection, recognizing that those circles which are filled correspond to the Northern hemisphere, and those that are open correspond to the Southern hemisphere. If the vector sum of all of the arrows is zero, then materials with this point group cannot be pyroelectric. In cases where not all of the components cancel, then a net polarization (and pyroelectric coefficient) can be observed for this orientation.

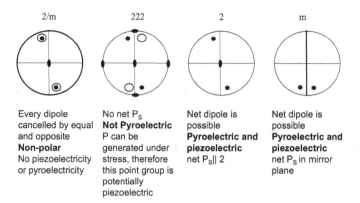

Fig. 28.9 Stereographic projections for several point groups and the link to pyro- and piezo-electricity. P_S refers to the spontaneous polarization. Redrawn from R. E. Newnham, *Structure–Property Relations*, Springer-Verlag, New York (1975). Fig. 31. Adapted with permission by Springer.

28.4 Structural Origins of Ferroelectricity

Consider the definition of a ferroelectric material as a compound that has a spontaneous polarization that can be reoriented between crystallographically defined states by application of an electric field. This means that it must be possible in a ferroelectric material to identify, on a unit cell level, a spontaneous dipole: a separation in space between the centers of positive and negative charge. Furthermore, more than one possible orientation for the polarization must be possible and accessible by a realizable electric field that does not cause dielectric breakdown. The process of polarization reversal will be described further below.

There are numerous key crystal structures for ferroelectricity, including the perovskite structure, polyvinylidene difluoride, $Bi_4Ti_3O_{12}$, and $NaNO_2$. The origin of the polarization differs for these cases, and it is interesting to examine the various possibilities. Consider, for example sodium nitrite, $NaNO_2$. The dipole moment in this compound is a result of the bent NO_2^{1-} molecule. As shown in Fig. 28.10, the O–N–O bond angle is ~114.3°. Thus, the NO_2^{1-} molecule itself has a dipole, where the N^{5+} ion acts as the positive charge and the center of negative charge is in between the O^{2-} ions. At elevated temperatures, the orientation of this molecule is randomized, so there is no net dipole. Below the ferroelectric transition temperatures (the Curie temperature, written as T_C) the NO_2^{1-} molecules align to produce a net polarization. This makes $NaNO_2$ an *order–disorder* ferroelectric. That is, the local dipole exists at all temperatures, but orders below T_C.

In contrast, many of the perovskite ferroelectrics have a *displacive* character to the phase transition. Perovskites have the general chemical formula ABX_3, where A is a large ion, B is a smaller cation that prefers octahedral coordination, and X is typically O, F, or S. Consider, for example, the crystal structure of $PbTiO_3$. The Ti is octahedrally coordinated to O, the Pb is 12-coordinated, and the O has a distorted octahedral coordination of two Ti and four Pb neighbors. Above T_C, this has the prototype

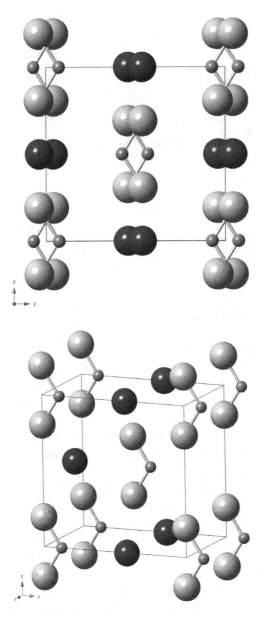

Fig. 28.10 Crystal structures of (top) paraelectric and (bottom) ferroelectric $NaNO_2$. The bent NO_2^{1-} molecules are held together by the large Na ions. At high temperatures the NO_2^{1-} molecules and Na^+ are disordered among the possible positions shown; they order in the ferroelectric phase. In the ferroelectric phase, the material can be switched between different orientations for the spontaneous polarization.

perovskite structure, with Pb^{2+} ions on the corners of the unit cell, O^{2-} on the face centers, and Ti^{4+} at the unit cell center. At T_C, the Ti *displaces* towards one of the O atoms, as shown in Fig. 28.11. As it does so, a spontaneous polarization develops parallel to the Ti displacement. At the same time, the unit cell elongates parallel to the

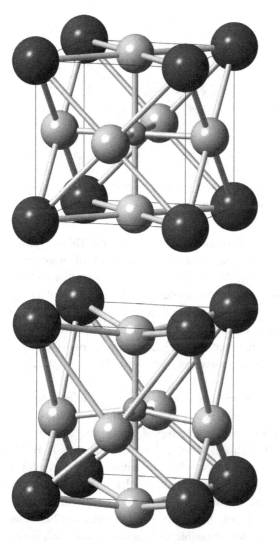

Fig. 28.11 PbTiO$_3$ crystal structures (top) paraelectric cubic phase, with Pb atoms at the cell corners, O at the face centers, and Ti in the center of the cell (bottom) ferroelectric phase with the Ti displaced along the c axis towards the O neighbor at the top of the cell.

spontaneous polarization, and contracts laterally, so that it is tetragonal rather than cubic. This points to the fact that the shape of the unit cell and the polarization are intimately linked. Bear this in mind, as it is important to an understanding of the origin of piezoelectricity.

It is also critical to note that because the Ti has six O neighbors, there are six possible orientations for the spontaneous polarization. That is, for the unit cell shown in Fig. 28.11, the Ti could move up, down, forward, back, left, or right. The polarization and the spontaneous strain of the unit cell will follow the Ti displacement.

In addition to the ferroelectric distortions just described, perovskite single crystals can also show a number of non-ferroelectric distortions associated with tilting of the

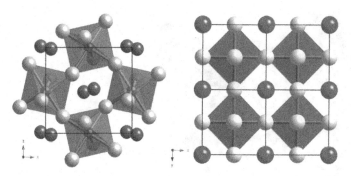

Fig. 28.12 Examples of (left) a tilted perovskite where the B-site cation–O–B-site cation angle is not 180°, and an (right) untilted perovskite.

octahedral units (with or without distortion of the octahedra). This is typically a result of the perovskite crystal structure collapsing around smaller A-site ions. An example of a tilted perovskite is given in Fig. 28.12. Many possible tilt transitions are known. When such a tilt transition occurs without the simultaneous development of a polarization, the material is referred to as ferroelastic, rather than ferroelectric, since the domain states differ only in the spontaneous distortion. Good examples of perovskite ferroelectrics in which tilt transitions are important are $BiFeO_3$ and $Na_{1/2}Bi_{1/2}TiO_3$. In the case of $BiFeO_3$, the spontaneous polarization has been attributed to displacement of the Bi atom, rather than the Fe.

There are many other ferroelectric materials that are related to the perovskite structure. $Bi_4Ti_3O_{12}$, for example, is an example of one of the Aurivillius phases, in which the perovskite periodicity is interrupted along one axis by insertion of a Bi_2O_2 layer. The general chemical formula for such materials is $Bi_2M_{n-1}R_nO_{3n+3}$; here M is a large ion, while R tends to be a transition metal that adopts octahedral coordination. The parameter n corresponds to the number of octahedra in the thickness of a perovskite block. In Aurivillius compounds with an even number of octahedra in the perovskite block, the ferroelectric polarization lies along either the $\pm a$ or $\pm b$ axis, producing four possible polarization states. $Bi_4Ti_3O_{12}$ has three layers in the perovskite block before interruption with the Bi_2O_2 (see Fig. 28.13). Aurivillius compounds with an odd number of octahedra in the perovskite block have the preponderance of their polarization in the a–b plane, with a small component parallel to c, yielding eight possible polarization directions. The spontaneous polarization is believed to be related to displacements of the Ti and O from their positions in the prototype phase.

In many hydrogen-bonded inorganic ferroelectrics, such as potassium dihydrogen phosphate, KH_2PO_4 or KDP, the ferroelectricity arises as a result of alignment of H along the bonds between oxygen atoms. As described in Chapter 11 for ice, H often forms one short covalent/ionic bond with oxygen and one longer hydrogen bond. The H can often move back and forth along the bond line. At elevated temperatures in KDP (see Fig. 28.14), the H are disordered amongst the available sites between the phosphate tetrahedra. At low temperatures, the H order amongst the available sites, with two close to each phosphate group. There is a simultaneous displacement of the

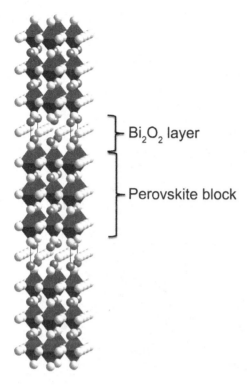

Fig. 28.13 Crystal structure of $Bi_4Ti_3O_{12}$ showing the interleaving of Bi_2O_2 and perovskite-like layers. Bi is the dark circle, and the TiO_6 octahedra are shaded. The unit cell is outlined in black.

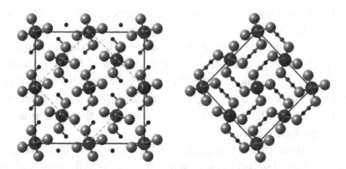

Fig. 28.14 Crystal structures of KDP, where the large filled ions are K, the small black ion is H, the small light ion is P, and the larger light ions are O. (Left) Orthorhombic ferroelectric phase with ordered H positions. (Right) Tetragonal paraelectric phase. In the tetragonal crystal, two possible H positions are shown in between phosphate groups; these are only 1/2 occupied in a random arrangement, so that there is one H between two oxygens. The outline of the tetragonal unit cell is shown in the dashed gray line on the orthorhombic cell to illustrate the relation between the two structures.

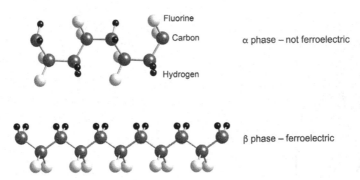

Fig. 28.15 Two different crystal structures for PVDF. When all of the fluorines are located on one side of the chain as in the β-phase, a spontaneous polarization develops.

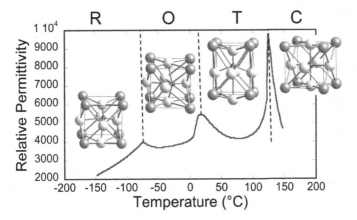

Fig. 28.16 Dielectric permittivity of BaTiO$_3$ as a function of temperature, showing the importance of the phase transitions.

potassium and phosphate ions in opposite directions along the ferroelectric c axis. These displacements are the cause of the spontaneous polarization.

The crystal structure of the polymeric ferroelectric polyvinylidene fluoride PVDF is shown in Fig. 28.15. Here, the dipole arises in the β-form of PVDF because all of the electronegative F are on "one side" of the chain, while the more electropositive H are on the other. Because the other chains align with the same orientation, a net polarization develops. PVDF does not have a clear Curie temperature, the alpha structure can be envisaged as a paraelectric phase.

The perovskite crystal structure is commercially important in the fields of dielectric and piezoelectrics. Modified BaTiO$_3$ is used in ~ 3×10^{12} multilayer ceramic capacitors annually (as of 2015). The dielectric permittivity of BaTiO$_3$ is high over a wide temperature range due to a sequence of phase transitions, as shown in Fig. 28.16. In the figure, R refers to the rhombohedral phase. In this region, the Ti displaces along one of the $\langle 111 \rangle$ directions of the original cubic cell. This moves the Ti closer to three neighboring O, towards one of the cube corners. Since there are eight corners on a cube, there are eight possible orientations for the spontaneous polarization. O refers to the orthorhombic phase; in this phase, the Ti displaces along one of the original cubic

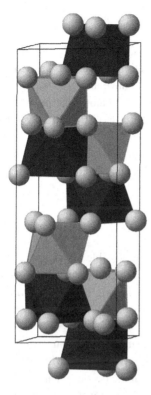

Fig. 28.17 Crystal structure of $LiNbO_3$, with the Li octahedra shown with the darker color, and the Nb octahedra shown in the lighter color. It is clear that both octahedra are distorted.

$\langle 110 \rangle$ axes. There are 12 possibilities here, and hence there are 12 possible polarization directions. In the tetragonally distorted version, the Ti displaces towards one of the neighboring O atoms along one of the cubic $\langle 100 \rangle$ directions. As was described for $PbTiO_3$, this yields six possible polarization directions. It is apparent from the figure that whenever the material is poised on the brink of a structural instability (i.e. at a phase transition), the polarizability, and hence the ε_r goes through a maximum. Piezoelectricity in perovskites will be discussed in Section 28.6.

Another commercially important ferroelectric is $LiNbO_3$. This can be regarded as a derivative of the corundum structure, as shown in Fig. 28.17, in which the cations order along the c axis as Li–Nb–☐, Li–Nb–☐. The off-centering of the Li ions in the octahedral site is the origin of the spontaneous polarization. In this material, the polarization is symmetry constrained to appear along the c axis, so only "up and "down" polarization states are possible.

In examining all of these ferroelectrics, a number of common structure features emerge.

(1) Many ferroelectric oxides contain d^0 ions with a tendency towards covalency, such as Ti^{4+}, Zr^{4+}, and Nb^{5+}. The resulting orbital mixing tends to produce somewhat asymmetric coordination polyhedra. Indeed, when Ti^{4+} is dissolved in water, it coordinates to six adjacent water molecules, with one of them being

much closer than the others – a coordination analogous to that observed in tetragonal $BaTiO_3$.

(2)   Atoms with lone-pair electron configurations, such as Pb^{2+}, Bi^{3+}, Sn^{2+}, Te^{4+}, S^{4+}, and I^{5+}, also prefer asymmetric coordination geometries, and tend to favor ferroelectricity. Moreover, these ions tend to have a large polarizability.

(3)   All of the atom displacements should be <1 Å along the polar direction with respect to the polar phase. This allows the polarization to be reoriented more easily (e.g. at lower electric fields).

(4)   Atom displacements from the prototype (or zero displacement) phase should exceed ~0.1 Å or the average amplitude of the thermal displacement for the atom.

28.5   Domains and Domain Walls

Because ferroelectric materials have more than one possible orientation of the polarization, it is not surprising that in different regions of the solid, the spontaneous polarization will be pointed in different directions. This is the origin of the *domain structure* of ferroelectric materials. The plane where the dipole orientation changes from one domain to the next is the *domain wall*.

In ferroelectric materials, domain walls are usually quite thin – only a couple of lattice spacings. To a first approximation, the lattice can be treated as quasi-continuous across the domain wall, with only small changes in the displacements corresponding to the ferroelectric polarization in the two domains. Domain walls are named for the angle made by the polarization across the wall. Thus, in a tetragonally-distorted perovskite, both 90° domain walls and 180° domain walls exist. In contrast, in rhombohedral perovskites, 180°, 71°, and 109° walls occur. Schematic illustrations of two types of domain walls in perovskites are given in Fig. 28.18. Permissible domain walls are those in which there is mechanical compatibility across the wall, so that excessive strains do not arise.

It is important to note that domain walls can be moved by electric fields, stresses, or combinations of the two. Note that 180° domain walls have no difference in spontaneous strain across the wall, since both domains will be elongated in the same way. Thus, domain walls of this type cannot be moved with a uniform applied stress, but can be moved by an electric field. Non-180° domain walls, sometimes called twin walls, have different spontaneous strains in the two domains. Thus, such a domain wall can be moved by an applied stress or by an applied electric field.

Domain wall motion is the process by which the spontaneous polarization is reoriented, producing the characteristic polarization–electric field hysteresis loop shown in Fig. 28.19. The underlying materials science that underpins the hysteresis loops is as follows: on cooling from the high temperatures used in processing the sample, the spontaneous polarization in a ferroelectric is typically aligned differently in different regions of the material; the net result is a net polarization near zero as-produced. When a high electric field is applied, the polarization tries to align with the applied electric field by motion of existing domain walls and nucleation of domains with orientations that are better aligned with respect to the electric field. This produces a large increase in

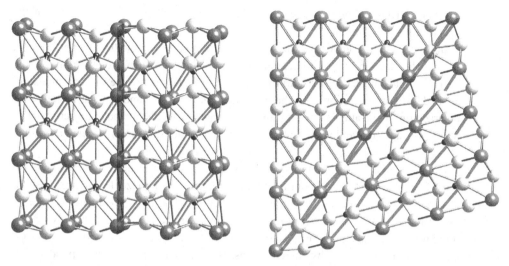

Fig. 28.18 Schematic representations of 180° and 90° domain walls in a tetragonal ferroelectric perovskite. For the 180° domain wall marked by the plane, the polarization is "up" on one side of the wall, and "down" on the other. In the 90° wall, the polarization is pointed "up" in the left-hand domain, and to the "left" in the domain on the right, following the directions of the Ti displacement. The distortions have been exaggerated to facilitate viewing. It is also worth noting that while the walls have been shown as essentially one unit cell in width here, non-180° walls, in particular, are believed to be a few unit cells wider.

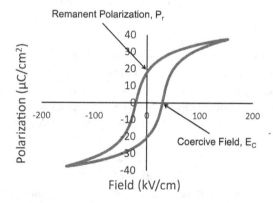

Fig. 28.19 Characteristic ferroelectric polarization–electric field hysteresis loop. The loop is traversed in a counter-clockwise fashion as the electric field is cycled.

the polarization. Once all of the domains are aligned as well as possible with respect to the field, further increases in the field lead to additional distortions of the material primarily via electronic and ionic polarizability. This accounts for the finite slope up at the tips of the hysteresis loop. When the field is removed, not all of the domains reorient, so that there is a remanent polarization in the sample. Unless the sample is a correctly oriented single crystal, the remanent polarization will be smaller than the spontaneous polarization.

In order to return the net polarization to zero, a reverse electric field can be applied, so that the domains begin to reorient to oppose the remanent polarization. At the coercive field, the net polarization is zero. If even larger reverse electric fields are applied, a net polarization with the opposite sign develops, and a negative remanent polarization can be attained. As is shown, for large-amplitude alternating electric fields, the polarization curve does not retrace itself, but demonstrates *hysteresis* – hence the characteristic open "loop".

Domain walls can also move short distances at lower electric fields, and so provide an important contribution to the dielectric and piezoelectric response of most ferroelectrics. Even at these modest electric fields, there is some hysteresis in the response, which produces domain wall losses in ferroelectric materials. In the microwave region, there are drastic reductions in the permittivity of $BaTiO_3$ capacitors, accompanied by increased losses. These frequencies correspond to the piezoelectric resonances of individual domains, and/or to motion of domain walls. Under dc bias, the domain walls are clamped, and marked reductions in both ε_r' and ε_r'' are observed. Similar loss phenomena occur in magnetic materials, and will be discussed in Chapter 29.

28.6   Pyroelectricity and Piezoelectricity

Pyroelectricity is a change in polarization, ΔP_i, with a change in temperature, ΔT:

$$\Delta P_i = \pi_i \Delta T,$$

where π_i is the pyroelectric coefficient. Consider, for example, $BaTiO_3$. As was discussed above, the material is tetragonal, with a spontaneous polarization at room temperature, and undergoes a phase transition to the cubic paraelectric phase on heating. As this occurs, the spontaneous polarization is lost. Thus, the polarization changes on heating, and the material is *pyroelectric*. A schematic of this behavior is shown in Fig. 28.20.

All ferroelectric materials are pyroelectric. In addition, some non-ferroelectric polar materials are also pyroelectric. An excellent example of the latter is the mineral tourmaline. One can sometimes recognize it in a shop window because the surfaces will be coated with dust. This is because the pyroelectric effect associated with normal day–night temperature changes leads to the development of surface charges on the crystal that attract dust particles from the air. Of more industrial importance is the use of pyroelectric materials for thermal imaging applications.

The *piezoelectric* effect entails a linear coupling between electrical and mechanical energies. Numerous piezoelectric coefficients are in use, depending on the electrical and mechanical boundary conditions imposed on the part under test. Each of the piezoelectric coefficients can be defined in terms of a direct and a converse effect; the two sets of

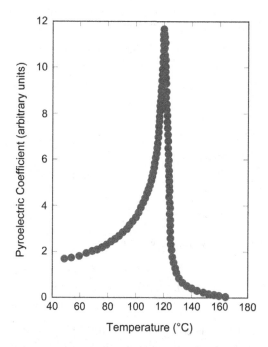

Fig. 28.20 Pyroelectric response of $BaTiO_3$ as a function of temperature.

coefficients are related by thermodynamics. For example, the piezoelectric charge coefficient, d_{ijk}, can be defined via:[1]

$$P_i = d_{ijk}\sigma_{jk},$$

where P_i is the induced polarization and σ is the applied stress. The same d_{ijk} coefficients are used in the converse piezoelectric effect, where

$$x_{ij} = d_{kij}E_k.$$

E is the applied electric field and x is the induced strain.

The piezoelectric coefficients are third-rank tensors, hence the piezoelectric response is anisotropic. The number of non-zero coefficients is governed by crystal symmetry, as described by Nye.[2] In most single crystals, the piezoelectric coefficients are defined in terms of the crystallographic axes; in polycrystalline ceramics, by convention the poling axis is referred to as the "3" axis.

The piezoelectric coupling coefficients describe how efficiently the material can convert between electrical and mechanical energies. For any vibration mode, a coupling coefficient can be defined as:

[1] IEEE Standard on Piezoelectricity, ANSI/IEEE Std 176–1987.
[2] J. F. Nye, *Physical Properties of Crystals: Their Representation by Tensors and Matrices*, Clarendon Press, Oxford (1979).

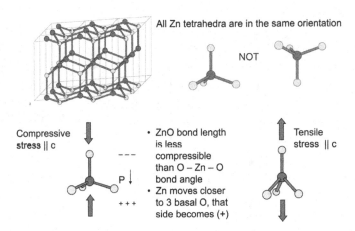

All Zn tetrahedra are in the same orientation

NOT

Compressive stress || c

• ZnO bond length is less compressible than O – Zn – O bond angle
• Zn moves closer to 3 basal O, that side becomes (+)

P ↓

+ + +

– – –

Tensile stress || c

Fig. 28.21 Piezoelectricity in ZnO with the wurtzite crystal structure.

$$(coupling\ coefficient)^2 = \frac{mechanical\ energy\ output}{electrical\ energy\ input}.$$

Piezoelectricity is observed in many materials with the wurtzite crystal structure, such as AlN and ZnO. The structural basis of this is shown in Fig. 28.21. In the wurtzite crystal structure, the cation-centered tetrahedra are all oriented in the same fashion within a single crystal. That is, all of them make triangular pyramids with the base down, in the figure shown. *When a stress is applied to the material, as a rule of thumb, it is easier to change the bond angle than it is to change the bond length.* The net result is that the legs of the tetrahedra splay out when a compressive stress is applied parallel to the *c* axis. This moves the cation closer to the base of the tetrahedron, making that side more positive. This stress-induced change in the polarization is the piezoelectric effect. The opposite sign of charge would develop on electroded surfaces if the sign of the stress was reversed. This mechanism for piezoelectricity is commercially of considerable importance in AlN thin-film bulk acoustic resonators used for frequency filtering in cell phones.

Perhaps the most important piezoelectric single crystal is quartz. Quartz is valued not so much for the magnitude of the piezoelectric coefficients, but for the excellent temperature stability and high quality factor Q that can be achieved. The crystal structure of α-quartz was discussed in Chapter 19. The Si^{4+} cations are in tetrahedral coordination, and the oxygens in a bent linear configuration. Interconnected chains of SiO_4 polyhedra spiral along the *c* axis in the trigonal crystal. Both right- and left-handed forms of quartz exist, depending on the handedness of the spiral. It is important in piezoelectric applications that a single handedness be utilized. In contrast to the perovskites discussed earlier, quartz is not a ferroelectric material. Thus, there is no spontaneous polarization that can be reoriented between crystallographically defined states by an applied electric field. The piezoelectric properties are set during the crystal growth.

There have been multiple explanations proposed for the origin of the piezoelectric effect in α-quartz. Recent X-ray diffraction measurements under applied electric fields support the model in which the inverse piezoelectric coefficient is associated with a rotation of the SiO_4 tetrahedra, rather than a deformation of the tetrahedral building blocks. The strong covalent contribution to the bonding is believed to effectively "stiffen" the tetrahedral unit, making it difficult to distort the O–Si–O bond angle. The rotation also seems likely given the existence of the $\alpha-\beta$ phase transformation. The net result is that for electric fields applied along the [100] or [110] crystallographic directions (i.e., perpendicular to the spiral axis), there is a change in the lattice parameters that is responsible for the observed d_{111}.

Quartz is used as the timing element in watches, for computers, and in many global positioning system devices. For these applications, one of the most valuable features of quartz is that it is possible to orient crystals so that the temperature coefficient of resonance frequency is effectively zero. This is achieved because of the presence of the $\alpha-\beta$ phase transformation in quartz. Essentially, in creating temperature independent cuts in quartz, the temperature dependence of the elastic constants is used to counterbalance the thermal expansion in the material. This is achieved by cutting crystals of different orientations. Quartz is also notable for its extremely high quality factor, of about a million.

The perovskite structure is remarkably stable, and can incorporate most of the atoms on the periodic table. As a result, a wide variety of useful properties can be engineered in perovskite ferroelectrics. As an example, the $PbZrO_3$–$PbTiO_3$ (PZT) solid solution shown in Fig. 28.22 is the basis of many important piezoelectric ceramics. Close to the middle of this complete substitutional solid solution, there is a nearly temperature independent morphotropic phase boundary between rhombohedrally and tetragonally distorted phases. As was discussed for the case of $BaTiO_3$, the polarizabilities become high when a material is poised between two different phases. As a result, the dielectric and piezoelectric coefficients are maximized at the morphotropic phase boundary

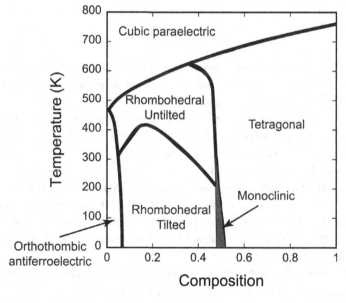

Fig. 28.22 PZT phase diagram.

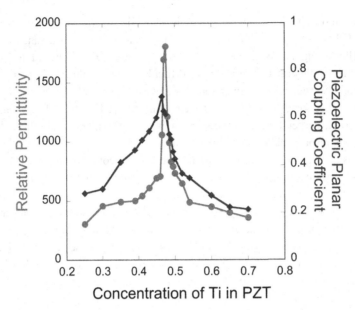

Fig. 28.23 Effect of composition on the dielectric and piezoelectric properties of $PbZr_{1-x}Ti_xO_3$ ceramics.

(see Fig. 28.23). Moreover, these high piezoelectric coefficients are retained over a wide temperature range. In addition, the energy penalty associated with making domain walls drops at the morphotropic phase boundary, so that motion of domain walls can help contribute to the dielectric, piezoelectric, and mechanical properties. PZT ceramics are widely used for applications including sonar systems, precise positioning devices, buzzers, and spark igniters, among others.

28.7   Problems

(1) Draw a stereographic projection for the point group of α-quartz. The space group is $P3_221$. On your diagram, show the location of all of the symmetry elements. Could such a material be pyroelectric? Explain your reasoning.

(2) Write an equation for the piezoelectric effect, defining all of the terms. Explain, based on your understanding of structure and bonding, why the following materials are piezoelectric: $(-CH_2-CF_2-)_n$ and $Pb(Zr_{1-x}Ti_x)O_3$.

(3) Describe (do not just list) the mechanisms that contribute to the dielectric constant of solids. Explain, based on your understanding of crystal structures and polarizability, why α-quartz has a low dielectric constant, and $BaTiO_3$ has a high dielectric constant.

(4) Explain the structural origins of a spontaneous polarization in a material of your choice. Describe how the existence of this polarization affects the dielectric and piezoelectric properties.

(5) Perovskite structured $BaTiO_3$ has a spontaneous electric polarization of 26 μC/cm^2 at room temperature that is oriented along the crystal "c" axis. The lattice dimension along the a direction is 4.0 Å.

(a) What is the fractional electronic free charge/unit cell on a "c" surface when the crystal is in equilibrium? Recall that a polarization is a charge/area.

(b) If the dipole moment were completely attributed to the displacement of a Ti^{4+} ion from the center of the unit cell along the "c" axis ($c \sim 4$ Å), what would the displacement be at room temperature?

(c) What is the electric field level that would be required to induce a similar density of surface charge on an NaCl crystal ($\varepsilon_r = 10$ for NaCl; assume properties are linear even at high fields).

(6) Single crystal tourmaline has a pyroelectric coefficient of 4 $\mu C/m^2$ °C along the polar axis. The dielectric constant along this axis is 7.1. Calculate the polarization induced by a 1°C change in temperature. How large an electric field would have to be applied to generate a comparable polarization?

(7) Bismuth titanate ($Bi_4Ti_3O_{12}$) is a ferroelectric material that is the basis of a large family of materials of interest for nonvolatile memories. The crystal structure at high temperatures is orthorhombic with space group *Fmmm*, with $a = 5.51$ Å, $b = 5.4487$ Å, $c = 32.84$ Å. The position parameters are:

Bi   (0, 0, 0.06700) and (0, 0, 0.211)
Ti   (0, 0, 0.5) and (0, 0, 0.372)
O   (0.25, 0.25, 0), (0.25, 0.25, 0.25), (0, 0, 0.436), (0, 0, 0.308), and
     (0.25, 0.25, 0.128).

(a) Plot the unit cell.

(b) Give the coordination of the Ti and the two types of Bi atoms.

(c) In the ferroelectric phase, the space group is *A1m1*. Along what axes can there be a finite value for the pyroelectric coefficient? Detail your reasoning.

(8) The low-frequency relative permittivity of water varies from 88.00 at 0 °C to 55.33 at 100 °C. Explain this behavior. Over the same range in temperature, the index of refraction (at 589.3 nm) goes from roughly 1.33 to 1.32. Why is the change in the refractive index so much smaller than the change in ε_r?

(9) The perovskite crystal structure is known for the variety of transformations it can support. The prototype structure is cubic (e.g. $SrTiO_3$ at room temperature). It has space group *Pm3m* (#221), with $a = 3.905$ Å. Sr is at (0, 0, 0) and Ti is at (0.5, 0.5, 0.5), and O at (0.5, 0.5, 0).

(a) Plot the unit cell of this structure showing the bonds of the atoms involved. Replot it showing the connectivity (and shape) of the Ti polyhedra.

(b) What are the coordinations of each ion? For the Ti itself, give each of the Ti–O bond lengths.

(c) $BaTiO_3$ is a ferroelectrically-distorted perovskite (at $T < 130$ °C). The structural distortion is quite small. At room temperature, it is tetragonal with space group *P4mm* (#99), with $a = 3.99$ Å, $c = 4.03$ Å. Ba is at (0, 0, 0) and Ti is at (0.5, 0.5, 0.514), and O at (0.5, 0.5, –0.025), and (0, 0.5, 0.485). Compare this structure to that of cubic $SrTiO_3$. (Hint: it helps to draw slightly more than one unit cell.) In your comparison be sure to include the Ti–O distances in tetragonal $BaTiO_3$.

(d) Is this a displacive or a reconstructive phase transformation? What does this imply about the kinetics of this phase transformation?

(e) Why should the cubic phase be favored at high temperatures?

(f) A ferroelectric, by definition, is a material with a spontaneous dipole which can be reoriented between crystallographically defined states by an applied electric field. In your $BaTiO_3$ unit cell, estimate the positions of the center of positive and negative charge in the unit cell. Also note that there were six possible directions in which the Ti could have displaced; as a result these are the six possible polarization directions in tetragonally distorted $BaTiO_3$. Note that two additional structural distortions (to orthorhombically and rhombohedrally distorted versions) also occur at lower temperatures.

(g) $GdScO_3$ is an orthorhombically distorted perovskite which is not ferroelectric. The space group is orthorhombic (*Pnma*, space group 62) with $a = 5.742$ Å, $b = 7.926$ Å, and $c = 5.482$ Å). Gd sits at (0.44058, 0.75, 0.48392), Sc sits at (0, 0, 0.5), and O at (0.4494, 0.25, 0.1201) and (0.1956, 0.5623, 0.1927). Plot a unit cell of this, and compare the structure to that of cubic $SrTiO_3$. It may be easiest to see this if you show the octahedra around the Sc.

(10) Would you expect the following materials to be pyroelectric? If so, along which axes? Show your reasoning.

(a) $LiNbO_3$ – space group $R3c$.

(b) Rutile – space group $4_2/mnm$.

(11) (a) Using drawings, explain the structural origin of ferroelectricity in one material.

(b) How many directions are there for the spontaneous polarization? Justify your reasoning.

(c) Why is this material piezoelectric?

29 Magnetism

29.0 Introduction

Basic electromagnetic theory points to a series of analogies between dielectrics and magnetism, as shown in Table 29.1. A magnetic field, $\vec{H}$, produces a magnetization, $\vec{M} = \frac{magnetic\ dipole\ moment}{unit\ volume}$, in much the same way that an electrical field, $\vec{E}$, produces a polarization, $\vec{P} = \frac{electric\ dipole\ moment}{unit\ volume}$. While electric dipoles are produced by a separation of positive and negative charge, magnetic dipoles are produced by *moving electric charge*. The fundamental unit of the magnetic moment is the Bohr magneton, $\mu_B = 9.27 \times 10^{-27}\ \mathrm{Am}^2$. This chapter describes the origin of the magnetic dipoles on atoms, the forces that direct the alignment of magnetic dipoles from one atom to the next, and magnetic domains and domain walls. The crystal structures of a number of important magnetic materials are also discussed.

29.1 Origins of Magnetic Dipoles

Moving electric charges produce a magnetic field as described by the right hand rule. That is, if the fingers of the right hand are pointed in the direction of the current, the thumb points in the direction of the resulting magnetic field. There are several possible means by which this can be developed.

(1) The spin on an electron. Each electron can be thought of as spinning on its own axis. As described by the fourth quantum number, there are two spins states possible in each orbital, "spin up" and "spin down."

(2) The orbital motion of the electrons around the atom. Given that electrons are moving around the nucleus, a magnetic dipole moment is produced.

(3) Charge moving in a loop of wire. Electromagnetics involve coils of wire through which a current is passed to induce a magnetic field on a macroscopic scale.

Magnetic dipole moments are often illustrated schematically using arrows.

In practice, in solids, the orbital contribution to the magnetic moment is typically quenched. Thus, the electron spin contribution often dominates. In atoms with filled electron orbitals, each spin up electron is paired with a spin down electron, producing no net magnetic moment. Thus, incompletely filled electron shells are very important to net magnetic moments on atoms.

Table 29.1 Analogy between dielectrics and magnetism

Dielectric	Magnetic
Polarization, $\vec{P}$	Magnetization, $\vec{M}$
Electric field, $\vec{E}$	Magnetic field intensity, $\vec{H}$
Permittivity of free space, $\varepsilon_o = 8.85 \times 10^{-12}$ F/m	Permeability of free space, $\mu_o = 4\pi \times 10^{-7}$ H/m
Dielectric displacement, $\vec{D} = \varepsilon_o \vec{E} + \vec{P} = \varepsilon_o \varepsilon_r \vec{E}$	Magnetic induction, $\vec{B} = \mu_o \vec{H} + \mu_o \vec{M} = \mu_o \mu_r \vec{H}$
Relative dielectric susceptibility, χ	Relative magnetic susceptibility, χ_m
Relative dielectric permittivity, $\varepsilon_r = \chi + 1$	Relative magnetic permeability, $\mu_r = \chi_m + 1$

Table 29.2 Electron configurations and net spins for several 3d transition metal ions in the high spin state

Ion		Electron configuration	Net spin
Ti^{3+}, V^{4+}	$3d^1$	↑	$1\mu_B$
V^{3+}, Cr^{4+}	$3d^2$	↑↑	$2\mu_B$
Cr^{3+}, Mn^{4+}	$3d^3$	↑↑↑	$3\mu_B$
Mn^{3+}, Cr^{2+}, Fe^{4+}	$3d^4$	↑↑↑↑	$4\mu_B$
Fe^{3+}, Mn^{2+}	$3d^5$	↑↑↑↑↑	$5\mu_B$
Fe^{2+}, Co^{3+}	$3d^6$	↑↓↑↑↑↑	$4\mu_B$
Co^{2+}, Ni^{3+}	$3d^7$	↑↓↑↓↑↑↑	$3\mu_B$
Ni^{2+}	$3d^8$	↑↓↑↓↑↓↑↑	$2\mu_B$
Cu^{2+}	$3d^9$	↑↓↑↓↑↓↑↓↑	$1\mu_B$
Zn^{2+}, Cu^{1+}, Ga^{3+}	$3d^{10}$	↑↓↑↓↑↓↑↓↑↓	$0\mu_B$

Table 29.3 Electron configurations and net spins for several 4f rare-earth ions

f^0	f^1	f^2	f^3	f^4	f^5	f^6	f^7	f^8	f^9	f^{10}	f^{11}	f^{12}	f^{13}	f^{14}
Y^{3+} La^{3+} Ce^{4+}	Ce^{3+} Pr^{4+}	Pr^{3+}	Nd^{3+}	Pm^{3+}	Sm^{3+}	Eu^{3+} Sm^{2+}	Gd^{3+} Eu^{2+} Tb^{4+}	Tb^{3+}	Dy^{3+}	Ho^{3+}	Er^{3+}	Tm^{3+}	Yb^{3+}	Lu^{3+}
$0\mu_B$	$1\mu_B$	$2\mu_B$	$3\mu_B$	$4\mu_B$	$5\mu_B$	$6\mu_B$	$7\mu_B$	$6\mu_B$	$5\mu_B$	$4\mu_B$	$3\mu_B$	$2\mu_B$	$1\mu_B$	$0\mu_B$

Recall that chemical bonding, particularly for ionic or covalent bonds, tends to produce filled shells. Metallic bonding does not necessarily produce electron pairing. Unpaired electron spins are common in transition metals and rare-earth ions, where there are incompletely filled d and f shells. The electron spins in the orbitals are filled in a manner consistent with Hund's rules and Pauli's exclusion principle. Tables 29.2 and 29.3 show the electron configurations and number of unpaired spins for the high spin state of 3d and 4f ions. The configurations for other ions can be calculated from the data given in Chapter 9.

As was described in Chapter 14, sometimes the energy level splits for different d and f orbitals are large enough that low spin states are observed. In this case, all of the low-energy states are filled with one spin first, then the opposite spins in the low-energy states are filled, before high-energy states are utilized.

29.2   Types of Magnetic Alignment

There are several different classes of magnetic response to applied magnetic field. Several of these, including possible arrangements of the magnetic dipoles, are illustrated in Fig. 29.1.

In materials where there are no unpaired electron spins, the induced magnetization *opposes* the applied magnetic field. The origin of the negative response is a field-induced

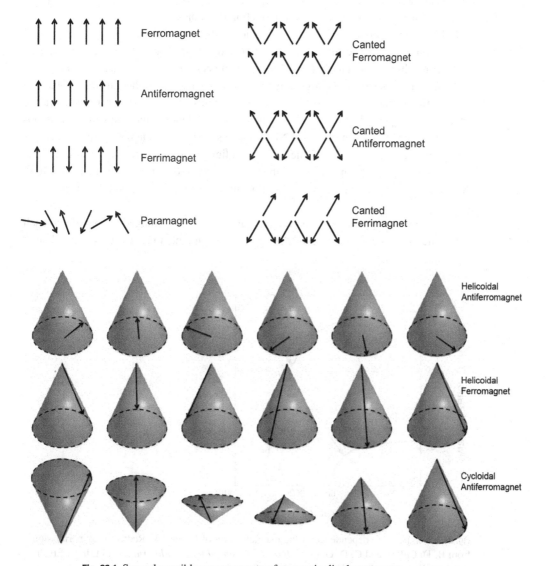

Fig. 29.1 Several possible arrangements of magnetic dipole moments.

change in the orbital motion of electrons consistent with Lenz's law. The χ_m is small, negative (typically $\sim -10^{-5}$) and weakly dependent on temperature; such materials are referred to as *diamagnetic*. Diamagnetism is present in all materials, but is masked if stronger magnetic effects are present. Particularly large diamagnetic responses are observed in aromatic compounds like benzene (where circulating currents due to the resonating multiple bonds can run around the carbon ring), and in superconducting materials, as described in Chapter 26.

Paramagnetic materials have electron spins on individual atoms, but these spins are randomly oriented in space. An applied magnetic field acts to align the spins parallel to the field, producing a positive χ_m, often on the order of 10^{-3} to 10^{-6}. Higher temperatures favor disorder of the spins, reducing the amount of field-induced spin alignment. Thus, χ_m decreases with temperature, $\chi_m = \frac{constant}{T}$. A second contribution to the paramagnetic response can arise due to conduction electrons.

In *ferroelectric materials*, the net magnetic moments on different atoms align parallel to each other below the magnetic transition temperature (the Curie temperature, T_C). The net result is a spontaneous magnetization in each magnetic domain. The magnetic susceptibility is positive and large in magnitude ($\sim 10^3$), as an applied magnetic field can reorient the magnetization direction. The susceptibility also shows a strong temperature dependence, $\chi_m = \frac{constant}{T - T_C}$. Figure 29.2 shows typical data for the temperature dependence of the magnetization of Fe and Ni. It can be seen that the magnetization is largest at low temperatures, where all of the spins are aligned. As temperature increases, some spins within any domain flip to another orientation, reducing the net magnetization. Alignment is lost at the transition temperature – above this point the material is paramagnetic.

Antiferromagnetic solids occur when magnetic spins are arranged antiparallel to each other, such that on a unit cell level, the net magnetic moment is zero. This occurs below

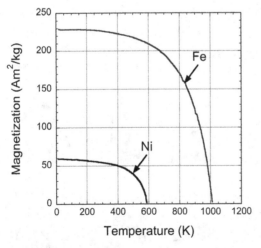

Fig. 29.2 Temperature dependence of the magnetization of Fe and Ni. Redrawn, by permission, from B. D. Cullity and C. D. Graham, *Introduction to Magnetic Materials*, 2nd Edn, IEEE Press and John Wiley and Sons, Hoboken, NJ (2009). Figure 4.5, page 120.

the Néel point. The χ_m tends to be modest in magnitude ($\sim 10^{-3}$, except near the Néel temperature) and positive.

Ferrimagnetic materials have incomplete cancellation of the magnetic dipoles at a unit cell level. The magnetic atoms in these materials occupy different sublattices in the crystal structure. The spin alignment is antiferromagnetic in character, but a net magnetization can result because the dipole moments on the sublattices differ in strength. Here, again, there is a spontaneous magnetization.

Symmetry also acts on the currents that produce magnetic dipoles. To visualize this, many texts introduce a new symmetry operator, the time reversal operator, which acts to make the currents responsible for the magnetic moments "run backwards," as if time had been reversed. To denote the time reversal operation, a prime is added to other symmetry descriptors, so that m' is a mirror plane which also reverses time. This is discussed in detail in many crystal physics textbooks.[1,2] When this operator is added to the other symmetry elements, the number of magnetic space groups rises to 1421,[3] and there are 90 magnetic point groups.

Symmetry acts on physical properties as described by Neumann's law. The effect of several symmetry operators on the current loop response for magnetization, and the resulting magnetic dipole is shown in Fig. 29.3. As shown there, the symmetry operators operate on the current loop. The right hand rule is then used to assess the orientation of the magnetic dipole moment.

29.3   Direct Exchange

When atoms with magnetic spins interact with each other directly, the coupling of the spins is governed by the direct exchange interaction. This mechanism is quantum mechanical in origin. Without explaining in detail, the potential energy describing the coupling of the two atoms with spins S_i and S_j is: $w_{ij} = -2JS_i \cdot S_j$. J is the exchange integral. When J is positive, the potential energy is lowered when the spins are parallel, i.e. when the spins couple ferromagnetically. A negative J value leads to antiferromagnetic coupling of the spins. The sign of J is a function of the atomic separation distance relative to the size of the electron orbitals responsible for the magnetic spins (typically those from d or f orbitals). This is shown in Fig. 29.4. Of the 3d transition metal elements, the only ones for which J is positive are Co, Ni, and Fe. Because Mn has a negative exchange integral, it is antiferromagnetic. In practice, antiferromagnetic and ferromagnetic coupling is often engineered in superlattices by controlling the interatomic separation distances.

It is not straightforward to predict the number of unpaired electron spins per atom in metals. In the formation of the band structure, electrons can be transferred, for example,

[1] J. F. Nye, *Physical Properties of Crystals*, Oxford Science Publications, Oxford (1985).

[2] R. E. Newnham, *Physical Properties of Materials: Anisotropy | Symmetry | Structure*, Oxford University Press, Oxford (2005).

[3] S. C. Miller, W. F. Love, and PEC Research Associates, *Tables of Irreducible Representations of Space Groups and Co-Representations of Magnetic Space Groups*, Pruett Press, Boulder, Colorado (1967).

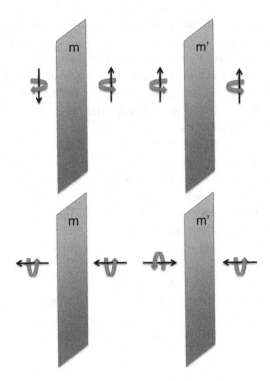

Fig. 29.3 Effect of mirror planes, m, and mirrors with time reversal symmetry, m' on the current loops (curved arrows) and the resulting dipole moment (shown with a straight arrow).

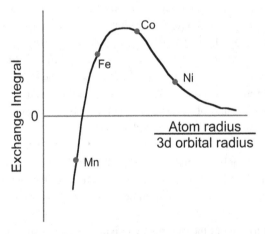

Fig. 29.4 Illustration showing schematically the exchange interaction as a function of a normalized separation distance for a few 3d transition metal atoms (the Bethe–Slater curve).

between levels that originated as 4s and 3d orbitals. The number of spins per Fe in metallic iron can be determined from neutron diffraction experiments to be $\frac{2.2\,\mu_B}{Fe}$, Co is $\frac{1.72\,\mu_B}{Co}$, and there are $\frac{0.60\,\mu_B}{Ni}$ at 0 K. In the ferromagnetic state, the electron configuration of the Fe is $d^{7.4}s^{0.6}$. Of the 7.4 d electrons on average per atom, 4.8 have one sign of

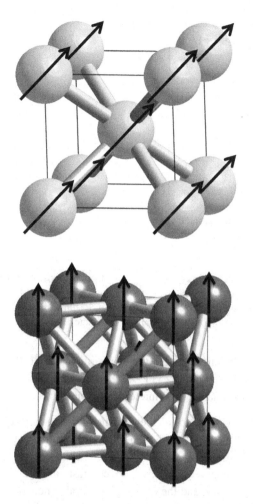

Fig. 29.5 Crystal structures of (top) BCC Fe and (bottom) FCC Ni showing the arrangement of magnetic spins.

spin, 2.6 of the other. The net gives the $\frac{2.2\,\mu_B}{Fe}$. Even for the same atom, the number of magnetic spins depends on the host compound. As a result, while Fe in 8-coordination has $\frac{2.2\,\mu_B}{Fe}$, FeAl (with the CsCl structure: each Fe has eight Al neighbors) has no magnetic moment until the material is distorted to form some Fe–Fe connections. In Fe₃Al, there are two different Fe sites. In one, the Fe has four Fe neighbors and four Al neighbors. Fe atoms on this site have $\frac{1.8\,\mu_B}{Fe}$. On the other site, each Fe atom has eight Fe neighbors; these atoms have $\frac{2.2\,\mu_B}{Fe}$.

Figure 29.5 shows the magnetic structures of Fe, and Ni, where the arrows indicate the direction of the magnetic dipole moment of the atoms. In α-Fe, the bonds are oriented along the ⟨111⟩ directions, and the ferromagnetic moments in each domain are directed along the ⟨100⟩. For large magnetic fields, the magnetization can be rotated to other orientations, but with some difficulty, and when the field is removed, the magnetization will rotate back to one of the ⟨100⟩ directions. Figure 29.6 shows the magnetization curves for Fe. It is clear that it is easier to fully align the

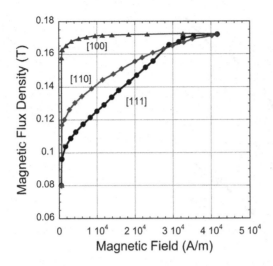

Fig. 29.6 Magnetization curves for iron single crystals in different orientations. Lower magnetic fields are required to align the magnetization for the [100] direction, since that is the preferred orientation for the magnetic spins. Redrawn, by permission, from Soshin Chikazumi, *Physics of Magnetism*, John Wiley and Sons, New York (1964). Figure 7.6, page 136.

magnetization along the $\langle 100 \rangle$ directions. In Ni, the bonds run along $\langle 110 \rangle$, while the spontaneous magnetization is oriented along $\langle 111 \rangle$.

The direct exchange mechanism has several interesting implications.

- Ferromagnetic alloys can be made of elements which are not themselves ferromagnetic. For example, MnBi and Cu_2MnSn are both ferromagnetic, whereas elemental Mn is an antiferromagnet. In the alloys the Mn atoms are far enough apart that the exchange integral is positive. In contrast, in Mn the Mn–Mn separation distances are smaller, and the exchange integral is negative.
- The exchange integral depends on separation distance, and does not require periodicity. Thus, it is possible to make amorphous metals ferromagnets.
- The strength of the interaction determining magnetic alignment drops rapidly with distance. Thus only near, next near, and next next near neighbors matter.

29.4   Superexchange

Superexchange is an important mechanism controlling the coupling of magnetic spins through an intermediary atom. It is also more readily understood in an intuitive way than is direct exchange. As an example, consider the case of wustite, FeO. Wustite has the rocksalt crystal structure at room temperature, but distorts slightly to a rhombohedral structure below the Néel temperature. The Fe^{2+} ion has an electron configuration of [Ar] $3d^6$, that is, the two electrons lost on ionizing the ion were the two from the 4s shell. This leads to four unpaired spins per atom. Two Fe^{2+} atoms interact through the oxygen 2p orbitals, as shown in Fig. 29.7. The electrons from the oxygen can interact with the

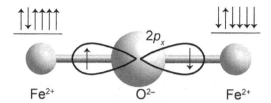

Fig. 29.7 The superexchange interaction illustrated for FeO. The spin alignment on Fe^{2+} ions is mediated through the oxygen $2p_x$ orbital.

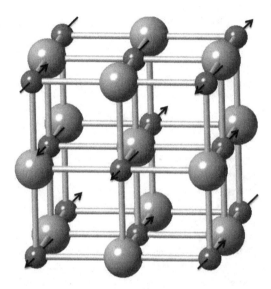

Fig. 29.8 Portion of the structure of FeO showing antiferromagnetic ordering resulting from superexchange. The magnetic dipole moments are directed along $\langle 111 \rangle$ axes. The unit cell has a slight rhombohedral distortion in the antiferromagnetic phase, so that the Fe–O–Fe bond angle is $89.42°$.

3d shell of the Fe^{2+} either if it is excited to a higher energy level, or if there is some degree of covalency in the bonding. Hund's rules will still be obeyed in the filling of the d shell with the electron from the oxygen atom. Thus, for the Fe^{2+} ion on the left-hand side of the figure, this means that the spin of the electron from the oxygen must be spin down, since all of the spin-up slots are occupied. Within the O^{2-} $2p_x$ orbital shown, the Pauli exclusion principle must be obeyed. Thus, the other electron in the orbital will be spin-up. Since this electron will spend some time interacting with the Fe^{2+} ion on the right-hand side, a slot is required in the d shell of that ion for a spin-up electron. Thus, the two Fe^{2+} will have net spins arranged antiferromagnetically, producing no net magnetization.

This superexchange interaction is strongest when the cation–anion–cation bond angle is $180°$, so that the cations interact through one of the oxygen 2p orbitals. The strength of the

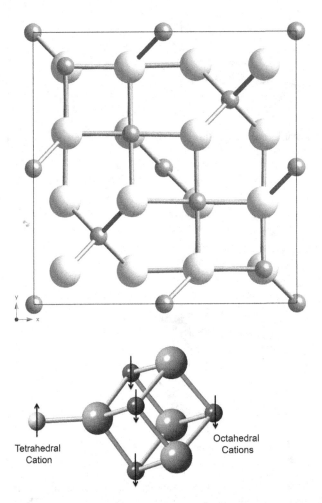

Fig. 29.9 (Top) Spinel crystal structure. Only half the depth of the unit cell is shown, so that the rocksalt-like and diamond-like blocks are more easily visible. (Bottom) Coordination of oxygen atom showing superexchange coupling between the tetrahedral and octahedral cations.

interaction drops when the cation–anion–cation bond angle approaches $90°$. In wustite, the rocksalt structure produces antiferromagnetic coupling on all cation sublattices, as shown in Fig. 29.8. Comparable behavior is observed in the isomorphs NiO and MnO.

To achieve a net magnetic moment, it is preferable to utilize a more complex crystal structure. Consider, for example, the structure of the spinel ferrites $AFe_2O_4 \equiv A^{2+}O \cdot Fe_2^{3+}O_3$. The spinel structure is shown in Fig. 29.9. The oxygens are approximately cubic close-packed, with one tetrahedral site and two octahedral sites occupied by cations. The O coordination is four, as shown.

The tetrahedral cation–O–octahedral cation bond angle is close to $125°$, while the octahedral cation–O–octahedral cation bond angle is close to $90°$. Superexchange strongly couples the spins of the tetrahedral and octahedral cations; a weaker secondary interaction favors antiferromagnetic coupling amongst the octahedral cations.

For $NiFe_2O_4$, the Ni^{2+} preferentially occupies one of the two octahedral sites.

Tetrahedral	Octahedral	Octahedral
Fe^{3+}	Ni^{2+}	Fe^{3+}
$5\mu_B \uparrow$	$2\mu_B \downarrow$	$5\mu_B \downarrow$

This produces a net spin of $2\mu_B \downarrow$ per formula unit. The spins from the strongest magnetic ions, Fe^{3+}, thus cancel each other out, and serve primarily to align the spins of the Ni^{2+}. $NiFe_2O_4$ is thus ferrimagnetic. Since there are eight formula units per unit cell, the magnetization is $\frac{8(2\mu_B)}{cell\ volume}$. This predicts that the spontaneous magnetization, $M_S = \frac{16\left(9.27 \times 10^{-24}\ Am^2\right)}{\left(8.37 \times 10^{-10}\ m\right)^3} \approx 2.5 \times 10^5$ A/m. The experimental value is close to 3×10^5 A/m, so the agreement is reasonably good.

The magnetization of spinel ferrites thus depends explicitly on the *site occupancy* of the cations, that is, which ions occupy the tetrahedral and octahedral sites. There are four basic types of spinels.

- So-called *normal spinels* have 2+ cations in the tetrahedral sites, and 3+ cations in the octahedral sites. Examples include $MgAl_2O_4$ and $ZnFe_2O_4$. The O has a variable atomic position parameter ($3/8 + u$) in spinels which allows small adjustments in cation–anion bondlengths. The relative sizes of the octahedral and tetrahedral interstitial sites depend on the oxygen parameters. If the oxygen coordinate is 3/8, the oxygens are cubic close-packed. In the mineral spinel ($MgAl_2O_4$), $u = 0.012$, and the Al–O bond length is $a\sqrt{\left[3\left(\frac{1}{8} + u\right)^2\right]} \cong \frac{\sqrt{3}}{8}a(1 + 8u) = 1.92$ Å. If the cubic close-packing were perfect, with $u = 0$, the Al–O and Mg–O distances would be 2.02 and 1.75 Å, respectively, and it would be difficult to understand why the larger Mg^{2+} ion is in the tetrahedral site.
- The distortion of the oxygen close-packing is related to the structure. In $MgAl_2O_4$, Al octahedra share edges along the $\langle 110 \rangle$ directions. As in many other structures, cation repulsion causes the shared edge to shorten. In spinel, the shared edges are 2.57 Å, much shorter than the unshared edges (2.86 Å). Similar shortening occurs in the ferrites.
- In *inverse spinels* the tetrahedral sites are occupied by one of the 3+ cations, while the octahedral sites are filled by the 2+ cation and the other 3+ cation. In the absence of other considerations, the relative size of cations would lead to this configuration being preferable: 3+ ions are typically smaller than 2+ ions, and so preferentially occupy the smaller site. Many of the spinel ferrites, including $NiFe_2O_4$, Fe_3O_4, and $MnFe_2O_4$ tend to have this configuration.
- In *random spinels*, there is no particular preference for either of the ions to occupy one of the sites. As a result, the 2+ and 3+ cations randomly fill the available sites. $MgFe_2O_4$ often adopts this configuration.
- γ-Fe_2O_3 is a *vacancy spinel*. The chemical formula can be re-written as: $\frac{4}{3}$ $Fe_2^{3+}O_3 \equiv Fe_{8/3}\square_{1/3}O_4$, where the vacancies, $\square$, preferentially occupy the octahedral sites.

Table 29.4 Octahedral site preference energies in spinels[4]

Cation	Octahedral site preference energy (kJ)	Site preference
Si^{4+}	150	
Ge^{4+}	110	
Cd^{2+}	70	
Zn^{2+}	53	Tetrahedral
Mn^{2+}	45	
Co^{2+}	20	
Mg^{2+}	20	
Fe^{2+}	18	
Fe^{3+}	0	No preference
Ga^{3+}	−4	
Cu^{2+}	−6	
Ni^{2+}	−28	
Co^{3+} (high spin)	−30	
Al^{3+}	−36	Octahedral
V^{3+}	−55	
Ti^{4+}	−50	
Mn^{3+}	−95	
Cr^{3+}	−160*	

* The numbers in this table differ somewhat from those that appear in Navrotsky, *Physics and Chemistry of Earth Materials*, Cambridge University Press (1994). In particular, the number for Cr^{3+} appears to be close to −120 kJ in that reference.

Which of these distributions is observed depends on the octahedral site preference energies given in Table 29.4. Positive octahedral site preference energies correspond to cases where there is an energy cost to locating that atom at an octahedral site. It is not surprising that Si^{4+}, which prefers tetrahedral coordination with oxygen, has a strongly positive energy value. In contrast, some transition metals can lower their free energy more in octahedral coordination, according to the crystal field stabilization energy described in Chapter 14. These atoms have negative octahedral site preference energies. Atoms near the zero point readily occupy either site.

It is clear from the table that Zn^{2+} has a strong preference for the tetrahedral site, as is true for many d^{10} ions. It has no dipole moment because its spins are paired. When a solid solution is made with $Ni_{1-x}Zn_xFe_2O_4$, the Zn^{2+} displaces some Fe^{3+} into octahedral sites:

Tetrahedral	Octahedral	Octahedral
$x\,Zn^{2+} + (1-x)\,Fe^{3+}$	$(1-x)\,Ni^{2+} + x\,Fe^{3+}$	Fe^{3+}
$x\,(0\mu_B) + (1-x)\,5\mu_B \uparrow$	$(1-x)2\mu_B \downarrow + x\,(5\mu_B)\downarrow$	$5\mu_B \downarrow$
$(5-5x)\mu_B \uparrow$	$(2+3x)\,\mu_B \downarrow$	$5\mu_B \downarrow$

[4] Data are from H.S.C. O'Neill and A. Navrotsky, "Cation distributions and thermodynamic properties of binary spinel solid solutions," *Am. Mineralogist*, **69**, 733–753 (1984).

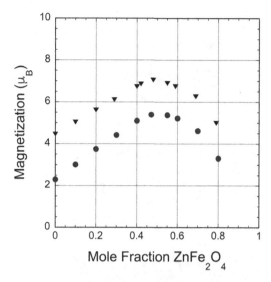

Fig. 29.10 Magnetization of two ferrite spinel solid solutions. The experimental points show measured data, the dashed lines show extrapolations to pure $ZnFe_2O_4$ as a normal spinel under the assumption that magnetic coupling is retained. Experimental data diverge substantially from the extrapolated line when there are not enough magnetic ions on the tetrahedral site to drive superexchange. In this case, a weaker antiferromagnetic coupling takes over for the Fe on the octahedral sites. Data from J. Smit and H. P. J. Wijn, *Ferrites*, John Wiley and Sons, New York (1959).

The net magnetic moment is $[(5 - 5x) - (7 + 3x)]$ μ_B/formula unit $= (2 + 8x)$ μ_B/formula unit. Thus, addition of a non-magnetic ion *increases* the net magnetization, as shown in Fig. 29.10. The magnetization rises as more Zn^{2+} is added until the point where there are no longer enough magnetic ions on the tetrahedral site to drive the superexchange process. The Curie temperature for the $M_{1-x}Zn_xFe_2O_4$ spinels drops as x increases, as the Zn decreases the strength of the alignment.

The spinel ferrites typically have high magnetic permeabilities and reasonably high electrical resistivities. The latter is important, since some magnets are used in applications where the magnetic field is cycled (e.g. in transformers). In more electrically conducting magnets such as Fe, the change in magnetic field generates an electrical current, I, and Joule heating (I^2R power losses), where R is the resistance. The spinel ferrites have much higher electrical resistivities, so the resulting currents and power losses are very much lower for alternating magnetic fields.

Other complex structures with interesting magnetic properties include the ilmenite, magnetoplumbite, and garnet structures illustrated in Fig. 29.11 and Fig 29.12. The structure of ilmenite, $FeTiO_3$, is a derivative of the corundum structure in which two different cations are arranged into layers in the octahedral sites. Fe^{2+} has a magnetic moment of 4 μ_B per atom, while Ti^{4+} has no moment. The magnetic alignment is aligned in a ferromagnetic fashion within each Fe layer, but adjacent layers are aligned antiparallel, producing no net magnetization. Chemical substitutions which shift the balance of moments can produce a net magnetization. For example, $NiMnO_3$ also has the ilmenite structure, but in this case, both cation sublattices have a net moment. Mn^{4+}

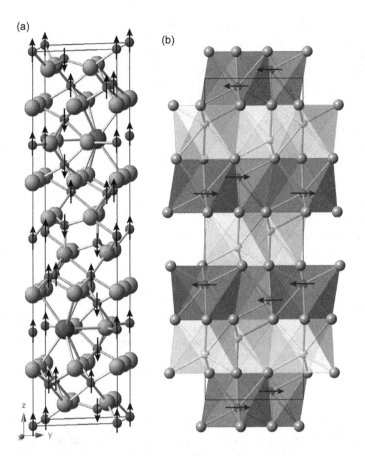

Fig. 29.11 Crystal structures of (a) magnetoplumbite, $BaFe_{12}O_{19}$ and (b) ilmenite, $FeTiO_3$.
In the magnetoplumbite structure, the large dark atoms are the Ba, the intermediate size are O, and the smallest are the Fe atoms. The orientations of spins are shown as arrows. In the ilmenite structure, layers of FeO_6 and TiO_6 octahedra are aligned perpendicular to c. The Fe atoms couple ferromagnetically within a layer, but adjacent layers have spins in opposite orientations. Only one half of the unit cell is shown along c.

has a magnetic dipole moment of 4 μ_B per atom, while Ni^{2+} has 4 μ_B per atom; the two sublattices couple antiferromagnetically. Thus, $NiMnO_3$ is a ferromagnet with a magnetization of $2\mu_B$/formula unit.

The magnetoplumbite structure can be thought of as a derivative of the spinel structure. The large Ba^{2+} ions form a close-packed array with the O^{2-} anions, with a mixed sequence of hexagonal and cubic close-packing. Trivalent iron ions occupy three types of sites within the magnetoplumbite structure: octahedral, tetrahedral, and an unusual five-coordinated trigonal bipyramid site. The magnetization aligns parallel to the long axis of the unit cell, in the [001] direction.

The prototype garnet crystal structure has a cubic unit cell with sites for tetrahedral, octahedral, and cubic cation coordinations. This is useful, as rare earth ions, which can have up to seven unpaired spins, are slightly too large to fit comfortably in the spinel crystal structure. As an example, consider $Gd_3Fe_5O_{12}$. The site occupancy in this

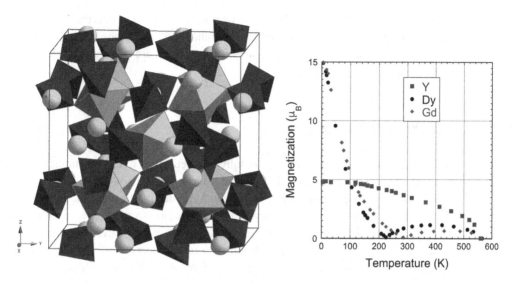

Fig. 29.12 (Left) Garnet crystal structure, showing the tetrahedra and octahedra around the Fe atoms. The 8-coordinated atoms are shown as spheres. (Right) Temperature dependence of the magnetization for several ferromagnetic garnets, with either Y, Gd, or Dy on the 8-coordinated site. Data on right replotted, by permission, from S. Chikazumi, *Physics of Magnetism*, John Wiley and Sons, New York (1964). Figure 5.21, p. 104.

compound is $Gd_3^{VIII}Fe_2^{VI}Fe_3^{IV}O_{12}$, where the Roman numeral superscripts are the coordination numbers. Each oxygen is tetrahedrally coordinated to two Gd, one octahedral Fe and one tetrahedral Fe. The tetrahedral and octahedral Fe couple antiferromagnetically through a superexchange interaction, leaving a net of 5 μ_B from the five Fe^{3+} ions/formula unit. A weaker antiferromagnetic coupling between the Gd spin and the *net* spin from the Fe^{3+} aligns the magnetic moment from the rare-earth ions. Gd^{3+} has 7 μ_B of unpaired spin, leading to a total magnetic moment of $\frac{3(7\mu_B)-5\mu_B}{Gd_3Fe_5O_{12}}$. The magnetic moments are aligned along the crystallographic [111] axis. Because the f electrons are buried deeper in the atom than the d electrons, the Gd spins disorder at lower temperatures than do the spins from the Fe. This leads to a compensation temperature where the net magnetization direction changes sign, as shown in Fig. 29.12. In the isostructural $Y_3Fe_5O_{12}$, the Y atom has no unpaired spin, so the net magnetization is due only to the net Fe spin, and no compensation temperature is observed.

At elevated temperatures, paramagnetism is favored thermodynamically, due to the entropy associated with randomization of the spins. Magnetic transition temperatures are a measure of the amount of energy required to disrupt the magnetic ordering. The temperature at which ferri- or ferro-magnetism is lost is the Curie temperature, T_C, while antiferromagnetic alignment disappears at the Néel temperature, T_N. Thus, as the strength of the exchange or superexchange interactions rises, so do the Néel or Curie temperatures. For superexchange interactions, the transition temperatures rise with the degree of covalency in the bonding. This also implies that the transition temperatures of oxides tend to be higher than that of fluorides. Exceptions can occur in ions with large spin–orbit interactions, including Co^{2+} in octahedral fields, or Ni^{2+} in tetrahedral fields. Magnetic transition temperatures are also usually lower for compounds with $\leq d^4$ electrons.

29.5   Domains and Domain Walls

A magnetic domain is a region that is fully magnetized (e.g. the net moments from adjacent unit cells are aligned). In different volumes of the solid, the magnetization directions may differ, producing the domain structure of the magnet. In magnets, there is a tendency to try and achieve closure domains, so that there is not a large magnetic field extending out of the material, as shown in Fig. 29.13. Because it takes a finite energy to make a domain wall (all surfaces are energetically expensive), it is necessary to balance the energy caused by incomplete closure with the energy to make a wall.

One of the significant differences between ferroelectric and ferromagnetic materials pertains to domain walls. In ferroelectrics, it is generally energetically expensive to rotate the polarization away from one of the preferred axes. This leads to domain wall widths of only a few unit cells or less (typically ~0.4–4 nm). In contrast, the energy cost for rotating the magnetization away from the preferred magnetic axis is much smaller, leading to wider domain wall widths (~10–100 nm). Schematic illustrations for the change in the orientation of a magnetic dipole changes across a domain wall for Bloch and Néel walls are shown in Fig. 29.14.

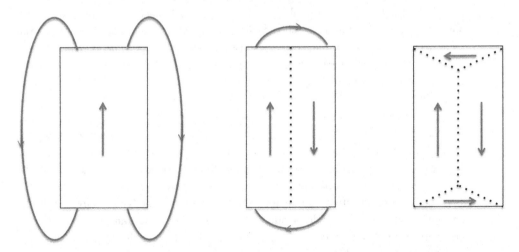

Fig. 29.13 A variety of domain configurations. The magnet is the solid rectangle. Magnetizations within a domain are shown with arrows. Curved field lines are shown outside of the magnet. (Left) A single domain crystal with a considerable amount of magnetic field outside of the magnet. (Middle) A domain wall, shown as a dotted line, reduces the amount of magnetic field outside of the sample. (Right) Closure domains.

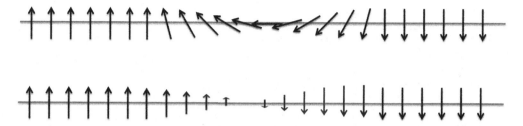

Fig. 29.14 Two types of magnetic domain walls. (Top) Néel domain wall. (Bottom) Bloch domain wall. In the first, the magnetic dipole moment rotates in the plane, in the other the rotation is out of the plane of the page.

The width of the domain wall is governed by several competing factors. (1) The anisotropy energy tends to make the magnetization line up along specific crystallographic axes. The magnetocrystalline anisotropy quantifies the fixed, average energy cost to rotate the magnetization from the preferred direction. The preferred direction is referred to as the magnetic easy axis. A large magnetocrystalline anisotropy tends to make domain walls thin. (2) Exchange (or superexchange) energy is a short-range force which causes spins to line up with each other. This tends to make the domain walls wide, since the magnetization direction tries to be as close to parallel as possible in adjacent unit cells. For the most part, the stronger the exchange (or superexchange) energy, the harder it is to disrupt the magnetic ordering, and the wider the domain walls will be. (3) Magnetostriction is the change in shape of a magnet due to a change in magnetization. It is easy to picture the converse of this. A change in the atomic spacing may cause changes in the exchange or superexchange interaction. This could lead to one magnetic orientation becoming more favorable. The induced distortion may be easier in some directions than others, and so may produce another driving force for a preferred orientation for the magnetization.

Because the domain walls already involve rotation of the magnetization from one orientation to the next, movement of a domain wall requires only modest rotations of the magnetization. Ferro- or ferrimagnetic domain walls can be moved by an applied magnetic field, increasing the magnetic permeability. Interactions with pinning centers produce magnetic losses at low fields. Grain boundaries, dislocations, or porosity act to pin magnetic domain walls, reduce the magnetic permeability, and increase the magnetic losses. Thus, large-grained materials are sought in ceramic ferrites where large permeabilities are required.

In ferro- and ferrimagnetic materials, an applied magnetic field acts to align the magnetic dipole moments. The process is conceptually similar to aligning electrical dipoles in a ferroelectric (see Chapter 28). At modest magnetic fields, the applied field tends to rotate the magnetization direction into the easy orientation most closely aligned with the applied magnetic field. Thus, the net magnetization increases. At higher fields, the magnetization may rotate off the preferred crystallographic axis, and saturate when all of the moments are aligned. When the magnetic field is reduced to zero, some of the magnetic dipoles remain aligned, producing a net magnetization. As shown in Fig. 29.15, a reverse magnetic field is required to

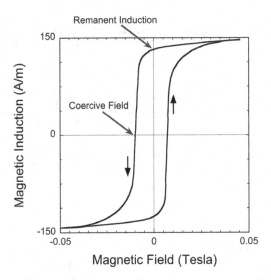

Fig. 29.15 Magnetic hysteresis loop of iron. The coercive field and remanent induction are labeled. The other arrows show the direction that the loop has traversed on increasing and decreasing the magnetic field.

reduce the net magnetization to zero at the coercive field. For large ac magnetic fields, the magnetization data will trace a hysteretic response.

29.6   Hard and Soft Magnets

Magnets for which it is easy to reorient the magnetization direction are referred to as soft magnets, whereas those where it is hard to reorient the magnetization are hard magnets. Hard magnets are useful for permanent magnets; $BaFe_{12}O_{19}$, $SmCo_5$, and $Nd_2Fe_{14}B$ are good examples. Figure 29.16 shows the crystal structures of $SmCo_5$ and $Nd_2Fe_{14}B$. $SmCo_5$ belongs to the hexagonal crystal system. Its hard magnetic properties arise from an unusually large magnetocrystalline anisotropy, which places a high energy burden on rotating the magnetization away from its preferred crystallographic axis. Structurally, the Co_5 units form trigonal bipyramids, that look like a slice out of a hexagonally close-packed structure. A higher magnetic moment can be achieved in a related structure by replacing some of the Sm atoms by a pair of Co atoms. The crystal structure of $Nd_2Fe_{14}B$ is a complex one. Dy-doped $Nd_2Fe_{14}B$ is used as a magnet in the motors of electric vehicles.

Intermediate hardness values, with very square hysteresis loops, are preferred for memory magnets, so that the magnetization is reorientable, but not by stray fields. γ-Fe_2O_3, FePt, CoPt, CrO_2, and $Fe_{0.8-x}Co_xFe_{0.2}$ have all been used in this application, as are complex stacks of metallic films. The crystal structures of most of these compounds have already been discussed: γ-Fe_2O_3 is a defect spinel, FePt and CoPt are ordered derivatives of the FCC structure shown in Fig. 29.17, and CrO_2 has the rutile structure.

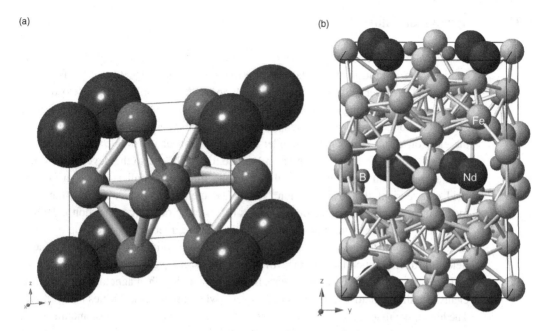

Fig. 29.16 Crystal structures of two hard magnetic materials: (a) SmCo$_5$, where Co$_5$ trigonal bipyramids are clearly visible, and (b) the complex tetragonal structure of Nd$_2$Fe$_{14}$B.

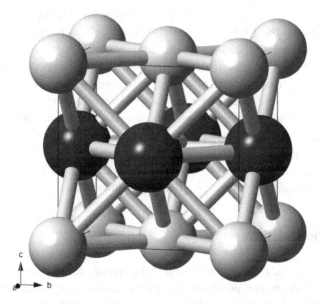

Fig. 29.17 Crystal structure of CoPt. The tetragonal distortion associated with alternating layers of Co and Pt increases the magnetic anisotropy relative to normal FCC metals. In this material, the magnetization is directed along the c axis. Co atoms are light, Pt atoms are dark.

29.7   Magnetostriction

Magnetostriction describes the change in shape of a material when a magnetic field is applied. The saturation strain that can be produced due to magnetostriction is on the order of 10 parts per million (10 ppm) in many magnets. The origin of magnetostriction is primarily the coupling between the magnetic spin and the shape of the electron cloud around an atom. Consider, for example, atoms with magnetic dipole moments in which the electron cloud is elongated parallel to the direction of the dipole moment due to spin–orbit coupling. In the ferromagnetic phase, this will lead to the development of a spontaneous strain associated with the magnetic moment. Application of an electric field can reorient the magnetization, and hence the strain. Typically, the transition metals have modest magnetostrictive strains because the orbital contribution is largely quenched. In contrast, the rare-earth ions have much larger spin–orbit coupling. The large resulting distortions in the electron clouds can be reoriented by applied magnetic fields, producing much larger values of the saturation magnetostrictive strain.

There are industrial uses for materials with low and high magnetostrictive coefficients. Small magnetostriction strains are useful when high permeabilities are required. The higher the magnetostriction, the more difficult domain wall motion is, and the less it can contribute to the properties. It is observed that higher permeabilities can be achieved in materials where domain wall motion is easy. In the spinel ferrites, the total strain that can be induced by a magnetic field is ~1 ppm, a very small value for the magnetostriction. At compositions where the magnetostriction is reduced to ~ zero, the relative permeability can be as high as 40 000.

In contrast, magnetostrictive transducers are designed to produce large strains in response to applied magnetic fields. A family of materials was developed by the Naval Ordinance Laboratory (NOL) for this purpose, including Ga–Fe and Tb–Fe based alloys. As was described for the case of piezoelectric materials, larger magnetostrictive strains can be achieved if the materials are poised on the brink of a structural instability. Terfenol-D has the composition $Tb_{0.3}Dy_{0.7}Fe_2$. Unusually large magnetostrictive responses of ~2000 ppm are observed for this composition because at room temperature the material sits at a phase boundary between rhombohedrally distorted structures, with the easy magnetization axis along $\langle 111 \rangle$ and tetragonally distorted structures with the magnetically easy axis along $\langle 100 \rangle$. The key compounds have a crystal structure corresponding to one of the Laves phases, described in Chapter 10. Underwater sound projectors based on the magnetostrictive effect are of interest for sonar systems.

29.8   Magnetocaloric Materials

Magnetocaloric materials exhibit reversible temperature changes in response to changing magnetic fields. For applications near room temperature, the rare-earth metal gadolinium has been shown to be the best magnetocaloric material for use in refrigeration devices. Figure 29.18 shows the adiabatic temperature changes for Gd, Dy, and $ErAl_2$ with magnetic transitions ranging from 14 to 294 K. All four possess second-order phase transformations.

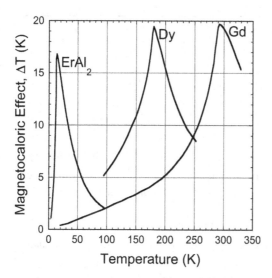

Fig. 29.18 Magnetocaloric effects in $ErAl_2$, Dy, and Gd measured in demagnetization experiments between zero and 10 Tesla. Adapted from V. K. Pecharsky and K. A. Gschneider, Jr., "Advanced material for magnetic cooling," *Material Matters*, **2.4**, 4 (2007), with permission from Sigma-Aldrich Co. LLC.

Further improvements in magnetocaloric behavior have been achieved using magnetic materials with first-order phase transformations. In such materials, the change in entropy associated with magnetic ordering is accomplished by a change in entropy associated with a crystallographic transformation. For example, the material may undergo a phase transformation from a paramagnetic or antiferromagnetic state to a ferromagnetic one. The contribution from the structural effect can be larger than that of the magnetic transition. $Gd_5Si_xGe_{4-x}$ alloys are typical of the giant magnetocaloric materials.

29.9   Problems

(1) Explain what is meant by paramagnetism, diamagnetism, ferromagnetism, antiferromagnetism, and ferrimagnetism. Illustrate your answer with possible spin arrangements.

(2) Explain what is meant by superexchange, illustrating for the cases of FeO and Fe_3O_4.

(3) Draw a picture showing the magnetization as you cross a domain wall in a ferromagnetic material. Explain the factors that control the width of the domain wall.

(4) Calculate the saturation magnetization for Fe_3O_4. The lattice parameter is 837 pm. Assume the material is an inverse spinel.

(5) The magnetic susceptibility data for $SrRuO_3$ are given below. Are the data consistent with the Curie law? If so, evaluate the Curie constant and Curie temperature.

χ_m	Temp (K)	χ_m	Temp (K)	χ_m	Temp (K)
33.7×10^{-3}	187	5.70×10^{-3}	313	2.28×10^{-3}	531
24.8×10^{-3}	196	4.22×10^{-3}	361	2.01×10^{-3}	579
17.5×10^{-3}	212	3.61×10^{-3}	393	1.77×10^{-3}	636
13.6×10^{-3}	228	2.97×10^{-3}	446	1.61×10^{-3}	686
7.74×10^{-3}	276	2.64×10^{-3}	481	1.46×10^{-3}	738

(6) The experimental value for the magnetic moment of $Li_{0.5}Fe_{2.5}O_4$ spinel is 2.6 Bohr magnetons (μ_B) per formula unit. How do you justify this result from the net spins of the ions involved? What position(s) in the crystal lattice does Li^+ occupy? What about Fe^{3+}?

(7) Illustrate the similarities and differences between domain walls in ferroelectrics and ferrimagnets. Explain the factors that control the domain structure in both cases.

(8) **(a)** Calculate the saturation magnetization for γ-Fe_2O_3. Assume that the vacancies in the spinel structure are on the octahedral sites and that the lattice parameter is ~8.5 Å.

    (b) How would you expect the magnetization to change as a function of temperature?

(9) To a first approximation MnO, can be treated as a cubic rocksalt-structured material. Given this, explain why Mn contributes to the magnetic moment of $MnFe_2O_4$, while MnO has little useful magnetization.

(10) The magnetic susceptibility of a Gd^{3+} containing salt was measured as a function of temperature (see below). Are the results consistent with the Curie–Weiss law? If yes, calculate the Curie constant.

    Molecular weight of salt = 851 g/mole, density = 3 g/cm^3.

T (K)	100	142	200	300
χ_m (cm^3/mole)	6.9×10^{-5}	5×10^{-5}	3.5×10^{-5}	2.4×10^{-5}

30 Mechanical Properties

30.0 Introduction

This chapter describes the response of solids to mechanical stresses, where the stress is the force per unit area.[1] Some materials undergo *plastic deformation*, a permanent change in shape in response to applied stresses, as in the case of metallic wire that is bent. Ductility is a measure of a material's ability to deform plastically. Other solids deform while the stress is applied, then recover their original shape as the stress is removed via *elastic deformation*. Whether or not a given material deforms elastically or plastically depends also on the load levels. All materials fracture under large tensile stresses, but the fracture process can take a variety of forms. *Brittle fracture* is frequently catastrophic, with rapid and unstable crack development. *Cleavage* is one particular form of brittle fracture, in which failure occurs along particular crystallographic planes. Elastic and plastic deformation are illustrated in Fig. 30.1.

Numerous descriptions of the elastic properties of solids are in use. *Elastic stiffness* can be defined, per Hooke's law, as $\sigma_{ij} = c_{ijkl}\varepsilon_{kl}$, where σ_{ij} is stress, ε_{kl} is strain, and c_{ijkl} is the elastic stiffness, or, alternatively, as $\varepsilon_{ij} = s_{ijkl}\sigma_{kl}$, where s_{ijkl} is the *compliance*. Two directions are needed to specify stress (the direction of the force and the normal to the face on which the force acts); thus, stress is a second-rank tensor. *Compressibility* describes the relative change in volume of a material caused by a change in isostatic pressure. The compressibility is expressed as $K = -\frac{1}{V}\frac{dV}{dP}$, where V is volume, and P is pressure. In tensor notation, $K = s_{1111} + s_{1122} + s_{1133} + s_{2211} + s_{2222} + s_{2233} + s_{3311} + s_{3322} + s_{3333}$.

As was described in Chapter 23, this means solids with less symmetric crystal structures require many different coefficients to fully specify the elastic response (up to 21). As symmetry increases, the number of independent elastic constants decreases. For the Curie group symmetries $\infty\infty m$ and $\infty\infty$, the independent elastic constants are s_{1111} and s_{1122}. In many texts on strengths of materials, high symmetry is assumed, and Young's modulus E is used: $E = 1/s_{1111}$. Likewise, Poisson's ratio, v, is $v = -\frac{s_{1122}}{s_{1111}}$. The rigidity ratio (also called the shear modulus), G, describes the shear strain induced by a shear stress. For the highest symmetry cases, $G = \frac{E}{2(1-v)} = \frac{1}{s_{2323}} = \frac{1}{2(s_{1111}-s_{1122})}$.

---

[1] Portions of this chapter are adapted (with permission) from R. E. Newnham, *Structure–Property Relations*, Springer-Verlag, New York (1975).

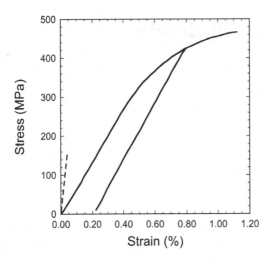

Fig. 30.1 Comparison of typical stress–strain curves for alumina and aluminum. The arrows show the loading and unloading of stress. Al_2O_3 behaves elastically up until the failure point, whereas aluminum shows hysteresis in the strain on unloading. For the aluminum sample shown, about 0.6% of the strain is elastic, while around 0.2% is plastic.

Another subject pertinent to the mechanical characteristics of materials is tribology, the study of friction and wear. This is a complex subject often involving a number of different materials; in addition to the two contacting solids, special surfactant coatings and a mixture of lubricants can be involved. The lifetime of many engineering components is ultimately determined by friction and wear. Hard materials have low wear rates, but often fail by brittle fracture. Some softer materials are mechanically tougher, but have less wear resistance. To reduce wear, the best approach is to coat a tough material with a hard surface film. For this reason, steel crankshafts and bearings are often carburized to improve wear resistance. A thin carbide coating creates the necessary hardness to extend the lifetime of such automotive components.

In addition to friction and wear, lubricants, grinding, and polishing are discussed in this chapter.

30.1   Plastic Deformation and Slip

Plastic deformation describes the case wherein the unloading of stress produces a permanent change in a material's shape. This can occur, for example, due to slip in metals, or changes in chain configurations in polymers. The plasticity of a pure metal can be predicted from the ratio of the bulk modulus K to the shear modulus G. High K/G implies ductility, as in thallium, where $K/G = 6.52$. Beryllium, on the other hand, which is extremely brittle, has a low K/G value of only 0.64.[2]

---

[2] S. F. Pugh, *Phil. Mag.*, **45**, 823 (1954).

Fig. 30.2 Slip in a face-centered cubic metal develops along close-packed planes. The slip shown is for a shear stress applied to the top and bottom layers of atoms. Slip produces deformation, as shown on the right.

Slip is an important deformation mechanism in some solids, including close-packed metals. Under shear stress (Fig. 30.2), two planes glide past one another to the next equilibrium site, producing shear strain.

In hexagonal or cubic close-packed metals, slip occurs on the most densely packed planes. These planes are strongly bonded within the plane, are flat, and have the smallest number of out-of-plane bonds that must be broken and reformed when the crystal slips. The non-directional nature of the metallic bond is also helpful in accommodating changes in the atom positions during slip.

Metallic deformation generally takes place via the slip plane and slip direction most nearly parallel to the applied stress. Together, the slip plane and the slip direction make up a slip system. In cubic close-packed (face-centered cubic) crystals, the close-packed planes are the four {111} planes. Within a given slip system, the slip direction is the close-packed direction; for (111) the slip directions are [110], [011], and [101]. In FCC metals, there are 12 slip systems (four slip planes and three slip directions in each) so that slip takes place rather easily. As a result, FCC metals tend to be comparatively more ductile.

Magnesium and other hexagonal close-packed metals typically slip on {001} planes. The slip directions are $\langle 110 \rangle$ (see Fig. 30.3), providing only three slip systems in most HCP metals. As a result, these metals are much less ductile than FCC metals. When c/a is reduced to about 1.59 (rather than the ideal value of 1.633), slip occurs along other planes as well.

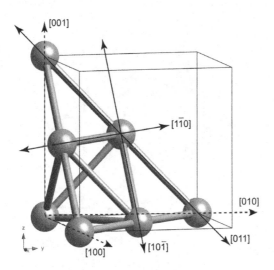

Fig. 30.3 Slip directions in a cubic close-packed metal.

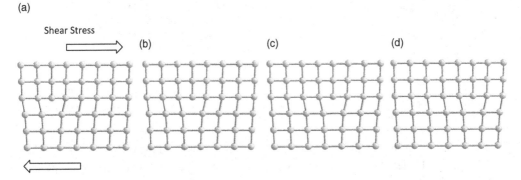

Fig. 30.4 Motion of an edge dislocation in response to a shear stress. In response to the stress, a row of bonds going into the page is broken to move the dislocation. Parts (a)–(d) show sequential steps.

Body-centered cubic metals like α-Fe are not close-packed, and as a result, the slip systems are not as well defined. Because the {110} plane is *almost close-packed*, this plane tends to be important for slip. Each atom has six near neighbors in the (110) plane but four are slightly closer than other two.

The *von Mises criterion* for slip says that a polycrystalline metal will tend to be ductile if there are at least five independent slip systems. With fewer than five, polycrystalline materials tend to fail in a brittle fashion.

As was described in Chapter 16, slip does not occur along an entire plane of atoms at a time, since this requires many bonds to be broken simultaneously. Instead, slip occurs via the process of dislocation motion, as shown in Fig. 30.4. Thus, the yield stress (the stress at the onset of the plastic deformation process) increases when it is more difficult to move dislocations. With more dislocations in the sample, and hence more dislocations interacting, the metal *work hardens*. It is readily demonstrated that a copper wire is easily bent initially but, after repeated plastic deformations, the wire becomes

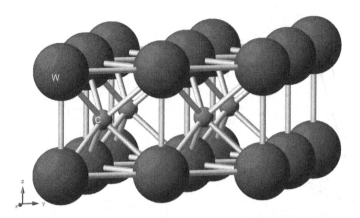

Fig. 30.5 Tungsten carbide has a hexagonal crystal structure with carbon atoms bonding the tungsten layers together.

progressively stiffer and more brittle. Work hardening is a consequence of generation of new dislocations in response to the applied stress. As the dislocation density rises, atoms are displaced from their equilibrium lattice sites, and dislocation motion becomes more difficult. Thus, the likelihood of slip decreases. The material can be restored to a more ductile state by heat-treating the part at temperatures where dislocations can be eliminated.

Dislocation motion also can be impeded by making solid solutions within a metal. The interstitial solid solution of steel, in which C is inserted into interstices in the Fe lattice, impedes dislocation motion, since the strong Fe–C bonds have to be broken for the dislocation to propagate. Thus steels are significantly harder than pure Fe. Yield stress and fracture strength also rise considerably. Tungsten carbide (WC) embedded in cobalt metal is one of the "cemented carbides" used in cutting steels. Cemented carbide dies and cutting tools are commonly referred to as tungsten carbide, although they are actually composites in which the WC grains are bonded to one another by a metallic phase. It is the combination of the rigid carbide network and a tough metallic substructure that provides these cemented carbides with their outstanding mechanical properties. The metal improves the ability to withstand brittle fracture, but degrades the hardness, stiffness, and resistance to corrosion. The crystal structure of WC (Fig. 30.5) is an unusual one, with the interstitial carbon atoms bonded to six tungsten atoms arranged in a trigonal prism, rather than the more common octahedral or tetrahedral interstices found in other metal carbides. The radii of the tungsten and carbon are 1.37 and 0.77 Å, respectively. Each tungsten atom is bonded to six carbons at a distance of 2.20 Å, and to eight tungstens: six at 2.91 and two at 2.84 Å. In the BCC structure of elemental tungsten, the bond lengths to the eight near neighbors are 2.74 Å, only slightly shorter than the metal–metal bonds in WC. Strong covalent bonds from the interstitial carbon atoms reinforce the metal–metal bonding to form an excellent abrasive.

The importance of substitutional solid solutions is illustrated for a series of Cu alloys in Fig. 30.6. As ions of different sizes and bonding characteristics are substituted into Cu, they produce local distortions in the lattice (see Fig. 30.7). The more distorted the lattice, the more difficult it is for slip to occur, and the higher the yield stress. This is

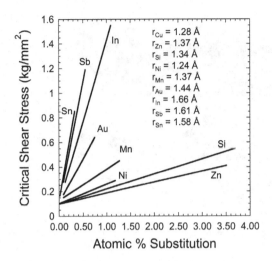

Fig. 30.6 Critical shear stress for plastic deformation as a function of doping in Cu alloys. The dopant atoms introduce deformations into the lattice which impede motion of dislocations and make the material harder. Replotted, by permission, from data in *ASM Speciality Handbook : Copper and Copper Alloys*, ed. J. R. Davis, Materals Park, OH (2001), p. 38, Fig. 11.

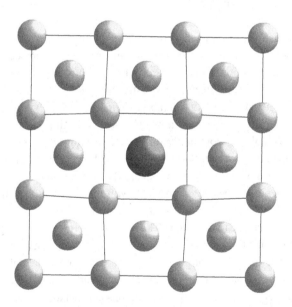

Fig. 30.7 Local distortions in a metal lattice caused by substitutional solid solution with an atom of the wrong size.

referred to as solution hardening. Thus, elements like Sn produce a higher yield stress when alloyed with Cu than does Zn.

If you think about it, solid solutions changed human history. Military equipment – battle-axes, helmets, shields, and swords – date back to about 3000 BC; the oldest

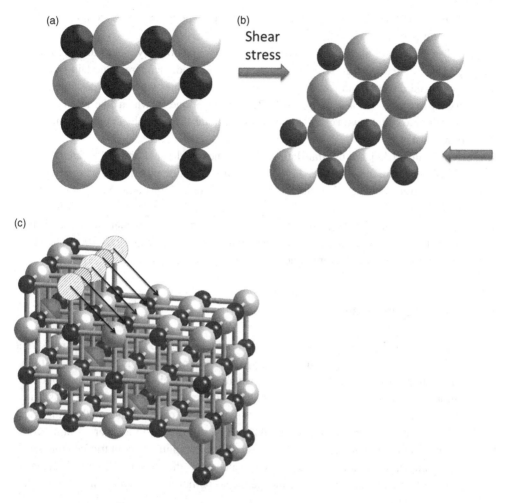

Fig. 30.8 Simple slip is difficult in rocksalt-structured oxides. While the unstressed structure (a) has anions (open) and cations (filled circles) in contact with each other, a large shear stress (b) would bring like ions into contact. (c) In rocksalt single crystals, slip can occur on the {110} planes.

known examples were made from copper. Cu–Sn bronzes are one of the oldest alloys known to man; bronze weapons held an edge better than Cu ones. Cu–Sn alloys containing 10% Sn are prepared in wrought form and hammered into shape. Likewise, the interstitial solid solution of steel produces a significant additional military advantage. These are just two examples of the ways materials science has changed history.

Metals have much less resistance to shear stresses than do most oxides. In oxides, the repulsive forces between next near nearest neighbors inhibits the slippage of one layer past another (see Fig. 30.8). The repulsive forces prevent slip and cause catastrophic rupture. No such effect is present in pure metals where neighbors are alike, and the bonding is non-directional.

The absence of slip in most ionically bonded solids not only makes them brittle but gives them great compressive strength as well. That said, there are, in certain symmetry planes and directions in some oxide single crystals, planes where it is possible for the crystal to shear like a deck of cards. This plane is one in which like atoms are not brought into contact with each other when the crystal slips, and the ion movement meets little resistance.

In halite, {011} are the translation glide planes, and [011] and $[01\bar{1}]$ are the glide directions within the plane (see Fig. 30.8(c)). In translation gliding, the preferred planes and directions are such that the Na^+ ions glide over Cl^- ions and vice versa, partially avoiding electrostatic repulsion. For the (110) plane, this condition is met for the [110] direction, but not for the [001]. When stressed in the latter orientation, halite cleaves rather than plastically deforming. [110] gliding takes place on the (100) and (111) planes of NaCl, with the same line of reasoning holding true.

30.2   Elastic Deformation

The elastic moduli describe the elastic deformation produced by an applied stress. That is, $\sigma_{ij} = c_{ijkl}\varepsilon_{kl}$, where σ_{ij} is the stress, ε_{kl} is the strain, and c_{ijkl} is the elastic stiffness. Alternatively, one can write $\varepsilon_{ij} = s_{ijkl}\sigma_{kl}$, where s_{ijkl} is the elastic compliance. Notice the unfortunate truth for English speakers that the convention is to use c for *s*tiffness, and s for *c*ompliance. It is also notable that some introductory courses in strength of materials teach that there is one Young's modulus, a Poisson ratio, and a shear modulus for a given material. In many crystals, this is a gross oversimplification: in triclinic materials, for example, there are 21 independent values for the elastic stiffness. Increasing symmetry reduces the number of independent coefficients to two for materials in the cubic crystal systems.

Because a material's response to strain results in the deformation of bonds, it makes sense that, the stronger the bonds are, the harder it will be to strain the solid. That is: *strong bonds are stiff bonds*. Thus, it is not surprising that stronger bonds, higher charges, and shorter interatomic separation distances all produce higher elastic stiffnesses. To a reasonable approximation, the elastic stiffnesses vary as d^{-n}, where d is the separation distance, and n typically ranges between 4 and 6. Data for the alkali halides are shown in Fig. 30.9. One can develop a reasonable understanding of the elastic stiffness, then, by thinking of the type of bonding, and the way in which the bonds are arranged in space.

Stiffness rises with bond density, bond strength, and density, and falls with increasing bond length. A relatively simple analogy helps to visualize why this is the case. The bonds that hold together the atoms in the solid can be visualized as springs. Short, strong bonds are stiff springs. They are difficult to perturb with an applied stress, producing a high stiffness. In contrast, weak bonds act like soft springs, which can readily be deformed with modest stresses. The more stiff springs there are in a given volume (i.e. the higher the bond density), the stiffer the material will be.

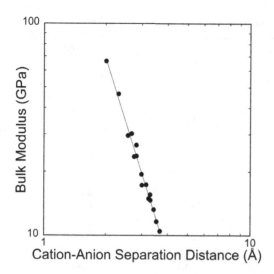

Fig. 30.9 Relation between bulk modulus and interatomic separation distance in the alkali halides.

Table 30.1 compares the elastic stiffnesses, c_{1111}, of a several families of cubic crystals. There are a number of trends which inform an intuitive understanding of the elastic constants.

(1)   Covalently bonded crystals are generally stiffer than ionically bonded solids. The covalent bond is very strong, and highly directional. In contrast, the ionic bond tends to be somewhat weaker and is non-directional, so that small perturbations in atomic arrangements are less energetically costly. As a result, NaF has a smaller stiffness coefficient than Ge, even though its bond length is shorter (~2.02 Å compared with 2.41 Å).

(2)   As bond length increases for a given structure and bond type, the stiffness decreases. Thus the stiffness of diamond far exceeds that of Si, which is in turn larger than that of Ge.

(3)   The elastic stiffnesses increase with the melting temperature. That is, bonds which are hard to perturb thermally are also hard to perturb with an elastic stress. Recall that some materials include multiple bond types. For example, in many polymers, there are strong covalent bonds in the backbone of the chain, and weaker van der Waals or hydrogen bonds between the chains. The melting temperature and the elastic moduli of bulk samples are typically determined by the weaker of these bonds.

(4)   The elastic stiffnesses for a given material typically decrease as temperature increases. As temperature rises, thermal vibrations increase in amplitude, inter-atomic separation distances lengthen, and the bond strength is degraded.

(5)   Anomalies in elastic moduli occur at phase transitions. It is a general rule of thumb that to make a material particularly responsive to an external stimulus, it is helpful to poise the material at the brink of a structural instability. In the case of elastic properties, it is common to see at least one of the elastic moduli soften at phase transitions. An excellent example of this is observed for $SrTiO_3$. At room temperature, $SrTiO_3$ has the ideal perovskite crystal

Table 30.1 Elastic stiffness constants for several families of cubic materials

Material	c_{1111} (GPa)	c_{1122} (GPa)	c_{2323} (GPa)	Structure
C (diamond)	1040	170	550	Diamond
Si	164.9	63.5	79.5	Diamond
Ge	129	48	67.1	Diamond
Al	108	62	28.3	FCC
Cu	169	122	75.3	FCC
Ni	247	153	122	FCC
Ag	123	92	45.	FCC
Pd	224	173	71.6	FCC
Pt	347	251	76.5	FCC
Fe	230	135	117	BCC
Na	7.59	6.33	4.20	BCC
K	3.71	3.15	1.88	BCC
Rb	2.96	2.44	1.60	BCC
Cs	2.47	1.48	2.06	BCC
Nb	245	132	28.4	BCC
Ta	262	156	82.6	BCC
W	517	203	157	BCC
NaF	97.0	24.2	28.1	Rocksalt
NaCl	49.1	14.0	12.7	Rocksalt
NaBr	34.6	5.2	5.0	Rocksalt
NaI	30.2	9.0	7.36	Rocksalt
KF	65.0	15.0	12.5	Rocksalt
KCl	39.89	7.25	6.25	Rocksalt
KBr	34.68	5.80	5.07	Rocksalt
KI	27.4	4.3	3.70	Rocksalt
RbCl	36.4	6.3	4.8	Rocksalt
CsF	4.2	15.4	7.58	Rocksalt
CsCl	36.6	9.0	8.07	Rocksalt
CsBr	30.7	8.4	7.49	Rocksalt
MgO	294	93	155	Rockslat
CaO	224	60	80.6	Rocksalt
SrO	170	46	55.6	Rocksalt
BaO	121	44	34.4	Rocksalt
GaP	152	63	71.6	Zincblende
GaAs	118	53.5	59.4	Zincblende
GaSb	88.4	40.3	43.4	Zincblende
InP	102	58	46.0	Zincblende
InAs	84.4	46.4	39.6	Zincblende
InSb	66.0	35.8	30.1	Zincblende
$NiCr_2O_4$	252	140	83.5	Spinel

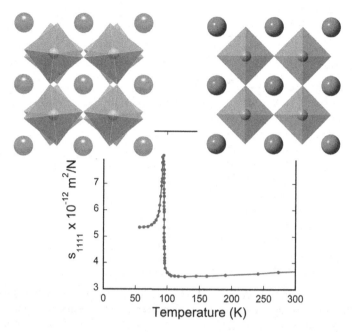

$s_{1111} \times 10^{-12} \ m^2/N$

Temperature (K)

Fig. 30.10 Effect of phase transition on the elastic compliance of $SrTiO_3$. Elastic constant data from *Landolt–Börnstein Numerical Data and Functional Relationships in Science and Technology New Series, Vol. 16 Ferroelectrics and Related Substances*, ed. K.–H. Hellwege and A. M. Hellwege, Springer-Verlag, New York (1981).

structure, and its elastic properties show a "normal" temperature dependence. At lower temperatures, however, the Sr is too small to occupy the 12-coordinated site, and the structure slumps through cooperative rotation of octahedra at ~102 K, as shown in Fig. 30.10. The elastic compliance goes through a maximum at the phase transition temperature, because the TiO_6 octahedra have large thermal oscillations at this temperature. Above this temperature, the cubic prototype structure is stable, and the compliance is small (stiffness is large). Below the phase transition temperature, the compliance is higher because applied mechanical stresses can move domain walls between different twin states, in which the rotation has occurred along different $\langle 001 \rangle$ axes of the prototype structure. That is, the material is ferroelastic (the elastic counterpart of ferroelectric or ferromagnetic materials, as described in Chapters 28 and 29).

(6)   As a first approximation, Young's modulus of an alloy can be treated using a mixing law between the end members. Systems with eutectic phase diagrams often show Young's moduli below the linear average of the end members. In contrast, materials that form intermediate compounds often have a positive deviation from the linear average, with maxima at the intermetallic points.

(7)   Motion of point, line, or area defects can lead to plastic deformation. Point defects can migrate under stresses, changing the shape of the material. Similarly, at low stress values, sections of pinned dislocations will bow out. At higher stresses, the dislocations can break away from pinning centers, producing a permanent shape change. Finally, grain boundary sliding is also an irreversible process.

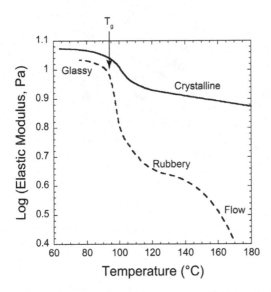

Fig. 30.11 Temperature dependence of the elastic modulus of two forms of polystyrene. The crystalline form has the higher elastic modulus over the whole temperature range. An amorphous thermoplastic shows glassy, rubbery, and flow regimes with increasing temperature. Adapted, with permission, from A. V. Tobolsky, *Properties and Structure of Polymers*, John Wiley and Sons, New York (1962).

(8)   Many polymeric materials make transitions between elastic and plastic deformation as a function of temperature. Consider, for example, the case of an amorphous thermoplastic subject to a tensile stress, as shown in Fig. 30.11. At high temperatures, the chains can slide past each other comparatively easily in the *flow* regime. Once this has happened, there is no restoring force for the material to return to its original shape, and the deformation is plastic. As temperature decreases, the material response becomes *rubbery*. The elongation of the material is large as the chains uncoil (without slipping). As the stress is released, the chains recoil, recovering the original shape of the sample. It is important to note that in this regime, the elastic stress-strain curve is *not* linear. The material has a low stiffness when the chains can uncoil. However, once the chains are largely aligned parallel to the applied stress, the stiffness rises substantially, since deformation of bonds in the backbone is now required. Furthermore, it is important to note that above the glass transition temperature, T_g, the polymer is viscoelastic, such that both the rate at which the stress is applied and the time under stress influences the amount of elongation. As the temperature decreases further, a major increase in stiffness is observed when the material drops below T_g. Below this temperature, thermal agitation of the chains drops substantially, and the chains are essentially "frozen in" to their positions. Thus, below this temperature, the material shows much smaller strains for the same stress; the material is *glassy* and tends to fail in a brittle fashion. Cross-linking the polymer produces covalent bonding between chains, and can lead to the disappearance of the flow and rubbery regimes, since chains can no longer readily move with

respect to each other. In much the same way, large sidegroups on the polymer decrease chain slippage and produce more brittle behavior.

(9) Many polymers are viscoelastic, such that the rate at which stress is applied is also important to the mechanical response. This occurs because when chains try to move, it takes time. Thus, the strain rate matters. As a result some materials (think silly putty) will shatter if the strain rate is very high due to crack propagation. At intermediate strain rates, the material behaves more like a rubber, so that large strains can be induced and recovered. At much lower strain rates, the same material deforms plastically by chains flowing past each other. Polymers are also often viscoplastic in that once chains move past each other, the deformation will not be completely recovered on unloading.

(10) Polymers which have bulky sidegroups tend to be stiff and brittle because it is difficult to move the sidegroups through the matrix. Polystyrene and polymethylmethacrylate are examples of this type of behavior.

(11) Materials such as polyvinylchloride can be made compliant and tough by adding plasticizers such as dioctylphthalate that weaken the intermolecular forces and reduce the crystallinity.

30.3 Anisotropy of Elastic Properties

All crystals are elastically anisotropic, meaning that the elastic properties differ in different directions. Again, the spring analogy for the bonding is useful here. When a tensile stress is applied parallel to the strongest bonds in the solid, then in order to deform the material, the strong, stiff bonds must be elongated. This is difficult to do, and the stiffness is large. Consider the case of the rocksalt crystal structure under stress, as shown in Fig. 30.12. When stress is applied parallel to the [100] direction, the bonds directly oppose the applied stress, and the elongation is modest. Because the atomic arrangements are identical in the [100], [010], and [001] directions, the elastic stiffness is the same in those three directions. However, as shown in the figure, stresses applied in other orientations (for example along [110]) can deform the material by shear without requiring large changes in the bond length. Thus, in most materials with the rocksalt crystal structure, the elastic stiffness is larger along $\langle 100 \rangle$ than along $\langle 110 \rangle$. Some exceptions to this general rule do occur. If, for example, the cation is much smaller than the cation, then the anions may be nearly in contact with each other at zero applied stress. Periclase (MgO) is a good example of this. When stress is applied parallel to $\langle 110 \rangle$, shear deformations are difficult, as they would require the anions to be pushed closer together. In this case, the stiffness is *higher* $\parallel \langle 110 \rangle$.

For cubic crystals, an anisotropy factor, A, can be defined as: $A = \frac{2c_{2323}}{c_{1111} - c_{1122}}$. If $A < 1$, then the cubic crystal is stiffest along $\langle 100 \rangle$, and least stiff $\parallel \langle 111 \rangle$. If $A = 1$, then the crystal is elastically isotropic, and the stiffness c_{1111} is the same for any orientation of the sample. When $A > 1$, the crystal is most stiff $\parallel \langle 111 \rangle$, and least stiff $\parallel \langle 100 \rangle$. Along other orientations, the stiffness lies between the values along $\langle 100 \rangle$ and $\langle 111 \rangle$. Looking back at Table 30.1, it is clear that the alkali metals have large values of A. This makes sense, since the bonds in BCC metals are parallel to $\langle 111 \rangle$, and materials are typically

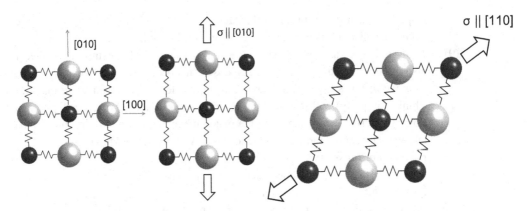

Fig. 30.12 Illustration of the spring model for describing the elastic constants of solids. In general, stiffness is high when the stress is working directly against the strongest bonds in the solid, and smaller when the stress can change a bond angle, rather than a bond length. Thus, in many materials with the rocksalt structure, the stiffness is higher $\parallel \langle 100 \rangle$ compared with $\langle 110 \rangle$.

stiffest parallel to the strongest bonds. In the rocksalt structures (excepting MgO and the Li salts, where the cations are small), the strong bonds are $\parallel \langle 100 \rangle$, so A is less than one. FCC metals have bonds arranged in the $\langle 110 \rangle$ directions, so the stiffness is expected to be highest in those directions. Since $\langle 110 \rangle$ is closer to $\langle 111 \rangle$ than $\langle 100 \rangle$, it is not surprising that the A values exceed one. Other crystal structures do not have continuous chains of strong bonds in a particular orientation (e.g. the diamond and zincblende structures). As a result, stresses in any direction will tend to produce some elongation of bonds and some changes in bond angles. In such materials, the anisotropy is small and the A values are closer to one.

Crystalline polymers show very large anisotropies in the elastic properties, due to the strong anisotropy in the bonding. Typically, the elastic modulus is approximately two orders of magnitude higher parallel to the chains, where the bonding is predominantly covalent, than perpendicular to the chains, where weak van der Waals or hydrogen bonds hold chains together. Typical numbers might be around 10 GPa perpendicular to the chains, and ~300 GPa parallel to the chains, for measurements at 4 K.

A stiffness surface is a plot of the elastic stiffness as a function of orientation. Examples of stiffness surfaces are given in Fig. 30.13 for NaCl, diamond, and polyethylene.

30.4   Toughness and Cleavage

To be durable, a material must be hard and tough, to withstand abrasion and impact, and strong so it does not fracture under applied stress.[3] In many durable materials, final

---

[3] Portions of this section are reproduced by permission from R. E. Newnham, *Structure-Property Relations*, Springer-Verlag (1975).

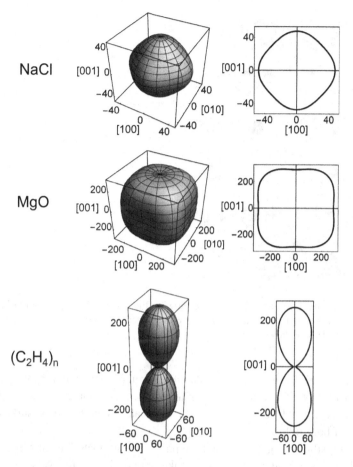

Fig. 30.13 Plots of elastic anisotropy for NaCl, diamond, and polyethylene. Data for the first two come from the *Landolt–Bornstein Tables Volume II: Elastic Piezoelectric, Pyroelectric, Piezooptic, Electroopic Constants, and Nonlinear Dielectric Susceptibilities of Crystals*, eds. Hellwege and Hellwege, Springer-Verlag, New York (1979). Stiffness constant for polyethylene were taken from Odajima and Maeda, "Calculation of the elastic constant and the lattice energy of the polyethylene crystal," *J. Pol. Sci. C.*, **15**, 55–74 (1966).

mechanical failure occurs by fracture, a phenomenon that depends on the microstructure, rather than the crystal structure, of the solid. Since microstructure is outside the scope of this book, this section will concentrate on failure by cleavage, where there is a strong link to crystal structure. Hardness was described in Chapter 4. Toughness means resistance to fracture, and can be defined as the energy required to fracture the sample, and is quantified as the integral of the stress–strain curve. Toughness is an irreversible property, since testing leaves the material in a permanently altered condition.

Fracture energy is a measure of toughness. Bars are often fractured in three-point bending, and the force–displacement curves are integrated to obtain the work of fracture. This provides a measure of the work required to propagate a crack over a given cross section, and is expressed in J/m^2.

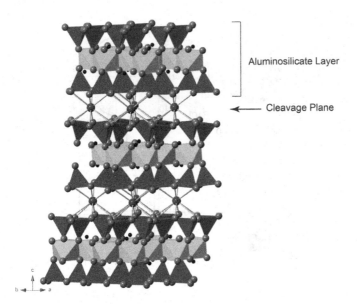

Aluminosilicate Layer

← Cleavage Plane

Fig. 30.14 Layer structure of muscovite mica showing the strongly bonded aluminosilicate layers held together by long, weak K–O ionic bonds. Cleavage occurs when a plane of the weak K–O bonds is broken.

Cleavage is the fracture of materials along preferred crystallographic planes. It is sensible that this should occur either on planes which have bonds that are easily broken, or along planes which have fewer bonds to break per unit area of surface created. Examples of both follow.

The layer silicates, gypsum, and graphite are all materials in which strongly bonded layers are held together by weaker bonds between layers. For example, muscovite mica, $KAl_2(AlSi_3O_{10})(OH)_2$, is layer silicate in which the strong Al–O (~1.9 Å) and Si–O (~1.62 Å) bonds are confined to the (001) plane. The aluminosilicate layers are joined by weak K–O bonds (~3.2 Å long). Because of this disparity in bond strengths, the K–O bonds are readily broken during cleavage to produce thin transparent layers; the cleavage planes are illustrated in Fig. 30.14. In former times such cleavage plates were used as stove windows and as high-breakdown-strength parallel-plate capacitors.

Gypsum, the main component of common wallboard, also crystallizes in a layer-like structure, as was shown in Fig. 21.12. Ca ions bond to $(SO_4)^{-2}$ sulfate groups in layers parallel to the monoclinic (001) plane. The $CaSO_4$ layers in gypsum ($CaSO_4 \cdot 2H_2O$) are bonded together by water molecules. The protons of the water are attracted to oxygens of the sulfate groups forming hydrogen bond bridges across the (010) plane. Cleavage takes place through these weak hydrogen bonds.

The easy cleavage in graphite is why graphite can be used as a lubricant. The structure of graphite, and its close relative, hexagonal boron nitride, is shown in Fig. 8.17. Covalently bonded layers are held together by long, weak van der Waals bonds. The latter are readily broken by stress, producing the cleavage.

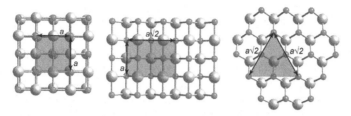

Fig. 30.15 Different surfaces of NaCl, showing how many bonds are broken for forming different surfaces. The lattice parameter is a. For each of the figures, both atoms on the top surface and ones on the plane below are shown, so that it is easier to see how many completed bonds are present. In rocksalt, each atom should have 6-coordination. In order to cleave the crystal bonds holding these planes to the next have to be broken.

In other materials that cleave, there are no planes of weak bonding. Instead, cleavage occurs where there are the fewest bonds to be broken to create the new surface area. For example, rocksalt crystals cleave into cubes on $\{100\}$ faces, while diamond and fluorite cleave into octahedrons on $\{111\}$ faces. To see why this is so, consider the case of rocksalt. Figure 30.15 shows several surfaces of the rocksalt crystal structures. On $\{100\}$ surfaces, there are five bonds made for each surface atom – four to other atoms in the same plane, and one to the plane underneath. There are four atoms on the corners, each of which is shared by four other area units. Thus, the corner atoms have $4\left(\frac{1}{4}\right)$ bonds broken. The four atoms on the edges of the area units are shared between two area units. Since there are four atoms on the edges, they contribute $4\left(\frac{1}{2}\right)$ bonds broken. For the atom at the center of the area unit, there is one broken bond. Thus, there are $\frac{4\left(\frac{1}{4}\right)+4\left(\frac{1}{2}\right)+1\ bonds\ broken}{a^2} = \frac{4\ bonds\ broken}{a^2}$ for cleavage on the $\{100\}$ plane.

For the $\{110\}$ plane, each surface atom has four completed bonds – two in the plane, and two to atoms in the plane beneath. Two bonds are missing, pointed 45° out of the plane. For the area shown shaded in Fig. 30.15, this means that there are $\frac{4\left(\frac{1}{2}\right)+4(1)+4\left(\frac{1}{2}\right)\ bonds\ broken}{\sqrt{2}a^2} = \frac{4\sqrt{2}\ bonds\ broken}{a^2}$. The three terms correspond to the top, middle, and bottom rows of atoms. For the $\{111\}$ planes, cleavage would require $\frac{4\sqrt{3}\ bonds\ broken}{a^2}$. Comparison of these numbers makes it clear that the $\{100\}$ surfaces are the preferred cleavage faces, because they require much less energy to separate. As a result, rocksalt cleaves into cubes.

For the diamond and fluorite structures, the order is reversed, with fewer broken bonds per unit area on $\{111\}$ planes than on $\{110\}$ or $\{100\}$. Therefore diamond and fluorite structure crystals cleave into octahedral shapes. Simple arguments based on counting the number of broken bonds generally allow cleavage planes to be identified, in the absence of more quantitative models.

Materials with high toughness are those in which crack propagation is difficult. There are a number of ways in which this can be engineered that hinge on crystal structure. This will be illustrated for fiber toughening, a stress-induced phase transformation that induces volume increase, and crack diversion.

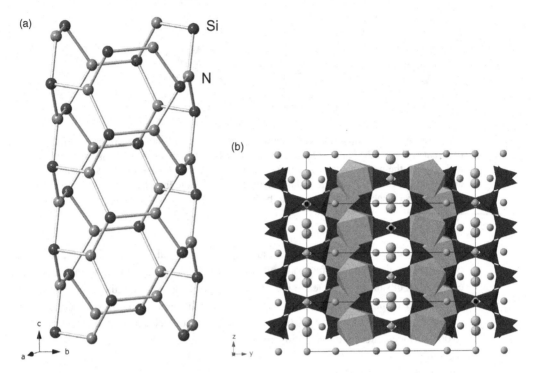

Fig. 30.16 Sections of the crystal structures of (a) β-Si_3N_4, silicon nitride, showing the central tube and (b) nephrite jade, a double-chain silicate. In both cases, the crystal structure favors a fibrous morphology that produces high fracture toughness.

Fiber toughening is often accomplished by making composite materials with fibrous reinforcements (e.g. a fiberglass-reinforced polymer for canoes). However, materials with fibrous morphologies resulting from their crystal structure can also show some amount of fiber toughening. Both silicon nitride and nephrite jade have crystal structures in which there is stronger bonding along a single axis, as shown in Fig. 30.16. When random polycrystalline samples are prepared, propagation of a crack can follow along the bond between an acicular crystal and the matrix. In order to fail the part, it is then necessary to pull out the fibrous grain. This mechanism is particularly important in structural ceramics like silicon nitride, which are used in engine components like turbines, cam rings, and bearings at high temperatures. Failure of any component in an aircraft engine, for example, clearly needs to be avoided, so high toughness is desirable. In the same way, dense polycrystalline chain silicates have surprisingly large fracture energies despite the presence of cleavage. Nephrite jade, $Ca_2(Mg,Fe)_5Si_8O_{22}(OH)_2$, is an amphibole double-chain silicate, as described in Chapter 19. The long, slender amphibole crystals are interwoven like cloth, causing crack pinning, crack deflection, and frictional effects associated with pullout of acicular grains. The result is a fracture energy >100 J/m^2, enabling jade to be carved into very delicate shapes. Many

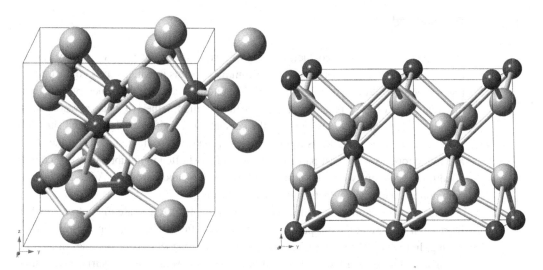

Fig. 30.17 Comparisons of the crystal structures of (left) baddelyite, with its unusual 7-coordination of the Zr atoms and (right) tetragonal partially stabilized zirconia.

glass-ceramics which have crystallites with fibrous morphologies are utilized to make dinnerware with high fracture toughness.

Partially stabilized zirconia is an excellent example of a material which undergoes a phase transformation as a crack relieves stress locally. The stable phase of ZrO_2 at room temperature is baddelyite, an unusual compound with Zr in 7-coordination which leads to poor packing efficiency and comparatively lower density ($\rho = 5.827$ g/cm^3). When ceramics are prepared in a partially stabilized form of zirconia, the material is doped to lead to a denser metastable form of ZrO_2 (for which the density would be 6.09 g/cm^3 if all of the cations were Zr) in which the Zr atom has four near neighbors and four slightly more distant neighbors (see Fig. 30.17). As a crack propagates through the material, a phase transformation to the larger volume phase takes place, closing the crack tip.

Microcracking can also be a toughening mechanism, as is done in machinable glass ceramics such as MACOR,[4] in which small mica particles are precipitated randomly from a glass. Mica cleaves easily parallel to the silicate sheets. As a crack moves into the material and reaches a mica particle, it is readily deflected by easy cleavage due to the layer structure. If many small particles are available, the crack can be diverted into many small microcracks, dissipating all of the available mechanical energy, without cracking the materials through.

In polycrystalline materials, the situation is more complicated than for a single crystal. Grain boundaries often have different mechanical characteristics than the grains, given that the bonding is incomplete, and that the chemistry of the grain boundary can differ from that of the grain. Interested readers are encouraged to follow up on this topic in other textbooks.[5]

---

[4] MACOR is the tradename of a product produced by Corning, Inc.

[5] See, e.g., David J. Green, *An Introduction to the Mechanical Properties of Ceramics*, Cambridge University Press, New York (1998).

30.5   Friction and Wear

When surfaces come into contact, rubbing can cause the surfaces to wear away at each other.[6] *Wear* and *friction* processes have common origins in which small contact points undergo cold-welding, or where surface penetration and cutting take place. Three different types of wear phenomena have been identified: adhesive wear, abrasive wear, and chemical wear. In *adhesive wear*, the contacting asperities cold-weld and shear off below the interface. This leads to the transfer of material from one surface to the other, and to the formation of wear particles. It is found most often in unlubricated, dry-contact conditions between two metal surfaces of about the same hardness. Adhesive wear is common in electrical contacts and conveyor bearings.

Abrasive wear is more common when hard and soft materials are in contact, or when lubrication is present. Hard asperities penetrate the softer surface and remove material by cutting. In magnetic recording, for example, the hard magnetic oxide particles in the tape act as a very fine abrasive. In the aerospace industry, wear-resistant surfaces are often made of hard ceramics, cemented carbides, and cermets. Such materials are also of interest for gyroscopes and for high-temperature environments where no wear can be tolerated.

Chemical wear can also play an important role. Alumina (Al_2O_3) wears away boron carbide (B_4C) despite the greater hardness and higher melting point of the carbide. Localized heating occurs when the alumina particles are rubbed against the boron carbide surfaces. If air is present, chemical oxidation takes place: $B_4C + 4O_2 \rightarrow 2B_2O_3 + CO_2$. Alumina then abrades the softer B_2O_3 phase, removing the harder boron carbide by chemical reaction.

The field of chemomechanical polishing, or chemomechanical planarization, is widely used in the semiconductor industry to produce flat surfaces over complex topologies. This relies on a combination of mechanical wear and chemical reactions to prepare extremely smooth surfaces over large wafers.

Abrasion is closely related to hardness, so Moh's hardness scale provides some insight into the wear process. In principle, materials with the same Moh's hardness should not abrade each other. However, if the surfaces slide past each other at high velocities, frictional heating develops. In much the same way, a person might rub their hands together to warm them by friction on a cold morning. Higher temperatures, in turn, increase the wear rates. This is why it is possible to wear diamond away with a high-speed rotating glass rod. Higher wear rates are observed in materials which have low thermal conductivities and high thermal expansions, as these tend to keep temperatures high, and generate thermal expansion-induced stresses (see Chapter 24).

Both crystal structure and surface roughness influence friction. The coefficient of friction μ is defined by the ratio of the frictional force F to the weight of the sample W: $\mu \equiv F/W$. Friction is lowest for smooth surfaces, whereas rough surfaces interfere with each other's motion to a greater extent. Fluid films can lubricate the surfaces and significantly lower the frictional forces, as is discussed in the next section.

---

[6] Portions of this section are reproduced by permission from R. E. Newnham, *Structure–Property Relations*, Springer-Verlag (1975).

Friction is anisotropic, and is generally lowest in planes of greatest packing density when sliding along close-packed directions. Planes with high packing density are often relatively flat. For example, in corundum, the coefficient of friction is smallest for sliding on the (0001) plane, where the oxygens are nearly hexagonally close packed; the sliding direction is [110]. In diamond, friction is lowest for the {111} planes, where each of the C atoms has three out of its four neighbors, and the surfaces are reasonably flat (see Fig. 30.18). These planes are the best-bonded of the faces on diamond.

The coefficient of friction is inversely proportional to cohesive energy, elastic modulus, and hardness (see Chapter 5). For the rocksalt series of MgO, LiF, KCl, and KBr, the bond strength decreases from MgO to KBr. As a result, the elastic stiffness and hardness decrease in this order; in contrast, friction increases. Materials with short, strong bonds have low coefficients of friction.

Asperities on surfaces wear away due to frictional forces. The frictional force is proportional to normal load and independent of the apparent contact area. However, most theories of friction are based on the idea that the true area of contact is proportional to load, making friction proportional to the shear area, $F = Sa$. The shear strength S can be controlled by surface conditions: a few layers of organic materials on a metal surface reduce frictional forces by orders of magnitude. The measured frictional coefficients are strongly dependent on the nature of the bonding at the surface, as well as the surface composition. FeO and Fe_3O_4 are superior lubricants to Fe_2O_3 and cause a large variation in the frictional forces of hardened steel with oxygen pressure.

Friction coefficients range over several orders of magnitude. For dry, unlubricated surfaces, a typical value for μ is about 0.5. The friction coefficients for oil-based lubricants are very low (0.001–0.01) but they degrade at relatively low temperatures, typically 100 °C. MoS_2 ($\mu \sim 0.2$) and graphite ($\mu \sim 0.1$) are better lubricants at high temperatures. Brake linings require substantial friction coefficients over temperature ranges of several hundred degrees. The heat generated during rapid braking can be a severe problem. Good thermal conductivity and high melting points are an advantage. Automobile brakes are made from phenolic resins as binders for a variety of powders and fibers that provide different functions such as vibration and noise control, changes in the friction coefficient, and wear resistance. Among the powders are SiC, graphite, and MoS_2. Much more refractory brake linings are needed for jet planes.

30.6   Lubrication

As all engineers know, friction and wear can be greatly reduced with a suitable lubricant. Lubricating oils for automobiles are obtained from refining heavy distillates, about 2% of the petroleum production. The Society of Automotive Engineers has classified oils by index number according to a variety of tests for viscosity, flash point, and sludge resistance. Improvements in performance are achieved through the use of additives such as detergents and antioxidants.

Most lubricants are colloids, an important class of industrial materials whose characteristics lie between those of liquids and solids. Paints, cosmetics, grease, some medicines, adhesives, asphalt, and many types of biological tissue are colloidal in nature.

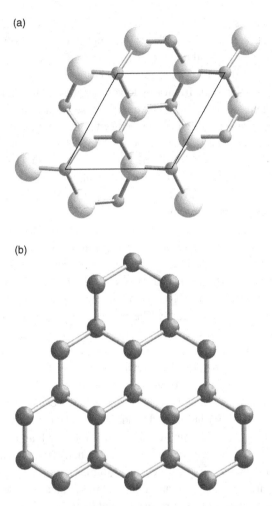

Fig. 30.18 Frictional forces are low on flat, well-bonded surface planes. (a) Basal (0001) plane of α-Al₂O₃ showing the [110] direction for which friction coefficients are minimized. Al are shown as smaller dark circles, oxygen is larger. The unit cell is outlined in black. (b) Strongly bonded (111) plane of diamond, with excellent wear resistance.

The change in viscosity with mechanical stress is a key feature of this class of materials. Non-drip paint, custard, and magic putty are examples of liquids which become "thicker" or "thinner" when stirred.

Colloidal gels like these are made up of long chain-like molecules dispersed in a solvent. The polymer chains contain covalently-bonded backbones with ionic segments that attract one another. As explained in Fig. 30.19, these attractive forces can occur within the same chain or between neighboring chains. The changes in cross-linking that result under stirring lead to large changes in viscosity.

Molecules like this are sometimes called *amphiphilic*, meaning that they can bond to both ionic and covalent liquids. *Surfactants* such as soap are amphiphilic, with ionic

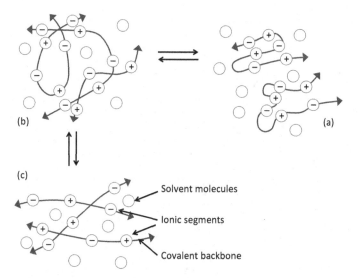

Fig. 30.19 Polymer chains with covalent backbones and ionized segments with weak attractive forces. (a) When the attraction takes place between ionic segments in the same chain, the chains curl up and act as separate molecules. In this state, the viscosity is low, but increases under mechanical agitation as the ionic segments in adjacent chains attract each other (b). These cross-linked chains can be separated under further agitation to lower the viscosity (c).

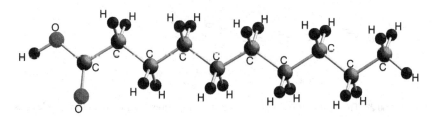

Fig. 30.20 Carboxylate surfacants have ionic COO⁻ groups at one end and covalent methyl termination at the other. In keeping with the "like likes like" rule, the ionic end likes water and the covalent end likes oils and fats.

groups at one end and covalent groups at the other (Fig. 30.20). The negative end of the carboxylate surfactant (obtained when the H is lost from the COOH group, leaving COO⁻) is balanced by positively charged cations such as Na⁺. Ionic terminations are *hydrophilic* (water-loving) and covalent *hydrophobic* (water-hating). When dissolved in water, surfactant molecules try to shelter their hydrophobic tails from the water molecules by immersing them in nearby oil globules, or by forming micelles or surface layers (Fig. 30.21). A layer of surfactant molecules such as this lowers the surface tension. When immersed in oil, the surfactants form "reverse" micelles with the ionic

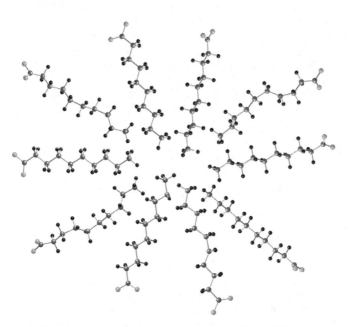

Fig. 30.21 Surfactant molecules immersed in water tend to form globular micelles or surface coatings. The figure shows a cross section of such a globe.

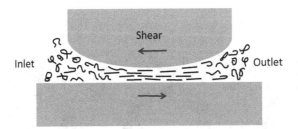

Fig. 30.22 The behavior of a lubricant under shear stress. As lubricant molecules pass into the contact region, they must disentangle and align in the shear direction, a process that requires time. Under steady-state motion, a continuous gradient of molecular order is established in the contact region. The degree of order may be different on the upper and lower surfaces. Molecules exiting the contact region will return to their unconfined configurations during a characteristic time. Adapted, with permission, after C. Drummond and J. Israelachvili, "Dynamic behavior of confined branched hydrocarbon lubricant fluids under shear," *Macromolecules*, **33** (13), 4910–4920 (2000). Fig. 14. Copyright American Chemical Society.

portions of the chain shielded from the covalent oil molecules. Micelles are very common in living systems.

The transient effects in a lubricated interface are shown in Fig. 30.22. As two surfaces begin to slide past each other, the friction starts high and, as sliding continues, decreases to a lower, steady-state value as *shear-thinning* takes place.

Dilatency is the second effect. In order to move a flat surface sideways it is first necessary to move it upwards. The movement may be small, but it should be sufficiently large to allow molecular reorganization of the lubricant.

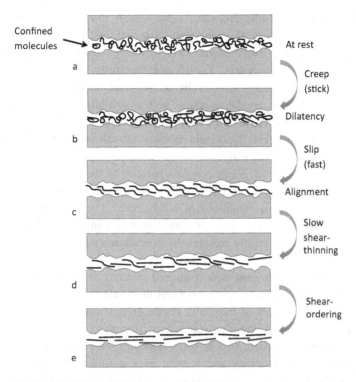

Fig. 30.23 (a) Confined lubricant molecules trapped between two surfaces at rest. (b) Under shear stress, the molecules begin to orient in the dilatency stage. (c) Slip begins as the molecules align, followed by shear thinning (d), and a phase transformation (e) to ordered domains within the contact regions. Adapted, with permission, after Fig. 14 of C. Drummond and J. Israelachvili, "Dynamic behavior of confined branched hydrocarbon lubricant fluids under shear," *Macromolecules*, **33** (13), 4910–4920 (2000). Fig. 14. Copyright American Chemical Society.

Figure 30.23 illustrates the changes in a lubricant film as it adapts to steady-state sliding conditions. The molecules proceed from a disordered state through the dilatency stage to thin ordered layers that reduce friction.

Solid lubricants provide a low-shear interface between bearing surfaces. The need for solid lubricants has increased during the past few decades. Automotive and aircraft engines operate at higher temperatures for longer periods of time to achieve improved efficiency and fuel economy. Under these conditions, liquid lubricants are susceptible to evaporation and decomposition. Likewise, solid lubricants are required in some vacuum systems, where the need for low outgassing precludes liquid lubricants with high vapor pressure components. Thus, solid lubricants are used when other lubricants degrade or decompose at high temperature or under high radiation flux, and when fluids congeal at low temperature or volatilize in high vacuums. Solid lubricants also find applications in space vehicles, metalworking operations, automotive components, engine oil additives, and electrical contacts. Freedom from maintenance is another attractive feature, especially when bearing surfaces are inaccessible or run unattended. Solid lubricants are effective at high loads and low speeds, but lack the self-healing and heat-dissipative qualities of fluid lubricants.

To minimize friction, solid lubricants must possess low shear strength in at least one crystallographic direction. Three important classes of solid lubricants are: (1) materials with layer structures, (2) soft materials which undergo plastic deformation, and (3) polymers with low surface energy. Factors contributing to low shear strength include crystal structure, intercalated gases, and chemical interactions with the surfaces.

Materials with lamellar crystal structures such as graphite and molybdenite (MoS_2) are effective lubricants because of the slippage between planes. The crystal structure of molybdenite is shown in Fig. 30.24, and its friction coefficient is compared with other engineering materials in Table 30.2.

The lubricating ability of graphite is less because of the π bonds between layers. Graphite depends on the presence of adsorbed surface layers to retain lubricity, especially in the near-vacuum conditions of outer space. The most effective treatment involves the fluorination of graphite to form substoichiometric graphite fluoride CF_x ($x = 0.3$ to 1.1). Fluorine greatly enlarges the interplanar spacing from 3.35 Å to 6–8 Å (see Fig. 30.24).

Another solid-state lubricant with a layered structure is boric acid, H_3BO_3. This material forms spontaneously on the surfaces of borides exposed to humid air. The structure consists of layers of triangular borate groups linked together by hydrogen bonds. The widely spaced layers provide low friction up to 170 °C.

Many polymers have been used as solid lubricants. Polytetrafluoroethylene $(CF_2)_n$, has a very low surface energy and is used as a self-lubricating coating for metal components. Polyimides have great potential in outer space applications because of their low wear rates, the low frictional coefficients in vacuum, and their high thermal stability.

Finally, metals and inorganic salts with modest melting points can undergo plastic deformation, making them useful as solid state lubricants. Lead (Pb) has long been used as a lubricant in motor bearings, and eutectic mixtures of calcium and barium difluoride can be used as high temperature lubricants (>500 °C).

Surface films can be used to reduce friction. For example, friction can be reduced by reacting a metal with a gas to form a layer-like surface structure. Solid layer compounds such as these can be used at higher temperatures than oils (Table 30.3).

In addition to low shear strength, a good lubricant must have thermal stability, low vapor pressure, and oxidation resistance. It must form an adherent film to the substrate, a self-healing film with mobile absorbate atoms to provide endurance. To avoid damaging metallic surfaces, the hardness of the lubricant is limited to about 5 on Mohs scale.

30.7   Grinding and Polishing

More than 90% of the multibillion-dollar gemstone business involves the cutting, faceting, and polishing of diamonds.[7] Webster's dictionary defines *polishing* as "to make smooth and glossy, usually by friction." Top and bottom views of the 58 facets on a standard brilliant cut for engagement rings are shown in Fig. 30.25. Shape, color, and

---

[7] Portions of this section are reproduced, by permission from R. E. Newnham, *Structure–Property Relations*, Springer-Verlag (1975).

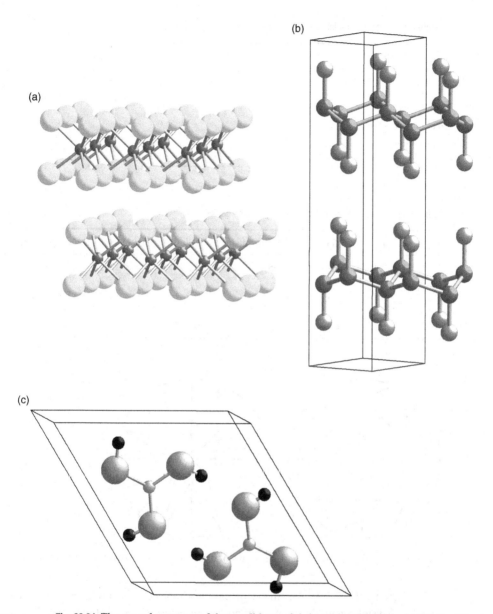

Fig. 30.24 The crystal structure of three solid-state lubricants. (a) Molybdenite, MoS_2. Mo atoms are shown as small spheres and S atoms as the larger spheres. The structure can be envisaged as a stacking of S/Mo/S layers. MoS_2 is an excellent solid lubricant because of the weak van der Waals bonds between layers. (b) Graphite fluoride with C layers separated by F atoms. (c) Boric acid, H_3BO_3. B is the small light sphere, O the larger sphere, and H the small black sphere.

clarity are of the utmost importance in selecting the rough diamond. To shape the preform, the rough stone is sawed using a thin phosphor bronze disk impregnated with diamond powder and lubricated with oil and additional diamond dust. Further shaping is done on a lathe by rubbing one diamond against another, followed by grinding and

Table 30.2 Friction coefficients for several engineering materials

Material	μ
Graphite	0.1
MoS_2	0.2
Polyethylene $(CH_2-CH_2)_n$	0.7
Teflon $(CF_2-CF_2)_n$	0.1
Steel FeC_x	0.8
Wood	0.5

Table 30.3 Friction-lowering surfaces layers on metals

Metal	μ	Gas	Film formed	μ with surface layer	Temperature limit
Mo	2.0	H_2S	MoS_2	0.2	800 °C
U	1.2	H_2S	US_2	0.4	700 °C
B	1.0	N_2	BN	0.7	1100 °C
Ti	1.2	I_2	TiI_2	0.3	400 °C
Cr	2	Cl_2	$CrCl_3$	0.2	700 °C

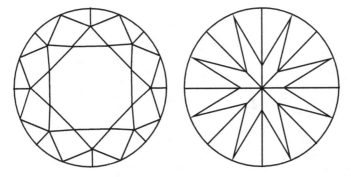

Fig. 30.25 (left) Top and (right) bottom views of a brilliant-cut diamond.

polishing on a horizontal lap made of cast iron coated with olive oil and diamond powder. Crystal orientation is very important in determining faceting times. The very hard {111} surfaces are generally avoided because they polish so slowly.

Hardness is one important factor that determines grinding or lapping rates. However, quantitative prediction of grinding rates is challenging, since they depend on many additional factors, such as the speed of lapping, the nature of the abrasive, and the forces involved. Nevertheless, the relative grinding rates of various crystal faces can sometimes be predicted based on an understanding of crystal structure.

Consider, for example, the case of diamond. The different faces of diamond polish at different rates, with the relative rate also depending on the direction of motion, as shown in Table 30.4. A good rule of thumb in assessing relative grinding rates is that rougher surfaces experience greater friction and hence have faster grinding rates.

Table 30.4 Relative grinding times required to remove a given thickness from a diamond crystal by polishing on various planes in different directions

Plane	Direction	Time
(111)	Any	$>10^3$
(100)	[010]	3
(100)	[011]	22
(110)	[001]	1
(110)	$[\bar{1}10]$	51

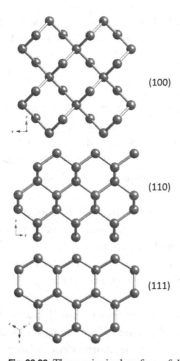

(100)

(110)

(111)

Fig. 30.26 Three principal surface of diamond. Redrawn from R. E. Newnham, *Structure–Property Relations*, Springer-Verlag, New York (1975). Fig 85, p. 201. Adapted with permission by Springer.

The carbon atoms on the (111), (110), and (100) surfaces of diamond (shown in Fig. 30.26) have different amounts of bonding, and hence these surfaces polish at different rates. The better bonded the atoms on a surface, the more energetically expensive it is to polish it. As a result, the {111} surfaces of diamond are exceedingly difficult to polish. The surfaces are parallel to the puckered hexagonal rings, which are comparatively smooth, and for which the surface atoms are held in place by three out of four bonds. Figure 30.26 shows that the (110) face has grooves parallel to $[\bar{1}10]$, the slow-grinding direction in this plane. Polishing proceeds much more rapidly in the [001] direction on this plane, because of the greater surface roughness perpendicular to the grooves. Motion in the [010] and [001] directions give the fastest polishing rates for the (100) face. These directions are rotated 45°

with respect to the grooves parallel to [011] of $[0\bar{1}1]$; thus there is always surface roughness to increase the polishing rates.

30.8   Weathering

Silica glass can be made extremely strong. Anhydrous, flaw-free glass can withstand tensile stresses of ≥ 14 GPa, about 10 times the strength of common metal alloys. This makes it one of the strongest materials known. In most engineering applications, however, the surface of the glass is exposed to chemicals and abrasives that create small surface flaws and promote their growth. The flaws present a difficult challenge for the design engineer. These small cracks not only reduce the measured mechanical strength, but also influence its behavior over time. Trans-oceanic optical fibers and other stress-bearing systems can fail after several years as small surface cracks grow in size, and act as the failure sites in brittle fracture.

The hydration of silica thus provides a good example of how environmental conditions can weaken the surface of a solid. Under normal conditions, oxygen ions are present on the outermost surfaces of silica glass because they are larger, more highly charged, and more polarizable than the silicon ions. Water molecules are then attracted to the surface through the formation of hydrogen bonds. This is described in more detail in Chapter 18.

Water weakens the surface of silica through a chemical reaction; the reaction is accelerated considerably in the presence of an applied stress. As pictured in Fig. 30.27, the reaction proceeds in four stages. Water molecules approach the surface of silica, and hydrogen bonds begin to form with the surface oxygens. Hydroxyl ions $(OH)^-$ are formed on the surface, converting the surface oxygens from bridging to non-bridging. This has the effect of weakening the glass surface through the formation of tiny surface flaws. Brittle substances like glass tend to fracture because small cracks and fissures enlarge under tensile stresses. Water continues to react with the new surfaces formed, further weakening the glass. As a result of this type of *stress corrosion cracking*, the mechanical strength of silica is lower in humid environments than in dry conditions. The rate of crack propagation can rise up to a million times in humid environments.

The susceptibility of oxide glasses and ceramics to failure under tensile stress can be reduced by placing the surface in compression. This can be done either thermally or chemically. The thermal process is called "tempering" and is carried out by cooling the surface of the glass more quickly than the bulk. Chemical strengthening takes place by replacing small modifier ions with larger ones by ion exchange. Replacing Li^+ with Na^+ or Na^+ with K^+ greatly increases the strength of the glass. Likewise, β-spodumene $(LiAlSi_2O_6)$ and nepheline glass-ceramics are strengthened in a similar way. Magnesium aluminosilicate glass-ceramics are strengthened by replacing surface Mg^{2+} with two Li^+ ions.

Under normal atmospheric conditions, the surface of silica is covered with hydrated silanol groups (–SiOH). The hydroxyl groups have a strong affinity for water, linking the surface hydroxyls to a monolayer of water. Heating the hydrated surface leads to a reversible reaction in which the water layer is removed. Further heating (to about 300 °C) removes about half the hydroxyl groups, liberating water

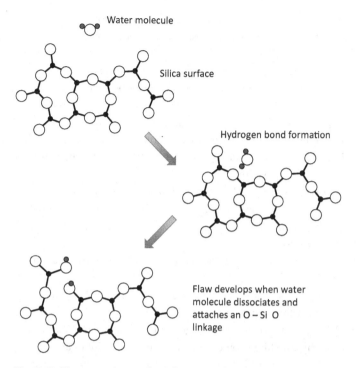

Water molecule

Silica surface

Hydrogen bond formation

Flaw develops when water molecule dissociates and attaches an O – Si O linkage

Fig. 30.27 The strong interaction of water and silica leads to the formation of hydrogen bonds, molecular dissociation, broken Si–O bonds, the creating of hydroxyl groups, and the development of surface flaws which weaken the mechanical strength of glass.

and leaving behind a strained siloxane surface. The reaction is complete near 600 °C. On cooling to room temperature, the siloxane surface readily reacts with moisture to form the hydrated silanol surface. A water-friendly surface such as this can be converted from hydrophilic to hydrophobic by chemical reaction with trimethylchlorosilane (CH_3SiCl).

30.9   Problems

(1) Why is Cu ductile, while MgO is brittle?

(2) Draw a curve showing the temperature dependence of the elastic modulus in an amorphous thermoplastic polymer. Explain why it has the shape it does.

(3) Give an example of a material that undergoes plastic deformation, and explain the mechanism that is responsible.

(4) Describe the factors that control cleavage. Provide examples of crystal structures that illustrate the ideas.

(5) Define slip. Describe the structure–property relations that govern which planes will be slip planes in crystals. Illustrate your points.

(6) Consider Fe in the BCC crystal structure. Would you expect it to be stiffer along the $\langle 100 \rangle$ or $\langle 111 \rangle$ directions? Detail your reasoning.

(7) Contrast reconstructive and displacive phase transitions.
     Describe materials in which a phase transition influences:
     (a) the elastic stiffness, and
     (b) the magnitude of the thermal expansion coefficient.

(8) Diamond is extremely hard, while graphite can be used as a solid-state lubricant.
     Discuss why this should be, based on the structures and bonding.

(9) (a) Explain why Fig. 30.9 shows the trend that it does.
     (b) Pick one alkali halide. Would you expect the elastic stiffness to be the same
          in the [100], [110], and [001] directions? If not, which would be highest?
          Justify your reasoning.

(10) BN is polymorphic; a cubic form comparable to sphalerite exists; in addition a
     hexagonal form is characterized by (space group 186: $P6_3mc$ with $a = 2.504$ Å,
     $c = 6.6612$ Å) and B at (1/3, 2/3, 1/4) and N at (1/3, 2/3, 3/4) and all symmetry-
     related positions.
     (a) Draw a unit cell of this structure showing the bonds that connect the B and
          N ions.
     (b) What is the coordination number for the B and N atoms?
     (c) Contrast this structure to that of graphite.
     (d) Why is hexagonal boron nitride a useful solid-state lubricant?

(11) Explain why graphite should be a good solid-state lubricant.

Appendix A Crystallographic Symbols

$\vec{a}, \vec{b}, \vec{c}$	Unit cell vectors
a, b, c	Unit cell lengths
α, β, γ	Unit cell angles
V	Unit cell volume
$\vec{r}$	Atomic position vector
r	Length of position vector
x, y, z	Atomic coordinates in cell fractions
$[uvw]$	Lattice direction = zone axis
$\langle uvw \rangle$	All symmetry-equivalent directions
(hkl)	Miller indices for a plane
$\{hkl\}$	Symmetry-equivalent planes
hkl	Indices of a Bragg reflection
d_{hkl}	Interplanar distance
$\vec{a}*, \vec{b}*, \vec{c}*$	Reciprocal lattice vectors
$a*, b*, c*$	Magnitudes of reciprocal lattice vectors
$\alpha*, \beta*, \gamma*$	Reciprocal lattice angles
$V*$	Volume of reciprocal lattice cell

Appendix B Shannon–Prewitt Ionic Radii

Ion	Coord. #	Spin state	Crystal radius (Å)	Ionic radius (Å)
Ac^{3+}	6		1.26	1.12
Ag^{1+}	2		0.81	0.67
Ag^{1+}	4		1.16	1.02
Ag^{1+}	5		1.23	1.09
Ag^{1+}	6		1.29	1.15
Ag^{1+}	7		1.36	1.22
Ag^{1+}	8		1.42	1.28
Ag^{2+}	4		0.93	0.79
Ag^{2+}	6		1.08	0.94
Ag^{3+}	4		0.81	0.67
Ag^{3+}	6		0.89	0.75
Al^{3+}	4		0.53	0.39
Al^{3+}	5		0.62	0.48
Al^{3+}	6		0.675	0.535
Am^{2+}	7		1.35	1.21
Am^{2+}	8		1.4	1.26
Am^{2+}	9		1.45	1.31
Am^{3+}	6		1.115	0.975
Am^{3+}	8		1.23	1.09
Am^{4+}	6		0.99	0.85
Am^{4+}	8		1.09	0.95
As^{3+}	6		0.72	0.58
As^{5+}	4		0.475	0.335
As^{5+}	6		0.6	0.46
At^{7+}	6		0.76	0.62
Au^{1+}	6		1.51	1.37
Au^{3+}	4		0.82	0.68
Au^{3+}	6		0.99	0.85
Au^{5+}	6		0.71	0.57
B^{3+}	3		0.15	0.01
B^{3+}	4		0.25	0.11
B^{3+}	6		0.41	0.27
Ba^{2+}	6		1.49	1.35
Ba^{2+}	7		1.52	1.38
Ba^{2+}	8		1.56	1.42
Ba^{2+}	9		1.61	1.47
Ba^{2+}	10		1.66	1.52

(*cont.*)

Ion	Coord. #	Spin state	Crystal radius (Å)	Ionic radius (Å)
Ba^{2+}	11		1.71	1.57
Ba^{2+}	12		1.75	1.61
Be^{2+}	3		0.3	0.16
Be^{2+}	4		0.41	0.27
Be^{2+}	6		0.59	0.45
Bi^{3+}	5		1.1	0.96
Bi^{3+}	6		1.17	1.03
Bi^{3+}	8		1.31	1.17
Bi^{5+}	6		0.9	0.76
Bk^{3+}	6		1.1	0.96
Bk^{4+}	6		0.97	0.83
Bk^{4+}	8		1.07	0.93
Br^{1-}	6		1.82	1.96
Br^{3+}	4		0.73	0.59
Br^{5+}	3		0.45	0.31
Br^{7+}	4		0.39	0.25
Br^{7+}	6		0.53	0.39
C^{4+}	3		0.06	−0.08
C^{4+}	4		0.29	0.15
C^{4+}	6		0.3	0.16
Ca^{2+}	6		1.14	1
Ca^{2+}	7		1.2	1.06
Ca^{2+}	8		1.26	1.12
Ca^{2+}	9		1.32	1.18
Ca^{2+}	10		1.37	1.23
Ca^{2+}	12		1.4	1.34
Cd^{2+}	4		0.92	0.78
Cd^{2+}	5		1.01	0.87
Cd^{2+}	6		1.09	0.95
Cd^{2+}	7		1.17	1.03
Cd^{2+}	8		1.24	1.1
Cd^{2+}	12		1.45	1.31
Ce^{3+}	6		1.15	1.01
Ce^{3+}	7		1.21	1.07
Ce^{3+}	8		1.283	1.143
Ce^{3+}	9		1.336	1.196
Ce^{3+}	10		1.39	1.25
Ce^{3+}	12		1.48	1.34
Ce^{4+}	6		1.01	0.87
Ce^{4+}	8		1.11	0.97
Ce^{4+}	10		1.21	1.07
Ce^{4+}	12		1.28	1.14
Cf^{3+}	6		1.09	0.95
Cf^{4+}	6		0.961	0.821
Cf^{4+}	8		1.06	0.92
Cl^{1-}	6		1.67	1.81
Cl^{5+}	3		0.26	0.12

(*cont.*)

Ion	Coord. #	Spin state	Crystal radius (Å)	Ionic radius (Å)
Cl^{7+}	4		0.22	0.08
Cl^{7+}	6		0.41	0.27
Cm^{3+}	4		1.11	0.97
Cm^{4+}	4		0.99	0.85
Cm^{4+}	8		1.09	0.95
Co^{2+}	4	HS	0.72	0.58
Co^{2+}	5		0.81	0.67
Co^{2+}	6	LS	0.79	0.65
Co^{2+}	6	HS	0.885	0.745
Co^{2+}	8		1.04	0.9
Co^{3+}	6	LS	0.685	0.545
Co^{3+}	6	HS	0.75	0.61
Co^{4+}	4		0.54	0.4
Co^{4+}	6	HS	0.67	0.53
Cr^{2+}	6	LS	0.87	0.73
Cr^{2+}	6	HS	0.94	0.8
Cr^{3+}	6		0.755	0.615
Cr^{4+}	4		0.55	0.41
Cr^{4+}	6		0.69	0.55
Cr^{5+}	4		0.485	0.345
Cr^{5+}	6		0.63	0.49
Cr^{5+}	8		0.71	0.57
Cr^{6+}	4		0.4	0.26
Cr^{6+}	6		0.58	0.44
Cs^{1+}	6		1.81	1.67
Cs^{1+}	8		1.88	1.74
Cs^{1+}	9		1.92	1.78
Cs^{1+}	10		1.95	1.81
Cs^{1+}	11		1.99	1.85
Cs^{1+}	12		2.02	1.88
Cu^{1+}	2		0.6	0.46
Cu^{1+}	4		0.74	0.6
Cu^{1+}	6		0.91	0.77
Cu^{2+}	4		0.71	0.57
Cu^{2+}	4		0.71	0.57
Cu^{2+}	5		0.79	0.65
Cu^{2+}	6		0.87	0.73
Cu^{3+}	6	LS	0.68	0.54
Dy^{1+}	2		0.04	−0.1
Dy^{2+}	6		1.21	1.07
Dy^{2+}	7		1.27	1.13
Dy^{2+}	8		1.33	1.19
Dy^{3+}	6		1.052	0.912
Dy^{3+}	7		1.11	0.97
Dy^{3+}	8		1.167	1.027
Dy^{3+}	9		1.223	1.083
Er^{3+}	6		1.03	0.89

(cont.)

Ion	Coord. #	Spin state	Crystal radius (Å)	Ionic radius (Å)
Er^{3+}	7		1.085	0.945
Er^{3+}	8		1.144	1.004
Er^{3+}	9		1.202	1.062
Eu^{2+}	6		1.31	1.17
Eu^{2+}	7		1.34	1.2
Eu^{2+}	8		1.39	1.25
Eu^{2+}	9		1.44	1.3
Eu^{2+}	10		1.49	1.35
Eu^{3+}	6		1.087	0.947
Eu^{3+}	7		1.15	1.01
Eu^{3+}	8		1.206	1.066
Eu^{3+}	9		1.26	1.12
F^{1-}	2		1.145	1.285
F^{1-}	3		1.16	1.3
F^{1-}	4		1.17	1.31
F^{1-}	6		1.19	1.33
F^{7+}	6		0.22	0.08
Fe^{2+}	4	HS	0.77	0.63
Fe^{2+}	4	HS	0.78	0.64
Fe^{2+}	6	LS	0.75	0.61
Fe^{2+}	6	HS	0.92	0.78
Fe^{2+}	8	HS	1.06	0.92
Fe^{3+}	4	HS	0.63	0.49
Fe^{3+}	5		0.72	0.58
Fe^{3+}	6	LS	0.69	0.55
Fe^{3+}	6	HS	0.785	0.645
Fe^{3+}	8	HS	0.92	0.78
Fe^{4+}	6		0.725	0.585
Fe^{6+}	4		0.39	0.25
Fr^{1+}	6		1.94	1.8
Ga^{3+}	4		0.61	0.47
Ga^{3+}	5		0.69	0.55
Ga^{3+}	6		0.76	0.62
Gd^{3+}	6		1.078	0.938
Gd^{3+}	7		1.14	1
Gd^{3+}	8		1.193	1.053
Gd^{3+}	9		1.247	1.107
Ge^{2+}	6		0.87	0.73
Ge^{4+}	4		0.53	0.39
Ge^{4+}	6		0.67	0.53
H^{1+}	1		−0.24	−0.38
H^{1+}	2		−0.04	−0.18
Hf^{4+}	4		0.72	0.58
Hf^{4+}	6		0.85	0.71
Hf^{4+}	7		0.9	0.76
Hf^{4+}	8		0.97	0.83
Hg^{1+}	3		1.11	0.97

(cont.)

Ion	Coord. #	Spin state	Crystal radius (Å)	Ionic radius (Å)
Hg^{1+}	6		1.33	1.19
Hg^{2+}	2		0.83	0.69
Hg^{2+}	4		1.1	0.96
Hg^{2+}	6		1.16	1.02
Hg^{2+}	8		1.28	1.14
Ho^{3+}	6		1.041	0.901
Ho^{3+}	8		1.155	1.015
Ho^{3+}	9		1.212	1.072
Ho^{3+}	10		1.26	1.12
I^{1-}	6		2.06	2.2
I^{5+}	3		0.58	0.44
I^{5+}	6		1.09	0.95
I^{7+}	4		0.56	0.42
I^{7+}	6		0.67	0.53
In^{3+}	4		0.76	0.62
In^{3+}	6		0.94	0.8
In^{3+}	8		1.06	0.92
Ir^{3+}	6		0.82	0.68
Ir^{4+}	6		0.765	0.625
Ir^{5+}	6		0.71	0.57
K^{1+}	4		1.51	1.37
K^{1+}	6		1.52	1.38
K^{1+}	7		1.6	1.46
K^{1+}	8		1.65	1.51
K^{1+}	9		1.69	1.55
K^{1+}	10		1.73	1.59
K^{1+}	12		1.78	1.64
La^{3+}	6		1.172	1.032
La^{3+}	7		1.24	1.1
La^{3+}	8		1.3	1.16
La^{3+}	9		1.356	1.216
La^{3+}	10		1.41	1.27
La^{3+}	12		1.5	1.36
Li^{1+}	4		0.73	0.59
Li^{1+}	6		0.9	0.76
Li^{1+}	8		1.06	0.92
Lu^{3+}	6		1.001	0.861
Lu^{3+}	8		1.117	0.977
Lu^{3+}	9		1.172	1.032
Mg^{2+}	4		0.71	0.57
Mg^{2+}	5		0.8	0.66
Mg^{2+}	6		0.86	0.72
Mg^{2+}	8		1.03	0.89
Mn^{2+}	4	HS	0.8	0.66
Mn^{2+}	5	HS	0.89	0.75
Mn^{2+}	6	LS	0.81	0.67
Mn^{2+}	6	HS	0.97	0.83

(*cont.*)

Ion	Coord. #	Spin state	Crystal radius (Å)	Ionic radius (Å)
Mn^{2+}	7	HS	1.04	0.9
Mn^{2+}	8		1.1	0.96
Mn^{3+}	5		0.72	0.58
Mn^{3+}	6	LS	0.72	0.58
Mn^{3+}	6	HS	0.785	0.645
Mn^{4+}	4		0.53	0.39
Mn^{4+}	6		0.67	0.53
Mn^{5+}	4		0.47	0.33
Mn^{6+}	4		0.395	0.255
Mn^{7+}	4		0.39	0.25
Mn^{7+}	6		0.6	0.46
Mo^{3+}	6		0.83	0.69
Mo^{4+}	6		0.79	0.65
Mo^{5+}	4		0.6	0.46
Mo^{5+}	6		0.75	0.61
Mo^{6+}	4		0.55	0.41
Mo^{6+}	5		0.64	0.5
Mo^{6+}	6		0.73	0.59
Mo^{6+}	7		0.87	0.73
N^{3-}	4		1.32	1.46
N^{3+}	6		0.3	0.16
N^{5+}	3		0.044	−0.104
N^{5+}	4		0.27	0.13
Na^{1+}	4		1.13	0.99
Na^{1+}	5		1.14	1
Na^{1+}	6		1.16	1.02
Na^{1+}	7		1.26	1.12
Na^{1+}	8		1.32	1.18
Na^{1+}	9		1.38	1.24
Na^{1+}	12		1.53	1.39
Nb^{3+}	6		0.86	0.72
Nb^{4+}	6		0.82	0.68
Nb^{4+}	8		0.93	0.79
Nb^{5+}	4		0.62	0.48
Nb^{5+}	6		0.78	0.64
Nb^{5+}	7		0.83	0.69
Nb^{5+}	8		0.88	0.74
Nd^{2+}	8		1.43	1.29
Nd^{2+}	9		1.49	1.35
Nd^{3+}	6		1.123	0.983
Nd^{3+}	8		1.249	1.109
Nd^{3+}	9		1.303	1.163
Nd^{3+}	12		1.41	1.27
Ni^{2+}	4		0.69	0.55
Ni^{2+}	4		0.63	0.49
Ni^{2+}	5		0.77	0.63
Ni^{2+}	6		0.83	0.69

(cont.)

Ion	Coord. #	Spin state	Crystal radius (Å)	Ionic radius (Å)
Ni^{3+}	6	LS	0.7	0.56
Ni^{3+}	6	HS	0.74	0.6
Ni^{4+}	6	LS	0.62	0.48
No^{2+}	6		1.24	1.1
Np^{2+}	6		1.24	1.1
Np^{3+}	6		1.15	1.01
Np^{4+}	6		1.01	0.87
Np^{4+}	8		1.12	0.98
Np^{5+}	6		0.89	0.75
Np^{6+}	6		0.86	0.72
Np^{7+}	6		0.85	0.71
O^{2-}	2		1.21	1.35
O^{2-}	3		1.22	1.36
O^{2-}	4		1.24	1.38
O^{2-}	6		1.26	1.4
O^{2-}	8		1.28	1.42
OH^{1-}	2		1.18	1.32
OH^{1-}	3		1.2	1.34
OH^{1-}	4		1.21	1.35
OH^{1-}	6		1.23	1.37
Os^{4+}	6		0.77	0.63
Os^{5+}	6		0.715	0.575
Os^{6+}	5		0.63	0.49
Os^{6+}	6		0.685	0.545
Os^{7+}	6		0.665	0.525
Os^{8+}	4		0.53	0.39
P^{3+}	6		0.58	0.44
P^{5+}	4		0.31	0.17
P^{5+}	5		0.43	0.29
P^{5+}	6		0.52	0.38
Pa^{3+}	6		1.18	1.04
Pa^{4+}	6		1.04	0.9
Pa^{4+}	8		1.15	1.01
Pa^{5+}	6		0.92	0.78
Pa^{5+}	8		1.05	0.91
Pa^{5+}	9		1.09	0.95
Pb^{2+}	4		1.12	0.98
Pb^{2+}	6		1.33	1.19
Pb^{2+}	7		1.37	1.23
Pb^{2+}	8		1.43	1.29
Pb^{2+}	9		1.49	1.35
Pb^{2+}	10		1.54	1.4
Pb^{2+}	11		1.59	1.45
Pb^{2+}	12		1.63	1.49
Pb^{4+}	4		0.79	0.65
Pb^{4+}	5		0.87	0.73
Pb^{4+}	6		0.915	0.775

(cont.)

Ion	Coord. #	Spin state	Crystal radius (Å)	Ionic radius (Å)
Pb^{4+}	8		1.08	0.94
Pd^{1+}	2		0.73	0.59
Pd^{2+}	4		0.78	0.64
Pd^{2+}	6		1	0.86
Pd^{3+}	6		0.9	0.76
Pd^{4+}	6		0.755	0.615
Pm^{3+}	6		1.11	0.97
Pm^{3+}	8		1.233	1.093
Pm^{3+}	9		1.284	1.144
Po^{4+}	6		1.08	0.94
Po^{4+}	8		1.22	1.08
Po^{6+}	6		0.81	0.67
Pr^{3+}	6		1.13	0.99
Pr^{3+}	8		1.266	1.126
Pr^{3+}	9		1.319	1.179
Pr^{4+}	6		0.99	0.85
Pr^{4+}	8		1.1	0.96
Pt^{2+}	4		0.74	0.6
Pt^{2+}	6		0.94	0.8
Pt^{4+}	6		0.765	0.625
Pt^{5+}	6		0.71	0.57
Pu^{3+}	6		1.14	1
Pu^{4+}	6		1	0.86
Pu^{4+}	8		1.1	0.96
Pu^{5+}	6		0.88	0.74
Pu^{6+}	6		0.85	0.71
Ra^{2+}	8		1.62	1.48
Ra^{2+}	12		1.84	1.7
Rb^{1+}	6		1.66	1.52
Rb^{1+}	7		1.7	1.56
Rb^{1+}	8		1.75	1.61
Rb^{1+}	9		1.77	1.63
Rb^{1+}	10		1.8	1.66
Rb^{1+}	11		1.83	1.69
Rb^{1+}	12		1.86	1.72
Rb^{1+}	14		1.97	1.83
Re^{4+}	6		0.77	0.63
Re^{5+}	6		0.72	0.58
Re^{6+}	6		0.69	0.55
Re^{7+}	4		0.52	0.38
Re^{7+}	6		0.67	0.53
Rh^{3+}	6		0.805	0.665
Rh^{4+}	6		0.74	0.6
Rh^{5+}	6		0.69	0.55
Ru^{3+}	6		0.82	0.68
Ru^{4+}	6		0.76	0.62
Ru^{5+}	6		0.705	0.565

(*cont.*)

Ion	Coord. #	Spin state	Crystal radius (Å)	Ionic radius (Å)
Ru^{7+}	4		0.52	0.38
Ru^{8+}	4		0.5	0.36
S^{2-}	6		1.7	1.84
S^{4+}	6		0.51	0.37
S^{6+}	4		0.26	0.12
S^{6+}	6		0.43	0.29
Sb^{3+}	4		0.9	0.74
Sb^{3+}	5		0.94	0.8
Sb^{3+}	6		0.9	0.76
Sb^{5+}	6		0.74	0.6
Sc^{3+}	6		0.885	0.745
Sc^{3+}	8		1.01	0.87
Se^{2-}	6		1.84	1.98
Se^{4+}	6		0.64	0.5
Se^{6+}	4		0.42	0.28
Se^{6+}	6		0.56	0.42
Si^{4+}	4		0.4	0.26
Si^{4+}	6		0.54	0.4
Sm^{2+}	7		1.36	1.22
Sm^{2+}	8		1.41	1.27
Sm^{2+}	9		1.46	1.32
Sm^{3+}	6		1.098	0.958
Sm^{3+}	7		1.16	1.02
Sm^{3+}	8		1.219	1.079
Sm^{3+}	9		1.272	1.132
Sm^{3+}	12		1.38	1.24
Sn^{4+}	4		0.69	0.55
Sn^{4+}	5		0.76	0.62
Sn^{4+}	6		0.83	0.69
Sn^{4+}	7		0.89	0.75
Sn^{4+}	8		0.95	0.81
Sr^{2+}	6		1.32	1.18
Sr^{2+}	7		1.35	1.21
Sr^{2+}	8		1.4	1.26
Sr^{2+}	9		1.45	1.31
Sr^{2+}	10		1.5	1.36
Sr^{2+}	12		1.58	1.44
Ta^{3+}	6		0.86	0.72
Ta^{4+}	6		0.82	0.68
Ta^{5+}	6		0.78	0.64
Ta^{5+}	7		0.83	0.69
Ta^{5+}	8		0.88	0.74
Tb^{3+}	6		1.063	0.923
Tb^{3+}	7		1.12	0.98
Tb^{3+}	8		1.18	1.04
Tb^{3+}	9		1.235	1.095
Tb^{4+}	6		0.9	0.76

(*cont.*)

Ion	Coord. #	Spin state	Crystal radius (Å)	Ionic radius (Å)
Tb^{4+}	8		1.02	0.88
Tc^{4+}	6		0.785	0.645
Tc^{5+}	6		0.74	0.6
Tc^{7+}	4		0.51	0.37
Tc^{7+}	6		0.7	0.56
Te^{2-}	6		2.07	2.21
Te^{4+}	3		0.66	0.52
Te^{4+}	4		0.8	0.66
Te^{4+}	6		1.11	0.97
Te^{6+}	4		0.57	0.43
Te^{6+}	6		0.7	0.56
Th^{4+}	6		1.08	0.94
Th^{4+}	8		1.19	1.05
Th^{4+}	9		1.23	1.09
Th^{4+}	10		1.27	1.13
Th^{4+}	11		1.32	1.18
Th^{4+}	12		1.35	1.21
Ti^{2+}	6		1	0.86
Ti^{3+}	6		0.81	0.67
Ti^{4+}	4		0.56	0.42
Ti^{4+}	5		0.65	0.51
Ti^{4+}	6		0.745	0.605
Ti^{4+}	8		0.88	0.74
Tl^{1+}	6		1.64	1.5
Tl^{1+}	8		1.73	1.59
Tl^{1+}	12		1.84	1.7
Tl^{3+}	4		0.89	0.75
Tl^{3+}	6		1.025	0.885
Tl^{3+}	8		1.12	0.98
Tm^{2+}	6		1.17	1.03
Tm^{2+}	7		1.23	1.09
Tm^{3+}	6		1.02	0.88
Tm^{3+}	8		1.134	0.994
Tm^{3+}	9		1.192	1.052
U^{3+}	6		1.165	1.025
U^{4+}	6		1.03	0.89
U^{4+}	7		1.09	0.95
U^{4+}	8		1.14	1
U^{4+}	9		1.19	1.05
U^{4+}	12		1.31	1.17
U^{5+}	6		0.9	0.76
U^{5+}	7		0.98	0.84
U^{6+}	2		0.59	0.45
U^{6+}	4		0.66	0.52
U^{6+}	6		0.87	0.73
U^{6+}	7		0.95	0.81
U^{6+}	8		1	0.86

(cont.)

Ion	Coord. #	Spin state	Crystal radius (Å)	Ionic radius (Å)
V^{2+}	6		0.93	0.79
V^{3+}	6		0.78	0.64
V^{4+}	5		0.67	0.53
V^{4+}	6		0.72	0.58
V^{4+}	8		0.86	0.72
V^{5+}	4		0.495	0.355
V^{5+}	5		0.6	0.46
V^{5+}	6		0.68	0.54
W^{4+}	6		0.8	0.66
W^{5+}	6		0.76	0.62
W^{6+}	4		0.56	0.42
W^{6+}	5		0.65	0.51
W^{6+}	6		0.74	0.6
Xe^{8+}	4		0.54	0.4
Xe^{8+}	6		0.62	0.48
Y^{3+}	6		1.04	0.9
Y^{3+}	7		1.1	0.96
Y^{3+}	8		1.159	1.019
Y^{3+}	9		1.215	1.075
Yb^{2+}	6		1.16	1.02
Yb^{2+}	7		1.22	1.08
Yb^{2+}	8		1.28	1.14
Yb^{3+}	6		1.008	0.868
Yb^{3+}	7		1.065	0.925
Yb^{3+}	8		1.125	0.985
Yb^{3+}	9		1.182	1.042
Zn^{2+}	4		0.74	0.6
Zn^{2+}	5		0.82	0.68
Zn^{2+}	6		0.88	0.74
Zn^{2+}	8		1.04	0.9
Zr^{4+}	4		0.73	0.59
Zr^{4+}	5		0.8	0.66
Zr^{4+}	6		0.86	0.72
Zr^{4+}	7		0.92	0.78
Zr^{4+}	8		0.98	0.84
Zr^{4+}	9		1.03	0.89

From http://v.faculty.umkc.edu/vanhornj/shannonradii.htm

Index